Forestry and Agricultural Management

Forestry and Agricultural Management

Editor: Malcolm Fisher

Callisto Reference
www.callistoreference.com

Callisto Reference,
118-35 Queens Blvd., Suite 400,
Forest Hills, NY 11375, USA

Visit us on the World Wide Web at:
www.callistoreference.com

ISBN: 978-1-63239-878-9 (Hardback)

The publisher's policy is to use permanent paper from mills that operate a sustainable forestry policy. Furthermore, the publisher ensures that the text paper and cover boards used have met acceptable environmental accreditation standards.

Printed in China.

Cataloging-in-Publication Data

Forestry and agricultural management / edited by Malcolm Fisher.
p. cm.
Includes bibliographical references and index.
ISBN 978-1-63239-878-9
1. Forests and forestry. 2. Agriculture. 3. Forest management. 4. Agricultural administration.
5. Agriculture--Economic aspects. I. Fisher, Malcolm.
SD131 .F67 2017
630--dc23

Table of Contents

Preface

The main aim of this book is to educate learners and enhance their research focus by presenting diverse topics covering this vast field. This is an advanced book which compiles significant studies by distinguished experts. This book addresses successive solutions to the challenges arising in the area of application, along with it; the book provides scope for future developments.

This book on forestry and agricultural management talks about methodological practices followed in pastures and cropland for the cultivation of trees, shrubs and plants. Methods employed in this field follow traditional as well as modern land-use management systems. Agroforestry allows for rearing of animals alongside crop cultivation and forest maintenance. The contents included in this text deal with the advanced concepts that promote forest cultivation and conservation. The aim of this book is to present researches that have transformed this discipline and aided its advancement. From theories to research to practical applications, case studies related to all contemporary topics of relevance to this field have been included in this book. It is appropriate for students seeking detailed information in this area as well as for experts.

It was a great honour to edit this book, though there were challenges, as it involved a lot of communication and networking between me and the editorial team. However, the end result was this all-inclusive book covering diverse themes in the field.

Finally, it is important to acknowledge the efforts of the contributors for their excellent chapters, through which a wide variety of issues have been addressed. I would also like to thank my colleagues for their valuable feedback during the making of this book.

Editor

Particle size for improvement of peptide production in mixed-culture solid-state fermentation of soybean meal and the corresponding kinetics

Junjun Guan*, Guohao Yang, Haicheng Yin, Feng Jia, Jinshui Wang

College of Biological Engineering, Henan University of Technology, Lianhua Street, Zhengzhou New & High-tech Industrial Development Zone, Zhengzhou 450001, China

Email address:
junjunguan@163.com (Junjun Guan)

Abstract: Experiments were undertaken to study particle size and its distribution for improvement of peptide production by mixed-culture (*Bacillus sublitis*, *Saccharomyces sp.* and *Lactococcus lactis*) solid-state fermentation (SSF) of soybean meal and the corresponding kinetics. A particle size of 1-1.4mm of soybean meal gave the highest finial peptide yield, and as the proportion of 1-1.4mm/1-2mm in natural soybean meal up to 61.30%/58.58%, the distribution of particle size was also appropriate for peptide production. Logistic model fitted the data most accurately and could be used for growth kinetic profiles during the course of fermentation, and the higher growth rate was calculated for the substrate with optimal particle size distribution. Hence the mechanism could be inferred that appropriate particle size did improve the according rate of microbial growth, thus leading to the higher peptide yield in soybean meal SSF within a limited fermentation time.

Keywords: Particle Size, Peptide, Soybean Meal, Kinetics, Solid-State Fermentation

1. Introduction

The need for animal feed will inevitably increase in the future. As a main protein source for animal feedstuffs, soybean meal often has several antinutritional factors including carbohydrates [1], and antigenic proteins [2], which might cause hypersensitivity. Microbial fermentation [3, 4] is a more economical approach to reduce the antinutritional factors of soybean meal, and the use of different strains for soybean meal fermentation, not only can produce microbial enzymes to reduce anti-nutritional factors in soybean meal, but also can accumulate some other beneficial metabolites such as peptides that exert specific positive effects on cultured animal cells [5]. Therefore, improvement of peptide production from soybean meal fermentation will provide a higher nutritional value for animal growth and also be interested by more and more researchers now.

Solid-state fermentation (SSF) is often used to process soybean meal for the lower cost of production. In SSF process, the moist solid substrate is used and without free-flowing water, and thus the adherence and penetration of microorganisms are clearly associated with the physical properties of the substrate such as the crystalline or amorphous nature, the accessible area, surface area, porosity, particle size, etc. In all the above factors, particle size should play a major role due to the dependence of other factors on it [6]. Hence it is of interest to know how particle size affects the production based on the substrate availability during SSF. The effect of particle size on the growth and product formation in SSF was studied by many [7-9]. But until now almost no report has been available regarding the effects of the particle size and its distribution on the changes in the peptide production during mixed-culture SSF of soybean meal. Therefore an attempt was made in the present study, with particular reference to the effects of the particle size and its distriburtion on peptide production and growth profiles of soybean meal SSF.

2. Materials and Methods

2.1. Microorganisms and Culture Preparation

Bacillus sublitis (ACCC 01746), *Saccharomyces sp.*

(CICIM Y0362) and *Lactococcus lactis* (ACCC 11092) were provided by Henan Engineering Laboratory for Collection and Selective Breeding of Industrial Microorganisms (Zhengzhou, China). The strains were cultured on nutrient agar (peptone beef extract agar) plates for *Bacillus sublitis,* potato dextrose agar plates for *Saccharomyces sp.*, and MRS agar plates for *Lactococcus lactis*. The plates were incubated at 30 °C for 24 h for *Bacillus sublitis* and *Saccharomyces sp.* And 7 days for *Lactococcus lactis* until colonies produced, and were kept at 4 °C as cultures until use.

2.2. Solid Substrate and Chemicals

Soybean meal was produced by Henan Sunshine Oils and Fats Group (Zhengzhou, China), analysed for protein (46%, N×6.25) by Kjedhal method, subjected to a sieving procedure using sieves mesh-size of 10, 14, 18, 40 and 80, and classified into six different diameter sizes: <0.18mm(P1), 0.18-0.425mm(P2), 0.425-1mm(P3), 1-1.4mm(P4), 1.4-2mm(P5), and >2mm(P6), according to which of the sieves mesh-size that the soybean meal retained. The soybean meal without treatment of sieving procedure was called natural soybean meal and expressed by "natural" in this study. The soybean meal with different particle sizes and its natural were used alone or in combination as solid substrates. All chemicals and medium ingredients were purchased from Shanghai Chemical Co. (Shanghai, China).

2.3. Solid-State Fermentation (SSF)

Two culture loops of *Bacillus sublitis* (B), *Saccharomyces sp.* (S) and *Lactococcus lactis* (S) were respectively transferred to 250 mL conical flasks containing 50 mL of liquid soybean meal dextrose medium made of 4% soybean-meal powder (particle size <0.15mm, passing through 100-mesh screen) and 1% dextrose, in order to adapt the strains for the solid substrate. The flasks were shaken at 150 rpm for *Bacillus sublitis* and statically placed for *Lactococcus lactis* and *Saccharomyces sp.* At 30 °C for 8 h. The solution was used as a stock solution for SSF.

A 250-mL conical flask, which was capped with rubber stopper perforated with a syringe needle for gas release, was used for SSF. A total inoculum of 10% (v/w) was used to inoculate 50 g of substrate with the inoculation ratio of B:S:L = 1:1:1 (v:v:v). Solid substrate with an original moisture content of 12% was adjusted to a moisture content of 50% by using distilled water, and the initial pH was maintained at a natural level as reported [15] and actually tested at 6.8±0.1 in this study. No nutrients or minerals were added to the substrate, due to soybean meal as the natural medium that contains the organic carbon, nitrogen, phosphorus, vitamins and metal ions to allow microbial growth, and the case that a supplement such as wheat bran might relatively reduce the total content of protein in soybean meal owing to the lower content of protein in supplement and thus would affect the nutrient value for application in animal feed. In addition, for this scheme the effect of particle size of soybean meal on the SSF could be investigated clearly without the interference of the supplement. The substrate was sterilized at 121 °C and 103 kPa in an autoclave for 20 min. And, SSF was done by statically incubating the substrate at 30 °C in a thermostat incubator within a water reservoir to maintain the relative humidity and moisture level of the substrate, and the samples were withdrawn every 3 h. Content of peptide of the fermented soybean meal was measured. All treatments were repeated three times.

2.4. Peptide and Dry Matter Assays

The peptide content in sample was determined by Lowry method [16]. 1 g sample was put into 50 mL of 15% (w/w) TCA solution and stirred and incubated for 5 min at 25 °C, and then the mixture was filtered. After the filtrate was centrifuged at 4,000×g for 30 min, the obtained supernatant, containing peptide (< 10 kDa) from fermented soybean meal, could be analyzed by Lowry method. The content of small peptide, expressed as a percentage (dwb), was calculated by dividing the weight of TCA-soluble peptide by the dry matter mass in the original sample.

The dry matter mass was determined by weighing after oven-drying at 105 °C overnight.

2.5. Calculation

The linear, exponential and logistic equations have typically been used to characterize growth profiles in SSF [17]. The integrated forms of these equations are expressed as follows:
linear,

$$X = Kt + X_0 \tag{1}$$

exponential,

$$X = X_0 e^{\mu t} \tag{2}$$

and logistic,

$$X = \frac{X_m}{1 + (\frac{X_m}{X_0} - 1)e^{-\mu t}} \tag{3}$$

in which, X, biomass concentration; X_0, X at time 0; t, fermentation time; K, growth rate; μ, constant of growth rate; X_m, carrying capacity. For the measurement of biomass (X) in the construction of growth profiles, some different units have been used by researchers [17], and in this study the content of peptide (dwb), produced by microorganisms in SSF, was used for the growth profile, due to the purpose of this study, relation of peptide production and particle size distribution, which could also reflect the microbial growth indirectly.

For the logistic equation, the maximum growth rate (dX/dt) can be obtained through the second derivative of Eq. (3) at,

$$\frac{d^2X}{dt^2} = 0$$

and the solution to this equation is,

$$t_{X_m/2} = \frac{\ln(\frac{X_m}{X_0} - 1)}{\mu} \quad (4)$$

at which, $X = X_m/2$. The point ($t_{Xm/2}$, $X_m/2$) is the inflexion of the Eq. (3) for a sigmoidal curve, where the maximum rate occurred.

Eqs. (1) - (3) were fitted to triplicate sets of data using linear and non-linear regression modules of the software, Polymath (Version 6.0, USA). The correlation coefficients R^2 were used to judge whether the equation represents correctly the data and for comparing various equations representing the same dependent variable. The goodness of fit was evaluated with the root-mean-square deviation (Rmsd).

2.6. Statistical Analysis

Analysis of variance (ANOVA) and least significant difference (LSD) were used in order to determine the differences among treatments ($P < 0.05$) by SAS (version 9.1).

3. Results and Discussion

3.1. Effect of Particle Size of Substrate on the Peptide Yield of Soybean Meal SSF

During SSF process, the availability of surface area plays an important role in microbial attachment, mass transfer, subsequent microbial growth and metabolite production, and is most associated with the particle size of the substrate [10]. To study the effect of the particle size of substrate on the peptide yield in SSF of soybean meal for 36h, different particle sizes (<0.18mm, 0.18-0.425mm, 0.425-1mm, 1-1.4mm, 1.4-2mm, >2mm, and natural) were used in the cultivation and the results are shown in Figure 1. The maximum of peptide content ($P<0.05$) was obtained on the soybean meal with particle size of 1-1.4mm. An increase or decrease in the particle size led to reduction of peptide yield, but no significant difference ($P>0.05$) was found between particle sizes of 1-1.4mm and 1.4-2mm, and between particle sizes of 0.425-1mm and >2mm. Fermentation of the fine (<0.425mm) soybean meal resulted in significantly lower content ($P<0.05$) of peptide. Moreover, it was interesting that SSF with natural substrate had a high yield of peptide, only lower than treatments with particle sizes of 1-1.4mm and 1.4-2mm (Figure 1), which should be related to its high content (42%) of particle size of 1-2mm (Table 1) and would be concerned with particle size distribution of mixture.

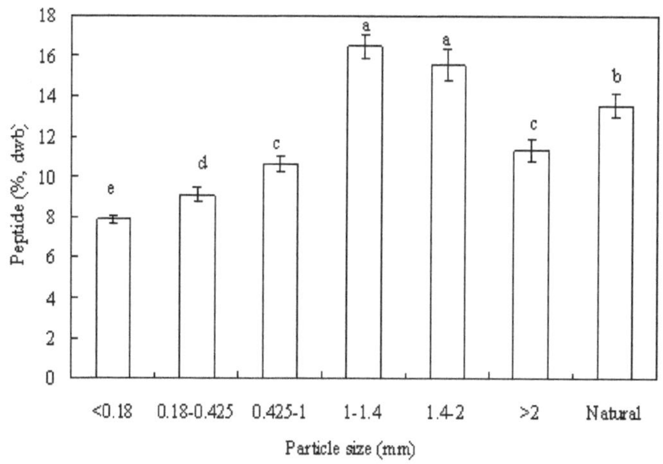

Figure 1. *Effect of particle sizes on peptide contents in mixed-culture SSF of soybean meal for a fermentation time of 36h. N = 3. Means followed by the same letter are not significantly different at P = 5%.* $LSD_{0.05} = 0.9152$.

Table 1. *Particle size distribution of different substrates.*

Substrates	Particle size distribution (%)						
	P1	P2	P3	P4	P5	P6	*1-2mm*
Natural	3.00	17.00	30.00	22.00	20.00	8.00	*42.00*
Natural + P4 (w : w)							
M1(1:0.25)	2.25	12.75	22.50	37.60	15.00	6.00	*52.60*
M2(1:0.5)	2.00	11.30	20.00	48.00	13.30	5.30	*61.30*
M3(1:1)	1.50	8.50	15.00	61.00	10.00	4.00	*71.00*
M4(1:2)	1.00	5.70	10.00	74.00	6.70	2.70	*80.70*
Natural + P4 + P5 (w : w : w)							
M5(1:0.1:0.1)	2.50	14.17	25.00	26.66	25.00	6.67	*51.66*
M6(1:0.2:0.2)	2.14	12.14	21.43	30.00	28.58	5.71	*58.58*
M7(1:0.3:0.3)	1.87	10.63	18.75	32.50	31.25	5.00	*63.75*
M8(1:0.5:0.5)	1.50	8.50	15.00	36.00	35.00	4.00	*71.00*

3.2. Effect of Particle Size Distribution of Mixture on the Peptide Content during SSF of Soybean Meal

Natural mixed with other substrates with some particle size was investigated during the course of SSF and the selection of supplement was based on the results of previous batches. Substrate of P4 (1-1.4mm) was hence chosen and P5 (1.4-2mm) was also considered for comparison. The particle size distributions of different mixtures are shown in Table 1, and effects of mixtures on changes of peptide content with fermentation time can be seen from Figure 2. The increase of proportion of P4 or P4+P5 in the mixture could enhance the final yield of peptide. For the total fermentation process, the yield of peptide increased significantly ($P<0.05$) as the proportion of P4 and P4+P5 was above 61.30% and 58.58% respectively, compared to natural. However, for the similar type treatments such as M1, M2, M3 and M4, when the proportion of P4 and P+P5 was above 71.00% and 63.75% respectively, there was no significantly difference in peptide production between those mixtures (M3 and M4, M7 and M8) used as substrates. Moreover, no significantly difference between P4, M4 and M8, which indicated that the proportion of P4 and P4+P5 in mixture reaching 80.70% and 71.00% respectively had a similar peptide yield to substrate only with 1-1.4mm particle size.

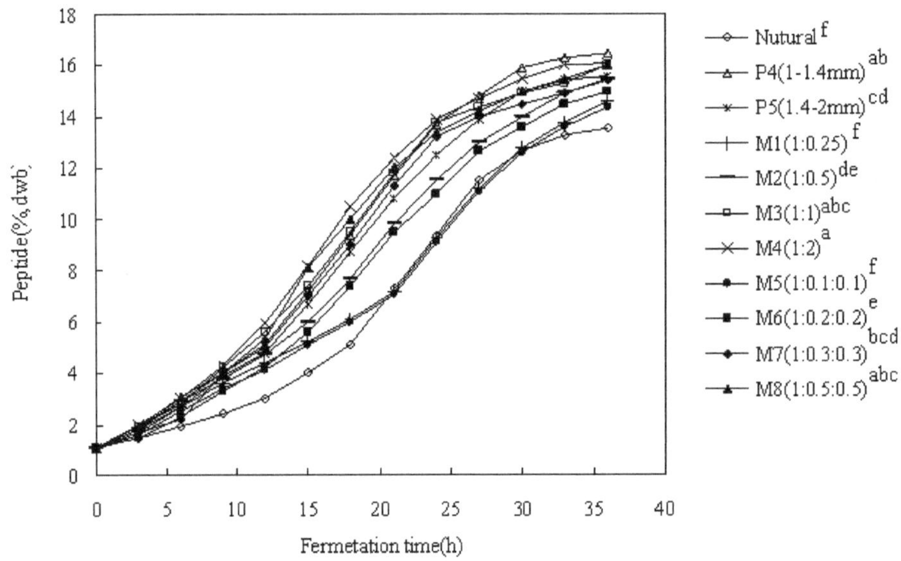

Figure 2. Time profiles of peptide content at different substrates during mixed-culture SSF of soybean meal. Different letters represent significant differences at P = 5%. LSD0.05 =0.562. All different substrates are described in Table 1.

3.3. Growth Kinetic Profiles

Table 2. *Kinetic analysis of growth profiles using liner, exponential and logistic equation*[*].

S	Linear			Exponential			Logistic				
	K (%·h^{-1})	R^2	Rmsd	μ (h^{-1})	R^2	Rmsd	X_m(%,dwb)	μ (h^{-1})	$t_{X_m/2}$ (h)	R^2	Rmsd
N	0.33	0.91	0.38	0.076	0.87	0.46	18.15ab	0.11e	*24.9*a	0.985	0.160
P4	0.47	0.97	0.28	0.083	0.62	0.96	17.59abc	0.16bc	*16.9*bcd	0.998	0.067
P5	0.43	0.97	0.25	0.081	0.66	0.86	16.82abc	0.15cd	*17.7*bc	0.998	0.057
M1	0.36	0.96	0.24	0.077	0.86	0.47	17.93a	0.12e	*22.7*a	0.992	0.110
M2	0.41	0.98	0.21	0.080	0.73	0.71	16.88abc	0.14d	*19.0*b	0.998	0.062
M3	0.45	0.97	0.27	0.082	0.56	0.98	16.46bc	0.17ab	*15.5*cd	0.998	0.070
M4	0.47	0.97	0.28	0.083	0.50	1.07	16.60abc	0.18a	*14.7*d	0.999	0.035
M5	0.35	0.96	0.27	0.077	0.87	0.45	17.96a	0.12e	*22.4*a	0.995	0.090
M6	0.39	0.97	0.25	0.079	0.77	0.66	16.79abc	0.14d	*19.0*b	0.999	0.049
M7	0.43	0.96	0.28	0.081	0.61	0.92	16.19c	0.16bc	*16.4*cd	0.997	0.081
M8	0.46	0.97	0.26	0.082	0.54	1.00	16.34c	0.17ab	*15.5*d	0.998	0.061

[*] S,Substrates; N,Natural; All different substrates are described in Table 1. Values followed by the same lowercase letter within a column are not significantly different at P = 5%.

Growth kinetic profiles of different substrates including natural, substrate with single particle size, and mixture, were analyzed by fitting the data in Figure 2 to three empirical growth model (linear, exponential and logistic) and results shown in Table 2. The higher R^2 values were found in linear (0.91-0.97) and logistic (0.985-0.998) equations, indicating that these two models could represent correctly the data but logistic equation seemed to represent given sets of data much more correctly by further comparison. Therefore, exponential model would not be considered because the lower R^2. Furthermore, the smallest Rmsd values (0.035-0.160) were found in logistic equation, also showing that this model fitted the data most accurately. So, logistic model would provide good answer for questions under investigation for mixed-culture SSF of soybean meal. This model fitting data from SSF with appropriate distribution of particle size (containing higher proportion of 1-1.4mm or 1-2mm) of mixtures (M3, M4, M7 and M8) or 1.4mm particle size of substrate had a smaller X_m value and a higher μ value. On the other hand, as the proportion of P4 and P4+P5 was above 61.30% and 58.58% respectively, the growth rate constant (μ) increased significantly ($P<0.05$).

3.4. Discussion

It is known that the size of particle is directly associated with the specific surface area and porosity of the solid substrate. And in general, bigger particles, with less surface area, cause a trend towards poorer accessibility to the microorganism, but inter-particular porosity is greater, while smaller particles, with larger surface area, are preferred for better accessibility to the microorganism, but the porosity is less. These two opposing factors probably interact to determine the products corresponding to optimum microbial growth [11]. Furthermore, smaller particles are favorable for anaerobic fermentation, due to forming a higher solid medium density and a less void fraction. But, too small particles may diminish the heat transfer and carbon dioxide exchange rates because of medium compaction [12]. On the other hand, larger particles provide better aeration/respiration opportunities [13], which does encourage the growth of the aerobic organism. Thus, an appropriate particle size is required for a particular process in SSF [14]. In this study, three microorganisms were combination of aerobe (*Bacillus sublitis*) and facultative anaerobe (*Saccharomyces sp.* And *Lactococcus lactis*), and final results showed that appropriate particle size could make three microorganisms grow as quickly as possible, but otherwise microbial growth would be inhibited to some extent, thus leading to reduce the peptide yield. So, an optimal particle size of 1-1.4mm should to be chosen for soybean meal SSF, which represented a compromise between the accessibility of nutrients and the availability of oxygen. In addition, particle size of 1.4-2mm might be considered, mainly owing to no statistically significant difference with particle size of 1-1.4mm.

Results also suggested that the yield of peptide could be improved by adjusting the proportion of P4 (1-1.4mm) or P4+P5 (1-2mm) of natural substrate. But it was found that in the all process of SSF, P4 (1-1.4mm) treatment had overall higher peptide yield ($P<0.05$) than P5 (1.4-2mm), other than previous batches, partly due to the different data obtained at fermentation time of 36h and 0-36h, respectively, thus leading to the smaller value of $LSD_{0.05}$ for comparison analysis of the latter, and this result also showed that the 1-1.4mm was the optimal particle size of the substrate in practice for all process of soybean meal SSF. In addition, numerically, SSF with M4 seemed to produce higher yield of peptide than that with P4, although statistically, no significant difference between both substrates. And the reason for these results might need further study to investigate the relation between growth profiles and particle size distributions.

A smaller X_m value and a higher μ value indicated a lower carrying capacity or biocapacity of substrate, which was the maximal load of substrate for peptide production and microbial growth that should be related to living environment, namely the particle size distribution in this study, and a higher growth rate, under those distributions. However, statistically, there was no significant difference between most X_m values obtained, partly due to a limited effect of particle size distribution on X_m and maybe other factors needing further study. In addition, the increase of the growth rate constant (μ) implied that the particle size distribution affected growth rate significantly, which was also showed from the data of the time to maximal growth rate ($t_{Xm/2}$): less time for higher proportion of P4 and P4+P5 (Table 2). Therefore, the mechanism could be inferred that appropriate particle size could improve the corresponding rate of microbial growth.

4. Conclusions

In all, the optimal particle size of soybean meal for peptide production by SSF was 1-1.4mm, and when the proportion of 1-1.4mm/1-2mm in natural soybean meal was increased to 61.30%/58.58% or more, the distribution of particle size was also appropriate for peptide production. Logistic model fitted the data most accurately, and showed a higher growth rate for the substrate with optimal particle size distribution in SSF. Hence the mechanism could be inferred that appropriate particle size could improve the corresponding rate of microbial growth, thus causing an increase of peptide yield in soybean meal SSF within a limited fermentation time.

Acknowledgements

We are grateful to Zhengzhou Siwei Biotechnology Co., Ltd. For supplying the soybean meal sample. The results reported in this paper are from research carried out with financial support from Zhengzhou key scientific and technological project, no. 20120664, "Key technology of fermented soybean meal processed", and no. 121PCXTD518, "Technology innovation team of biotechnology and biomass resources transformation and security".

References

[1] Choct M, Dersjant-Li Y, McLeish J, Peisker M. "Soy oligosaccharides and soluble non-starch polysaccharides: a review of digestion, nutritive and anti-nutritive effects in pigs and poultry." *Asian-Aust J Anim Sci*, 23, pp. 1386-1398, 2010.

[2] Song YS, Pérez VG, Pettigrew JE, Martinez-Villaluenga C, de Mejia EG. "Fermentation of soybean meal and its inclusion in diets for newly weaned pigs reduced diarrhea and measures of immunoreactivity in the plasma." *Anim Feed Sc. Tech*, 159, pp. 41-49, 2010.

[3] Hong K-J, Lee C-H, Kim SW: Aspergillus oryzae GB-107 fermentation improves nutritional quality of food soybeans and feed soybean meals. *J Med Food* 2004, 7: 430-435.

[4] Refstie S, Sahlström S, Bråthen E, "Baeverfjord G, Krogedal P: Lactic acid fermentation eliminates indigestible carbohydrates and antinutritional factors in soybean meal for Atlantic salmon (*Salmo salar*)." *Aquaculture*, 246, pp. 331-345, 2005.

[5] Franěk F, Hohenwarter O, Katinger H. "Plant Protein Hydrolysates: Preparation of Defined Peptide Fractions Promoting Growth and Production in Animal Cells Cultures." *Biotechno Prog*, 16, pp. 688-692, 2000.

[6] Ramana Murthy MV, Karanth NG, Raghava Rao KSMS: "Biochemical Engineering Aspects of Solid-State Fermentation. in: *Advances in Applied Microbiology*." *Volume 38*. Edited by Saul N, Allen IL. Academic Press; pp. 99-147, 1993.

[7] Dilipkumar M, Rajamohan N, Rajasimman M. "Inulinase production in a packed bed reactor by solid state fermentation." *Carbohydr Polym*, 96, pp. 196-199, 2013.

[8] Smail T, Salhi O, Knapp JS. "Solid-state fermentation of carob pods by *Aspergillus niger* for protein production: effect of particle size." *World J Microb Biot*, 11, pp. 171-173, 1995.

[9] Zadrazil F, Puniya AK. "Studies on the effect of particle size on solid-state fermentation of sugarcane bagasse into animal feed using white-rot fungi." *Bioresour Technol*, 54, pp. 85-87, 1995.

[10] Prakasham RS, Rao CS, Sarma PN. "Green gram husk—an inexpensive substrate for alkaline protease production by *Bacillus sp.* in solid-state fermentation." *Bioresour Technol*, 97, pp. 1449-1454, 2006.

[11] Muniswaran PKA, Charyulu NCLN. "Solid substrate fermentation of coconut coir pith for cellulase production." *Enzyme Microb Technol*, 16, pp. 436-440, 1994.

[12] Molaverdi M, Karimi K, Khanahmadi M, Goshadrou A. "Enhanced sweet sorghum stalk to ethanol by fungus Mucor indicus using solid state fermentation followed by simultaneous saccharification and fermentation." *Ind Crops Prod*, 49, pp. 580-585, 2013.

[13] Pandey A. "Recent process developments in solid-state fermentation." *Process Biochem*, 27. pp. 109-117, 1992.

[14] Ellaiah P, Srinivasulu B, Adinarayana K. "Optimisation studies on neomycin production by a mutant strain of Streptomyces marinensis in solid state fermentation." *Process Biochem*, 39. pp. 529-534, 2004.

[15] Wang D, Guang X, Deng X, Wei Z, Ding Y. "Study on fermenting soybean meal by a multi-strain mixtur." *Cereal Feed Ind*, (4), pp. 36-39, 2011.

[16] Lowry OH, Rosebrough NJ, Farr AL, Randall RJ. "Protein measurement with the folin phenol reagent." *J Biol Chem*, 193, pp. 265-275, 1951.

[17] Viccini G, Mitchell DA, Boitl SD, Gern JC, Rosa ASD, Costa RM, Dalsenter FDH. "Analysis of growth kinetic profiles in solid-state fermentation." *Food Technol Biotechnol*, 39, pp. 271-294, 2001.

2

Effects of wastewater irrigation on the growth of two bean spices and soil chemical properties under greenhouse conditions

Hamed Fathi[1], Mojan Mirzanejad[2]

[1]Department of soil Sciences, college of Agriculture, Gorgan University of Agriculture Sciences and Natural Resources Gorgan, Iran
[2]Islamic Azad University, Rasht, Iran

Email address:
hamed_fathi_dokht@yahoo.com (H. Fathi)

Abstract: This study has the objectives to evaluate the short-term effect of different dilutions of the wastewater on the chemical properties of sandy loam and sandy clay soils in two bean species under greenhouse condition. The effect of the irrigation with different wastewater dilutions on the bean crops and their chemical properties were studied. The results showed that the irrigation with wastewater reduced the pH in the sandy loam soil significantly in comparison with sandy clay soil, and this could be due to lower buffering capacity of sandy loam soil as compared to sandy clay soil registering a change in pH increasing. Available N, P and K status, OC and metallic cation content were significantly higher in sandy clay than in sandy loam soils. Generally, sewage water irrigation improved chemical properties and fertility status of both soils. The metallic cations contamination of the bean crops were observed below the maximum permissible limit. However, several folds have been built of metallic cations in soils, fronds and grain of the bean crops indicating that it is possible to increase the metallic cations contamination beyond maximum permissible limit if the waste water is continuously used for irrigating for a long time.

Keywords: Wastewater, Bean, Chemical Properties, Greenhouse Conditions

1. Introduction

The demand for water is continuously increasing in arid and semi-arid countries. Therefore, water of higher quality is preserved for domestic use while that of lower quality is recommended for irrigation. The use of industrial or municipal wastewater in agriculture is a common practice in many parts of the world (Ensink *et al*., 2002;Kizilogluet *al*., 2008). Municipal wastewater is less expensive and considered an attractive source for irrigation in these countries (Al-Rashed and Sherif, 2000; Mohammad and Mazahreh, 2003), and any sources of water which might be used economically and effectively should be considered to promote further development. The total area irrigated with wastewater was conservatively estimated at around 9000 ha (Scott *etal*., 2004).Iran, as a semi-arid country, suffers from shortages in water supplyfor domestic, industrial and agricultural purposes. Water resources in Iran are very limited because of naturally arid and semi-arid conditions. Hence, limited water supplies require careful management for successful agricultural production. The interest in reusing wastewater for irrigation is rapidly growing in Iran as in most countries. Wastewater is a valuable source for plant nutrients and organic matter needed for maintaining fertility and productivity of arid soils (Horswell *et al*., 2003). The reuse of wastewater for irrigation may potentially create environmental problems if not properly treated and managed (Bahri and Brissaud, 1996; Kiziloglu *et al*.,2008). Application of treated wastewater for irrigation of plants and crops is gradually becoming a common practice worldwide (Angelakis *et al*., 2003). Treatment of wastewater, as a segment of water management, usually produces a liquid effluent of suitable quality that can be used for irrigation purposes with minimum impacts on human health or the environment (Qishlaqi *et al*., 2008; Hofstedt, 2005). In contractual wastewater treatment, noteworthy portions of heavy metals remain in the treated effluent if special advanced treatment is not conducted. Thus, long term effects of irrigation with wastewater might include pollution of ground water and soil with heavy metals (Tyagi *et al.*, 2009). Other impacts of treated wastewater in agriculture include the health impacts of possible contamination of crops by pathogenic bacteria and heavy metals (Jiménez, 2006).Also Change in the basic physicochemical properties of soil (soil quality degradation) is another negative impact of wastewater application in agricultural lands (Qishlaqi *et al.*, 2008; Mahmoud *et al.*, 2010). This study has the objectives to evaluate the short-term effect of different dilutions of wastewater on chemical properties of sandy loam and sandy clay soils and assess the chemical build up in two bean spices on pots under greenhouse conditions.

2. Materials and Methods

The experiment was conducted in the greenhouse of department of Soil Science, College of Agriculture, Shiraz University. The municipal wastewater used in this experiment was from the wastewater treatment plant of Marvdasht located near city of Shiraz. The experiment was laid out in factorial whit completely randomized design with three wastewater purification variants: fresh water irrigation (T1), primary treated wastewater and fresh water in equal proportions (T2), primary treated wastewater irrigation (T3) on two bean species (Red bean& Green bean) and two texture soil (sandy loam (SL) and sandy clay (SC) with three replicates). Primary treatment wastewater is the removal of settable organic and inorganic solids by sedimentation and the removal of floating materials by skimming. Approximately 25–50%of the incoming biological oxygen demand (BOD5), 50–70% of the total suspended solids, and 65% of the oil and grease are removed during the primary treatment but colloidal and dissolved constituents are not affected (Pescod, 1992; Kiziloglu *et al.*, 2008). Some characteristics of primary treated wastewater and fresh water are shown in Table 1. The primary treated wastewater and fresh water used for preparing various dilutions have total suspended solids (TSS) and pH within safe range for irrigation (Wu *et al.*, 2009) (Table 1). The soil samples were drawn prior to experimentation and at harvest of plant crops and analyzed for chemical properties following standard methods of analysis (Table 2). Soil pH and EC were measured by pH meter and EC meter from saturated soil paste extracted by vacuum pump. Organic carbon (OC) contents in soil were determined by the wet digestion method (Walkley and Black 1934), total N by the Kjeldhahl method and available P according to Olsen et al. (1954). Available K+ was extracted by vacuum pump, from saturated paste and analyzed by flame emission spectrophotometer, metallic cation were extracted by DTPA and analyzed by atomic absorption spectrophotometer (Page *et al.*, 1982). The dry ashing method was used for analyzing the heavy metal contents in crop samples. Each determination was carried out by accurately measuring a sample of 1 g of a ground sample in a crucible. The crucible with its content was placed in a muffle furnace and ashed at 550° C for 7 h. The ash was digested with 5 ml of 20% (V/V) Analar HCl solution. The residue was filtered into 50-ml volumetric flask using What man filter paper No.41 and the solution was made to the mark with deionised water. Atomic absorption spectrophotometry was used for the heavy metal determination. The sandy loam and sandy clay soils used were non-saline, mildly alkaline and calcareous in nature (Table 2).Also; the electrical conductivity (EC) was slight to moderately saline. The metallic cations content were below the maximum permissible limits as prescribed by Chen (2004).The status of available N and P was medium, while K was high in both the soils. The DTPA extractable Fe, Cu and Mn were high, while Zn was low in both the soils. Data were analyzed statistically the statistical software called SPSS (Abedi-Koupai, *et al.*, 2006).

Table 1.*Characteristics of well and sewage water.*

Charactries	unite	tab water	wastewater
TSS	mg L^{-1}	4.5	85
PH	-Log[H^+]	7.78	7.8
EC	ds m^{-1}	0.51	1.46
Na	me L^{-1}	0.8	10.2
SAR	-------	1.02	6.8
BOD	mg L^{-1}	-------	35

Table 2.*chemical properties of soil prior to experimentation.*

Parameter	unit	sandy loam	sandy clay
EC	ds m^{-1}	0.57	0.72
PH	-Log[H^+]	7.5	8.1
CCE	%	44.5	39.8
availbale N	%	0.056	0.061
availbale p	mg kg^{-1}	18.3	20.5
availible K	mg kg^{-1}	305	348
Zn	mg kg^{-1}	0.68	0.73
Cu	mg kg^{-1}	0.61	0.78
Mn	mg kg^{-1}	4.35	5.87
Fe	mg kg^{-1}	8.7	10.13
Pb	mg kg^{-1}	Trace	Trace
Ni	mg kg^{-1}	Trace	Trace
Cd	mg kg^{-1}	Trace	Trace
OM	%	1.1	1.32

3. Results and Discussion

3.1. Effect on Soil Properties

3.1.1. Chemical Properties

The effect of soil texture, wastewater dilutions and vegetable crops on some of the chemical properties of the soils was analyzed (Tables 3A & 3B). The decrease in pH was more in the sandy loam soil as compared to sandy clay soil, and this could be due to lower buffering capacity of sandy loam soil as compared to sandy clay soil registering a change in pH (Gelsomino, *et al.*, 2006). A decrease in pH of the soil with the use of wastewater dilutions for irrigation might be due to lower pH of wastewater being used for irrigation. Significantly highest value of EC was observed for sandy clay soil irrigated with wastewater as such as compared to rest of the treatment combinations indicating a higher build-up of salts in moderately fine textured soil. This could be ascribed to its high salt content with a mean value of 1.46 $dS.m^{-1}$ as against 0.51 $dS.m^{-1}$ of well water and hence contributing more dissolved salts in the soil. Similar results of pH were reported by Malla, *et al.*, 2007 and Ganeshamurthy, *et al.*,2008) and of EC were reported by (Malla, et al., 2007). On the other hand, the soils on which red bean grown soils recorded a small quantify EC than green bean grown soils. Available N, P and K status, OC and metallic cation content values were significantly higher in sandy clay than in sandy loam soils. It was further enhanced with wastewater irrigation as such or in its dilution indicating improvement in the fertility status of the soils except buildup of metallic cations in the soil. The increase in N, P and K content of soil with prolonged irrigation was also reported by (Masto, *et al.*,2009). Sandy loam soil contained a higher value of $CaCO_3$content which decreased on application of wastewater as such or in its dilution.

3.1.2. Metallic Cations Build up

The table 3B shows the Average of heavy metal concentration in soil of different irrigation treatments when vegetable crops were irrigated with waste water as such with reference to the control. Considering the mean values of the content buildup of metallic cations irrespective of the crops grown on them, a higher content buildup of Zn (2.58 ppm) and Ni(0.43 ppm) was recorded for sandy clay soil while, the buildup of Cd (1.4 ppm) was comparatively higher in sandy loam soil. Similarly, irrespective of the textural class of the soils buildup of Zn (2.58 ppm), Cu (2.48 ppm), Fe (17.78 ppm), Mn (10.93 ppm), Cd (1.4 ppm) and Ni (0.43 ppm) was relatively higher in soil where green bean was grown. These may be ascribed to differential removal of cations from the soil by the vegetable crops under study (Malla, R., &Totawat, K. L. 2006). The increase in DTPA extractable metallic cations was also reported by (Rana, *et al.*, 2010).

Table 3A.*Effect of soil texture, sewage dilutions and vegetable crops on chemical properties of the soils.*

species	Texture	Treatment	PH	EC $ds\ m^{-1}$	OM	Caco3	N	P	K
					----------------	-%----------------		--------mg kg^{-1}-------	
G	SL	T1	7.62	0.62	1.22	39.8	0.0598	16.7	288
G	SL	T2	7.48	0.856	1.47	36.74	0.073	20.9	283
G	SL	T3	7.21	1.12	1.83	33.23	0.823	26.8	278.9
G	SC	T1	8.14	0.87	1.41	35.8	0.065	18.88	327.6
G	SC	T2	7.95	1.13	1.72	32.98	0.83	24.86	324.2
G	SC	T3	7.78	1.62	2.18	31.27	0.102	29.36	320.74
R	SL	T1	7.59	0.61	1.21	39.6	0.051	15.9	286.5
R	SL	T2	7.43	0.89	1.4	35.68	0.069	20	282.3
R	SL	T3	7.2	1.2	1.78	32.97	0.81	25.1	9
R	SC	T1	8.16	0.84	1.38	36.2	0.062	17.23	327
R	SC	T2	7.93	1.09	1.675	33.85	0.079	22.36	322
R	SC	T3	7.81	1.91	2.12	32.1	0.989	27.45	319

G: Green BeanR: Red BeanSL: Sandy LoamSC: Sandy Clay

Table3B. *Effect of soil texture, sewage dilutions and vegetable crops on chemical properties of the soils.*

species	Texture	Treatment	Fe	Zn	Cu	Mn	Cd	Ni
			-------------------------------	---mg kg^{-1}---	-------------------------------			
G	SL	T1	6.95	0.573	0.498	4.1	T	T
G	SL	T2	10.23	1.23	1.11	6.88	0.8	0.158
G	SL	T3	12.79	1.95	1.65	8.45	1.4	0.346
G	SC	T1	9.112	0.678	0.711	5.64	T	T
G	SC	T2	14.23	1.632	1.84	9.23	0.65	0.2
G	SC	T3	17.78	2.58	2.48	10.93	1.1	0.43
R	SL	T1	6.9	0.58	0.493	3.96	T	T
R	SL	T2	9.65	1.12	0.98	6.5	0.65	0.146
R	SL	T3	11.3	1.79	1.58	8.1	1.3	0.324
R	SC	T1	8.88	0.61	0.683	5.4	T	T
R	SC	T2	14.7	1.55	1.72	9	0.48	0.187
R	SC	T3	16.88	2.43	2.42	10.88	0.91	0.394

G: Green BeanR: Red BeanSL: Sandy LoamSC: Sandy ClayT: Trace

3.2. Effect on Plant Composition

3.2.1. Metallic Cations Buildup in Grain and Fronds

The content buildup of metallic cations in the fronds and grain of crops when irrigated with waste water as such with reference to the control is given in Tables 4A and 4B, and figure 1. Considering the mean values of the content buildup of metallic cations in the fronds of crops irrespective of the type of crop grown, comparatively higher buildup of Zn (97.6 ppm), Cu (9.7 ppm), Fe (266.4 ppm) and Mn (179.9 ppm) was observed when crops were grown on sandy clay soil. Irrespective of the textural class of the soil, content buildup of Fe, Cu, Mn, and Zn was higher in the fronds of kidney bean. Similar to the content buildup of metallic cations in fronds, comparatively higher buildup of Zn (42.2 ppm), Cu (29.9 ppm), Fe (258 ppm) and Mn (75.3 ppm) was recorded in the grain of crop grown on sandy clay soil. Irrespective of the textural class of the soils, content buildup of Fe, Cu, Mn, and Zn was higher in the fronds of kidney bean. The variations in the content buildup of metallic cations in the soils and vegetable crops may be attributed to a greater variation in the initial values of the metallic cations in the soil prior to experimentation and preferential absorbance of a particular cation by different vegetable crops under study (Sharma and Shukla, 2013). Even if the concentration of heavy metal in sewage effluents is low, the extent of build-up of metals in sewage/waste water-irrigated soils often depends on the period of its application (Malla, *et al.*, 2007). In spite of higher content buildup of metallic cations in the vegetable crops, they were all within the safe limit prescribed by (WHO, 1996).

Table 4A. *Metalliccations build up in grain in sandy loam soil.*

SPICES	Part of plant	Irrigation Treatment	N%	P	K	Cl	Ca	Ni
			----------mg kg^{-1}----------					
G	Grain	T1	8.46 E	0.453 F	0.451 F	2.987 E	N.D	N.D
G	Grain	T2	9.01 ED	0.519 E	0.534 E	3.198 E	N.D	N.D
G	Grain	T3	9.54 CD	0.567 D	0.583 D	3.421 D	N.D	N.D
R	Grain	T1	10.2 BC	0.588 C	0.609 C	2.9 C	N.D	N.D
R	Grain	T2	10.86 AB	0.675 B	0.723 B	3.8 B	N.D	N.D
R	Grain	T3	11.54 A	0.732 A	0.787 A	4.7 A	N.D	N.D
		Metallic cations build up in fronds in sandy loam soil						
G	Fronds	T1	4.83 D	0.323 F	2.164 F	6.328 E	N.D	N.D
G	Fronds	T2	5.33 C	0.386 E	2.264 D	6.249 D	N.D	N.D
G	Fronds	T3	5.95 B	0.438 C	2.352 E	6.153 F	N.D	N.D
R	Fronds	T1	4.08 E	0.42 D	2.921 C	6.8 C	N.D	N.D
R	Fronds	T2	6.04 B	0.501 B	3.056 B	7.2 B	N.D	N.D
R	Fronds	T3	7.168 A	0.576 A	3.175 A	7.6 A	N.D	N.D

N.D: Not detection

Table 4B. *Metallic cations build up in grain in sandy clay soil.*

SPICES	Part of plant	Irrigation Treatment	N%	P	K	Cl	Ca	Ni
			----------mg kg^{-1}----------					
G	Grain	T1	9.06 F	0.49 C	0.48 C	3.4 C	N.D	N.D
G	Grain	T2	9.69 E	0.563 B	0.57 BC	3.63 BC	N.D	N.D
G	Grain	T3	10.3 D	0.61 B	0.62 BC	3.9 AB	N.D	N.D
R	Grain	T1	11.4 C	0.6 B	0.8 BC	3.638 BC	N.D	N.D
R	Grain	T2	12.6 B	0.71 A	0.85 B	3.884 AB	N.D	N.D
R	Grain	T3	14.1 A	0.748A	1.2 A	4.173 A	N.D	N.D
		Metallic cations build up in fronds in sandy loam soil						
G	Fronds	T1	5.2 D	0.35 D	2.3 C	7.2 A	N.D	N.D
G	Fronds	T2	5.74 C	0.418 CD	2.40 C	7.11 A	N.D	N.D
G	Fronds	T3	6.4 B	0.48 C	2.5 C	7 A	N.D	N.D
R	Fronds	T1	4.3 E	0.8 B	3.6 B	7.704 A	N.D	N.D
R	Fronds	T2	6.3 B	0.88 AB	3.9 AB	7.607 A	N.D	N.D
R	Fronds	T3	7.2 A	0.9 A	4.2 A	7.49 A	N.D	N.D

N.D: Not detection

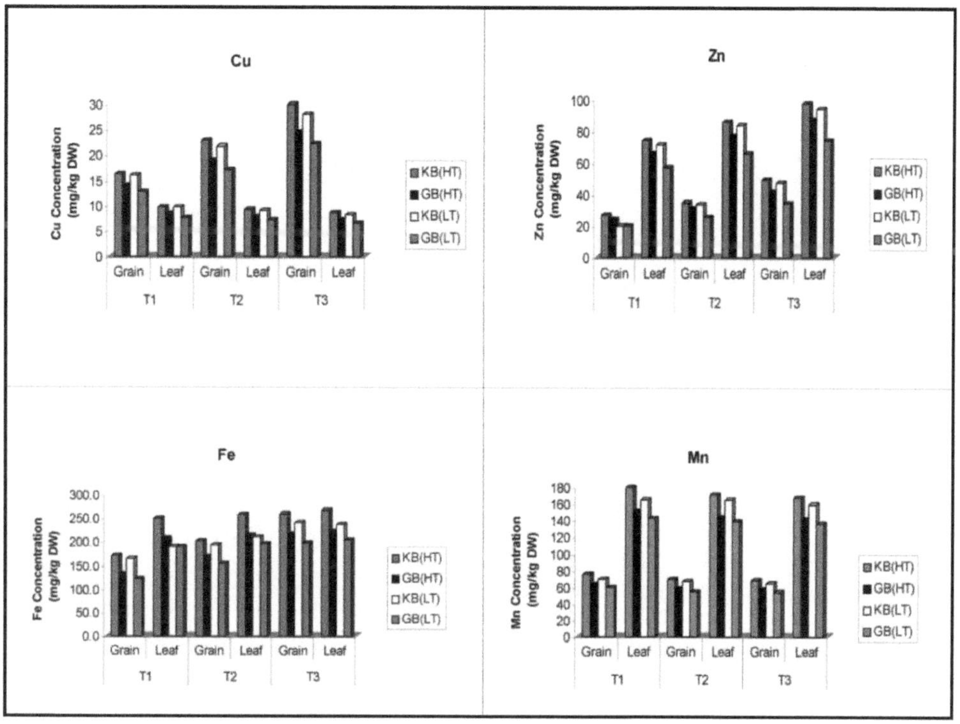

Figure 1. *The content builds up of metallic cations in the dry weight of fronds and grain of two beans (KB: Red Bean GB: Green Bean HT: Sandy clay LT: Sandy loam).*

4. Conclusions

Generally speaking, sewage water irrigation improved chemical properties and fertility status of both soils. The metallic cations contamination of vegetable crops was observed to be below the maximum permissible limit. However, several folds buildup of metallic cations in soils, fronds and grain of vegetable crops under study indicated that there might be an increase in metallic cations contamination beyond maximum permissible limit if such waste water is continuously used for irrigation on long term. Thus timely monitoring of sewage water, soil and crops is essential to maintain soil quality for consumption of safe foods grown on them. Establishment of at least a primary sewage treatment plant is suggested where raw sewage is being diverted for irrigation.

References

[1] Abedi-Koupai, J., Mostafazadeh-Fard, B., Afyuni, M., &Bagheri, M. R. (2006).Effect of treated wastewater on soil chemical and physical properties in an arid region. Plant Soil and Environment, 52(8), 335.

[2] Al-Rashed, M. F., & Sherif, M. M. (2000).Water resources in the GCC countries: an overview. Water Resources Management, 14(1), 59-75.

[3] Angelakis, A. N., Bontoux, L., &Lazarova, V. (2003).Challenges and prospectives for water recycling and reuse in EU countries. Water Science & Technology: Water Supply, 3(4).

[4] Bahri, A., &Brissaud, F. (1996).Wastewater reuse in Tunisia: assessing a national policy. Water Science and Technology, 33(10), 87-94.

[5] Chen, Y., Wang, C., Wang, Z., & Huang, S. (2004). Assessment of the contamination and genotoxicity of soil irrigated with wastewater. Plant and Soil, 261(1-2), 189-196.

[6] Ensink, J. H., Van der Hoek, W., Matsuno, Y., Munir, S., &Aslam, M. R. (2002).Use of Untreated Wastewater in Peri-Urban Agriculture in Pakistan.

[7] Ganeshamurthy, A. N., Varalakshmi, L. R., &Sumangala, H. P. (2008). Environmental risks associated with heavy metal contamination in soil, water and plants in urban and periurban agriculture. J HorticulturSci, 3(1), 1-29.

[8] Gelsomino, A., Badalucco, L., Ambrosoli, R., Crecchio, C., Puglisi, E., &Meli, S. M. (2006). Changes in chemical and biological soil properties as induced by anthropogenic disturbance: A case study of an agricultural soil under recurrent flooding by wastewaters. Soil Biology and Biochemistry, 38(8), 2069-2080.

[9] Hofstedt, C. (2005). Wastewater use in Agriculture in Andhra Pradesh, India: An evaluation of irrigation water quality in reference to associated health risks and agricultural suitability.

[10] Horswell, J., Speir, T. W., & Van Schaik, A. P. (2003). Bio-indicators to assess impacts of heavy metals in land-applied sewage sludge. Soil Biology and Biochemistry, 35(11), 1501-1505.

[11] Jiménez, B. (2006). Irrigation in developing countries using wastewater. International Review for Environmental Strategies, 6(2), 229-250.

[12] Kiziloglu, F. M., Turan, M., Sahin, U., Kuslu, Y., & Dursun, A. (2008). Effects of untreated and treated wastewater irrigation on some chemical properties of cauliflower (< i> Brassica olerecea</i> L. var. botrytis) and red cabbage (< i> Brassica olerecea</i> L. var. rubra) grown on calcareous soil in Turkey. Agricultural water management, 95(6), 716-724.

[13] Kiziloglu, F. M., Turan, M., Sahin, U., Kuslu, Y., & Dursun, A. (2008). Effects of untreated and treated wastewater irrigation on some chemical properties of cauliflower (< i> Brassica olerecea</i> L. var. botrytis) and red cabbage (< i> Brassica olerecea</i> L. var. rubra) grown on calcareous soil in Turkey. Agricultural water management, 95(6), 716-724.

[14] Mahmoud, A., Olivier, J., Vaxelaire, J., &Hoadley, A. F. (2010).Electrical field: A historical review of its application and contributions in wastewater sludge dewatering. Water research, 44(8), 2381-2407.

[15] Malla, R., & Totawat, K. L. (2006).Effect of Municipal Sewage on Soil Properties and Chemical Build-up in Vegetable Crops Grown on Haplustepts of Sub-humid Southern Plains of Rajasthan. Journal of the Indian Society of Soil Science, 54(2), 226-231.

[16] Malla, R., Tanaka, Y., Mori, K., & Totawat, K. L. (2007).Effect of short-term sewage irrigation on chemical build up in soils and vegetables.

[17] Masto, R. E., Chhonkar, P. K., Singh, D., &Patra, A. K. (2009).Changes in soil quality indicators under long-term sewage irrigation in a sub-tropical environment. Environmental geology, 56(6), 1237-1243.

[18] Mohammad, M. J., & Mazahreh, N. (2003).Changes in soil fertility parameters in response to irrigation of forage crops with secondary treated wastewater. Communications in soil science and plant analysis, 34(9-10), 1281-1294.

[19] Page AL (1982). Methods of soil analysis, part 2: chemical and microbiological properties. Agronomy monograph No. 9 (part 2). ASA/SSSA, Madison, Wisconsin

[20] Pescod, M. B. (1992). Wastewater treatment and use in agriculture.

[21] Qishlaqi, A., Moore, F., & Forghani, G. (2008).Impact of untreated wastewater irrigation on soils and crops in Shiraz suburban area, SW Iran. Environmental monitoring and assessment, 141(1-3), 257-273.

[22] Olsen, S. R., C. V. Cole, F. S. Watanabe, and L. A. Dean.1954.Estimation of available phosphorous in soil by extraction with sodium bicarbonate. USDA Circ. 39, U. S. Gover. Prin. Office Washington, DC.

[23] Rana, L., Dhankhar, R., &Chhikara, S. (2010). Soil Characteristics Affected by long term Application of Sewage Wastewater. International Journal of Environmental Research, 4(3).

[24] Scott, C. A., Faruqui, N. I., & Raschid-Sally, L. (2004).1.Wastewater Use in Irrigated Agriculture: Management Challenges in Developing Countries.

[25] Sharma, D., & Shukla, A. K. (2013). Analysis of heavy metals in soil and vegetables irrigated with wastewater in Meerut city region. Int. J. Adv. Life Sci, 6(4), 287-292.

[26] Tyagi, R. D., Surampalli, R. Y., Yan, S., Zhang, T. C., Kao, C. M., & Lohani, B. N. (2009). Sustainable sludge management: production of value added products. ASCE Publications.

[27] Walkley, A., & Black, I. A. (1934).An examination of the Degtjareff method for determining soil organic matter, and a proposed modification of the chromic acid titration method. Soil science, 37(1), 29-38.

[28] World Health Organization, 1996. Water Quality Monitoring - A Practical Guide to the Design and Implementation of Freshwater Quality Studies and Monitoring Programmers.

[29] Wu, L., Chen, W., French, C., & Chang, A. C. (2009). Safe application of reclaimed water reuse in the southwestern United States (Vol. 8357).UCANR Publications.

3

Municipal solid waste characterization and it is assessment for potential compost production

Habib Fathi[1], Abdolhossinpari Zangane[2], Hamed Fathi[3, *], Hossein Moradi[4]

[1]Academic center for Education, Culture and Research, Environmental Research Instituterasht, Iran
[2]Zanjan University Department of chemistry Zanjan, Iran
[3]Department of soil Sciences, college of Agriculture, Gorgan University of agriculture sciences and Natural resources Gorgan, Iran
[4]Higher educational complex of saravan, Iran

Email address:
hamed_fathi_dokht@yahoo.com (H. Fathi)

Abstract: The increasing amount of municipal solid waste (MSW) generations and their improper disposal have great social costs, and is an issue of increasing concern. There are strict mandatory targets in many nations to reduce the amount of MSW entering the landfill. At present, the source reducing, recycling, reusing, waste combusting, and also composting the MSW are the major strategies which are being advised spread across the world. Municipal solid waste management (MSWM) practices in Zanjan, a city with population of about 350,000 persons, are generating about 300 tons of MSW daily those are consisting of the collection, transportation and land filling of the wastes. The wastes are not recycled and composted but just mainly land filled. Therefore, this research is conducted to study the generation, characterization and management strategies of solid wastes in Zanjan city. The samplings of MSW were performed in all four seasons during 2010 and the contents of its main components are determined. The qualitative and quantitative results indicated that the composting of MSW is feasible and can be adopted as a proper MSWM approach in the studied area. Recycling of paper, plastic and glass along with composting of MSW are highly recommended.

Keywords: Municipal Solid Waste Management, Composting, Recycling, Zanjan, Solid Waste

1. Introduction

Solid waste is a serious environmental problem in both developing and developed nations. Most developing countries have recently started to improve their municipal solid waste management practices. Inadequate management of solid waste leads to problems that impair human and animal health and ultimately result in economic, environmental and biological losses (Sharholy et al., 2008). In these countries MSW is generally disposed of in low-lying areas without taking any precautions or operational controls. Therefore, MSWM is one of the major environmental problems confronting many developing countries. MSWM involves activities associated with waste generation, its storage, collection, transfer and transport, processing and disposal. In most Iranian cities, the MSWM system comprises only for four activities, i.e., waste generation, collection, transportation, and disposal (Ziaee et al., 2012). The management of MSW requires proper infrastructure, maintenance and upgrade for all activities. This is very expensive and complex due to the continuous and unplanned growth of urban centers. Providing the desired level of public service in the urban centers is often attributed to the poor financial status of the managing municipal corporations (Ahsan., 1999, Mor., 2006, Raje et al., 2001, Siddiqui et al., 2006, Kumar Kaushal et al., 2012). Per capita generation and characteristics of household waste depends on household income levels and area characteristics. MSW generation in Asia in 1998 was 0.76 million ton/day (Jin et al., 2006) with a growth rate of 2-3 percent annually in developing countries as compared to 3.2 - 4.5 percent in developed countries. The per capita MSW generation rate is varied in different parts of Asia ranging from 0.88 kg/day in Iran (Damghani et al., 2008) to 0.95 kg/day in Turkey (Metin et al., 2003) and 1.21 kg/day in China (El-Salam, M. M. A. 2013).

The MSW disposal problems can be reduced by waste minimization, reuse and recycling of its components and converting the organic components of the waste into a valuable product and reused without adversely affecting the environment. This can be achieved by converting the waste into biogas through anaerobic digestion and into plant nutrients by way of composting (Mbuligwe and Kassenga2004, Jha l et al., 2013). Compost characteristics are essentially dependent upon the raw materials and the factors that affect the progress of the process. The factors affecting the composting process can be divided into two groups: those depending on the formulation of the composting mix, such as nutrient balance, pH, particle size, porosity and moisture; and those dependent on the process management, such as O_2 concentration, temperature and water content. Nutritional balance is mainly defined by the Carbon/Nitrogen ratio. Microorganisms require an energy source (degradable organic Carbon) and Nitrogen for their development and activity (Hubbeet al., 2010).

This study was conducted in ZanjanProvince which occupies about 1.3% of Iranian territory and has an area of 22,164 square kilometers. The province is located in the northwestern part of Iran between latitudes 36^0 27' and 37^0 15' N and longitudes 47^0 17' and 48^0 54' E. Zanjan Province has a population of nearly 1,077,254 people which is about 1.5% of Iran's population. Zanjan city is the capital of the province with a population of about 350,000 people. The municipality of Zanjan is responsible for all aspects of solid waste management. Zanjan has a main municipality and is composed of 3 sub municipalities. MSWM practices in Zanjan, generating about 300 tons of MSW daily, consist of collection of waste, waste transportation and land filling. The waste is mainly land filled and recycling and composting are not yet introduced. This paper presents an overview of current MSWM system in Zanjan city, and provides several recommendations for its improvements.

2. Materials and Methods

To perform household wastes management studies in any city, it is necessary to know the constituents of wastes; that is physical and chemical analysis, measurement of weight, volume, quality and quantity of wastes produced in different seasons. Hence in this method, based on the most recent instructions available, the wastes from different localities of Zanjan city were collected during seven continues days in the mid of each seasons during 2010. After analysis and evaluation of MSW, the results were used in studying the quality of wastes from Zanjan City.

2.1. Sample Collection and Segregation

MSW samples were collected through seven working days in each season and from each of four different communities (i.e. High class districts, Middle economic class, Low class and commercial districts).This means a total number of 20 samples were collected and segregated per day from four different socioeconomic localities. The samples were then segregated manually into different physical components such as; organic material, paper, plastics, rubber, wood, glass, metals, textiles and PET. Each of these recyclable materials was weighted to determine its fraction in the total solid waste sample collected. The remaining material was a uniform mixture of soil, mud, sand and other inert materials that were not manually separable, and is termed mixed organic materials. 1.5 kg of this mix from each sample was collected in polyethylene bags, brought to the laboratory, and analyzed for moisture content immediately. The remaining samples were stored and analyzed for other important parameters.

2.2. Laboratory Sample Analyses

Representative samples (organic components) were first grounded to homogenous powder in a miller apparatus and then the following analyses were carried out.

Moisture content: Samples of 100 g were taken in triplicate, dried to a constant weight in an oven at 105°C for 24 h, cooled in a desiccators and the difference in weight was recorded (Vesilind et al., 2003). Moisture content was calculated using Eq. (1)

$$M_n = ((W_w-W_d)/W_w) \times 100 \qquad (1)$$

M_n = moisture content (%) of material

W_W = wet weight of the sample, andW_d = weight of the sample after drying

Carbon/Nitrogen Ratio: Total Kjeldahl nitrogen and volatile solids (VS) were analyzed by standard methods (APHA, 1998). C/N ratio was calculated using Eq. (2)

$$\frac{C}{N} = \frac{(vs \times 0.58)\% drymatter}{(Kjeldhal - N)\% drymatter} \qquad (2)$$

pH: pH of the water extract (1:5) was measured using a pH meter (Metrohm model 780).

Statistical sample analyses**:** In this paper, SPSS software (version 11.5) was used for the statistical analysis of the data.

3. Results and Discussion

Waste production and composition depend on many factors, such as the stage of development; socio-economic, climatic and geographical conditions; and collection frequency (Collivignarelli et al., 2004). Data on quantity variation and generation are useful in planning for a collection and disposal system (Sharholy et al., 2008).The current state of Zanjan MSW management, its challenges and our recommendations for improvement of the system are discussed in this section.

3.1. Solid Waste Generation and Characteristics

A comparison of the average amounts of MSW components of Zanjan city in different seasons are reported in Table 1 & 2. The amount of waste produced in Zanjan is 300 tons per day. Qualitative study of the MSW sampled in different seasons, different social areas of the city (According to the culture of consumption and waste production that is associated with life)and different economic classes shows that maximum percent of organic materials in MSW are found in low economy class in spring and minimum organic contents belongs to samples from commercial areas in autumn. On the other hand, the maximum total dry recyclable waste in MSW is from commercial regions in summer, and minimum such contents are related to the lower economic classes and in autumn. The highest dispersal is related to organic materials.

Table 1.*Composition of municipal solid waste in different socioeconomic localities (high, middle, low and commercial localities).*

Socio-economic Localities	organic material(Percent %)	paper (Percent %)	Card board (Percent %)	rubber (Percent %)	plastic (Percent %)	PET (Percent %)	textiles (Percent %)	glass (Percent %)	iron (Percent %)	other metal (Percent %)	sand (Percent %)	wood (Percent %)	D(kg/m^3)
High Class	64.1	5.3	4.9	0.1	9.3	2.7	3.0	4.0	0.7	2.5	1.9	1.2	189
Middle Class	66.1	3.9	5.3	0.2	9.1	2.3	3.5	5.1	0.6	1.9	0.8	1.1	194
Low Class	67.5	3.7	4.5	0.9	7.7	1.9	2.9	3.1	0.9	1.8	2.6	2.2	275
Commercial	46.7	6.6	5.8	1.4	12.6	2.2	5.8	3.1	3.3	3.2	4.4	4.4	187
Mean	61.4	4.8	5.1	0.7	9.7	2.2	3.8	3.0	1.4	2.3	2.4	2.2	210

Table 2.*Composition of municipal solid waste in different seasons (by percent)*

Seasons	organic material	paper	Card board	rubber	plastic	PET	textiles	glass	iron	other metal	sand	wood	D(kg/m^3)
Autumn(October)	57.5	4.4	4.0	0.0	9.5	1.5	6.1	5.3	2.4	1.7	3.0	3.5	160
Winter(January)	62.9	4.9	5.7	0.7	11.4	2.1	2.5	3.1	0.4	2.6	2.4	1.2	206
Spring(April)	63.5	5.3	5.3	0.8	9.4	2.4	3.3	3.2	1.0	2.5	2.2	1.9	201
Summer(July)	61.5	4.7	5.3	1.4	8.5	3.0	3.2	3.7	1.9	2.5	2.0	2.2	276
Mean	61.4	4.9	5.1	0.7	9.7	2.2	3.8	3.8	1.4	2.3	2.4	2.2	210

3.2. Waste Handling and Separation/Processing at Source

The method of handling, storage and processing of solid wastes at the source plays an important role in public health, aesthetics and the efficiency of the MSW system (Abdoli, 1995). Source separation of wastes will not only bring economic benefits, but also make the recycling of other components more efficient (Aydin, 2004).Unfortunately, separation of household wastes is not yet considered in MSWM program in Zanjan city.

3.3. Collection of Solid Wastes

Collection, transportation and disposal of MSW are the critical problems of the MSWM in any city today (Kinman, 1987).The organic parts of MSW can be easily degraded, which causes problems in storage containers. In Zanjan, municipalities are fully responsible for waste collection and transportation directly, using their own infrastructure, and indirectly through private sector contracts. At present, both mechanical and manual methods are used to collect residential waste. Although mechanical methods of collection have been implemented in some parts of the city, direct lifting and carrying of waste to the collection vehicles is the most commonly used method for waste collection in most parts of Zanjancity .Almost all of the MSW generated is collected daily (including holidays) and are transported to Temporary Transfer Station (TTS) or disposal site.

3.4. Transport and Transfer

In recent years, the use of small TTS in different cities of Iran has become popular (Moghadam 2009) .At present, there is only one TTS in Zanjan city. The main reason for using a transfer stations is to decrease transport costs. The type of the transfer station in Zanjan is direct load. The waste that is collected by small vehicles is discharged into an open top trailer.

3.5. Final Disposal

Safe and reliable long-term disposal of waste is an important component of integrated solid waste management.Land filling, incineration and composting are three main methods of MSW disposal in the world.In Iran, land filling, open dumping and composting are the main methods of waste disposal (Moghadam 2009). In Zanjan, only open dumping method for waste disposal is adopted.

3.6. Evaluation of Composting by MunicipalSolid Waste in Zanjan

Composting is generally defined as the biological oxidative decomposition of organic constituents in wastes of almost any nature under controlled conditions. Since composting is a biological process of decomposing organic materials, it requires special conditions, particularly of temperature, moisture, aeration, pH, particle size and C/N ratio, related to optimum biological activity in the various stages of the process. The main products of aerobic composting are carbon dioxide, water, mineral ions and stabilized organic matter, often called humus. The process is accomplished through different phases(Sharma et al., 1997).

Amount of biodegradable materials in MSW of Zanjan is 75.2 percent, which comes to 225.6 tons per day and composting can be produced by utilizing almost 60 percent of MSW transferred to landfill.

The chemical composition of the bulk material affects some parameters, especially the C/N ratio. Very high C/N ratios delay the microbial metabolism, whereas low values cause the loss of nitrogen through ammonia volatilization (Sharma et al., 1997). Residual carbon to nitrogen ratio of Zanjan ranges between 14.22 to 19.53 and the average is 17.6. That is, its value is much lower than standards (Table 3). To obtain a high quality compost, the C / N ratio of the waste can be adjusted to an optimum level by adding cow manure, poultry manure, garden waste etc(Jilani 2007). Another parameter of significant importance is the percentage of water content in the feed material. It is to be noted that the higher water content may cause problems for the complete oxygenation of the material, whereas the lower values, on the other hand, may interrupt the process prematurely (Sharma et al., 1997).Relative Humidity interval in MSW of Zanjan city is 67.94 to 70.3 with a mean of 69.2. The standard rates of more than 50 were identified by Zucconin&deBestoldi (1987) as shown in table 3. As the mean relative humidity of MSW in Zanjan is high, it has to be reduced by aeration during the process of composting.

The parameters, such as pH, alkalinity and volatile acids, are closely inter-related in the composting process. In the beginning of the process, the formation of carbon dioxide and organic acids causes values of approximately 5-6, whereas, as the process progresses, the pH value reach even up to 8-8.5. This is mainly due to the decomposition of proteins, as well as elimination of the carbon dioxide (Sharma et al., 1997).

A pH of 6.7–9.0 supports good microbial activity during composting. Optimum values are between 5.5 and 8.0 (Bertoldi et al., 1983 and Miller., 1992). Usually pH is not a key factor for composting since most materials are within this pH range. However, this factor is very relevant for controlling N-losses by ammonia volatilization, which can be particularly high at pH >7.5. Elemental sulphur (S) has been used as an amendment for avoiding excessively high pH values during composting (Mari et al., 2005). Average pH in four different seasons was 5.45, which is almost close to the standards (Table 3).

Concerning biological aspects, it is necessary that the starting material is qualified physically, chemically and biologically. For example, the size and consistency of particles should be such that they confer suitable porosity to the bulking material, avoiding exaggerated packing and, thus, allowing free air passage, even in a wasting pile of nearly 2 m height. In order to avoid any eventual slowing of the microbiological degradation, the particle size must be in the cm range because the microbial transformation of a substance is directly proportional to the surface area available (Sharma et al., 1997). Most organic materials in MSW of Zanjan are in range of 8-40 mm (Table 4). Therefore in production of compost it is necessary to mill the organic components to have a more homogenous mater.

Table 3. *Average chemical composition of MSW in Zanjan and the standard values Suitable for composting*

MSWComponents	Autumn	Winter	Spring	Summer	Mean	*Standard values suitable forcomposting
Organic matter (% wet basis)	57.53	62.95	63.54	61.53	61.38	>20
C	44.50	38.46	37.90	55.35	44.05	No specs
N	3.20	2.30	2.01	2.94	2.61	>0.6
C/N	14.23	17.96	19.53	18.91	17.66	25-50:1
pH	5.25	5.49	5.515	5.55	5.45	5.5-8
Moisture	67.94	70.30	69.28	69.33	69.21	>50

* Standard (Zucconin&deBertoldi, 1987)

Table 4. *Composition of particle size in MSW*

Particle size(mm)	Autumn	Winter	Spring	Summer	Mean
8>	10.42	10.84	10.26	10.85	10.592
8-120	51.22	50.67	49.47	49.22	50.147
40-120	33.70	32.14	33.80	34.27	33.477
120<	4.62	6.32	4.48	5.53	5.237

3.7. Seasonal Change in MSW Characteristics

ANOVAs test was used to differentiate organic material contents and C / N ratio in samples collected in four different seasons with three degrees of freedom, mean squares of respectively 51.3 and 39.5 in the 5% significance level test (0.014 and 0.04) are significant. Relative humidity also with three degrees of freedom, mean squares of 6.6, and significant level of 5% is not significant.

4. Conclusion

The waste analysis of Zanjan showed that the percentage of organic waste in the MSW is substantially high, mainly due to the use of unprocessed food in the daily diet of inhabitants. At present, both mechanical and manual methods are used for the collection of residential wastes. Due to high organic components and its high moisture content, the MSW must be collected daily and transferred to disposal site, substantially increasing the cost of solid waste management.

Composting of MSW are feasible in the studied area and compost production can be chosen as a proper management methodutilizingabout60 percent of MSW generated, and substantially reducing the amount of waste land filled .Since composting is a biological process of decomposing organic materials, it requires special conditions, particularly of temperature, moisture, aeration, pH, particle size and C/N ratio. 61.38 percent of the total MSW produced in Zanjan (184.2 tons per day) is biodegradable and can be used for compost production. Average C/N ratio of the MSW was low (17.6) and must be adjusted to an optimum level by adding cow or poultry manure. Average pH values of MSW were 5.4 which are almost close to optimum values for composting. Mean relative humidity was high (69.2%) and may be reduced by aeration during the process of composting. In order to avoid any eventual slowing of the microbiological degradation during the process of composting, it is necessary to mill the organic components to have a more homogenous mater. Recycling of paper, plastic and glass along with composting of MSW are also advised.

Acknowledgment

Department of Environment in Zanjan province is acknowledged for their financial and scientific collaboration.

References

[1] Abbasi FM, Ahmad H, Perveen F, Inamullah, Sajid M, Brar DS 2010.Assesment of genomic relationship between *Oryza sativa* and*Oryzaaustraliensis*. Afr. J. Biotechnol. 9(12): 1312-1316.

[2] Abdoli MA, 1995. Solid waste management in Tehran. Waste Management and Research. 13: 519–531.

[3] Ahsan N, 1999. Solid waste management plan for Indian megacities.Environmental Protection. 19 (2): 90–95.

[4] APHA (American Public Health Association), AWWA (American Water Works Association), and WEF (Water Environmental Federation), Standard Methods for the Examination of Water and Waste Water, 20th Ed, APHA, Washington DC, 1998.

[5] Aydin GA, and Kocasoy G, 2004.Significance of Source Separation and Composting of Wastes of Istanbul: From Theory To Practice. Paper Presented at the ISWA World Congress. Rome. October 17–21.

[6] Damghani A M, 2008. Municipal solid waste management in Tehran: current practices, opportunities and challenges. Waste Management. 28: 929-934.

[7] El-Salam, M. M. A. (2013).Journal of Environmental and Occupational Science. Journal of Environmental and Occupational Science, 2(3), 131-140.

[8] Farrell.M and Jones.DL, 2009.Critical evaluation of municipal solid waste composting and potential compost markets.Bioresource Technology.100: 4301- 4310.

[9] Hubbe, M. A., Nazhad, M., & Sánchez, C. (2010).Composting as a way to convert cellulosic biomass and organic waste into high-value soil amendments: A review.BioResources, 5(4), 2808-2854.

[10] Jha, A. K., Li, J., Nies, L., & Zhang, L. (2013). Research advances in dry anaerobic digestion process of solid organic wastes. African Journal of Biotechnology, 10(64), 14242-14253.

[11] Jilani S, 2007. Municipal solid waste composting and its assessment for reuse in plant production.Pakistan Journal of Botany. 39: 271-277.

[12] Jin J, 2006.Solid waste management in Macao: practices and challenges. Waste Management . 26: 1045-1051.

[13] Kinman1987. Sanitary landfill disposal of urban SW. In: Workshop on SWM. New Delhi. Jan 16-17.

[14] Kumar Kaushal, R., Varghese, G. K., &Chabukdhara, M. (2012).Municipal Solid Waste Management in India-Current State and Future Challenges: A Review.International Journal of Engineering Science & Technology, 4(4).

[15] Mari I, 2005. Use of sulfur to control pH in composts derived from olive processing by-products. Compost Sci.Util.13: 281-287.

[16] Mbuligwe S. and Kassenga G, 2005. Feasibility and strategies for anaerobic digestion of solid waste for energy production in Dar-es-Salaam city Tanzania. Resources Conservation & Recycling. 42: 183-203.

[17] Metin, E, ErÖztÜrk, 2003. Solid waste management practices and review of recovery and recycling operations in Turkey. Waste Management 23, pp. 425-432.

[18] Miller FC, 1992.Composting as a Process Based on the Control of Ecologically Selective Factors. In: Meting, F.B., Jr.(Ed), Soil Microbial Ecology, Applications in Agricultural and Environmental Management. Marcel Dekker, Inc., New York, pp. 515-544.

[19] MoghadamMR and Mokhtarani B, 2009. Municipal solid waste management in Rasht city. Iran. Waste Management .29: 485-489.

[20] Mor S, 2006. Municipal solid waste characterization and its assessment for potential methane generation: a case study. Total Environment. 371 (1): 1–10.

[21] NareshKumar K, 2009. Characterization of Municipal Solid Waste (MSW) and a proposed management plan for Kharagpur,West Bengal, India. Resources, Conservation and Recycling. 53:166–174.

[22] Raje D V, 2001.An approach to assess level of satisfaction of the residents in relation to SWM system. Waste Management and Research. 19: 12–19.

[23] Sharholy.M, 2008.Municipal solid waste management in Indian cities– A review.Waste Management. 28: 459–467.

[24] Sharma VK, 1997. Processing of urban and agro-industrial residues by aerobic composting: Review.Energy convers. 38: 453-478.

[25] Siddiqui T, 2006.Sustainable development through integrated municipal solid waste management (MSWM) approach – a case study of Aligarh District. Paper presented at the National Conference of Advanced in Mechanical Engineering. New Delhi. India. AIME

[26] Vesilind P. A.,WorrellW. A., Reinhart DR, 2003.Solid waste engineering. Singapore: Thomson Asia Pvt. Ltd.,

[27] Ziaee, S., Omrani, G., Agha, M. M. A., &Mansouri, N. (2012).Qualitative and Quantitative Examination of Household Hazardous Waste in Tehran. Advances in Environmental Biology, 6(2), 676-683.

[28] ZucconiF, 1987. Compost Specification for the Production and Characterization of Compost from Municipal Solid Waste, In: Compost Production, Quality and the use.(Eds): M.de. Bestoldi, M.P.Ferranti, P.L., Hermite and F. Zucconi. Elsevier Applied Science Publishing Co., Inc., New York,

4

Assessment of microbiological proliferation and *in vitro* demonstration of the antimicrobial activity of the commonly available salad vegetables within Dhaka Metropolis, Bangladesh

Tasnia Ahmed[1], Nusrat Jahan Urmi[1], Md. Sakil Munna[1], Kamal Kanta Das[1], Mrityunjoy Acharjee[1], M Majibur Rahman[2], Rashed Noor[1,*]

[1]Department of Microbiology, Stamford University Bangladesh, Dhaka, Bangladesh
[2]Department of Microbiology, University of Dhaka, Dhaka, Bangladesh

Email address:
noor.rashed@yahoo.com (R. Noor)

Abstract: Present study mapped a complete pathogenic profile of the salad vegetables in Dhaka Metropolis, Bangladesh. In addition to a huge bacterial load found previously in lettuce, tomato, cucumber and carrot, current study further detected microbial contamination in chili, onion, capsicum and coriander samples. While *Vibrio* spp., *Salmonella* spp. and *Shigella* spp. fecal coliform and *Eshcherichia coli* were found to be absent within these vegetable samples; a colossal burden of *Aeromonas* spp. ($>10^6$ cfu/g) was observed in chili, capsicum, coriander, whereas *Staphylococcus aureus* (1.2×10^8 cfu/g) and *Klebsiella Pneumoniae* (10^4 cfu/g) were detected in onion. Fungal growth was also observed in all samples. Most of the pathogens from all 8 samples were resistant against erythromycin (15 μg), amoxicillin (30 μg) and ampicillin (10 μg) while susceptible against ciprofloxacin (5 μg), kanamycin (30 μg) and gentamicin (10 μg). Interestingly, lettuce and cucumber samples were found to exhibit the anti-bacterial activity against *Staphylococcus aureus* and *Aeromonas* spp.

Keywords: Salad Vegetables, Pathogens, Drug-Resistance, Antibacterial Activity, Food Safety

1. Introduction

Vegetables may undergo microbial spoilage by a huge array of pathogenic bacteria, fungi, viruses and parasites [1-2]. Raw vegetables may be bruised during processing and distribution resulting in the release of plant nutrients which may serve as the potential organic and inorganic substrates for microbial growth [3]. A variety of pathogenic bacteria including *Salmonella* spp., *Escherichia coli* O157:H7, *Bacillus anthracis*, *Mycobacterium* spp., *Brucella* spp., *Listeria monocytogenes*, *Yersinia enterocolitica*, *Clostridium perfringens*, and *Klebsiella* spp. can be introduced into vegetables from the manure used to promote their growth [4]. Moreover, pathogens existing in contaminated foods may harbor virulence genes and toxins/ enzymes, which further evoke the pathogenesis [5-7]. As a consequence, outbreaks of food borne diseases associated with vegetables in many countries are not unusual [4, 8-10].

Besides such spoilage threat, a major consideration in the medication of diseases has to be brought into the development of drug-resistance of the pathogens against commonly used antibiotics [11]. Such a situation drives the current interest on natural antimicrobial molecules in hope that they may be considered as anti-infective drug candidates. Many antimicrobial agents have previously been isolated from plant including secondary metabolites such as essential oil and terenoides, i. e., xanthones, benzophenones, coumarins and flavonoids [12]. These chemical substances can be even used as templates to produce more effective drugs through semi-synthetic and total synthetic procedures [13]. A new interest has occurred in the last decade in search of phytochemicals of native plants for pharmaceutical purposes as the plant-derived products have great potential as the source of pharmaceuticals [13].

Microbial study on vegetables in Bangladesh generally revealed the association of a huge pathogenic load. Several reports showed the growth and proliferation of pathogenic bacteria causing enteric diseases in salad vegetables [14]. Previously we also noticed an array of bacterial pathogens among the commonly consumed salad vegetables including lettuce (*Lactuca sativa*), tomato (*Solanum lycopersicum*), cucumber (*Cucumis sativus*) and carrot (*Daucus carota*) collected from Dhaka Metropolis [15]. Additional salad vegetables including green chili (*Capsicum annum*), onion (*Allium cepa*), capsicum (*Capsicum assum*) and coriander (*Coriandrum sativum*) were included in the study to portray a complete map of the overall microbiological traits of all salad vegetables found in Dhaka Metropolis. The drug resistance traits of the pathogenic isolates and the anti-microbial activity of all 8 types of salad samples were also examined to endow with an absolute insight on the food safety as well their possible role as natural antimicrobials.

2. Materials and Methods

2.1. Study Area, Sampling and Sample Processing

Samples of 8 categories of salad vegetables including lettuce (*Lactuca sativa*), tomato (*Solanum lycopersicum*), cucumber (*Cucumis sativus*), carrot (*Daucus carota*), chili (*Capsicum annum*), onion (*Allium cepa*), capsicum (*Capsicum assum*) and coriander (*Coriandrum sativum*) were randomly collected from different super shops, local markets and from street vans within a time frame of July 2012 to December 2012. Samples were collected early in the morning and quickly transported to the laboratory as soon as possible according to the method suggested by American Public Health Association [16]. For the identification and enumeration of pathogenic bacteria and fungi, 25 g of each sample was blended with 225 ml buffer peptone water (pH 7.2 ± 0.2) and serial dilutions were prepared up to 10^{-6} following the standard methods for plating purposes [17].

2.2. Microbiological Analysis

2.2.1. Enumeration of Total Viable Bacteria and Fungi

Total viable bacteria (TVB) were enumerated by spreading 0.1 ml of each sample suspension onto nutrient agar (Hi-Media Laboratories Pvt. Ltd., India). After incubation at 37 °C for 24 hours, plates were examined. For the estimation of fungal load, 0.1 ml of each sample was spread onto sabouraud dextrose agar (Hi-Media Laboratories Pvt. Ltd., India) followed by incubation at 25 °C for 48 hours.

2.2.2. Estimation of Total Fecal Coliform, Escherichia Coli and Klebsiella Pneumoniae

For enumerating total fecal coliform (TFC), 0.1 ml of each sample was spread onto membrane fecal coliform agar (Hi-Media Laboratories Pvt. Ltd., India). Plates were incubated at 44.5 °C for 24 hours. For the isolation of *Escherichia coli* and *Klebsiella pneumoniae*, 0.1 ml of suspension was spread over Mac Conkey agar (Hi-Media Laboratories Pvt. Ltd., India) for each sample and incubated at 37 °C for 18-24 hours. Presence of *E. coli* was further confirmed by the appearance of bluish-black colonies with green metallic sheen on the eosine methylene blue agar (Hi-Media Laboratories Pvt. Ltd., India). *Klebsiella* spp. was be identified by observing gummy pink color on Mac Conkey agar.

2.2.3. Isolation of Salmonella Spp., Shigella and Vibrio Spp

In case of *Salmonella*, *Shigella* and *Vibrio* spp, the absence of growth was assumptive of microbial cells of being in stressed condition or in viable but nonculturable (VBNC) state [18-19].

Therefore, enrichment technique was performed in selenite cystine broth (SCB) for *Salmonella* and *Shigella* spp. and in alkaline peptone water (APW) for *Vibrio* spp. For isolation of *Salmonella* and *Shigella* spp., one ml of homogenized sample suspension was transferred to SCB for enrichment followed by incubation at 37 °C for 6 hours and serial dilutions up to 10^{-6}, from which 0.1 ml was spread onto *Salmonella-Shigella* agar (Oxoid Ltd., Basingstoke, Hampshire, England) followed by incubation at 37 °C for 24 hours. In case of *Vibrio* spp., 1 ml of homogenized sample suspension was transferred to alkaline peptone water (APW) for enrichment and incubated at 37 °C for 6 hours. Serial dilutions up to 10^{-6} were made of the enriched broth, and 0.1 ml of suspension was spread onto TCBS agar, followed by incubation al 37 °C for 24 hours.

2.2.4. Estimation of Staphylococcus Aureus and Listeria Monocytogenes

For the estimation of *Staphylococcus aureus*, 0.1 ml of suspension was spread onto mannitol salt agar (Hi-Media Laboratories Pvt. Ltd., India) and the plates were incubated at 37 °C for 24 hours. The load of *Listeria monocytogenes* was enumerated by spreading 0.1 ml of suspension onto *Listeria* identification media (BD Diagnostic Systems, Europe) and plates were incubated at 37 °C for 24 hours. *Listeria monocytogenes* was identified as characteristic black gray color colonies surrounded by black halos on *Listeria* identification agar media [20].

2.2.5. Estimation of Clostridium Perfringens

One ml of each fresh salad vegetable sample blends were mixed in sterile normal saline in a ratio of 1:8 followed by heating at 80 °C for 15 minutes in order to kill vegetative cells of the microorganisms. From here, one ml of each samples were incubated for 4 hours at 37 °C with 9 ml of fluid thioglycolate broth. Then 0.1 ml of each enriched broth was subjected to 10-fold serial dilution for pouring on Clostridium isolation agar (Sigma-Aldrich, Inc., USA) plates, and was incubated at 37 °C within the anaerobic jar (2.5 L Anaero Jar, Oxoid Ltd., UK) for 48 hours.

2.3. Biochemical Tests for the Confirmative Identification

Finally, the standard biochemical tests were performed to confirm the identification of all the pathogenic isolates found in all 8 types of salad vegetable samples by the previously described methods [17,21].

2.4. Antibiotic Susceptibility Test

The pathogenic isolates were examined for antibiotic susceptibility traits (either drug resistant or sensitive) by disc diffusion assay on Mueller-Hinton agar (Difco Laboratories, Detroit, MI, USA) against commonly used antibiotics following the standard protocol [22-28]. Antibiotics used in the study included trimethoprime/sulfamethoxazole 25 µg, erythromycin 15 µg, amoxicillin 30 µg, ceftriaxon 30 µg, ciprofloxacin 5 µg, streptomycin 10 µg, ampicillin 10 µg, tetracycline 30 µg, chloramphenicol 30 µg, cefixime 5 µg, polymyxin B (300 units), kanamycin 30 µg, vancomycin 30 µg, gentamycin 10 µg, nalidixic acid 30 µg, azythromycin 15 µg and penicillin G 10 µg (Oxoid Ltd., Basingstoke, Hampshire, England).

2.5. Determination of Antibacterial Activity of the Salad Vegetable Samples

The investigation of the antibacterial activity of the vegetable samples was performed by using agar well diffusion method [29-30]. Briefly, vegetable blends were used directly on the Mueller-Hinton agar media. At first, the pathogenic bacterial suspensions (*Pseudomonas* spp, *Listeria* spp, *Aeromonas* spp, *Vibrio* spp, *Salmonella* spp, *Klebsiella* spp, *Staphylococcus aureus*, *E. coli*) with the equivalent turbidity standard of McFarland (0.5) were introduced evenly over the Mueller-Hinton agar media separately using cotton swab and wells were made in the Mueller-Hinton agar media by cork borer. From each of the crude vegetables blends, 100 µl samples were then introduced separately in the specified well with a positive control (antibiotic disc) and a negative control (normal saline). Presence of clear zone around the sample solution (if any) indicated the antibacterial potential of vegetable samples.

2.6. Statistical Analysis

All the experiments were performed in triplicate. Statistical analyses were performed by determining the P-value through *t-test*. Errors were also calculated.

3. Result and Discussion

Outbreaks of human disease associated with the consumption of raw fruits and vegetables often occur in developing countries and have become more frequent in developed countries over the past decade. Vegetable products are the most common food items in Bangladesh. However, vegetable borne diseases may put the overall public health at a serious risk. Multiplication of bacterial pathogens as well as their drug-resistance properties mainly accounts for the health related problems [31]. Along these lines, the present study portrays the (i) pathogenic load of all salad vegetables consumed in Bangladesh, (ii) the drug-resistance traits of the pathogens, and finally (iii) the anti-microbial activities of these salad samples.

3.1. Prevalence of Pathogenic Microorganism

Our previous study showed that *E. coli*, *Salmonella* spp., *Shigella* spp., *S. aureus*, *Vibrio* spp. and *Listeria* spp. were common in carrot (*Daucus carota*), tomato (*Solanum lycopersicum*), lettuce (*Lactuca sativa*) and cucumber (*Cucumis sativus*) samples [15]. The load of *Vibrio* spp., *Salmonella* spp. and *Shigella* spp. in those vegetables were nil before enrichment; however, after enrichment *Vibrio* spp. was estimated within a range of 2.0×10^4 to 8.3×10^7 cfu/g in carrot, lettuce and tomato samples, while *Salmonella* and *Shigella* spp. were found within a range of 1.0×10^3 to 3.1×10^7 cfu/g and 3.0×10^4 to 4.8×10^8 cfu/g, respectively in carrot, cucumber, lettuce, tomato samples. The similar results were observed in current study in case of the above mentioned vegetables. Moreover in the present study, we included chili (*Capsicum annum*), onion (*Allium cepa*), capsicum (*Capsicum assum*) and coriander (*Coriandrum sativum*) salad vegetables where VBNC state bacteria [15], i. e., *Vibrio*, *Salmonella* and *Shigella* spp. were found to be absent (Table 1). *E. coli* and *Pseudomonas* spp. were also absent in all 4 new samples. *Aeromonas* spp. was found in capsicum, chili and coriander within the range of 1.2×10^6 to 2.7×10^6 cfu/g. Staphylococci (1.2×10^8 cfu/g) in onion indicates health risk upon consumption of this raw vegetable. Moreover, the presence of pathogenic organisms such as *Aeromonas* spp. and *Klebsiella Pneumoniae*. Revealed the possibility of spreading enteric diseases to the consumers. *Clostridium* spp. was absent in all samples, while fungal growth was observed in all the vegetable samples tested.

Table 1. *Microbial load (cfu/g) of fresh salad vegetables.*

Samples	TVBC	Fungi	*Klebsiella* spp.	*Aeromonas* spp.	*Staphylococcus aureus*
Capsicum	6.3×10^6	6.6×10^6	0	1.2×10^6	0
Onion	2.6×10^7	1.3×10^8	1.0×10^4	0	1.2×10^8
Chili	1.9×10^7	1.4×10^8	0	2.2×10^6	0
Coriander	1.3×10^7	1.5×10^7	0	2.7×10^6	0

TVBC: Total Viable Bacterial Count.
All data were statistically analyzed and were found significant ($P < 0.1$).

Overall, our study revealed the presence of a huge range of microorganisms in the commonly consumed salad vegetables. The pathogens might be introduced from the crop land, organic fertilizers, irrigating water, packaging materials, transport vehicles etc. Besides, unhygienic personnel handling and processing of the vegetables and their storage in such a condition which favors microbial growth might also account for such spoilage of vegetables. The contaminating pathogens are responsible for various types of enteric diseases as well as serious intoxications in human health and hence the further detection of the virulent genes would be interesting.

3.2. Antibiotic Susceptibility Patterns of Pathogens Found in Salad Vegetables

Drug resistance is a serious problem in these days that is becoming more and more risky for the global public health. Our study of antibiogram revealed that most of the isolates were susceptible towards some antibiotics, whereas resistance towards several antibiotics indicated the risk of the emerging resistant isolates causing health hazards difficult to eradicate by those antibiotic therapy (Table 2).

Table 2. *Antibacterial susceptibility of the pathogenic isolates.*

Antibiotic	Disc Content	*E. coli*	*Klebsiella* spp.	*Salmonella* spp.	*Listeria* spp.	*Staphylococcus aureus*	*Shigella* spp.	*Aeromonas* spp.	*Vibrio* spp.
Trimethoprime/ Sulfamethoxazole	25 µg	S	S	S	S	R	S	R	ND
Erythromycine	15 µg	R	S	ND	ND	R	ND	R	ND
Amoxicillin	30 µg	S	R	R	ND	S	ND	S	R
Ceftriaxone	30 µg	S	R	ND	ND	S	S	I	I
Ciprofloxacin	5 µg	S	S	S	S	S	S	S	ND
Streptomycin	10 µg	I	R	ND	ND	S	ND	S	I
Ampicillin	10 µg	R	R	R	R	S	ND	S	R
Tetracycline	30 µg	S	S	ND	S	S	ND	S	R
Chloramphenicol	30 µg	S	S	S	ND	S	S	R	R
Cefixime	5 µg	S	R	ND	ND	S	ND	R	ND
Polymyxin B	300 units	S	S	ND	ND	R	ND	S	ND
Kanamycin	30 µg	ND	ND	S	S	S	ND	ND	I
Streptomycin	10 µg	ND	ND	ND	S	ND	ND	ND	ND
Vancomycin	30 µg	ND	ND	ND	S	ND	R	ND	R
Gentamycin	10 µg	ND	ND	S	S	ND	S	ND	ND
Nalidixic acid	30 µg	ND	ND	S	S	ND	S	ND	R
Azythromycin	15 µg	ND	ND	S	S	ND	ND	ND	R
Penicillin G	10 µg	ND	ND	ND	R	ND	ND	ND	R

ND= Not done, R= Resistant, S= Sensitive, I= Intermediate.

Majority of the isolates showed susceptibility against ciprofloxacin, tetracycline, kanamycin . However, most of the pathogenic isolates were resistance against ampicillin. Such resistance of the bacterial pathogens could be due to mechanistic-, epidemiologic- and genetic factors [32-34]. The resistance traits of the pathogens may result in the ineffective chemotherapy during food borne disease outbreaks.

Table 3. *Antimicrobial activity of the salad vegetables.*

Vegetable samples	Zone size of inhibition (mm) *Staphylococcus aureus*	*Aeromonas* spp.
Carrot	0	0
Lettuce	11mm	0
Tomato	0	0
Cucumber	11mm	12mm
Chili	0	0
Onion	0	0
Capsicum	0	0
Coriander	0	0

All data were statistically analyzed and were found significant (P <0.1).

3.3. Antibacterial Activity of Salad Vegetable Samples

Natural antimicrobials from the vegetables could be potential future candidates to be used as safe anti infective pharmaceutical products without causing any side effects like chemical medicines. Antimicrobial activity of different vegetables such as spinach, ghuniya, pumpkin, suran, cabbage, has been tested around the world in different times [13,30,35]. Along these lines, we determined the anti-microbial activity of the salad vegetables for the first time in Bangladesh (Table 3). Interestingly in present study, antibacterial activity was found in the lettuce and cucumber samples against *Staphylococcus aureus*. Apart from that cucumber samples also showed antibacterial activity against *Aeromonas* spp. (Table 3, Figure 1). However, the other vegetable samples (carrot, tomato, capsicum, onion, green chili, coriander) showed no antimicrobial activity against these bacterial isolates.

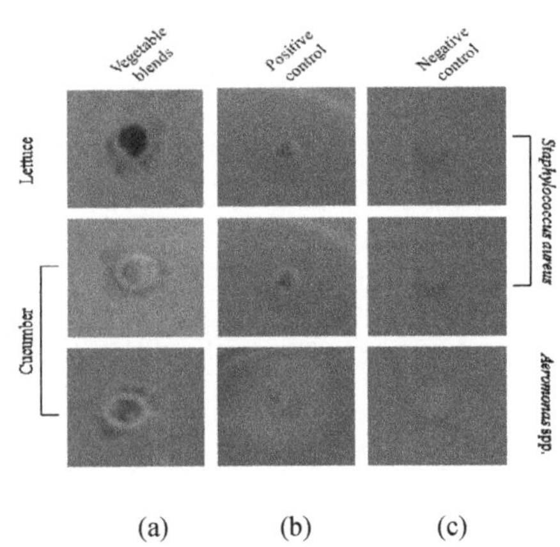

Figure 1. *Antibacterial activity of lettuce and cucumber. In the first column, lettuce and cucumber samples have been found to show antibacterial activity against Staphylococcus aureus and Aeromonas spp. In the second column, the first two images stand for positive controls using trimethoprime/sulfamethoxazole and the bottom image indicates the positive control using ciprofloxacin. In the third column, all the images are indicative of negative controls (normal saline).*

In cohort to our recent findings on food borne pathogens in within Bangladesh, the outcomes of the current study further added the insight of public health risk associated with fresh produce [36-39]. Moreover, the study might be further extended by identifying the genes responsible for the activity as well as to find out the way of using those gene products as anti-infective drugs to reduce the infection rates.

4. Conclusion

In the current study, we found that the common salad vegetable samples were largely populated with various microorganisms leading to serious public health hazards. The pathogenic bacteria present in the commonly consumed salad vegetables showed resistance against the regular antibiotics which is significant from the view point of public health. Our study thus imparted not only a complete picture on pathogenic profile of the salad vegetables but also presented a hopeful result on the antibacterial activity of lettuce and cucumber against microorganisms with pathogenic potential like *Staphylococcus aureus* and *Aeromonas* spp. Such antimicrobial trait of salad vegetables could further fortify the natural medication and an extended investigation could play role in preventing disease outbreaks caused by vegetable spoiling micro-flora. Nevertheless, our study contributed a detailed pathogenic profile of the salad vegetables which would be of significance for other countries/ regions with resource-poor settings.

Acknowledgements

Authors are acknowledging Stamford University Bangladesh for proving laboratory facilities, technical and financial support.

References

[1] L.R. Beuchat, "Ecological factors influencing survival and growth of human pathogens on raw fruits and vegetables." Micro. Infect., vol. 4(4), pp. 413-423, 2002.

[2] E.A. Szabo, and M.J. Conventry, Spoilage of processed foods; causes and diagnosis. Waterloo: AIFST Inc. (NSW Branch) Food Microbiology Group, 2001.

[3] J.A. Bartz, and C.I. Wei, The influence of bacteria postharvest physiology and pathology of vegetables. 2nd ed., New York: Marcel Dekker Inc., 2003, pp. 519-541.

[4] N.P. Aelice, "Manure and Microbes: Public and Animal Health Problem?" J. Dairy Sci., vol. 80 (10), pp. 2673-2681, 1997.

[5] D.R. Bhatta, A. Bangtrakulonth, P. Tishyadhigama, S.D. Saroj, J.R. Bandekar, R.S. Hendriksen, and B.P. Kapadnis, "Serotyping, PCR, phage-typing and antibiotic sensitivity testing of *Salmonella serovars* isolated from urban drinking water supply systems of Nepal." Let. Appl. Microbiol., vol. 44(6), pp. 588-594, 2007.

[6] A.J. Gubala, and D.F. Proll, "Molecula-Beacon Multiplex Real-Time PCR assay for detection of *Vibrio cholerae*." Appl. Environ. Microbiol., vol. 72(9), pp. 6424-6428, 2006.

[7] J.E. Jakee, E.I. Moussa, K.F. Mohamed, and G. Mohamed, "Using molecular techniques for characterization of *Escherichia coli* isolated from water sources in Egypt." Global Veterinaria, vol. 3(5), pp. 354-362, 2009.

[8] W.C.J. Cray, and W.H. Moon, "Experimental infection of calves and adult cattle with *Escherichia coli* O157:H7." Appl. Environ. Microbiol., vol. 61(4), pp. 1586-1590, 1995.

[9] J.A. Snowdon, D.O. Cliver, and J.C. Converse, "Land disposal of mixed human and animal wastes: A review." Waste Manag. Res., vol. 7, pp. 21-134, 1989.

[10] D. Starutch, "Survival of pathogenic microorganisms and parasites in excreta, manure sand ewage sludge." Rev. Sci. Tech., vol. 10(3), pp. 813-846, 1991.

[11] F.C. Tenover, "Mechanisms of Antimicrobial Resistance in Bacteria." American. J. Med., vol. 119(6A), pp. S3-S10, 2006.

[12] H. Belguith, F. Kthiri, A. Chati, A.A. Sofah, J.B. Hamida, and A. Ladoulsi, "Inhibitory effect of aqueous garlic extract (Allium sativum) on some isolated *Salmonella serovars*." African J. Microbiol. Res., vol. 4(5): pp. 328-338, 2010.

[13] A. Dubey, N. Mishra, and N. Singh, "Antimicrobial activity of some selected vegetables." Int. J. Appl. Biol. Pharma. Tech., vol. 1(3), pp. 994-999, 2010.

[14] V. Oni, A. Oni, and F. Esumeh, "Prevalence of Bacteria food poison from vegetable salads." Internet. J. Nutr. Well., vol. 10(1), 2010.

[15] F. Rahman, and R. Noor, "Prevalence of pathogenic bacteria in common salad vegetables of Dhaka metropilis." Bang. J. Bot., vol. 41(2), pp. 159-162, 2012.

[16] American Public Health Association, Standard Methods for the Examination of Water and Wastewater, 20th ed., Washington DC: American Public Health Association, 1998.

[17] J.G. Cappuccino, and N. Sherman, Microbiology- A Laboratory Manual, 4th ed., Menlo Park, California: The Benjamin/Cummings Publishing Co Inc, 1996, pp. 13-182.

[18] R.R. Colwell, "Non-culturable microorganisms in the environment. Washington DC, USA: American Society of Microbiology, 2000, pp. 325-342.

[19] J.D. Oliver, "The viable but nonculturable state in bacteria." J. Microbiol., vol. 43, pp. 93-100, 2005.

[20] C. Polcovnicu, L. Ionescu, and G. Bahrim, "Confirmation and identification of *Listeria* species from fresh lettuce." Romanian Biotechnol. Let., vol. 13(6), pp. 32-36, 2008.

[21] E.B. Alfrad, Bensons Microbiological Applications. New York: Mcgraw-Hill Book Company, 2007, pp 263-280.

[22] A.W. Bauer, W.M.M. Kirby, J.C. Sherris, and M. Tierch, "Antibiotic susceptibility testing by a standardized single disc method." American J. Clin. Patho., vol. 45(4), pp. 493-496, 1966.

[23] M.J. Ferraro, W.A. Craig, and M.N. Dudley, Performance standards for antimicrobial susceptibility testing. 11th ed. Pennsylvania, USA: NCCLS, 2001.

[24] S.K. Munshi, M.M. Rahman, and R. Noor, "Detection of virulence potential of diarrheagenic *Escherichia coli* isolated from surface water rivers surrounding Dhaka city." J. Bang. Acad. Sci., vol. 36(1), pp. 109-122, 2012.

[25] M. Acharjee, K. Fatema, F. Jahan, S.J. Siddiki, M.A. Uddin, and R. Noor, "Prevalence of *Vibrio cholerae* in different food samples in the city of Dhaka, Bangladesh." Int. Food Res. J., vol. 20(2). 2013.

[26] S. Dutta, M.R. Hasan, F. Rahman, M.S.A. Jilani, and R. Noor, "Study of antimicrobial susceptibility of clinically significant microorganisms isolated from selected areas of Dhaka city, Bangladesh." Bang. J. Med. Sci., vol. 12 (1), pp. 34-42, 2003.

[27] S.A. Khan, F. Feroz, and R. Noor, "Study of extended spectrum β-lactamase producing bacteria from urinary tract infection in Dhaka city, Bangladesh." Tzu. Chi. Med. J., unpublished.

[28] R. Noor, M.A. Uddin, M.A. Hoq, S.K. Munshi, M. Acharjee, and M.M. Rahman, "Microbiological study of vendor and packed fresh juices locally available in Dhaka city, Bangladesh." Int. Food Res. J., vol. 20(2), 2013.

[29] R.C. Jagessar, A. Mars, and G. Gones, "Selective antimicrobial properties of leaf extract against various micro-organisms using disc diffusion and agar well diffusion method." J. Nat. Sci., vol. 6(2), pp. 24-38, 2008.

[30] A. Hussain, S. Wahab, I. Zarin, and M.D.S. Hussain, "Antibacterial activity of the leaves of *Coccinia indica* (W. and A) W of India." Adv. Biol. Res., vol. 4(5), pp. 241-248, 2010.

[31] A.A. Butt, K.E. Aldrig, and C.V. Sanders, "Infections related to the ingestion of seafood Part I: Viral and bacterial infections." Lancet. Infect. Dis., vol. 4: pp. 201-212, 2004.

[32] P.M. Bennet, "Plasmid encoded antibiotic resistance: Acquisition and transfer of antibiotic resistance genes in bacteria." British J. Pharmacol., vol. 153(1), pp. 347-357, 2008.

[33] R. Canton, "Antibiotic resitance genes from the environment: A perspective through newly identified antibiotic resistance mechanisms in clinical setting." Euro. Soc. Clin. Microbiol. Infect. Dis., vol. 15(1), pp. 20-25, 2009.

[34] D.T. Hung, and B.B. Kaufman, "The Fast Track to Multidrug Resistance." Mol. Cell Biol., vol. 37(3), pp. 297-298, 2010.

[35] K.H. Kyung, and H.P. Fleming, "Antibacterial activity of cabbage juice against lactic acid bacteria." J. Food. Sci., vol. 59(1), pp. 125-129, 1994.

[36] F. Feroz, J.D. Senjuti, and R. Noor, "Determination of microbial growth and survival in salad vegetables through in vitro challenge test." International Journal of Nutrition and Food Science, vol. 2(6), pp. 312-319, 2013.

[37] N. Sarker, S. Islam, M. Hasan, F. Kabir, M.A. Uddin, and R. Noor. "Use of multiplex PCR assay for detection of diarrheagenic *Escherichia coli* in street vended food items." American Journal of Life Sciences, vol. 1(6), pp. 267-272, 2013.

[38] N. Fatema, M. Acharjee, and R. Noor, "Microbiological profiling of imported apples and demonstration of bacterial survival capacity through in vitro challenge test." American Journal of Microbiological Research, vol. 1(4), pp. 98-104, 2013.

[39] T. Ahmed, M. Acharjee, M.S. Rahman, M. Meghla, J. Jamal, S.K. Munshi, and R. Noor. "Microbiological study of drinking water: qualitative and quantitative approach." Asian J. of Microbiol. Biotech. Env. Sc. vol. 15(4), pp. 23-30, 2013.

Analysis of price transmission of fresh tomato and pineapple in the rural and urban markets of Akwa Ibom Sate, Nigeria

Sunday Brownson Akpan[1], Ini-Mfon Vincent Patrick[1], Glory Emmanuel Edet[2], Daniel Etim John[2]

[1]Department of Agricultural Economics and Extension, Akwa Ibom State University, Nigeria
[2]Department of Agricultural Economics and Extension, University of Uyo, Akwa Ibom State, Nigeria

Email address:
brownsonakpan10@gmail.com (Sunday B. A.), inivipako@gmail.com (Ini-Mfon V. P.), gakanmkpo@yahoo.com (Glory E. E.), dannyj2004@yahoo.com (Daniel E. J.)

Abstract: The study investigated the dynamics of price transmission and market integration of fresh tomato and pineapple in the rural and urban markets of Akwa Ibom State, Nigeria. Average monthly prices (measured in naira per kilogram) of fresh tomato and pineapple in the rural and urban markets were used in the analysis. The data was obtained from the quarterly publications of the Akwa Ibom State Agricultural Development Programme (AKADEP). The data covered the period from January 2005 to September 2013. The trend analysis showed that, prices of fresh tomato and pineapple in the rural and urban markets have positive significant relationships with time and positive exponential growth rate. The Pearson correlation coefficient of each of the respective pair of rural and urban price of fresh tomato and pineapple revealed significant positive and symmetric relationships. The bivariate Granger causality test revealed bi-directional relationships between the rural and urban price of fresh tomato and pineapple in the State. The co-integration test revealed the presence of co-integration between the rural and urban price of fresh tomato and pineapple. The coefficients of market integration in the rural and urban price equation display varied degrees of long run market integration. The results of the error correction model (ECM) confirmed the existence of the short run market integration between the rural and urban prices of fresh tomato and pineapple in the study area. The urban price of fresh tomato adjusted faster to the stable state in the long run than its respective rural price. Also the rural price of pineapple adjusted faster to a stable state in the long run than its respective urban price. The index of market connection for fresh tomato and pineapple supported the existence of the short run market integration between the rural and urban markets. Based on the findings, it is recommended that, the Akwa Ibom State government should continue to provide marketing infrastructures in the rural areas to improve the symmetric nature of information among fresh tomato and pineapple markets in the state. Effort should also be channeled by all stake holders such as governments, trade unions and other organizations to reduce excessive externality costs associated with the marketing of fresh tomato and pineapple in the state. The government of Akwa Ibom State should established market information centers and awareness programmes on mass media to facilitate efficient information flow among fresh tomato and pineapple producers and consumers in the state.

Keywords: Market, Fresh Tomato, Price, Integration, Pineapple, Akwa Ibom, Nigeria

1. Introduction

Fruit is considered as the sweet and fleshy product of a tree or other plant that contains seed and can be eaten as food. Fruits and vegetables play significant roles in human nutrition, especially as sources of vitamins (C, A, B6, thiamine, niacin, E), minerals, and dietary fiber [26, 27, 30]. Fresh tomato and pineapple is among the most important fruits in Nigeria. Fruits are generally high in fiber, water, vitamins, and sugars among others. Several reports have shown that adequate intake of fruits and vegetables form an important part of a healthy diet; as low fruit and vegetable intake constitute a risk factor for chronic diseases such as cancer, coronary heart disease (CHD), stroke and cataract formation [29]. Scientific evidence indicates that frequent consumption of fruits and vegetables can prevent oesophageal, stomach, pancreatic, bladder and cervical cancers and that a diet high in fruits and vegetables could prevent 20% of most types of cancers [11].

Tomato (*Solanum lycopersicum*) and pineapple (*Ananas comosus*) are among fruit crops cultivated in most areas of Nigeria. These crops are characterized by problems of seasonality, bulkiness and high perishable nature. The persistence of these problems affects production and marketing of these agricultural commodities. Due to the difficulty in the handling of fresh tomato and pineapple among farmers, price variation is a regular phenomenon in Nigeria. Price variation in fresh tomato and pineapple differs among regions, state, local government, rural and urban centers. Price been the coordinating cell for agricultural production and marketing; it determines resource allocation and utilization in the agricultural sector. In the context of the marketing environment, price shows the level of efficiency and the working mechanism among agricultural product markets. In a poverty stricken society like Nigeria, price is one of the fundamental determinants of consumer preference and taste [4]. Many researches in Nigeria have related price volatility in agricultural commodities to several factors including variances in bargaining power among consumers, cyclical income fluctuation among sellers and consumers, seasonality of production, natural shocks such as flood, pests, diseases, and inappropriate response by farmers to price signals [12, 1, 4]. Instability in commodity prices among markets could be detrimental to the marketing system and the economy as a whole. It could cause inefficiency in resources allocation among sellers and consumers depending on the source of variability (that is, whether it is induce by supply or demand side or both). It could also increase poverty level among low income earners in the society [25]. On the other hand, a unified product price among spatial markets might not be a rational policy to pursue in a developing country like Nigeria. This is because of the deteriorating marketing infrastructures, increase in cost of externalities and the nature of most agricultural products which often resulted in significant differences in the total variable costs incurred by sellers and consumers in these markets.

Nigeria is a vast country with clear distinct regions, rural and urban areas. Prices of agricultural commodities in each of this region, rural and urban areas differ depending on the availability of infrastructures among others. Spatial price linkages are often interpreted as providing insight into the efficiency of infrastructures of markets. This is especially true in developing economy or society, where infrastructure issues such as road systems, market development, transportation, and so forth may be especially pertinent [15]. In separated markets, when there is significant price difference between homogenous goods, such that the differences exceeded the transfer cost; the arbitrage activities will be stimulated. The arbitrageur will purchase commodities from lower-price markets and resale in higher-price markets. This is a situation where spatial markets are not integrated. On the other hand, two markets are integrated when there is a significant long-run relationship between prices of homogenous goods due to the smooth transmission of price signals and information across the two markets [22, 4]. Market integration could be perfect if price changes in one market are fully and instantaneously reflected in the alternative markets [14].

Hence, understanding the direction and magnitude of fresh tomato and pineapple price transmission between the rural and urban markets in a state like "Akwa Ibom" will provide indispensable input to policy makers to formulate workable policies for the agricultural sector in the state. It will also, promote the achievement of the self food sufficiency drive and help in minimizing the menace of poverty among citizenry in the state. For instance, the extent of market integration has often been used to measure the success of market liberalization and structural adjustment policies in developing countries [18]. Therefore, such information can help government at all tiers to decide the extent to which price transmission can be considered as efficient across different geo-political zones in their domains.

Several empirical investigations have delved into agricultural price transmission and market integration of food stuffs in Nigeria. For instance, Amusa, [7] in her study of the trend analysis of agricultural food prices in Nigeria reported that, food items such as vegetable oil, Garri, brown beans, ripe plantains, fresh tomatoes, green vegetables, onion bulbs, shelled melon seeds, experienced increase and fluctuations in their prices. Also, Okoh *and* Egbon [22] examined the integration of Nigeria's rural and urban foodstuffs markets. The study concludes that, the rural and urban foodstuffs markets were well integrated. The result further suggested that, the urban market price drives the rural market price. The size of the adjustment coefficient for the rural foodstuffs price revealed that, the speed of adjustment from disequilibrium was moderate. Similarly, Ohen *et al.,* [20] studied the vertical and horizontal price linkages for live catfish in Nigeria. The price variables used in the analysis were non-stationary and therefore were made stationary by first difference. The Johansen co-integration analysis was used to test for the relationship between prices. Results indicated that producer and export prices were co-integrated. Furthermore, the Granger causality Wald test suggested that, the retailed prices do have a causal relationship with producer prices. The dynamic regression analysis of prices also revealed that, the markets for live catfish have strong price linkages and thus are spatially integrated. In western Nigeria, Adeoye *et al.,* [3] examined the price transmission and market integration of banana and plantain in Oyo state, Nigeria. Six market links rejected their respective null hypothesis of no Granger causality ($P>0.05$), two of the market links exhibited bi-directional Granger causality or simultaneous feedback relationship; while four market links exhibited uni-directional Granger causality at 5% and 10% level of significance. Urban plantain market occupies the leadership position in the commodity price formation and transmission in the markets investigated. The Index of market connection or concentration (IMC) indicated that, the markets exhibited low short run integration. In addition, Ojiako *et al.,* [21] studied the spatial integration and price transmission in selected cassava products' (Lafun) markets in Nigeria. The study employed Vector Error Correction (VEC) model methodology. The result revealed the presence of the long-run equilibrium following exogenous shocks in the market. In addition, the result discovered unilateral Granger causality that runs from the rural to the urban market. The impulse response analysis revealed that, the rural price was more responsive to shocks emanating from the urban market, the effect of which was computed as 95.6% using the forecast error variance decompositions. The study further discovered that, the effects of the rural prices' shock on urban price were very negligible at 3.2% after 10 weeks. The implication is that the rural market was the dominant market for determining the price of *lafun* in the short-run. The error correction model revealed significant causality link between the peripheral and central markets, suggesting a clear trend in price leadership. In the South-South region, Akpan *et al.,* [4] examined the price transmission and market integration of local and foreign rice in rural and urban markets of Akwa Ibom State. The findings showed that, price of local and foreign rice in rural and urban markets has constant exponential growth rate of 0.59% which suggests perfect co-movement for rural and urban prices of local and foreign rice in the study area. Also, the Pearson correlation coefficient matrix revealed that, the rural price of local and foreign rice has linear symmetric relationships with their corresponding urban prices. The Granger causality test revealed bi-directional relationship between rural and urban price of local and foreign rice. The results of the co-integration test revealed the presence of co-integration between the rural and urban prices of local and foreign rice as well as support the hypothesis of perfect price transmission between the two markets. The results of the error correction (ECM) model also confirm the existence of the short run market integration between the rural and urban prices of local and foreign rice in the study area. In addition, the result shows that, the price of local rice in both rural and urban markets adjusted faster than prices of foreign rice once there is an exogenous shock in the marketing process of rice. The index of market connection (IMC) supported the high short run market integration between prices in rural and urban markets for local and foreign rice commodities and the quick adjustment of rural price of local rice in relative to rural price of foreign rice.

From the reviewed literature, it is cleared that, fruit price transmission studies has not enjoyed good patronage from the Nigeria's researchers. Fruit is important in the dairy dietary need of individuals. Balance diet and self food sufficiency pursuit of all tiers of government cannot be achieved without the contributions of the fruit and vegetable food stuffs. Hence, this study is designed to specially provide dynamic information on fresh tomato and pineapple marketing in Akwa Ibom State, Nigeria. Study like this will provide effective policy variables needed to formulate workable marketing policies and assessing the efficiency of fruit price transmission as well as availability of social infrastructures in the southern Nigeria. The specific objectives were to:

- examine the trends of the monthly prices (₦/Kg) of fresh tomato and pineapple in the rural and the urban markets of the study area.
- assess the nature of price transmission of fresh tomato and pineapple in the rural and urban market of the study area, and examine the degree of long and short market integration of fresh tomato and pineapple in the study area.

1.1 Research Hypotheses

Ho: Price trend in fresh tomato and Pineapple do not differ in rural and urban markets of Akwa Ibom State.

Ho: Price transmission of fresh tomato and pineapple in the rural and urban markets is not symmetric in Akwa Ibom State.

Ho: There are no long and short run relationships between prices of fresh tomato and pineapple in the rural and urban markets in the study area.

2. Research Methodology

2.1. Study Area and Data Source

The study was conducted in Akwa Ibom State located in the coastal South-South region of Nigeria. The state is located between latitudes $4°32^1$ and $5°33^1$ north and longitudes $7°25^1$ and $8°25^1$ east. It has a total land area of areas of 7,246km^2. The mean annual temperature of the state lies between 26°C and 29°C and average sunshine of about 1,450 hours per year. The mean annual rainfall ranges from 2,000mm to 3,000mm, depending on the area. Akwa Ibom State has a population of about 3,902,051 [19]. The state is basically an agrarian society where crops like maize, okra, cassava, pineapple, tomato, yam and rice are cultivated in large quantities.

2.2. Source of Data

Secondary data were used in this study. The data came from the quarterly publication of the Akwa Ibom State Agricultural Development Programme (AKADEP). It consisted of the average monthly retailed price in Naira per kilogram of fresh tomato and pineapple from sampled markets in rural and urban areas of Akwa Ibom State. The study period covered January 2005 to September 2013. A total of 105 weeks' retailed average monthly prices (₦/Kg) of fresh tomato and pineapple from rural and urban markets were used in the study.

2.3. Analytical Techniques

The study applied series of statistical and econometric techniques to test for the relationship between rural and urban prices of tomato and pineapple in the study area. The tests applied include; the trend analysis, bi-variate correlation analysis, Granger causality tests, co-integration and Error Correction Model (ECM) as well as Index of Market Connection analysis. Each of the tests is explained as shown below:

2.4. The Trend Analysis of Monthly Retailed Prices of Fresh Tomato and Pineapple in Rural and Urban Markets of Akwa Ibom State in Nigeria

To investigate the nature of growth rate in prices of fresh tomato and pineapple in the study area, the exponential growth rate equation was specified as thus:

$$P_t = b_o e^{bt} e^{ut} \quad (1)$$

$$log_e P_t = b_0 + b_1 t + U_t \quad (2)$$

Where exponential growth rate (r) is defined as $(e^{b1} - 1) * 100$; where "e" is 2.7183. The quadratic exponential trend equation was also used to test for the impact of increasing time on prices of fresh tomato and pineapple in the rural and urban markets of the study area. The quadratic exponential trend equation was specified as thus:

$$log_e P_t = b_o + b_1 t_1 + b_2 t_2^2 + u_t \quad (3)$$

If $b_2 > 0$; the variable investigated had accelerated growth rate: when $b_2 < 0$; the variable has decelerated growth rate over time. In this study, "P_t" was represented by:

$FT_{rt} =$ *monthly Price of fresh tomato in rural market*

$FT_{ut} =$ *monthly Price of fresh tomato in Urban market*

$PN_{rt} =$ *monthly Price of pineapple in rural market*

$PN_{ut} =$ *monthly Price of pineapple in urban market*

t = time trend (1, 2... 105). Note prices are in nominal unit; and measured in Naira/Kilogram.

The exponential growth rate equation was adopted in this study to investigate the growth in prices of fresh tomato and pineapple because, several literature have supported continuous inflated prices of agricultural commodities for some years in Nigeria [2, 11, 24, 5].

2.5. Pearson Correlation Matrix of Average Monthly Retailed Prices of fresh tomato and pineapple in Rural and Urban Markets of Akwa Ibom State in Nigeria

To test for the symmetrical and linear relationship between rural and urban prices of tomato and pineapple in Akwa Ibom State, the Pearson correlation coefficients were estimated. The formula is as described below:

$$P_r = \frac{n\sum FT_{rt}FT_{ut} - (\sum FT_{rt})(\sum FT_{ut})}{\sqrt{\{n\sum FT_{rt}^2 - (\sum FT_{rt})^2\}\{\sum FT_{ut}^2 - (\sum FT_{ut})^2\}}} \quad (4)$$

The "Pr" is the correlation coefficient between urban and rural market prices of fresh tomato and pineapple in the study area. A significant relationship between the considered variables suggests perfect price transmission or the presence of market integration between the two markets; while insignificant association suggests otherwise.

2.6. Bilateral Granger Causality Test on Average Monthly Retailed Prices of Fresh Tomato and Pineapple in the Rural and Urban Markets of Akwa Ibom State in Nigeria

The Granger causality test is a statistical hypothesis test for determining whether one time series is useful in forecasting another. A time series *X* is said to Granger-cause *Y* if it can be shown, usually through a series of t-tests and F-test on lagged values of *X* (and with lagged values of *Y* also included), that those *X* values provide statistically significant information about future values of *Y* [16]. If a time series is a stationary process, the test is performed using the level values of two (or more) variables. If the variables are non-stationary, then the test is done using first (or higher) differences. To determine the number of lags (and to control for serial correlations) in a model, the Akaike, Schwarz or Hannan-Quinn information criteria are generally applied [8]. In this study, the bilateral Granger Causality tests were conducted on the average monthly retailed prices of fresh tomato and pineapple in urban and rural markets of Akwa Ibom State, southern Nigeria. The primary model in Vector Autoregressive Regression forms are represented as thus:

$$\begin{cases} \Delta LnFT_{rt} = \beta_0 + \beta_1 \sum_{i=1}^{n} \Delta LnFT_{rt-1} + \beta_2 \sum_{i=1}^{n} \Delta LnFT_{ut-1} + \varepsilon_{1t} \ldots (5) \\ \Delta LnFT_{ut} = \delta_0 + \delta_1 \sum_{i=1}^{n} \Delta LnFT_{ut-1} + \delta_2 \sum_{i=1}^{n} \Delta LnFT_{rt-1} + \varepsilon_{2t} \ldots (6) \end{cases}$$

$$\begin{cases} \Delta LnPN_{rt} = \gamma_0 + \gamma_1 \sum_{i=1}^{n} LnPN_{rt-1} + \gamma_2 \sum_{i=1}^{n} \Delta LnPN_{ut-1} + \varepsilon_{3t} \ldots (7) \\ \Delta LnPN_{ut} = \alpha_0 + \alpha_1 \sum_{i=1}^{n} \Delta LnPN_{ut-1} + \alpha_2 \sum_{i=1}^{n} \Delta LnPN_{rt-1} + \varepsilon_{4t} \ldots (8) \end{cases}$$

There is bilateral Granger causality from urban markets to rural market of fresh tomato, if $\beta_2 \neq 0$ and $\delta_2 = 0$. The market, which Granger-causes the other is tagged the exogenous market or the lead market. Market price exogeneity can be weak or strong. According to Hendry [17] the weak exogeneity occurs when the marginal distribution of P_{1t-1} is independent of the joint distribution of both P_{1t-1} and P_{2t-1}. On the other hand, strong exogeneity occurs when there is no statistically significant Granger-causality from the other variable. If we have two spatial prices, P_{1t-1} and P_{2t-1}, the price P_{1t-1} is weakly exogenous to P_{2t-1} if P_{1t-1} is tested to be weakly exogenous and P_{2t-1} is not weakly exogenous to P_{1t-1}. This implies that P_{1t-1} is causing P_{2t-1} to change and not vice-versa [9]. Similarly, there is Granger causality from the rural market price to urban market price of fresh tomato if $\beta_2 = 0$ and $\delta_2 \neq 0$. The causality is considered as mutual or bidirectional if $\beta_2 \neq 0$ and $\delta_2 \neq 0$. Finally, there is no link between monthly price of fresh tomato in rural markets and urban markets if $\beta_2 = 0$ and $\delta_2 = 0$. The same interpretations follow for equations 7 and 8 for pineapple. The variables are as defined previously in equation (3). A bidirectional Granger causality test indicates the presence of perfect price transmission between prices of rural and urban markets for fresh tomato and pineapple in the study area.

2.7. Co- Integration Test for Retailed Prices of Fresh Tomato and Pineapple in the Rural and Urban Markets of Akwa Ibom State in Nigeria

If the geographically separated markets are co-integrated, then there exists an equilibrium or long run relationship among these markets [14, 28, 13]. The study applied the Engle and Granger two-step technique and Johansen co-integration approach to examine the co-integration relationships between rural and urban market prices of fresh tomato and pineapple in the study area. If two prices (FT_{rt} and FT_{ut}) are perfectly integrated, then $\gamma_1 = 1$ *in equation* 9. In this case, price changes in rural market (FT_{rt}) are fully reflected in the urban (FT_{ut}) market. When $\gamma_1 \neq 1$ ($i.e. \gamma_1 1$ or γ_1 1), then the degree of integration needs to be determined by investigating the variance of γ_1 from the threshold mark of 1. Following the law of one price, the time dependent rural price equation for fresh tomato and pineapple is given as follows:

$$LnFT_{rt} = \gamma_0 + \gamma_1 \sum_{i=1}^{n} LnFT_{ut} + U_{1t} \quad (9)$$

$$LnPN_{rt} = \varphi_0 + \varphi_1 \sum_{i=1}^{n} LnPN_{ut} + U_{2t} \quad (10)$$

Also, the urban price equation of fresh tomato and pineapple was specified in the similar manner. Following the Granger Representation Theorem, we specify the error correction model (ECM) for the co-integrating series in the study. The general specification of the ECM that was estimated for the rural price equation for fresh tomato and pineapple in the study area is shown below:

$$\Delta \mathrm{LnFT}_{rt} = \gamma_0 + \gamma_1 \sum_{i=1}^{n} \Delta LnFT_{rt-1} + \gamma_2 \sum_{i=1}^{n} \Delta \mathrm{LnFT}_{ut} + \gamma_3 \sum_{i=1}^{n} \Delta \mathrm{LnFT}_{ut-i} + \gamma_4 ECM_{t-1} + U_{1t} \quad (11)$$

$$\Delta \mathrm{LnPN}_{rt} = \vartheta_0 + \vartheta_1 \sum_{i=1}^{n} \Delta LnPN_{rt-1} + \vartheta_2 \sum_{i=1}^{n} Ln\Delta PN_{ut} + \vartheta_3 \sum_{i=1}^{n} Ln\Delta PN_{ut-i} + \vartheta_4 ECM_{t-1} + U_{2t} \quad (12)$$

In this study, these equation series were replicated for the urban price of fresh tomato and pineapple respectively. The specified variables are as defined previously in equation (3), and the coefficient (γ_4) and (ϑ_4) of the ECM_{t-1} ($-1 < \gamma_4, \vartheta_4 < 0$) measures the deviations from the long-run equilibrium in period ($t-1$)) in both FT_{rt} and PN_{rt}. The specification of equation 11 and 12 was to test for the short run market integration and determine the adjustment coefficient of FT_{rt} and PN_{rt} to equilibrium level in the long run when there is exogenous shock in the system. This exercise was also done on the urban prices of fresh tomato and pineapple (i.e. FT_{ut} and PN_{ut}).

2.8. Index of Market Connection (IMC)

The index of market connection (IMC) was used to measure the degree of short run price transmission or price relationship between integrated markets. Following Oladapo and Momoh, [23] technique, the relationship between the fresh tomato and pineapple in the rural and urban market is given by the equations below:

$$FT_{rt} = \delta_0 + \delta_1 FT_{rt-1} + \delta_2(FT_{ut} - FT_{ut-1}) + \delta_3 FT_{ut-1} + \varepsilon_{1t} \quad (13)$$

$$PN_{rt} = \varphi_0 + \varphi_1 PN_{rt-1} + \varphi_2(PN_{ut} - PN_{ut-1}) + \varphi_3 PN_{ut-1} + \varepsilon_{2t} \quad (14)$$

Then IMC = δ_1/δ_3 for fresh tomato and = φ_1/φ_3 for Pineapple. Note, when the estimated IMC < 1, it implies high short run market Integration; IMC > 1 implies low short run market Integration; IMC = ∞ implies no market integration; IMC = 1 high or low short run market integration. Other variables are as defined in equations 3.

3. Results and Discussions

3.1. Augmented Dicker Fuller and Augmented Dicker Fuller –GLS Unit Root Test Results

The stationary status of prices of fresh tomato and pineapple was examined by the unit root tests. The Augmented Dicker Fuller (ADF) and Augmented Dicker Fuller –GLS unit root tests were used to ascertain the stationary of the specified price variables. The result of the tests is presented in Table 1 and 2. The result showed that, prices of fresh tomato in the rural and urban markets were non-stationary at levels but stationary (at 1% significance level) at first difference for the ADF and ADF-GLS equations that contain constant only. On the other hand, prices of pineapple in the rural and urban markets were stationary at level using the two unit root equations. However, when the two unit root equations were specified to include constant and deterministic trend, prices of fresh tomato and pineapple were stationary at levels.

Table 1. Result of the Augmented Dicker Fuller unit root test for prices of fresh tomato and pineapple used in the analysis.

Logged Variables	Augmented Dicker Fuller Test for unit root With Constant			Constant and Trend		
	Level	1st diff.	OT	Level	1st diff.	OT
$LnFT_{rt}$	-2.778	-14.61*	1(1)	- 4.782*	—	1(0)
$LnFT_{ut}$	-3.360	-14.72*	1(1)	-5.982*	—	1(0)
$LnPN_{rt}$	-4.712*	—	1(0)	-7.735*	—	1(0)
$LnPN_{ut}$	-4.357*	—	1(0)	-7.321*	—	1(0)
1% (CV)	-3.49	-3.50		-4.05	-4.50	

Note: OT means order of integration. Critical value (CV) is defined at 1% significant level and asterisk * represent 1% significance level. Variables are as defined in equation 3.

***Table 2**. Result of the Augmented Dicker Fuller – GLS unit root tests for prices of fresh Tomato and Pineapple used in the analysis.*

Logged Variables	Augmented Dicker Fuller –GLS Test for unit root With Constant			Constant and Trend		
	Level	1st diff.	OT	Level	1st diff.	OT
$LnFT_{rt}$	-2.121	-11.325*	1(1)	-4.749*	—	1(0)
$LnFT_{ut}$	-2.260	-10.791*	1(1)	-5.948*	—	1(0)
$LnPN_{rt}$	-3.809*	—	1(0)	-7.731*	—	1(0)
$LnPN_{ut}$	-3.454*	—	1(0)	-7.308*	—	1(0)

Note: OT means order of integration. Critical value (CV) is defined at 1% significant level and asterisk * represent 1% significance level. Variables are as defined in equation 3.

The result implies that, prices of fresh tomato cannot be specified at their levels without the risk of obtaining spurious regression. Also, the result indicates that, there is need to test for the presence of co-integration and short run relationship between the rural and urban prices of fresh tomato in the study area. Despite the stationary status of pineapple prices in the rural and urban markets, the study also investigated the short run properties by subjecting the price variables to further empirical tests. The nature of the stationary of the specified price variables is similar to the research results reported by several researchers on different food stuffs in Nigeria. They include; Okoh *and* Egbon [22], Ojiako *et al.,* [21] and Akpan et al., [4].

3.2. Descriptive Analysis of Monthly Prices of Fresh Tomato and Pineapple in Akwa Ibom State (from January 2005 to June 2013)

The descriptive statistics of the price variable used in the analyses is shown in Table 3. The average price of fresh tomato in rural and urban markets of Akwa Ibom State was ₦321.15/kg and ₦244.41/kg respectively. Also, the average price of Pineapple was ₦121.84/kg and ₦139.57/kg in rural and urban markets respectively. These findings revealed that, there might be significant deviations or differences between rural and urban price of fresh tomato and Pineapple in the study area. The standard deviations of 90.25 and 90.87 in prices of fresh tomato in the rural and urban markets respectively, indicates that, changes in both prices assume similar patterns during the study period. In addition, the coefficient of variability in price of fresh tomato in the rural and urban markets was 39.00% and 37.20% respectively.

***Table 3**. Descriptive Statistic of Price variables used in the model.*

Parameters	Fresh tomato Price Rural Market (₦/Kg)	Fresh tomato Price Urban Market (₦/Kg)	Pineapple Price Rural Market (₦/Kg)	Pineapple Price Urban Market (₦/Kg)
Mean	231.15	244.41	121.84	139.57
Median	217.00	241.02	120.00	137.01
Minimum	77.400	80.600	62.870	73.66
Maximum	480.54	507.94	267.08	250.00
Std. deviation	90.250	90.870	40.675	45.620
Co.Variation	0.390	0.372	0.334	0.327
Skewness	0.517	0.366	0.877	0.452
Kurtosis	0.158	-0.127	1.111	-0.756
AGR (%)	3.907	4.418	4.192	3.822

Note: Computed by authors, and prices are expressed in nominal terms. AGR means average growth rate.

For Pineapple, it stood at 33.40% and 32.70% for rural and urban markets respectively. The finding showed that, rural prices of fresh tomato and Pineapple exhibited higher variations compared to their respective urban prices. Furthermore, the average linear growth rate in the price of fresh tomato in the rural and urban markets were 3.907% and 4.418% respectively. Similarly, the price of Pineapple grew at the rate of 4.192% and 3.822% in the rural and urban markets respectively.

3.3. Exponential Trend Analysis of Prices of Fresh Tomato and Pineapple in Rural and Urban Markets of Akwa Ibom State (2005 to 2013)

The exponential trend equations for each of the price variables specified in equation 3 is presented in Table 4. The estimated regression result for each of the price variables is followed by the calculated exponential growth rate and the conclusion derived from the respective long run exponential trend equation. The result revealed that, trends in prices of fresh tomato and pineapple in rural and urban markets showed positive significant relationship with time in the study area.

***Table 4**. Exponential Trend Analysis of Monthly Average Prices of fresh tomato and Pineapple in Rural and Urban Markets in Akwa Ibom State (2005 to 2013).*

Variables	$LnFT_{rt}$	$LnFT_{ut}$	$LnPN_{rt}$	$LnPN_{ut}$
Constant	4.78(90.66)*	4.87(97.82)*	4.34(97.44)*	4.46(102.7)*
Time	0.012(12.66)*	0.011(12.86)*	0.008(10.60)*	0.008(11.31)*
F- cal.	160.38*	165.41*	112.40*	127.94*
R-square	0.609	0.616	0.522	0.554
EGR (%)	1.099	1.054	0.776	0.808
Nature of Growth Rate				
Constant	4.808(59.83)*	4.83(63.76)*	4.19(64.35)*	4.33(67.57)*
Time (b_1)	0.009(2.68)**	0.013(3.81)*	0.016(5.61)*	0.015(5.41)*
Time (b_2)	1.58e-05(0.46)	-2.0e-05 (-0.66)	-7.72e-05(-2.97)*	-6.7e-05 (-2.61)*
F- cal.	79.68*	82.462*	64.903*	70.968*
R-square	0.609	0.618	0.559	0.559
Inference	No effect	No effect	Significant GR	Significant GR

Note: Values in bracket represent t-values. The asterisks *, and **represent 1%, and 5% significance levels respectively. EGR means exponential growth rate (%), Variables are as defined in equation 3.

This implies that, changes in price of fresh tomato and pineapple in the rural and urban markets is influenced by time variable. Prices of fresh tomato in the rural and urban market have exponential growth rate of 1.099% and 1.054% respectively. In the similar way, about 0.776% and 0.808% exponential growth rate was identified in the rural and urban prices of pineapple. These results suggest that, there are noticeably dispersions between prices of fresh tomato in rural and urban markets in the study area; whereas the dispersion was more conspicuous in prices of pineapple in the two spatial markets. The result suggests that, there is possibility of market integration between the rural and urban prices of fresh tomato and pineapple in the study area. The nature of exponential growth in each price variable investigated revealed that, over time the price of fresh tomato in rural and urban markets showed insignificant decrease.

On the other hand, prices of pineapple in the rural and urban markets exhibited significant negative relationships with increase in time. This implies that, the prices of pineapple in rural and urban markets witnessed significant marginal deceases over increase period of time. These results are in consonance with the research report of Amusa, [7] who asserted that, agricultural commodity prices experienced increase and fluctuations over time in Nigeria. Akpan *et al.,* [4] also attested to this result on grain food stuffs in the southern Nigeria.

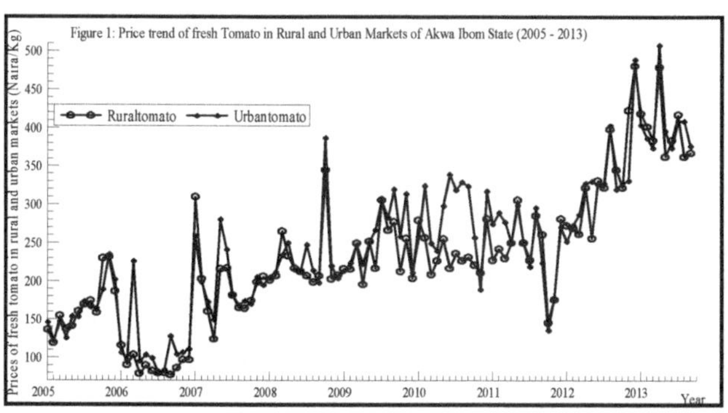

Figure 1: Price trend of fresh Tomato in Rural and Urban Markets of Akwa Ibom State (2005 - 2013)

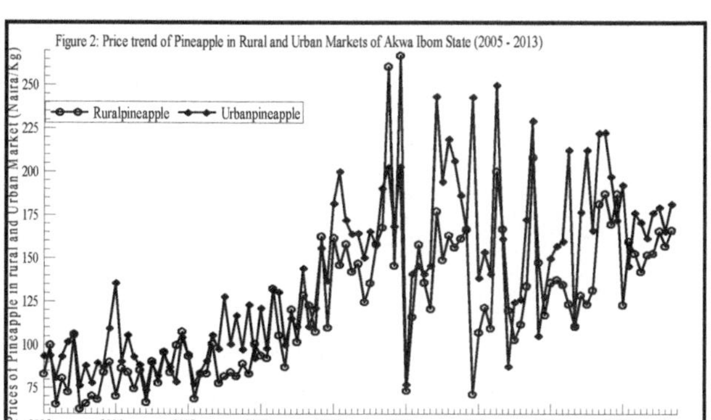

Figure 2: Price trend of Pineapple in Rural and Urban Markets of Akwa Ibom State (2005 - 2013)

The graphically representation of the linear price trend in fresh tomato and pineapple from January 2005 to September 2013, is shown in figure 1 and 2 respectively. The price trends show undulated fluctuations throughout the study period. In figure 1, the rural and urban price of fresh tomato seems to move together in most part of the year. However, both prices exhibited similar pattern of fluctuations throughout the study period. The result implies that, the rural and urban prices of fresh tomato move together in the period of investigation. Similarly, in figure 2, the linear trend in price of pineapple in rural and urban markets of Akwa Ibom State showed undulated fluctuations during the period of investigation. There were conspicuous dispersions between the rural and urban price of pineapple from 2010 to 2013 in the study area. On average, it seems prices of fresh tomato and pineapple assume similar pattern of fluctuations in most part of the period under consideration. Following this nature of fluctuations in the price of fresh tomato and pineapple in the rural and urban markets; it is suggested that, there is a strong evidence of symmetric price transmission mechanism or market integration between the two prices.

3.4. Pearson Correlation Matrix of Monthly Prices of Fresh Tomato and Pineapple (Expressed in ₦/Kg) in Rural and Urban Markets in Akwa Ibom State, Southern Nigeria

The linear and symmetric relationship between the average monthly price of fresh tomato and pineapple in the rural and urban markets of Akwa Ibom State was captured by the Pearson correlation coefficient. Table 5 presents the correlation matrix of rural and urban prices of fresh tomato and pineapple from January 2005 to September 2013 in Akwa Ibom State, Southern Nigeria. The result indicates that, prices of fresh tomato and pineapple in the rural market have positive significant (at 1% probability level) linear relationship with their corresponding prices in the urban markets. This means that, the price of fresh tomato in the rural market has a strong linear relationship with its own price in urban market in the study area. The same result was also obtainable for the pineapple market in the study area. These results further provide strong support for the existence of price integration or good price transmission mechanism between the rural and urban markets for vegetables (i.e. fresh tomato and pineapple) in Akwa Ibom State in southern Nigeria. This result implies that, factors that influence price of fresh tomato and pineapple marketing in the rural markets are similar to those in urban markets in the study area.

Table 5. *Pearson correlation matrix for prices of fresh tomato and Pineapple in Rural and Urban Markets in Akwa Ibom State (2005 - 2013).*

Variables	FT_{rt}	FT_{ut}	PN_{rt}	PN_{ut}
FT_{rt}	1.000	0.936 *	0.481*	0.561 *
FT_{rt}		1.000	0.506 *	0.616 *
PN_{ut}			1.000	0.785 *
PN_{ut}				1.000

Note: variables are as expressed in equation 3. * means statistically significant at 1% probability level; while values are bi-variate correlation coefficients.

3.5. Bilateral Granger Causality Test for Prices of Fresh Tomato and Pineapple in Rural and Urban Markets in Akwa Ibom State (2005 - 2013)

The causality relationship between the rural and urban prices of fresh tomato and Pineapple was tested in Akwa Ibom State. The result of the analysis is presented in Table 7. The result in Table 6 shows the optimal lag period used in the causality equation specified in equations 5 to 8. The asterisks below indicate the best (that is, minimized) values of the respective information criteria, AIC = Akaike criterion, BIC = Schwarz Bayesian criterion and HQC = Hannan-Quinn criterion.

Table 6. *The optimal Lag length for the causality equation.*

Lag	Loglikelihood	P(LR)	AIC	BIC	HQC
1	125.769	-	-2.094*	-1.577*	-1.885*
2	137.518	0.101	-2.010	-1.078	-1.633
3	145.384	0.471	-1.849	-0.503	-1.304
4	157.359	0.091	-1.769	-0.009	-1.057

Source: From data analysis.

The corresponding lag length indicates the best lag length for generating a more parsimonious causality equation for the specify price series. The result of the exercise indicated that lag 1 was more appropriate for the causality equations. This implies that the causality equations generated were done by using one period lagged of the variables involved. The estimated bilateral Granger causality estimates are presented in Table 7.

Table 7. *The vector autoregressive regression Granger causality estimates.*

Hypotheses	Lag	Sample size	F-Statistic	Prob.	Decision
$\Delta LnFT_{rt} \rightarrow \Delta LnFT_{ut}$	1	103	6.9398	0.000***	Rejected
$\Delta LnFT_{ut} \rightarrow \Delta Ln\, FT_{rt}$	1	103	4.6535	0.012**	Rejected
$Ln\, PN_{rt} \rightarrow LnPN_{ut}$	1	104	47.668	0.000***	Rejected
$LnPN_{ut} \rightarrow LnPN_{rt}$	1	104	49.775	0.000***	Rejected

Note: Variables are as defined in equation 3.→ means does not Granger cause.

The result presented in Table 7 suggests that, there is a bi-directional Granger causality association between the urban price and rural price of fresh tomato and pineapple from January 2005 to September 2013 in the study area. It means that, rural price of fresh tomato and pineapple is strongly endogenous to its respective urban price. The result also means that, the causality runs from urban market to rural market and vice versa. Alternatively, the result implies that, urban price of fresh tomato and Pineapple affected positively on their respective rural prices; and in the same manner the rural price of fresh tomato and pineapple influences their respective urban prices. In the similar way, the previous price in urban market significantly predicted the current price in the rural market for fresh tomato and pineapple in the study area. The presence of the bi-directional Granger Causality between the rural and urban prices of fresh tomato and pineapple indicated that, there is a perfect price transmission mechanism or market integration between the two markets in the study area. Following this result, the flow of markets information between the rural and urban markets concerning the fresh tomato and pineapple could be described as symmetric. The result further suggested that, the rural and urban prices of fresh tomato and pineapple might be tight together in the long run and the effect of transfer cost not significant. Based on the magnitude of the diagnostic statistics, it appears that, the rural market is the lead or the driving market in the short run for fresh tomato and urban market for Pineapple marketing in the state. This result can be attributed to the fact that, the bulk quantity of Pineapple consumed in the study area is brought from the neigbouring states. This commodity is mostly haboured in the urban markets for onwards distribution to other spatial markets. The result for the fresh tomato perhaps suggests the increasing importance of domestic production in the study area. In addition, the result suggests probably high degree of competitive market structure and strong endogeneity in the rural and urban prices of fresh tomato and Pineapple markets in the region. This result corroborates reports of Okoh *and* Egbon [22] and Adeoye *et al.,* [3] for different food stuffs in western Nigeria. Akpan *et al.,* [4] also confirm the similar relationship in grains in southern Nigeria.

3.6. The Engle Granger Two - Step and Johansen Co-Integration Test on Prices of Fresh Tomato and Pineapple in the Rural and Urban Markets of Akwa Ibom State, Nigeria

The result of the Engle and Granger two-step technique of co-integration regression tests for the residuals (ECM) generated in the long run equations is presented in the lower portion of Table 9 and Table 10. The results show that at 1% probability level of significance, the ADF test for the residual is significant (i.e. the estimated tau test from the residuals is greater than the critical tau value). Thus the Engle–Granger co-integration tests reject the null hypothesis of no co-integration for the four specified equations. Hence, there exist long run equilibria relationships between the rural and urban prices of fresh tomato and pineapple in the study area. Following the above results, the evidence of the long-run equilibria relationships between the rural and urban prices of fresh tomato and pineapple was strong, hence there was need to verify the result by conducting Johansen cointegration test. Table 8, presents results of the Johansen test meant to examine the long-run relationships between the rural and urban price of fresh tomato and pineapple. The null hypothesis of no cointegration between the rural and urban price of fresh tomato and pineapple were rejected at the conventional levels for both Trace and Max-Eingen statistics. There was no conflict between the Trace and Max-Eingen statistics, so the Johansen approach further confirms at least one cointegration equation between the rural and urban price of fresh tomato and pineapple in the study area.

Table 8. *Results of Johansen Cointegration Test.*

Markets	*Hypotheses*	*Trace Statistics*	*Hypotheses*	*Max.-Eigen Statistics*
$FT_{rt} \rightarrow FT_{ut}$	r = 0 r ≤ 1	14.544 (0.069)* 0.835 (0.361)	r = 0 r = 1	13.709(0.061)* 0.835 (0.361)
$PN_{rt} \rightarrow PN_{ut}$	r = 0 r ≤ 1	15.905(0.034)* 2.768 (0.096)*	r = 0 r = 1	13.137(0.075)* 2.768 (0.096)*

Source: Computed by authors. Trend assumption: linear deterministic trend. "r" indicates the number of co-integrating equation. Asterisk * represents rejection of the null hypothesis at 10% significance level. P-values derived from Mackinnon-Haug-Michelis, (1999) methodology. Critical values for the trace test are 15.495 and 3.841 respectively; while critical values for the Maximum Eingenvalue test are 14.265 and 3.841 respectively.

3.7. Regression Estimates for the Co-Integration Model and the Law of one Price (LOP) for Prices of Fresh Tomato and Pineapple in Rural and Urban Markets of Akwa Ibom State, Southern Nigeria

The result in Table 9 and Table 10 present the regression estimates to test the co-integration and law of one price in fresh tomato and Pineapple markets in the study area. Results in Table 9 contain the rural and urban price equation estimates for the fresh tomato in the study area. The R-square for the fresh tomato equations is 0.89. This implies that, about 89% of the total variability in the rural price of fresh tomato is explained by its respective urban prices. The result also applied to the urban price equation. The F-statistics for the two equations is significant at 1% probability level; thus confirming the significant of the estimated R-squares for the two equations. The result for the rural price of fresh tomato indicated that, the price transmission between the rural and urban markets exhibited near perfect long run integration. This is because; the integration coefficient (or inelastic coefficient) is approximately unity (i.e. 0.989). These results give a strong evidence of long run price integration between the rural and urban market prices of fresh tomato in the study area.

Table 9. *Long run Relationships between prices of Fresh Tomato and Pineapple in Rural and Urban markets of Akwa Ibom State (2005 – 2013).*

Variable	***LnFTrt= f LnFTut (Fresh Tomato)***	*Variable*	***LnFTut= f LnFTrt (Fresh Tomato)***
Constant	-0.006 (-0.032)	Constant	0.610 (3.620)*
$LnFt_{ut}$	0.989 (28.65)*	$LnFT_{rt}$	0.898 (28.65)*
F-cal	820.6511*	F-cal	820.657*
R^2	0.888	R^2	0.888
DW- test	1.321	DW- test	1.434
ADF test for errors from above equations			
ECM	-7.138***	-7.557***	

Note: the equation for the ADF test include constant and trend. Critical value at 1% = -4.05, Values in bracket represent t-values. The asterisk * represents 1% significance level. Variables are as defined in equation 3.

The result for the urban price equation of tomato also shows a strong indication of the law of one price. However, the coefficient of the long run market integration (inelastic coefficient) was not approximately unity. This means that, the long run market integration coefficient for the urban price equation is weaker than the rural price equation. The magnitude of market integration coefficient for the fresh tomato suggests that, the urban price impact on the rural price is more relative to the rural price impact on urban price in the study area. Hence, the market integration in fresh tomato is stronger when market information flow from urban to rural markets compared to the reverse case in the study area. Similarly, results in Table 10 contain the rural and urban price equation estimates for Pineapple in the study area. The R-square value is 0.67, and it implies that, about 67% of the total variability in the rural price of pineapple is explained by its respective urban prices. The same result applied to the urban price equation. The F-statistics for both equations is significant at 1% probability level; thus confirming the significant of the estimated R-square. The estimated rural price equation of pineapple indicated that, the price transmission between the rural and urban markets exhibited strong significant but not perfect long run integration. This is because; the integration coefficient (or inelastic coefficient) is far from unity (i.e. 0.809). These results give a good evidence of long run price integration between the rural and urban market prices of pineapple in the state. The result for the urban price equation of pineapple also revealed a good indication of the law of one price. However, the coefficient of the long run market integration (inelastic coefficient) was still far from unity.

Table 10. *Long run Relationships in prices of fresh Tomato and Pineapple in Rural and Urban Markets of Akwa Ibom State (2005 – 2013).*

Variable	***LnPNrt= f LnPNut (Pineapple)***	*Variable*	***LnPNut= f LnPNrt (Pineapple)***
Constant	0.796 (2.887)*	Constant	0.969 (3.548)*
$LnPN_{ut}$	0.8093 (14.37)*	$LnPN_{rt}$	0.8251 (14.37)*
F-cal	206.591*	F-cal	206.591*
R^2	0.667	R^2	0.667
DW- test	1.857	DW-test	1.764
ADF test for errors from above equations			
ECM	-8.523*		-9.671*

Note: the equation for the ADF test include constant. Critical value at 1% = -3.50, Values in bracket represent t-values. The asterisk * represents 1% significance level. Variables are as defined in equation 3.

Comparing the two price equations for pineapple, the long run market integration coefficient for the urban price equation is stronger than the rural price equation. Based on the magnitude of market integration coefficient for the rural and urban price of pineapple, it is suggested that, the rural price impact on the urban price is more relative to the reverse situation in the study area. Hence, the long run market integration in pineapple is stronger when market information flow from the rural to urban markets compared to the opposite case. The result provided additional evidence that, the rural market is the driver in the marketing of pineapple in Akwa Ibom State. However, the result for the fresh tomato supported urban market leadership role contrary to the previous report.

3.8. Error Correction Model of the price of fresh tomato and Pineapple in Rural and Urban markets in Akwa Ibom State (2005 – 2013)

Table 11. *ECM estimates for the rural and urban price of fresh tomato in Akwa Ibom State.*

$LnFT_{rt} = f(LnFT_{ut})$		$LnFT_{ut} = f(LnFT_{rt})$	
Variable	Coefficient and t-test	Variable	Coefficient and t-test
Constant	0.003 (0.240)	Constant	0.002 (0.13)
$\Delta LnFT_{rt-1}$	-0.084 (-0.82)	$\Delta LnFT_{ut-1}$	-0.034 (-0.34)
$\Delta LnFT_{ut}$	0.769 (14.40)*	$\Delta LnFT_{rt}$	0.856 (14.29)*
$\Delta LnFT_{ut-1}$	0.013 (0.14)	$\Delta LnFT_{rt-1}$	0.035 (0.32)
ECM_{t-1}	-0.507 (-4.59)*	ECM_{t-1}	-0.697 (-5.77)*
$R^2 = 0.708$; F-cal. = 59.31*; DW = 1.97; Log-likelihood = 68.80		$R^2 = 0.745$; F-cal. = 71.75*; DW = 2.02; Log-likelihood = 63.93	

Note: Values in bracket represent t-values. The asterisk * represent 1% significance level. Variables are as defined in equation 3.

Table 12. *ECM estimates for the rural and urban price of Pineapple.*

$LnPN_{rt} = f(LnPN_{ut})$		$LnPN_{ut} = f(LnPN_{rt})$	
Variable	Coefficient and t-test	Variable	Coefficient and t-test
Constant	0.002 (0.10)	Constant	0.005 (0.26)
$\Delta\ LnPN_{rt-1}$	-0.234 (-2.45)**	$\Delta\ LnPN_{ut-1}$	-0.235 (-2.31)**
$\Delta Ln\ LnPN_{ut}$	0.557 (7.24)***	ΔLn $LnPN_{rt}$	0.600 (7.61)***
$\Delta Ln\ LnPN_{ut-1}$	0.121 (1.21)	ΔLn $LnPN_{rt-1}$	0.069 (0.74)
ECM_{t-1}	-0.688 (-5.4)***	ECM_{t-1}	-0.605 (-4.7)***
$R^2 = 0.603$; F-cal. = 37.143***; DW = 2.013; Log-likelihood = 34.098		$R^2 = 0.541$; F-cal. = 28.710***; DW = 2.105; Log-likelihood = 32.707	

Note: Values in bracket represent t-values. The asterisk *** represent 1% significance level. Variables are as defined in equation 3.

The presence of the co-integration among the specified price variables demanded the specification of the Error Correction Model (ECM). Table 11 contains estimates of ECM generated from the rural and urban price of fresh tomato; while Table 12 contains estimates of the ECM generated from the rural and urban price of pineapple. For the rural and urban price equation of fresh tomato, the coefficient of the error correction terms is negative and statistically significant at 1% probability level respectively. The result validates the existence of the long-run equilibrium relationships between the rural and urban market prices of fresh tomato in the study area. The result further implies that, the rural and urban prices of fresh tomato are sensitive to departure from their equilibrium levels in the previous periods. The slope coefficients of the error correction term for the rural (-0.507) and urban (-0.697) price equation represents the speed of adjustment of respective price equation and also is consistent with the hypothesis of convergence towards the long-run equilibrium once the price equation is shocked.

The diagnostic tests for the ECM model of fresh tomato price equation revealed the R^2 values of 0.708 and 0.74 for the rural and urban price equation respectively. This means that the specified explanatory variables explained about 70.85% and 74.00% of the adjusted total variations in the rural and urban price of fresh tomato respectively in the study area. The F-statistics for the two equations are significant at 1% probability level, indicating that the two R^2 are significant and this implies that, the ECM of rural and urban price of fresh tomato has goodness of fit. The auto-correlation was not a serious problem in the two equations. The result revealed that, about 50.70% and 69.70% of the rural and urban price adjustments of fresh tomato take place respectively within every month due to exogenous shock. That is, it will take about 7 weeks 6 days for the rural price of fresh tomato; and 5 weeks 5 days for the urban price of fresh tomato; to fully adjust to the equilibrium position in the long run due to shock in the marketing system. The result suggests that, the urban price of fresh tomato adjusted faster than the rural price. By implication, movements in the rural price of fresh tomato is significantly detected by it respective urban price. Given the magnitude of adjustment, it seems the adjustment rate in the rural and urban price of fresh tomato in the study area is not instantaneous. It is suggested that, the rural price of fresh tomato is strongly endogenous to its urban price. The result is in consonance with the research report of Okoh *and* Egbon [22] and Akpan *et al.,* [4] on different food stuffs in Nigeria.

For the Pineapple market, the slope coefficient of the error correction term for the rural (-0.688) and urban (-0.605) price of pineapple is negative and statistically significant at 1% probability level. The coefficient of the error correction terms shows the speed of convergence to the long run equilibrium as a result of shock in the pineapple price equations. This indicates that, any disequilibrium in the long run of either rural or urban price of pineapple would be corrected in the short run. The significant value of the error correction term implies that, the urban price of pineapple will always react to bring stability in the rural price whenever there is significant variation in the rural price and vice versa. The result revealed that, about 68.80% and 60.50% of the rural and urban price adjustments take place respectively within every month due to exogenous shock in the system. That is, it will take about 5 weeks 5 days and 6 weeks 4 days for the rural and urban price of pineapple respectively, to fully adjust to equilibrium position in the long run due to shock in the system. The result revealed that, the rural price of pineapple adjusted faster than the rural price. By implication, movements in the rural price of pineapple is significantly detected by it respective urban price. The diagnostic statistics derived from the ECM with respect to pineapple price revealed R^2 values of 0.603 and 0.541 for the rural and urban price equation respectively. This means that, the explanatory variables contained in the pineapple price equation explain about 60.30% and 54.10% of the total variations in the rural and urban price of pineapple respectively in the study area. The F-statistics were significant at 1% probability level, indicating that, the equation has goodness of fit.

3.9. Discussion of the Long Run and Short Run Model Results for Fresh Tomato and Pineapple

The empirical results revealed that, the long run rural price cointegration coefficient for the fresh tomato converges to the postulate of the law of one price. This means that, this co-integration coefficient was approximately unity (0.989). This confirms the existence of high degree of long run market integration between the rural and urban prices of fresh tomato in the study area. This implies that, a shock on the urban price of fresh tomato is instantaneously transmitted to its corresponding rural price. This is conceivable when considering the flow of market information from urban to rural market. Also, the degree of market integration in the urban price equation of fresh tomato (0.898) was far from the threshold mark of unity. Although these also confers high degree of long run market integration, but is not efficient as compared to the previous result. This result is tenable when market activities flow from rural to urban areas. It therefore implies that, when fresh tomato marketing activities is stimulated by the rural market, the degree of market price transmission is slower than the reverse case in the study area.

In addition, the long run model for the rural and urban price of pineapple showed similar degrees of divergence from unity (i.e. 0.809 for rural price and 0.825 for urban price equation). The rural and urban price equations however confirm the presence of significant long run market integration between the urban and rural price of pineapple; but it failed to uphold instantaneous response of the urban price to rural price shock or vice versa. This means that, irrespective of the direction of flow of pineapple marketing activities in the long run, the degree of price change in the rural and urban markets will not be instantaneous. This result indicates that, there might be some mounted externality costs in the marketing of pineapple in the study area. It is suggested that, there are some levels of inefficiency in pineapple market information transmission in the long run between the urban and rural markets in the state. This result could be connected to the poor or inefficient marketing and social infrastructures in the rural areas of the state where most of the pineapple consumed is produced. Based on the result obtained from the rural and urban price analysis of fresh tomato and pineapple in the study area; it is established that, in the long run, the price of fresh tomato in the rural markets will synchronize faster with that of the urban market if marketing shock emanated from the urban areas. This is however the reverse case for pineapple. On the other hand, if the marketing shock came from the rural area, the price transmission of fresh tomato will not attain high level of efficiency as compared to pineapple in the study area. Similar results on the long run relationships among agricultural commodities have been established by Okoh *and* Egbon [22]; Ohen *et al.*,[20]; Ojiako *et al.*, [21] and Akpan *et al.*, [4].

For the short run model; the estimates of the rural price equation of fresh tomato showed significant contribution of the urban price to the total variations in the rural price. The result was similar in the urban price equation. In the same manner, the short run urban price equations of pineapple showed significant coefficient of the rural price variables. However, the short run market cointegration coefficients in the rural price equation of pineapple showed significant deviations from unity (i.e. 0.557). This result implies that, though there is significant short run market integration between the urban and rural prices of fresh tomato and pineapple irrespective of the direction of flow of market activities; but the degrees of short run market -integration varies between the rural and urban areas. The short run market integration in fresh tomato appears to be stronger when information or price movement flows from urban to rural markets than the reverse case. The case is reverse in pineapple marketing in the study area. The short run results are consistent with the opinion of Okoh *and* Egbon [22]; Ohen *et al.*,[20] and Akpan *et al.*, [4].

3.10. Index of Market Connection (IMC) for Fresh Tomato and Pineapple in Akwa Ibom State, Nigeria

The IMC was estimated for fresh tomato and pineapple markets in the study area. The estimates of IMC regression is shown in Table 13. The IMC were 0.961 and 0.095 for fresh tomato and pineapple market respectively.

Table 13. *Estimates of IMC Regression for fresh Tomato and Pineapple in Akwa Ibom State, Nigeria.*

$LnFT_{rt} = f(LnFT_{ut})$		$LnPN_{rt} = f(LnPN_{ut})$	
Variable	Coefficient and t-test	Variable	Coefficient and t-test
Constant	2.870 (0.33)	Constant	12.498 (1.43)
FT_{rt-1}	0.460 (4.86)***	PN_{rt-1}	0.069 (0.72)
$(FT_{ut} - FT_{ut-1})$	0.782 (14.40)***	$(PN_{ut} - PN_{ut-1})$	0.559 (8.08)***
FT_{ut-1}	0.501(5.15)***	PN_{ut-1}	0.724 (8.01)***
R^2 = 0.90; F-cal. = 300.50***; DW = 2.20		R^2 = 0.648; F-cal. = 61.36***; DW = 2.08	

Note: Values in bracket represent t-values. The asterisks * and *** represent 10% and 1% significance levels respectively. Variables are as defined in equation 3.

The IMC for fresh tomato and pineapple are less than unity. The result implies that, there is high short run market integration between rural and urban markets for fresh tomato and pineapple commodities in the state. The short run market integration was stronger in pineapple market relative to the fresh tomato market. This however confirms the ECM results discussed previously and further substantiates the presence of significant short run relationship and efficient price transmission mechanism between the rural and urban price of fresh tomato and pineapple in Akwa Ibom State.

4. Summary and Recommendations

The study used statistical and econometric techniques to analyze the dynamic of price transmission between the rural and urban prices of fresh tomato and pineapple in Akwa Ibom State, Southern Nigeria. Results revealed that, prices of fresh tomato and pineapple in rural and urban markets have positive relationship with time and exponential growth rates that is greater than unity in fresh tomato and less than unity in pineapple. The graphical analysis showed that, there were harmonies in the rural and urban price dispersion in fresh tomato and noticeably deviations in some months in the rural and urban price of pineapple. The result of the trend analysis suggested the prevalence of good price transmission mechanism between the rural and urban market for fresh tomato and perhaps less efficient for pineapple in the study area. Also, the Pearson correlation coefficient matrix revealed that, the rural price of fresh tomato and pineapple has linear symmetric relationships with their corresponding urban prices in the study area. The relationship was stronger in fresh tomato market than in pineapple market. The result connotes the existence of the symmetric market information flow between the rural and urban markets for fresh tomato and pineapple in the state. The Granger causality test revealed bi-directional relationship between the rural and urban price of fresh tomato and pineapple in the study area. This suggested that, the price transmission mechanism between the rural and urban markets for fresh tomato and pineapple is efficient; and has high tendency for market integration as well as strong endogeneity properties.

The results of the co-integration regression revealed the presence of the long run market integration relationship between the rural and urban prices of fresh tomato and pineapple as well as upheld the hypothesis of the perfect or fast price transmission between the two markets in the study area. The results also attested to the presence of varied long run market integrations between the rural and urban market of fresh tomato and pineapple depending on the direction of flow of the market information. This means that, the rural and urban price equation of fresh tomato and pineapple exhibit varied degrees of long run market integration as well as convergence to the law of one price. The results of the short run model or error correction model (ECM), confirm the existence of the short run market integration between rural and urban prices of fresh tomato and pineapple in the study area. Similarly, the short run market integration varied in magnitude based on the direction of flow of market information or price movement between the two markets. Based on the magnitude of the coefficient of error correction term, it was discovered that, the urban price of fresh tomato adjusted faster to the stable state in the long run than the rural price once there is exogenous shock in the marketing system in Akwa Ibom State. It was also discovered that, the rural price of pineapple adjusted faster than its corresponding urban price once there is exogenous shock in the system. The estimated index of market connection (IMC) supported the high short run market integration of fresh tomato and pineapple prices in the rural and urban markets of Akwa Ibom State.

Based on the discoveries of this study, it is recommended that, the Akwa Ibom State government should continue to make available marketing infrastructures especially in the rural areas to improve the symmetric nature of information among fresh tomato and pineapple markets in the state. Effort should also be channeled by all stake holders such as governments, trade unions, market unions and other organizations to reduce excessive externality costs (i.e. storage cost, transportation cost, toll gate, law enforcement charges) associated with the marketing of fresh tomato and pineapple in the state. This attempt will help to reduce the total variable cost and bring about insignificant price differential between the rural and urban markets in the state. The government of Akwa Ibom State should established market information centers and awareness programmes on mass media (such as radio, television and newspaper), to facilitate efficient communication and flow of information among fresh tomato and pineapple producers and consumers in the state.

References

[1] Adebusuyi, B. S. (2004). Stabilization of commodity market of interest to Africa: Paper presented at the workshop on constraints to growth in sub-Saharan Africa, held in Pretoria South Africa.

[2] Adenegan, K. O. and I. B. (2011). Adeoye. Price Analysis of Tomato in Rural and Urban Retail Markets of Oyo State. International Journal of Agricultural Economics & Rural Development. Volume 4 (2): Pp 90 – 96.

[3] Adeoye, I. B, Dontsop Nguezet P. M, Badmus M. A and Amao I. O. (2011). Price Transmission and Market Integration of Banana and Plantain in Oyo state, Nigeria. ARPN Journal of Agricultural and Biological Science; Vol. 6, No. 5, Pp 18-24.

[4] Akpan, S. B, Inimfon V. P, Samuel J. U. (2014). Analysis of Monthly Price Transmission of Local and Foreign Rice in Rural and Urban Markets in Akwa Ibom State, Nigeria (2005 to 2013). International Journal of Agriculture and Forestry; 4(1): 6-18.

[5] Akpan, S. B. (2007). Relative Price Variability and Inflation: A case study of Grain subsector in Nigeria. Unpublished Master Degree Thesis; University of Uyo.

[6] Akwa Ibom State Agricultural Development Programme (AKADEP), quarterly physical progress reports of various issues (from January 2005 to June 2013).

[7] Amusa, A. M. (1997).Trend Analysis of Agricultural Food Prices in Nigeria (1985-1995) (A Case Study of Ibadan). Unpublished M. Sc Project, Department of Agricultural Economics, University of Ibadan, Ibadan, Nigeria.

[8] Baumöhl, E., and Tomáš Výrost (2010). Stock Market Integration: Granger Causality Testing with Respect to Non-synchronous Trading Effects. Finance aúvěr *Czech Journal of Economics and Finance*, 60, no. 5, Pp 414-425.

[9] Chirwa, E. W. (199). "Food Marketing Reforms and Integration of Maize and Rice Markets in Malawi. University of Malawi". *Working Paper*, No. WC/05/99, 1999.

[10] Crawford, P. B, Obarzanek E., Morrison J & Sabry Z. I. (1994). Comparative advantage of 3-day food records over 24 recall and 5-day food frequency validated by observation of 9-and 10-year girls. J Am Diet Assoc 94 (6): 626-630.

[11] Famine Early Warning Systems Network (FEWSNET) (2008). Nigeria Food Security Outlook March to September 2008.

[12] Gilbert, C. (1999). Commodity risk management for developing countries: paper prepared for the third meeting of the international task force (ITF) held in Geneva 23-24 June 1999.

[13] Gonzalez-Rivera, G. & S. M. (2001). The Extent, Pattern and Degree of Market Integration: A Multivariate Approach for the Brazilian Rice Market. American Journal of Agriculture Economics, 2001; 83(3), 576-92.

[14] Goodwin, B. K. & T. C. Schroeder. (1991). "Cointegration tests and spatial price linkages in regional cattle markets." American Journal of Agricultur al Economics, 1991; 73(2): 452-464.

[15] Goodwin, B. K. (2005). Spatial and Vertical Price Transmission in Meat Markets. A Paper prepared for workshop on Market Integration and Vertical and Spatial Price Transmission in Agricultural Markets, University of Kentucky, April 21, 2005.

[16] Granger, C. W. J. (1969). Investigating causal relations by econometric models and Cross-spectral methods. Econometrica 37, 424-438.

[17] Hendry, D. F. (1986). 'The Role of Prediction in Eval uating Econometric Models', Proceedings of the Royal Society of London, 1986; 407, 25–34.

[18] Mushtaq K, Abbas F, & Ghafoor A. (2006). Testing the Law of One Price Rice Market Integration in Punjab, Pakistan. Pakistan Journal of Agricultural Science; 43(3): 213 – 216.

[19] National Population Council (NPC) of Nigeria, 2006 Population Census Data.

[20] Ohen, S. B., S.O. Abang & I. C. Idiong. (2007). Price Transmission and Market Integration: Vertical and Horizontal Price Linkages for Live Catfish in Nigeria. Journal of Agriculture & Social Sciences; Vol. 3, No. 1, 17–20.

[21] Ojiako, I. A., I. A., Chuma E, Godwin N. A & N. M. Nkang. (2012). Spatial Integration and Price Transmission in Selected Cassava Products' Markets in Nigeria: A Case of Lafun. World Applied Sciences Journal; 18 (9): 1209-1219.

[22] Okoh, R. N & Egbon P. C. (2005). The Integration of Nigeria's Rural and Urban Foodstuffs Markets. AERC Research Paper 151, Africa Economic Research Consortium, Nairobi, 2005.

[23] Oladapo, M. O., Momoh S, Yusuf S, & Awoyinka Y. (2007). Marketing Margin and Spatial Pricing Efficiency of Pineapple in Nigeria. Asian Journal of Marketing; 1(1), 14-22.

[24] Odozi, J. C., and Bolarin T. O., (2012). Governance options for price instability: A review of the food grain commodity in Nigeria. *Journal of Development and Agricultural Economics;* Vol. 4(4), pp. 93-100.

[25] Polaski, S. (2008). Rising Food Prices, Poverty and Doha Round. Carnegie Endowment for International Peace, 2008.

[26] Quebedeaux, B. and Bliss, F. A. (1988). Horticulture and human health. Contributions of fruits and vegetables. Proc. 1st Intl. Symp. Hort. and Human Health. Prentice Hall, Englewood, NJ.

[27] Quebedeaux, B. and Eisa, H. M. (1990). Horticulture and human health. Contributions of fruits and vegetables. Proc. 2nd Intl. symp. Hort. and Human Health. HortScience 25:1473-1532.

[28] Sexton, R. J., C. L. Kling, H. F. Carman. (1991)."Market integration, efficiency of arbitrage, and imperfect competition: methodology and application to U.S. celery." *American Journal of Agricultural Economics;* 73(3):568-580.

[29] Van Duyn, M. A & Pivonka E. (2000). Overview of the health benefits of fruit and vegetable consumption for the dietetics professional. J Am Diet Assoc 99 (10): 1241.

[30] Wargovich, M. J., (2000). Anticancer properties of fruits and vegetables. HortScience 35:573-575.

6

Yield response of rice in Nigeria

David Boansi

Department of Economic and Technological Change, Center for Development Research (ZEF), University of Bonn, Germany

Email address:
boansidavid@rocketmail.com (D. Boansi)

Abstract: To help bridge information gap in supply response studies for Nigeria and inform policy decision on how the demand-supply gap for rice in Nigeria could be bridged, the current study through the use of Johansen's Full Information Maximum Likelihood test estimated a yield response model for Nigeria using national level data for the period 1966-2008. The results suggest that, increasing yield levels for paddy rice in Nigeria and ensuring stability requires interplay of biophysical, socio-economic and structural forces. By estimates for the current study, bridging of the demand-supply gap can be realized through initiation of measures to address inefficiencies in the supply chain to ensure appropriate transmission of price increment, promotion of local rice consumption to ensure ready market for farmers in times of increasing output, addressing soil fertility challenges through efficient use of fertilizer and regular management of fertility of rice fields, and increasing farmers access to credit to help them meet cost of relevant inputs of production. The latter suggestion could to a greater extent incite appropriate response of farmers to both price and non-price incentives in the country. Diagnostic tests conducted indicate that the residual series is normally distributed, non-serially correlated and homoscedastic.

Keywords: Co-Integration, Yield, Error Correction Model, Nigeria, Prices

1. Introduction

With a population estimate of 174,507,539 persons and population growth rate of 2.54%, Nigeria happens to be not only the leading producer of rice in West Africa, but also among the leading importers of the commodity. Although endowed with a strong agricultural and natural resources base, as well as favorable climatic conditions for agricultural production, an amount of about ₦1 billion is spent daily by Nigeria on importation of rice [1]. Hindered by inconsistent policies on rice, improper methods of production used by farmers, high costs and scarcity of vital inputs of production, etc, local rice production for Nigeria has failed to catch up with the increasing pace of consumption. By estimates observed from the agricultural production database of FAO for Nigeria, the gap between domestic demand of rice for food (as against feed and other uses) has widened since the late 1990s.

Treated with benign neglect prior to independence due to self-sufficiency, rice has become a strategic and political commodity in Nigeria, attracting much attention due to its increasing role in the diet of the populace and its daily drainage of foreign exchange through imports. Rice used to be classified as a luxury food item prior to independence; it however now holds the status of a staple food, replacing cassava and yam among others [12]. The consumption of rice in spite of increasing prices induced through high tariff imposition has been increasing since the year 1976. Per capita consumption of rice in the country increased from as low as 3.4kg/year in 1976 to 20.9kg/yr in 2009 [15], an increase of over 500%. Production has however failed to catch up with the increasing demand for rice, leading to widening of the gap between domestic production and demand, increasing the role of imported rice in diets of the Nigerian populace, and making the country a net importer of the commodity in the process.

Various trade policies purposed on improving local rice production and marketing systems of the country have been adopted and applied by previous regimes to help reverse the net importer status of the country. Among such measures are imposition of tariffs, quantitative restrictions on imports through the use of quota, and outright ban on imported rice [12]. In spite of past and present efforts, the demand-supply gap persists. Like other West African countries, bridging of the rice demand-supply gap in Nigeria, besides the aforementioned trade policy measures, have been sought through expansion of area for production as against intensification (purposed on improving productivity). With area harvested of rice having generally depicted an increasing trend between the years 1976 and

2009, output of rough rice has more or less stagnated between the years 1989 and 2007. As shown in Figure1, the latter observation was accompanied by a general declining trend in yields between the two aforementioned years (1989 and 2007). Yield decreased from 2tonnes/ha in 1989 to 1.3tonnes/ha in 2007.

It is believed that coupling extensification (area expansion) measures with productivity-enhancing measures could produce better outcome in terms of output of rough rice, than sole expansion of harvested area. To achieve this however, there arises a need to identify and estimate the magnitude and effects of key determinants of rough rice yields and to inform policy decision on how beneficial implications of growth-enhancing determinants could be maximized and adverse implications from inhibitors minimized. Meeting this need is the goal of the present study. Magnitude and effects of such key determinants is sourced through the use of Johansen's Full Information Maximum Likelihood test.

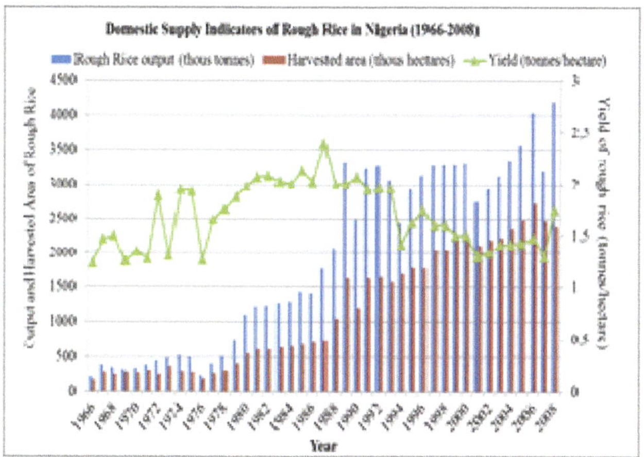

Data Source: IRRI (World Rice Statistics, FAO data)

***Figure 1.** Domestic supply indicators of rice in Nigeria*

1.1. Local, Regional and Continental Rice Yields

In a study by Norman and Kebe (2006) on "*African smallholder farmers: rice production and sustainable livelihoods*", it was revealed that rice production in most African countries lags well behind demand due to low productivity of farmers' fields. Most developing countries go after area expansion in pursuit of bridging their respective demand-supply gap for rice, placing quite minimal emphasis on productivity. Some important questions that need answering are "How long can we keep on expanding area under rice cultivation? How far can we get with this?" In his speech at the launch of a seminar organized by CGIAR Fund Office, Dr Robert Zeigler (Director General of the International Rice Research Institute (IRRI) placed an advise that

"Because of rapid population growth and diminished harvest due to climate change and other weather related stress, such as severe drought and floods, global demand for rice will outstrip supplies unless concerted action is taken now to boost yield growth and improve the management of water, land and other resources on which production depends" [36]

In spite of efforts in line with area expansion put forth so far by many developing countries, the demand-supply gap persists in majority of such countries, notably countries in West Africa. The problem therefore as identified by Norman and Kebe [29], and affirmed by Zeigler [36] is with yield as a component of output as against harvested area. With anticipated increases in population growth and corresponding increases in infrastructure development, there will come a time when expansion will no more be an option. Must we wait until that time is due? What then will we be protecting for posterity? This line of reasoning has over a decade now given most researchers a purpose to sacrifice much time, money and energy into finding alternative measures to help increase yields of major cereals and non-cereals on which farmers and most rural and sub-urban inhabitants depend for sustenance.

***Table 1.** Local, regional and continental rice yields*

Region / Country	YEAR (PERIOD)				
	1961- 1970	1971- 1980	1981- 1990	1991- 2000	2001- 2010
Côte d'Ivoire	1.00	1.21	1.17	1.34	1.88
Egypt	5.19	5.37	5.98	8.22	9.68
Ghana	1.11	0.92	1.11	1.87	2.19
Indonesia	1.91	2.75	3.96	4.35	4.68
Nigeria	1.26	1.69	2.08	1.69	1.52
Pakistan	1.65	2.35	2.45	2.73	3.15
Thailand	1.81	1.85	2.03	2.37	2.91
Vietnam	1.93	2.10	2.77	3.71	4.90
Western Africa	1.15	1.34	1.58	1.64	1.71
Northern Africa	5.15	5.27	5.93	8.14	9.58
Africa	1.82	1.84	1.93	2.17	2.37
Americas	2.08	2.25	2.76	3.70	4.82
Asia	2.11	2.55	3.30	3.80	4.19
World	2.11	2.52	3.23	3.73	4.12

Source: Author's computation (with FAO data from IRRI)

As seen from Table 1, the issue of low yield is seen as a problem not only with countries in West Africa, but some high esteemed countries in world rice trade as well. Analyzing systematically by regions and continents, rice yields for the West African sub-region are well below figures for the African continent for all the five decadal periods considered. Excluding the period 2001-2010 for Côte d'Ivoire, and 1991-2000 and 2001-2010 for Ghana, rough rice yields for the two countries are well below that for the sub-region and African continent for all the five decadal periods. Although generally above figures for the sub-region, rough rice yields in Nigeria (with the exception of average yield for the period 1981-1990, 2.08tonnes/ha) are lower than figures for the continent. Similarly, average yields of rough rice for the African continent are well below world averages for all the five decadal periods.

Yields however in Egypt are not just above figures for the North African sub-region, but almost twice world averages. Similarly, figures observed for Northern Africa are well above world averages, West African averages, averages for the African continent, and even those observed for Asia and the Americas. Although average yields observed for Asia are above world averages and those for Africa, noted countries in world rice production and trade like Pakistan and Thailand have average yield figures well below figures for the Asian continent, and world averages for all the five decadal periods. Excluding the period 1961-1970, average yields for Indonesia are well above average figures for the Asian continent and above world averages. In contrast however, average yield figures for Vietnam are well below figures for the Asian continent except for the period 2001-2010. Average yield figures for Vietnam are above world averages only in the periods 1991-2000 and 2001-2010. Similarly, average yield figures for the Americas, although above averages for Africa, Asia in the period 2001-2010, and world averages for the periods 1991-2000 and 2001-2010, they are generally below average yields in Asia, Northern Africa and world averages for majority of the period.

2. Literature Review

Agricultural supply response (be it output, acreage or yield) has been a fundamental issue and continues to attract much attention due to uncertainty in future food supply and crisis in the past (most notably, the commodity crisis of the year 2008 which impacted adversely on most households in the Sudan-Sahel region of West Africa). These concerns are held mostly due to the volatile nature of prices for agricultural commodities on the world market and to climatic influences on crop production.

Supply response studies observed so far in literature have been on either estimating response across a group of crops or for a specific crop of interest to the country under study. In either case, supply decision of farmers is assumed to be in line with economic theory; thus being influenced by price and non-price factors. Among the common non-price factors suggested in literature are access to capital, extension services, agro-climatic conditions, agricultural labor availability, area of land cultivated, status of rice farmers, and use of fertilizer [27, 7, 8, and 35]. Real producer price of rice, real producer price of competitive field crops like maize, world price of rice and maize with important indirect effects to producers, and price of relevant agro-chemicals like fertilizer [24, 26] are among the price-related drivers of supply identified so far in literature.

Defined as the variation of agricultural output as a result of variation in price and key inputs, supply response according to Molua [25] is explained by biophysical, socioeconomic and policy factors. For over three decades now, various theories have been developed and applied by economists to explain the dynamics of supply in agriculture and its key sub-sectors. Ideally, in line with economic theory, supply functions have been estimated on the assumption that the primary drivers of market supply of any given commodity are inputs, output and the state of technology [8]. Ahmed and Siddiqui [3] estimated supply response as a function of output and input prices, together with quasi-fixed inputs and supply shifters like technical change and policy intervention.

In a study on supply responsiveness of Indian farmers, Cummings [11] suggested that supply response for a given crop is equivalent to the response of acreage under cultivation to changes in economic and non-economic factors, and that estimates obtained from acreage response studies could be considered fairer reflections of supply response for a given commodity. Molua [24] however advised that positive signals observed from acreage response models will translate appropriately into output only on the employment of other vital complementary factors of production such as pesticides, fertilizer, high yielding varieties, other farm chemicals, improved cropping techniques and better farm management methods. Increasing production of a given crop can be achieved through expansion of the area under cultivation, increasing cropping intensity or increasing produce per unit area (yield) [13].

Beside these, other constraints have been noted to impact significantly on output of major staple crops, most importantly wheat, maize, rice, sorghum and millet for West Africa and other rice producing and importing regions. Among such constraints are exchange rate policies and pricing policies; fiscal policy, fertilizer policy, land policy, etc. [30].

3. Research Methodology

3.1. Analytical Framework

Co-integration analysis has so far been performed in supply response studies using the Engle-Granger two-step estimation technique [14], the residual-based approach of Phillips and Ouliaris [32], or the Johansen Full Information

Maximum Likelihood test [21]. For the purpose of the current study however, the Johansen Full Information Maximum Likelihood test is used. This estimation technique is chosen because, unlike the other two methods which assume a single co-integrating vector, the Johansen method allows for all possible co-integrating relationships and permits empirical determination of the number of co-integrating vectors [22]. In addition, short-run coefficients are estimated in such a way that they are guided by and consistent with long-run relationships.

The Johansen approach commences with the definition of a Vector Autogression given by

$$X_t = \prod_1 X_{t-1} + \prod_2 X_{t-2} + \ldots + \prod_p X_{t-p} + \mu_t \quad (1)$$

Where X_t is an $(n \times 1)$ vector of $I(1)$ variables, $\prod_1$ through $\prod_p$ represent $(m \times m)$ matrix of coefficients, and μ_t is $(n \times 1)$ vector of white noise errors. With the assumption that X_t is non-stationary, equation (1) can be re-parameterized or written in an error correction form as;

$$\Delta X_t = \Gamma_1 \Delta X_{t-1} + \Gamma_2 \Delta X_{t-2} + \ldots + \Gamma_{p-1} \Delta X_{t-p+1} - \prod X_{t-p} + \mu_t \quad (2)$$

Where $\Gamma_1 = \prod_1 - I$, $\Gamma_2 = \prod_2 - \Gamma_1$, $\Gamma_3 = \prod_3 - \Gamma_2$ and $\prod = I - \prod_1 - \prod_2 - \ldots - \prod_p$

Γi gives the short-run estimates while $\prod$ gives the long-run estimates. The matrix $\prod$ determines the extent to which a given system is co-integrated and is called the impact matrix. This implies that, information on the number of co-integrating relationships among the variables in X_t is given by the rank of the matrix $\prod$. If the rank of $\prod$ matrix r, is $0 < r > n$, there are r linear combinations of the variables in X_t that are stationary [22]. Thus $\prod$ can be decomposed into two matrices α and β, where α represents the error correction term and measures the speed of adjustment in ΔX_t and β contain r co-integrating vectors. Should there be variables which are $I(0)$ and are significant in the long-run co-integrating space but affect the short-run model, equation (2) is re-written as follows:

$$\Delta X_t = \Gamma_1 \Delta X_{t-1} + \Gamma_2 \Delta X_{t-2} + \ldots + \Gamma_{p-1} \Delta X_{t-p+1} - \prod X_{t-p} + vD_t + \mu_t \quad (3)$$

Where Dt represents the $I(0)$ variables

In identifying the number of co-integrating vectors, two likelihood ratio (LR) tests are used. These are the trace test and the maximum eigenvalue tests. The trace test is a joint test of the null hypothesis of r co-integrating vectors against the alternative that it is greater than r. The trace test statistic is given as

$$Jtrace(r) = -T\Sigma^{p}_{i=r+1} \ln (1-\lambda_i) \quad (4)$$

The maximum eigenvalue test on the other hand conducts separate tests on the individual eigenvalues for a null hypothesis that the number of co-integrating vectors is r, against an alternative of r+1. The test statistic for the maximum eigenvalue test is given as

$$Jmax\ (r, r+1) = -T \ln (1-\lambda_{r+1}) \quad (5)$$

The trace test according to Harris [18] shows more robustness to both skewness and excess kurtosis in the innovations than the maximum eigenvalue test. It is therefore selected over the maximum eigenvalue test in the present study. Kuwornu *et al* [22] suggested that in using the Johansen Full Information Maximum Likelihood test, the endogenous variables included in the Vector Autoregression (VAR) are all $I(1)$, also the additional exogenous variables which explain the short-run effect are $I(0)$.

In spite of the advantages held by the chosen method to co-integration analysis, the Johansen Full Information Maximum Likelihood test has a weakness of being heavily reliant on asymptotic properties and is sensitive to specification errors in limited samples. To ensure correct specification of the VAR or VECM (Vector Error Correction Model) and adherence to the appropriate standard Gaussian properties, there is a need to set appropriate lag length(s) and include variables that are likely to affect the short-run behavior of the model. Omitting important conditional variables could lead to residual misspecification [19]. In selecting the lag length, emphasis so far in literature has been placed on the Akaike information criterion (AIC), the Schwarz information criterion (SC) and the Hanna-Quinn information criterion (HQ).

3.2. Yield Response Model

This study estimated the yield response of paddy rice in Nigeria using the following implicit response function

$$LgYld = f(LgRR, LgOP, LgRM, LgFU, LgNRA) \quad (6)$$

Yield (LgYld) is considered a function of real producer price of rice (LgRR), aggregate output of paddy rice (LgOP), real producer price of maize (LgRM), total quantity of fertilizer used (LgFU), and nominal rate of assistance (NRA). All the variables except nominal rate of assistance (NRA) are expressed in the log form (LgX means X in log form). Data on yield, aggregate output and total quantity of fertilizer used were sourced from the World Rice Statistics of IRRI (FAO data for yield and output). Nominal producer price of rice and maize gathered from the agricultural production database of FAO were deflated into real price using 2005-based consumer price index (CPI) series for Nigeria from the World Rice Statistics of IRRI [20]. Data on nominal rate of assistance was gathered from the updated excel file of the World Bank on National and Global Agricultural Trade and Welfare Reduction Indexes, 1955 to 2010 [6].

4. Results and Discussion

4.1. Unit Root Test

As a vital step in co-integration analysis, the data gathered was tested for stationarity of the respective series and to ascertain the order of integration of the individual series. This, in co-integration analysis is considered a vital step in the data generation process and in the choice of estimator. For co-integration analysis to be deemed valid, Towsend [33] suggested that all the series considered must be integrated of the same order, usually I(1). Although several unit root tests have been suggested and applied in literature (most notably the Dickey-Fuller and Augmented Dickey-Fuller tests), the Phillips-Perron (PP) unit root test was used in the present study for verification of data. The PP test used can be viewed as a Dickey-Fuller statistic that has been made robust to serial correlation by using the Newey-West [28] heteroskedasticity-and-autocorrelation-consistent matrix estimator. Trend and intercept terms were included in specification for level, with only an intercept term being used in specification for first difference. The results are presented in Table 2.

The results of the unit root tests show that all the series are non-stationary at level except for nominal rate of assistance (NRA) which was found stationary at the 1% significance level. All the non-stationary series however became stationary on first difference at the 1% significance level. The results therefore underscore non-stationarity in the data set at level.

Table 2. *Results of unit root test*

Series	level	BW	First. diff	BW
LgYld	-3.367	4	-10.838***	0
LgRR	-2.056	0	-5.335***	5
LgOP	-2.455	0	-8.248***	4
LgRM	-3.299	2	-14.764***	40
LgFU	-1.364	3	-6.803***	3
NRA	-4.519***	2	-15.242***	16
Crit. Val.	-3.521		-2.935	

NB: 95% confidence level for critical values, ***1%, **5%

4.2. Results of Co-Integration

Having established the order of integration of the respective data series, an appropriate Vector Autoregression (VAR) for the study was defined, the optimum number of lags determined and the number of co-integrating equation(s) identified through the trace test. In defining the VAR, the variable NRA (nominal rate of assistance) was considered exogenous. In determining the optimum number of lags, a maximum of 3 lags was initially set due to the number of observations (1966-2008). From the output obtained, the Akaike information criterion (AIC) selected lag order three, whiles the Schwarz information criterion (SC) and Hannan-Quinn information criterion (HQ) selected lag order one. This study thus selected lag order one for the VAR model.

A test for number of co-integrating vector(s) based on trace statistic (assuming no deterministic trend) also indicated "one" co-integrating equation at the 0.05 level. By this, it can be concluded that the variables in the model are co-integrated. With co-integration having been confirmed, a dynamic error correction model or vector equilibrium correction model [23] instead of a VAR becomes more appropriate. Opting for a VECM instead of VAR, the statistical package EViews estimated the long-run elasticities with respect to real producer price of rice, aggregate output of paddy rice, real producer price of maize, and total quantity of fertilizer used. The normalized equation (variables in log) observed for yield of rough rice for Nigeria is given as

$$Yld = 0.479RR + 0.194OP - 1.042RM + 0.005FU + 4.861 \quad (7)$$

(0.087) (0.073) (0.157) (0.043) (0.993)

(5.483)*** (2.644)***(-6.642)***(0.114) (4.895)***

The dynamic error correction model for eq(7) in this study is presented as

$$\Delta Yld_t = \Gamma_0 + \Sigma^n_{i=0}\Gamma_{1i}\Delta Yld_{t-1} + \Sigma^n_{i=0}\Gamma_{2i}\Delta RR_t + \Sigma^n_{i=0}\Gamma_{3i}\Delta OP_t + \Sigma^n_{i=0}\Gamma_{4i}\Delta RM_t + \Sigma^n_{i=0}\Gamma_{5i}\Delta FU_t + \Gamma_{6i}NRA_t - \alpha EC_{t-1} \quad (8)$$

Where EC is the error correction term and "α" represents the rate at which deviations from the long-run equilibrium are restored (speed of adjustment). The negative sign on the error correction term ensures that any adjustment made or observed in the short-run is guided by and consistent with the long-run equilibrium relationship. A significant "α" validates the existence of long-run relationship between the variables. Having defined and estimated the error correction model, the estimates where tested for appropriate standard Gaussian assumptions. The residual series was found to be normality distributed, non-serially correlated and homoscedastic. Higher order test for serial correlation up to 20 lags still showed no sign of autocorrelation.

The results indicate that in the long-run, yield of paddy rice is dependent on real producer price of rice, aggregate output of paddy observed and real producer price of maize. In the short-run however, observed yield is dependent on lagged yield, real producer price of rice, aggregate output of paddy rice, real producer price of maize, and the quantity of fertilizer used. A total of about 59.82% of variation in yields of paddy rice for Nigeria is explained by variables in the implicit yield response function specified in the current study. Approximately 26% of total deviations in yield from the long-run equilibrium are restored in the current period, and this restoration is found significant at the 5% level.

Yields in Nigeria are reportedly low due to low application of vital agro-chemicals like fertilizer and pesticides, low mechanization of rice farms, over-reliance on rainfall, limited use and inappropriate management of irrigation facilities, limited availability of labour needed for carrying out vital cultural practices like weeding, pest control among others, broad use of genetically inferior varieties that exhibit low productivity, low producer price in the output market which limits incentive to invest appropriately in yield enhancing innovations and limited delivery of extension services [12, 34, and 31]. Pull in yields resulting from the inter-play of these biophysical, socio-economic and structural constraints have for over two decades now led not only to irregular supply of local rice, but also to yields well below climatic potential. Improvement observed in current yield, by estimate for the short-run model, is followed by significant declines in yields in the short-term, reflecting incidence of significant nutrient withdrawal in times of yield increment and subsequent ineffective replenishment.

Table 3. *ECM results for yield of paddy rice*

	Coeff	Std. Error	t-statistic
Δ ln Yld (-1)	-0.3768	0.1184	-3.1823***
Δln RR	0.2099	0.0845	2.4835**
Δ ln OP	0.3103	0.0739	4.2005***
Δ ln RM	-0.1993	0.0932	2.1389**
Δ ln FU	0.1558	0.0508	3.0644***
NRA	-0.0515	0.0368	-1.4015
EC_{t-1}	-0.2599	0.1232	-2.1092**
R^2	0.6584	J-Bera	0.994 (0.609)
Adj-R^2	0.5982	Q-stat (1)	1.664 (0.197),
S.E of reg.	0.1044	B-G LM	2.066 (0.160),
L-likelihood	38.3099	ARCH T.	0.014 (0.906)
AIC	-1.5273	Mean DV	0.0043
SC	-1.2348	S.D.DV	0.1647
HQ	-1.4208	SS resid	0.3704
D-W Stat.	2.3210		

NB: ***1%, **5%,

A 1% increase in yield observed in time t-1 may although be statistically significant by current estimates, along current levels, such improvement may not be sufficient enough to ensure significant increase in farmers' income, a condition vital for meeting future field improvement and soil fertility management needs. In the present study, a 1% increase in lagged yield is observed to trigger a 0.377% decrease in yields for the subsequent years. This decrease is found significant at the 1% level. Monitoring trends in harvested area and yields in the country, this observation could, beside the aforementioned constraints be attributed to major increases in area harvested following periods of yield increment, thereby creating much competition for limited nutrients available to the rice plants due to significant withdrawal. With their production decisions being influenced by current land tenure system of the country, most of the farmers invest a greater share of increases observed in their incomes as a result of major yield improvements on acquiring more lands, rather than investing in land improvement (knowing that the land is leased out to them only for a short-period of time). This consequently has significant adverse implication for future rice yields.

Having been identified as one of the constraints precluding the achievement of improved and stable yields, low producer price of rice in the output market has together with limited access to credit precluded most farmers from responding appropriately to non-price policy incentives purposed on increasing output. Among the vital inputs for rice production in the tropics are labour, water supply (irrigation) and application of agro-chemicals. These inputs account for a greater share of cost of production of local rice in West Africa (labor being prioritized for Nigeria due to more remunerative off-farm employment). In the presence of limited access to credit, the alternative means by which farmers could increase their financial base is primarily through increases in price of their produce and increases in the produce. Hindered by poor marketing structure however (as seen also in Ghana, see [5, 9]), increases in the initial low price of rice are usually poorly transmitted in the short-run and some farmers in areas with poor roads are to some extent precluded from sharing in such increments due to poor bargaining power and limited storage infrastructure (which course them to sell their produce at give-away prices to minimize post-harvest losses). In the long-run however, following accumulation of profits and improvements in the market structure to facilitate a relatively better transmission of price increment, farmers are able to meet vital cost of production better than in the short-run. This leads to anticipated high effect of own price increments in the long-run on yield than in the short-run.

Yield of rice is in the present study observed to increase by 0.210% and 0.479% respectively in the short-run and long-run for a 1% increase in real producer price of rice. Each of these effects is found significant at the 1% level. Using an autoregressive distributed lag model approach to estimate rice output supply response to changes in real prices for Nigeria, Ogazi [30] found insignificant short-run own-price coefficient of 0.043 and at 5% significance level a long-run coefficient of 0.271. In as much as the coefficients observed for yield in the present study compare favorably with the coefficients observed for output in the study by Ogazi, it is noted that own price coefficients in either study are inelastic. This observation reflects structural constraints faced by farmers mostly in access to and structure of the market. In addition, Ogazi observed an error correction term of -0.575 for the output response model, significant at the 1% level. A coefficient of -0.260 (significant at the 5% level) observed in the present study indicates that adjustments to long run equilibrium in output for Nigeria are faster than in yields.

In as much as most studies hold unto the dependence of output on yield, the current study is founded on the belief that in a country where productivity of farmers' fields is identified to be low, changes in output to a greater extent shape yields in both the long- and short-run. Effect of changes in output on yield could be either positive or negative depending on how well derived benefits from such increments are re-invested in the fields to improve and sustain future yields. Increment in output is under normal circumstances accompanied by nutrient withdrawal regardless of whether or not output is driven more by increases in area expansion than in yield. Should benefits from such increments be appropriately re-invested in the fields with a purpose of replenishing lost nutrient, a significant positive association could be observed between increments in output and yield in subsequent years. Failure however to ensure appropriate investment in the fields following output increments could lead to adverse implications.

In the present study, aggregate output of paddy rice had coefficients of 0.310 and 0.194 respectively for the short- and long-run. The coefficient observed for the short-run is significant at the 1% level, while that for the long-run is found significant at the 5% level. This implies that a 1% increase in output of rice leads to a 0.310% increase in yield in the short-run and 0.194% increase in the long-run. Although long-run coefficients are usually larger than the short-run coefficients, the respective coefficients observed in this study are economically and practically intuitive. In the short-run where most factors of production are considered fixed, increases in output could increase the capital and financial base of farmers to enable them invest appropriately in their fields and to meet other vital production costs, most importantly the cost of fertilizing and managing the fertility of rice fields. Majority of investments made in such situation are mostly purposed on increasing productivity (and in some cases expanding area of production in communities where land tenure pulls are minimum).

In the long-run however where almost all factors of production are considered variable, the relative share of increase in capital and financial base of farmers (due to increases in output) invested with productivity enhancement motive reduces due to increases in cost of other inputs of production and vital cultural practices. There usually is an increase in area harvested of rice in the long-run, and with this comes intensified competition by rice plants for limited nutrients in the soil due to low dosages of current fertilizer application and poor fertility status of current rice fields (amidst issues with pest and disease control, as well as water management). In as much as significant increases in financial base of farmers through increases in output (which also has some nutrient withdrawing implications) may help increase productivity through investment in productivity enhancing innovations, adverse effects from other forces in the long-run may be larger than in the short-run, thereby pulling (negating) more of the contribution from any investment made in the long-run than in the short-run.

By production theory, farmers are considered to be profit maximizers and cost minimizers. Their decisions are under normal circumstances made in such a way that, profits are maximized through better allocation of their resources, minimizing cost as well in the process. Facing price transmission challenges in the rice market (on the assumption that transmission is better in the maize market) amidst difficulty in accessing credits, as risk averse individuals and rationale beings, farms usually allocate resources in favor of ventures that yield higher returns at the expense of the low yielding ventures. Focusing on rice and maize as two competing field crops, ceteris paribus, farmers are expected to allocate more resources into rice production should conditions in the rice sub-sector (most importantly prices) be favorable than in the maize sub-sector. The opposite is expected should conditions be more favorable in the maize sub-sector.

By the coefficients observed in this study for real producer price of maize, it is inferred that a 1% increase in real producer price of maize leads to a 0.199% decrease in yield in the short-run and 1.042% decrease in the long-run. The short-run effect is found significant at the 5% level, while the long-run effect is significant at the 1% level. This observation is solely attributed to resource allocation in favor of maize production (thus withdrawal of resources from rice production into maize production). The nature of this re-allocation of resources however differs between the short-run and the long-run. In the short-run where most factors of production are considered fixed (most importantly land), the issue of withdrawal of resources from rice production into maize production is basically limited to labor and vital agro-chemicals. In the long-run, the degree of flexibility in production decisions is quite higher due to the ability to vary almost all vital inputs of production including land. By this, the withdrawal of resources from rice production goes beyond the labour and agro-chemical look for the short-run. In the long-run, land could even be diverted for maize production should economic and bio-physical conditions of production be more favorable. This explains the relatively low and inelastic coefficient observed in the short-run and the elastic coefficient observed in the long-run. This observation indicates that rice farmers in Nigeria respond more to price incentives for competitive field crops like maize than own-price incentives. This could be due to differences in efficiency of the supply chain for both commodities.

Characterized by large agro-ecological heterogeneity [2,4], rice fields in Nigeria respond differently to fertilizer application base on other bio-physical and water supply constraints. Heterogeneous environments tend to exhibit substantial spatial and temporal variability in soil properties [10, 17]. In as much as some fields respond positively to fertilizer application in both the long and short-run, some fields respond positively only in the short-run (following the law of diminishing returns), while others give totally mixed signals. By economic theory however, assuming a fixed area of production in the short-run (as farmers are believed to be constrained in regard to what production decision they can make), the "Flower-pot law" postulates that both marginal and average productivities initially increase, but eventually they both decrease. Information on these indicators is mostly useful in homogenous fields and in estimating responses at the farm/firm and to some extent regional level. At the national level, information on fertilizer use fails to provide details on regional and farm level differences in actual use but generally gives the aggregate figure for the country.

The aggregate provides no distinction between the amounts used during the main cropping season and that for the lean season in areas with two seasons of production (mostly in regions where the irrigated system of production dominates). Use of aggregate fertilizer for this yield response study is therefore purposed on identifying how changes in aggregate fertilizer used generally shape national yields of rice in the country, placing no major emphasis on the optimum needed due to agro-ecological heterogeneity and to likely differences in the true optimum for the respective agro-ecologies.

In this study, coefficients of 0.156 and 0.005 were respectively observed in the short-run and long-run for the variable "FU" (total quantity of fertilizer used). The short-run effect is found significant at the 1% level. The long effect however is not significant. This implies that, a 1% increase in total quantity of fertilizer used in the short-run leads to a 0.156% increase in average yield of paddy rice, but increases observed in the long-run are not significant. This observation could have been considered consistent with the "Flower pot law" should area of land under cultivation be deemed constant in both the long and short-run, other vital variables held constant in the long-run and the area of production be considered homogenous. With this however not being the case, the coefficients observed in this study reflect general inefficiency in use of fertilizer in Nigeria as quantity of fertilizer used leads to inelastic changes in yields in both the long and short-run regardless of the significant response in the short-run (where most inputs of production are deemed constant).

The insignificant effect of quantity of fertilizer used on national yields in the long-run is attributed to differences in the rate of increase in fertilizer application and area harvested of rice, increasing cost of production in the long-run which offsets benefits from increases in fertilizer application, increasing cost of water and soil fertility management, and to weather and other biological factors whose effect lies beyond the scope of this study. This suggests that improving yields in the country in the long-run as against the short-run requires relatively complex interaction of forces.

The effect of nominal rate of assistance (NRA) on yield was found to be insignificant. In as much as assistance from government, mostly reflected by decreases in taxation of farm incomes and increases in tariffs on imports is expected to incite positive response from farmers, it equally leads to secondary distortion in input prices thereby offsetting any beneficial implications of the variable. Its' being positive or negative (and significant or not) usually depends on the extent to which one effect supersedes the other and to how consistent the measure is. A noted problem with this variable for Nigeria is inconsistency. NRA which to some extent reflects the degree of trade restriction in a country has depicted multiple-peaked oscillations in short term intervals for Nigeria. This observation discourages farmers from investing appropriately in their fields due to anticipated fear of competition in the rice market; hence the observed insignificant response of farmers to nominal rate of assistance.

4.3. Limitations of Study

In as much as this study tried to estimate yield response for Nigeria using secondary data at the national level and could not cover all relevant variables suggested in literature due to the large model estimation problem of the selected estimation technique [23] and to data accessibility challenges (for climate related variables like precipitation and temperature), the effect of such vital variables could not be captured in the study. Secondly, effect of the variable "FU" (quantity of fertilizer used) on yield estimated in the present study although reflect inefficiency in fertilizer usage in the country, the observed coefficients may not necessarily hold in some agro-ecologies due to large agro-ecological heterogeneity of rice fields in the country.

In addition, no optimum quantity of fertilizer application was estimated because an optimum for one production system in a given ecology may not hold for another due to likely spatial and temporal variability in soil properties and other bio-physical forces. Future research in estimating yield response could improve on this study by analyzing response at the farm or regional level to help determine optimum for the respective agro-ecologies. In so doing, the effect of variables like farmer's expertize, adoption and use of improved production technologies and vital climate variables like precipitation and temperature could be captured (for example in assessment of effect of climate variability on yields).

5. Conclusion

To help bridge information gap in supply response studies for Nigeria and inform policy decisions on how the demand-supply gap could be bridged in the near future, the current study through the use of Johansen's Full Information Maximum Likelihood test estimated a yield response model for Nigeria using secondary data for the period 1966 to 2008. The results show that in the long-run, yield is significantly dependent on real producer price of rice, aggregate output of paddy rice and on real producer price of maize (a competitive field crop). In the short-run however, yield is dependent on lagged yield, real producer price of rice, aggregate output of paddy rice, real producer price of maize, and quantity of fertilizer used. A total of about 59.82% of variations in yield of paddy rice for Nigeria is explained by variables in the model estimated. A total of about 26% of deviations from the long-run equilibrium for yield is restored in the current period, and this restoration was found significant at the 5% level.

Interestingly, rice farmers in the country were found to respond more to changes in price of maize than own-price for rice. This observation was attributed to likely differences in efficiency of the supply chain for rice and maize, with transmission of price increment presumed to be higher in the maize market than in the local rice market. Inelastic response of yield to changes in real producer price of rice, as was also observed by Ogazi [30], shows that farmers in Nigeria are precluded from responding appropriately to own-price incentives due to structural constraints they face. In as much as improvement in yields is presumed to stimulate growth in output, increases in output in countries with low fertility of farmers' fields like Nigeria is found in the present study to shape (increase) yields through increases in capital and financial base of farmers. The effect was however found to be higher in the short-run than in the longer run, due to anticipated pulls from other production factors in the long-run.

Although rice farmers are observed to use fertilizer inefficiently in their cropping activities, observed effects in the present study may not hold in some agro-ecological zones due to presumed heterogeneity in rice fields, and to likely spatial and temporal variability in soil properties. The effect of nominal rate of assistance on yield was found to be insignificant. This was attributed to inconsistent nature of this variable in Nigeria, which discourages farmers from investing appropriately in their fields, due to anticipated fear of competition in the rice market.

The results generally suggest that, increasing yield levels for paddy rice in Nigeria and ensuring stability requires interplay of biophysical, socio-economic and structural forces. By estimates for the current study, bridging of the demand-supply gap through improvement in yields can be realized through initiation of measures to address inefficiencies in the supply chain to ensure appropriate transmission of price increment, promotion of local rice consumption to ensure ready market for farmers in times of increasing output (to help increase their capital and financial base), addressing soil fertility challenges through efficient use of fertilizer and regular management of fertility of rice fields, and increasing farmers access to credit to help them meet cost of relevant inputs of production. This latter suggestion could to a greater extent incite appropriate response of farmers to both price and non-price incentives in the country.

References

[1] Abubakar, S. (2013). Why Nigeria imports N1bn rice daily. DAILY TRUST http://allafrica.com/stories/201308080475.html

[2] Adedipe, N.O., Bakshi, J.S., Odegbaro, O.A., and Aliyu, A. (1996). Evolving the Nigerian Agricultural Research Strategy Plan: Agro-Ecological Inputs. The National Agricultural Research Project, Ibadan, 1-486pp.

[3] Ahmed, A.M. and Siddiqui, R. (1994). Supply response in Pakistan with "Endogenous" Technology. The Pakistan Development Review 33:4 Part II (Winter 1994) pp. 871-888

[4] Akpokodje, G., Lançon, F., and Erenstein, O. (2001). Nigeria's rice economy: State of the Art. West Africa Rice Development Association (WARDA), Bouake, Côte d'Ivoire

[5] Amanor-Boadu, V. (2012). Rice price trends in Ghana (2006-2011). Department of Agricultural Economics, Kansas State University. METSS-Ghana Research and Issue Paper Series No. 02-2012-June 2012

[6] Anderson, K. and Nelgen, S. (2012). Updated national and global agricultural trade and welfare reduction indexes, 1995 to 2010. World Bank, Washington DC

[7] Basorun, J.O. and Fasakin, J.O. (2012). Factors influencing rice production in Igbemo-Ekiti Region of Nigeria. Journal of Agriculture, Food and Environmental Science, 5(1):1-9

[8] Bingxin, Y. and Shenggen, F. (2009). Rice production response in Cambodia. Contributed paper prepared for presentation at the International Association of Agricultural Economists Conference, Beijing, China. August 16-22, 2009

[9] Boansi, D. (2013). Rice production in Ghana: Past, present and future. ISBN-13: 978-3659498220, LAMBERT Academic Publishing, Germany

[10] Cohen, M.J., Dunne, J.E.J., and Bruland, G.L. (2008). Spatial variability of soil properties in Cypress domes surrounded by different land uses. Wetlands 28:411-422. DOI:10.1672/06-182.1

[11] Cummings, J. (1975). The supply responsiveness of Indian farmers in the post-independence period: Major cereal and cash crops. Indian J. Agricultural Economics, 30(1): 25-40

[12] Daramola, B. (2005). Government policies and competitiveness of Nigerian rice economy. Paper presented at the 'Workshop on Rice Policy & Food Security in Sub-Saharan Africa' Organized by WARDA, Cotonou, Republic of Benin, November 07-09, 2005

[13] Defoer, T., Wopereis, M.C.S., Jones, M.P. Lançon, F., Erenstein, O., and Guei, R.G. (2004). Challenges and technical opportunities for rice-based production systems for food security and poverty alleviation in Sub-Saharan Africa. Proceedings of FAO Rice Conference, Rome, Italy, 12-13 February, 2004.

[14] Engle, R.F. and Granger, C.W.J. (1987). Co-integration and error correction: Representation, estimation and testing. Econometrica, 55:251-276

[15] FAOSTAT (2013). Agricultural production database. Food and Agricultural Organization of the United Nations, FAO, Rome, Italy. Accessed on June14, 2013 http://faostat.fao.org/site/703/default.aspx#ancor

[16] Fashola, O.O., Imolehin, E.D., and Wakatsuki, T. (2007). Water management practices for sustainable rice production in Nigeria. Nigeria Agricultural Journal, 38: 40-48. DOI: http://dx.doi.org/10.4314%2Fnaj.v38i1.3247

[17] Grunwald , S., Corstanje, R., Weinrich, B.E., and Reddy, K.R. (2006). Spatial patterns of labile forms of phosphorus in a subtropical wetland. Journal of Environmental Quality, 35:378-389 DOI:10.2134/jeq2005.0042

[18] Harris, R. (1995). Using co-integration analysis in econometric modelling. Oxford University Press, London

[19] Harris, R. and Sollis, R. (2003). Applied time series modelling and forecasting. Durham University. John Wiley & Sons

[20] IRRI. World Rice Statistics, International Rice Research Institute. Accessed on June 14, 2013 http://ricestat.irri.org:8080/wrs/

[21] Johansen, S. and Juselius K. (1990). Maximum likelihood estimation and inference on co-integration with application to the demand for money. Oxford bulletin of Economics and Statistics, 52:170-209

[22] Kuwornu, J.K.M., Izideen, M.P.M., and Osei-Asare, Y.B. (2011). Supply response of rice in Ghana: A co-integration analysis. Journal of Economics and Sustainable Development, 2(6):1-14

[23] Lütkepohl, H. and Krätzig, M. (2004). Applied time series econometrics. Cambridge University Press, New York

[24] Molua, E.L. (2010). Response of rice yields in Cameroon: Some implications for agricultural price policy. Libyan Agriculture Research Center Journal Internation, 1(3):182-194

[25] Molua, E.L. (2008). Turning up the heat on African Agriculture: Economic impact of climate change on agriculture in Cameroon. African J. Agricultural and Resource Economics, 2(1):45-64

[26] Mulwanyi, A., Hutagaol, P., and Sinaga, B.M. (2011). Impact of rice purchasing policy on welfare of both producers and consumers in Indonesia. J. ISSAAS, 17(2):48-57

[27] Mythili, G. (2008). Acreage and yield response for major crops in the pre- and post-reform periods in India: A dynamic panel data approach. Report prepared for IGIDR-ERS/USDA Project: Agricultural Markets and Policy. Indira Gandhi Institute of Development Research, Mumbai.

[28] Newey, W.K. and West, K.D. (1987). A simple positive semi-definite, heteroskedasticity and autocorrelation consistent covariance matrix. Econometrica 55(3):703-708

[29] Norman, J.C. and Kebe, B. (2006). African smallholder farmers: Rice production and sustainable livelihoods. International Rice Commission Newsletter (FAO), 55:33-44

[30] Ogazi, C.G. (2009). Rice output supply response to the changes in real prices in Nigeria: An autoregressive distributed lag model approach. Journal of Sustainable Development in Africa, 11(4):83-100

[31] Omonona, B.T. (2013). Fertilizer profitability across Nigeria's diverse agro-ecological zones and farming systems. At the Nigeria launch of 'Guiding Investments in Sustainable Agricultural Intensification in Africa (GISAIA)', Abuja. June 17, 2013

[32] Phillips, P.C.B. and Ouliaris, S. (1998). Testing for co-integration using principal component methods. Economics Dynamic and Control, 12:205-230

[33] Townsend, T.P. (2001). World cotton market conditions. Beltwide Cotton Conferences, Proceedings, Cotton Economics and Marketing Conference, National Cotton Council, Memphis, TN pp 401-405

[34] WARDA (2003). The Nigerian rice economy in a competitive world: Constraints, opportunities, and strategic choices. Strategy for rice sector revitalization in Nigeria

[35] Yeong-Sheng, T., Survani, D., Aswani, F.M.N., and Nurjihan, I. (2009). Acreage response of rice: A case study in Malaysia. MPRA Paper No 15300. University Library of Munich, Germany

[36] Zeigler, R.S.(2012) Advancing cutting-science to boost rice production, increase farmers' incomes and conserve the environment. International Rice Research Insitute. March 15, 2012 https://www.cgiarfund.org/sites/cgiarfund.org/files/Images/bbl_march15_2012_RZEIGLER.pdf

[37] Other source: Index Mundi, http://www.indexmundi.com/nigeria/demographics_profile.html

7

Voluntary feed intake and nutrient utilization of West African Dwarf sheep fed supplements of *Moringa oleifera* and *Gliricidia sepium* fodders

Adegun Maria Kikelomo

Department of Animal Production and Health Sciences, Ekiti State University, Ado-Ekiti, Nigeria

Email address:
patrickikelomo@yahoo.com

Abstract: *Moringa oleifera* (Mo) and *Gliricidia sepium* (Gs) fodders and their mixtures were fed as protein supplements to basal diets of *Panicum maximum* (Pm) and cassava peels. Thirty two (32) yearling West African Dwarf (WAD) rams were divided into four groups of eight animals balanced for body weight and allocated to dietary treatments in a Completely Randomized Design for fifteen (15) weeks. Animals in group 1 consumed the basal diet only while animals from group 2, 3 and 4 were fed basal diets with supplements of Mo, Gs and their mixtures respectively. Voluntary feed intake (VFI), nutrient digestibility and nitrogen utilization of the animals were evaluated. Animals fed basal diet only had reduced VFI (252.36±22.0g day^{-1}) while those on Mo and Gs mixtures had significantly higher ($p<0.05$) VFI value of 344.96±22.2g day^{-1}. Voluntary feed intake of basal diet was increased significantly ($p<0.05$) for animals in group 2 and 4 (260.20±30.00 and 270.46±20.6g day^{-1} respectively). Higher but significantly different ($p<0.05$) dry matter digestibility values of 63.90±3.06, 62.72±3.02 and 68.50±4.32% were obtained for animals in group 2, 3 and 4 respectively while the animals in group 1 had the least value of 48.39±2.03%. Similar trend exist in crude protein and crude fibre digestibility with animals in group 4 having significantly higher ($p<0.05$) values of 60.62±5.24 and 68.63±3.22% respectively. The result of nitrogen utilization indicate significantly lowest ($p<0.05$) value of 3.56±0.16g day^{-1} of digested nitrogen for animals in group 1. Higher nitrogen retention value ($p<0.05$) of 65.81±4.20% was obtained in sheep on Mo and Gs fodder combinations. Those on separate Mo and Gs fodders were statistically similar (58.55±3.02% and 58.00±5.30% respectively) but higher ($p<0.05$) than those on basal diet (43.70±3.02%). Supplementation of *Panicum maximum* and cassava peels basal diet with *Moringa oleifera* or *Gliricidia sepium* fodders improved the intake of basal diet and enhanced better nutrient utilization of WAD sheep. Mixtures of the two fodders in the same ratio were superior to either supplementation.

Keywords: *Moringa oleifera,* Fodder Trees, Voluntary Feed Intake, Foliage Mixtures, Supplementation

1. Introduction

The increased importance of fodder trees as a non-conventional feed resource has been recognized as one of the most effective means of improving forage supply in small holder ruminant production (1). They provide biomass that remain green and seem to be cheap protein supplements that can improve voluntary intake, digestibility and general performance of animals fed low quality feeds which characterize the pasture and agricultural by-products in the tropics (2).

Research efforts have been intensified in Nigeria to characterize the agronomic features, management requirements and feed value of some promising tree species. Agishi identified about 40 families of trees and shrubs as foliage species in Nigerian rangelands (3). Out of these trees, *Gliricidia sepium*, an exotic leguminous tree is found in abundance in South-Western Nigeria, but well known as fencing, staking and shade tree than as a source of fodder (4). Animals often refuse gliricidia leaves on the basis of smell without even tasting it (5). On the other hand, *Moringa oleifera* is one of several trees that have not received sufficient attention for animal use in Nigeria (6, 7). Moringa is called a “Miracle tree” because of its wide range of purposes (8). Its nutritional importance, fast growth rate coupled with low anti-nutrients components merit its use as a valuable fodder for ruminants in South

Western Nigeria (6).

Depending on availability, presence of toxic substance, adaptability, voluntary intake and quality, a range of browse feeding strategies can be developed and these may prompt farmers to mix foliage given to animals. High responses of these mixtures have been reported for small ruminants. For example, Bosman *et al.* fed West African Dwarf (WAD) goats with gliricidia and leucaena as mixture (9). Weight gain with gliricidia alone was $2g/W^{-0.75}$/day and for the mixture a weight gain of $8.2g/W^{-0.75}$ /day was obtained. In another study, foliage from *Leucaena leucocephala, Calliandra calothyrsus* and their mixtures were used as supplements to maize husks for goats. Daily weight gain was 28, 19 and 22g while dry matter intake was 335, 315 and 317g for animals supplemented with leucaena only, calliandra only and their mixtures respectively (10).

The disappearance of feed particles from the reticulo-rumen depends primarily on the rate of digestion. High fibre rations such as tropical pastures and crop by-products result in longer retention and lower intake (11). Voluntary intake is inversely related to the fibre content and directly proportional to chemical composition of herbages consumed which affect the survivability and productivity of animals (12).

This study was designed to evaluate the effects of fodders of *Gliricidia sepium* and *Moringa oleifera* and their mixtures as supplements to basal diet of *Panicum maximum* and cassava peels on voluntary feed intake, and nutrient utilization of West African Dwarf (WAD) rams in South-Western Nigeria.

2. Materials and Methods

2.1. Experimental Site and Animals' Management

The study was conducted at the Teaching and Research Farm, Ekiti State University, Ado-Ekiti, South Western Nigeria between April 2010 and April 2011. Ado-Ekiti lies between latitude 07° 37' 15" N and longitude 05° 13' 17" E with an average relative humidity of 72%. It experiences a tropical climate with a temperature range of 20-28°C and a bimodal rainfall distribution between April and October with peaks in June and September and a break in August. Dry season is between November and March. The average precipitation in this area is 1367mm.

Thirty two (32) yearling West African Dwarf (WAD) rams weighing 13.0±0.6kg were purchased from Otun-Ekiti, Ekiti State, Nigeria and quarantined for 30 days using the routine treatment developed at NAPRI (13).

2.2. Basal Diets

2.2.1. Cassava Peels

Cassava peels were obtained from the Garri Processing Unit of the Teaching and Research Farm, Ekiti State University, Ado-Ekiti. The fresh cassava peels were sun dried for about 2-4 days depending on the weather condition.

2.2.2. Panicum maximum (Guinea Grass)

Guinea grass was harvested from the University Campus premises with a sickle when the stems and leaves of the grass were still succulent. It was hereafter chopped into small pieces with a cutlass to prevent wastage by the animals. The grasses were harvested and wilted for a day before feeding to prevent scouring in the animals.

2.3. Supplemental Diets

Gliricidia sepium and Moringa oleifera fodders were harvested from gliricidia and moringa plantations at the Teaching and Research Farm, Ekiti State University. These fodders (leaves and twigs) were harvested daily and were left to wilt overnight before feeding to the animals.

2.4. Feeding Trial

The experiment lasted for fifteen weeks that comprise of one week for acclimatization to the feeding regime, twelve weeks for voluntary feed intake studies and two weeks for balance studies. The animals were weighed before the commencement of the feeding regime and divided into four groups of eight animals balanced for body weight and randomly allocated to dietary treatments in a Completely Randomized Design (CRD). The experimental diets were fed daily in two installments (morning and evening) at 08:00h and 17:00h. The basal diet of *Panicum maximum* plus cassava peels were supplemented with fodders of gliricidia, moringa or their mixture based on daily feed allowance of 4% of body weight. Equal fodder combinations of *Moringa oleifera and Gliricidia sepium* were thoroughly mixed manually to minimize animal preference/selection in Treatment 4. The rams had free access to fresh water daily. The dietary treatments were as follows

Treatment 1: *Panicum maximum* + Cassava peels

Treatment 2: *Panicum maximum* + Cassava peels + *Moringa oleifera*

Treatment 3: *Panicum maximum* + Cassava peels + *Gliricidia sepium*

Treatment 4: *Panicum maximum* + Cassava peels + *Moringa oleifera* + *Gliricidia sepium*

2.5. Digestibility Study

This was carried out immediately after the performance trial by transferring sixteen (16) WAD rams into wooden metabolic cages fitted with facilities for collection of faeces and urines. The quantity of feed offered, feed refusal, faeces and urine were determined for 7 days, after 7 days of adjustment to cages. Ten percent of the faeces and urine collected daily over the 7-day collection period were bulked. Nitrogen loss from urine and bacteria growth infestation were prevented by introducing $20cm^3$ of 10% H_2SO_4 into urine collection bottles and kept in freezing cabinet. Ten percent (10%) of faeces were weighed and used for moisture determination and the remaining oven-dried at 70°C for 36 hours and milled.

2.6. Chemical Analysis

Samples of experimental diets and faecal materials were collected for dry matter (DM) determination and proximate analysis. The samples were weighed and oven-dried at 105°C for 24 hours, weighed and ground to pass through a 2mm sieve. The milled experimental diets and faecal samples were analyzed for proximate analysis using the procedures described in AOAC (14). Gross energy (GE) contents of the experimental diets were determined against thermo-chemical grade benzoic acid standard using Gallenkamp ballistic bomb calorimeter (Cam Metric Ltd, Cambridge, UK). The nitrogen content of the urine was determined by the micro-Kjeldahl method as described by Pearson (15).

2.7. Statistical Analysis

The data collected were subjected to analysis of variance (ANOVA) using the general linear model (GLM) procedure of SAS. Significant differences among means were separated using the Duncan's Multiple Range Test (DMRT).

3. Results and Discussion

3.1. Results

3.1.1. Chemical Composition of the Experimental Fodders

The nutrient constituents of the basal diets; *Panicum maximum* and cassava peels, and their supplements; *Moringa oleifera, Gliricidia sepium* or combinations of *Moringa oleifera* and *Gliricidia sepium* are shown in Table 1. Crude protein (CP) of the fodders ranged from 3.96 g $100g^{-1}$ in cassava peels to 24.59 g $100g^{-1}$ in *G. sepium* fodder. Ash varied from 6.55 g $100g^{-1}$ in *G. sepium* fodder to 9.26 g $100g^{-1}$ in *P. maximum*. As for the fibre components, *M. oleifera*, *G. sepium* and their combinations were low in crude fibre with an average value of 7.52 g $100g^{-1}$ with *P. maximum* having the highest value of 41.16 g $100g^{-1}$. Similar trend exist with the Neutral Detergent Fibre (NDF) and Acid Detergent Fibre (ADF) values. NDF varied between 25.50 g $100g^{-1}$ in *M. oleifera* to 59.56 g $100g^{-1}$ in *P. maximum* while ADF ranged from 23.58 g $100g^{-1}$ in cassava peels to 78.56 g $100g^{-1}$ in *P. maximum*. All the fodders were high in organic matter with values ranging from 85.60g 100 g^{-1} in *P. maximum* to 91.56 g $100g^{-1}$ in *G. sepium*. The gross energy (GE) values of the experimental fodders ranged from 14.79 MJ kg^{-1} in *M. oleifera* to 22.06 MJ kg^{-1} in *P. maximum*.

Table 1. *Chemical composition of experimental fodders/fodder combinations fed to West African Dwarf (WAD) sheep (g $100g^{-1}$)*

	Experimental Diets				
Parameters	***Panicum maximum***	**Cassava peels**	***Moringa oleifera***	***Gliricidia sepium***	***Moringa oleifera+ Gliricidia sepium****
DM	82.52	80.32	92.22	92.17	92.20
Ash	9.26	8.40	7.48	6.55	7.02
CP	7.56	3.96	22.69	24.59	23.64
CF	41.16	19.60	6.49	8.55	7.52
EE	1.26	0.98	16.45	12.29	14.37
NFE	48.68	66.62	40.28	40.21	40.25
NDF	59.56	34.06	25.50	28.50	27.00
ADF	78.56	23.58	40.60	29.41	35.01
OM	85.60	89.56	90.06	91.56	90.81
GE (MJ kg^{-1})	22.06	17.83	14.79	16.19	15.49

* Nutrient contents of mixed fodder were calculated.

DM – Dry matter, CP – Crude protein, CF – Crude fibre, EE – Ether extract, NFE – Nitrogen Free extract, NDF – Neutral detergent fibre, ADF – Acid detergent fibre, OM – Organic matter, GE – Gross energy.

3.1.2. Voluntary Feed Intake and Nutrient Utilization of WAD Rams Fed Supplements of Moringa oleifera, Gliricidia sepium and their Combination

Table 2 shows the effect of dietary supplementation on voluntary feed intake and nutrient digestibility of WAD rams.

There were significant differences ($p<0.05$) in voluntary feed intake (VFI) of the diet in all the treatments. The VFI ranged from 252.36 g day^{-1} in animals fed the control diet to 344.96 g day^{-1} in the *M. oleifera* and *G. sepium* combination. Supplement intake ranged from 48.60 g day^{-1} in animals on *G. sepium* fodder to 74.50 g day^{-1} in animals fed mixture of *M. oleifera* and *G. sepium* fodder combination.

The apparent nutrient digestibility of dry matter (DM) were significantly different ($p<0.05$) in all the diets. Higher but not significantly difference ($p<0.05$) values of 63.90, 62.72 and 68.50% were obtained for *M. oleifera*, *G. sepium* and *M. oleifera* and *G. sepium* combination respectively while the control diet had the least DM digestibility of 48.39%. Similar trend exist in the CP and CF digestibility with the supplement with *M. oleifera* and *G. sepium* combination having the highest values of 60.62 and 68.63% for CP and CF digestibility respectively. The coefficients of digestibility of *M. oleifera* and *G. sepium* fodders fed separately as supplements to basal diets for CP (55.40 and 53.63%) and CF (60.32 and 62.46%) were not significantly different ($p<0.05$) but were significantly higher ($p<0.05$) than the coefficient of digestibility of CP (45.03%) and CF (52.00%) in the control diet.

***Table 2.** Voluntary feed intake and nutrient digestibility of WAD sheep fed supplements of Moringa oleifera and Gliricidia sepium fodder*

	Experimental Diets			
Parameters	**Control**	***Moringa oleifera***	***Gliricidia sepium***	***Moringa oleifera + Gliricidia sepium***
Intake				
Basal diet (g day^{-1})	252.36 ± 22.0^{c}	260.20 ± 30.0^{b}	254.62 ± 16.4^{c}	270.46 ± 20.6^{a}
Supplement diet(g day^{-1})	–	54.46 ± 3.20^{b}	48.60 ± 5.60^{c}	74.50 ± 4.80^{a}
Total (g day^{-1})	252.36 ± 22.0^{d}	314.66 ± 88.6^{b}	303.22 ± 20.5^{c}	344.96 ± 22.2^{a}
DM Digestilibility (%)	48.39 ± 2.03^{b}	63.90 ± 3.06^{a}	62.72 ± 3.02^{a}	68.50 ± 4.32^{a}
CP Digestilibility (%)	45.03 ± 4.62^{c}	55.40 ± 3.04^{b}	53.63 ± 1.02^{b}	60.62 ± 5.24^{a}
CF Digestilibility (%)	52.00 ± 2.80^{c}	60.32 ± 4.06^{b}	62.46 ± 2.82^{b}	68.63 ± 3.22^{a}

a,b,c Means with the same superscript(s) in the same row are not significantly different ($p > 0.05$).

3.1.3. Nitrogen Utilization of WAD Sheep Fed Moringa oleifera and Gliricidia sepium Fodders as Supplements to Basal Diet of Panicum maximum and cassava peels

The result of nitrogen utilization in Table 3 shows that nitrogen intake values of supplemental fodders ranged from 6.08±0.34 g day^{-1} in *M. oleifera* to 6.20±0.42 g day^{-1} in *G. sepium* and *M. oleifera* mixture which were significantly higher ($p<0.05$) than the nitrogen intake of sheep on control diet (4.79±0.62 g day^{-1}). Significant difference ($p<0.05$) exists among means of digested nitrogen (N). Sheep on fodder combination of *M. oleifera* and *G. sepium* had higher digested nitrogen (5.64±0.42 g day^{-1}) than those on separate *M. oleifera* and *G. sepium* fodders (5.39±0.40 g day^{-1} and 5.30±0.12 g day^{-1}) respectively. Sheep on control diet had significantly lower ($p<0.05$) value of 3.56±0.16 g day^{-1} of digested nitrogen.

Nitrogen balance showed significant differences ($p<0.05$) in animals in the different treatment groups. Sheep on fodder combination of *M. oleifera* and *G. sepium* had the highest nitrogen balance (4.08±0.20 g day^{-1}) while those on control diet had significantly lowest value (1.70±0.28 g day^{-1}). Nitrogen retention shows considerable variation ($p<0.05$). The highest N retention value of 65.81±4.20% was obtained in sheep on *M. oleifera* and *G. sepium* combination fodder supplement while the least value (43.70±3.02%) was obtained in sheep fed the control diet.

***Table 3.** Mean Nitrogen Utilization by WAD sheep fed Moringa oleifera and Gliricidia sepium fodders as supplements to basal diets of Panicum maximum and Cassava peels*

	Experimental Diets			
Parameters	**Control**	***Moringa oleifera***	***Gliricidia sepium***	***Moringa oleifera + Gliricidia sepium***
Nitrogen intake (g day^{-1})	4.79 ± 0.62^{c}	6.08 ± 0.34^{a}	6.12 ± 0.20^{a}	6.20 ± 0.42^{a}
Faecal nitrogen (g day^{-1})	1.23 ± 0.06^{a}	0.79 ± 0.02^{c}	0.82 ± 0.02^{b}	0.56 ± 0.03^{d}
Digested nitrogen (g day^{-1})	3.56 ± 0.16^{c}	5.39 ± 0.40^{b}	5.30 ± 0.12^{b}	5.64 ± 0.42^{a}
Urinary nitrogen (g day^{-1})	1.86 ± 0.32^{a}	1.83 ± 0.28^{a}	1.75 ± 0.52^{a}	1.56 ± 0.32^{b}
Nitrogen balance (g day^{-1})	1.70 ± 0.28^{c}	3.56 ± 0.46^{b}	$3.32 \pm .22^{b}$	4.08 ± 0.20^{a}
Nitrogen retention (%)	43.70 ± 3.02^{c}	58.55 ± 3.60^{b}	58.00 ± 5.30^{b}	65.81 ± 4.20^{a}

a,b,c Means with the same superscript(s) in the same row are not significantly different ($p > 0.05$).

4. Discussion

The crude protein (CP) contents of cassava peels and *P. maximum* fall within the range reported in the literatures, 2.8 – 6.5 g $100g^{-1}$ for cassava peels (17) and 5.9 – 12.4 g $100g^{-1}$ for *P. maximum* (18, 19). The CP values were below the requirement of 14 g $100g^{-1}$ for growing small ruminants (20). The nitrogen free extract (NFE) value of cassava peels is high and within the range of 62.5-72.9 g $100g^{-1}$ as reported by Adegbola and Asaolu (16). This means that cassava peel is an energy source suitable for ruminants but with a characteristically low protein content. *Panicum maximum* used in this study contained high NDF and ADF. Feeds with higher NDF (more than 35%) have lower digestibilities because NDF generally ferments and passes from the reticulo-rumen more slowly than other dietary constituents leading to greater filling effect over time than non-fibrous feed components (12).

The CP value of 24.59g $100g^{-1}$ reported in this study for *G. sepium* is above 14.70 g $100g^{-1}$ and 21.64 g $100g^{-1}$DM documented by Devendra and Asaolu *et al.* respectively (4, 17). These may be due to natural variations as a result of genetic background, environment, cultivation method, variations due to sample preparations and analysis as noted by Brisibe *et al.* (18). The CP of *M. oleifera* and *G. sepium* were above 20.0 g $100g^{-1}$. This agrees with the observation of Waldroup and Smith that multipurpose trees contain 20% CP and above in their leaves (19). The value of CP in *M. oleifera* and *G. sepium* were above the minimum requirement recommended for small ruminants (20). The fibre components of ADF and NDF of *M. oleifera* and *G. sepium* are generally low and similar. This is consistent with the findings of Okoli *et al.* that browses contain low to moderate ADF and NDF values which make them more available for degradation by rumen microbes (21).

The observed lower voluntary feed intake (VFI) in the

sheep fed the control diet could be traced to the level of CP and NDF. Minson observed that most of the tropical forage species and crop by-products have low intake due to the lower level of protein in their leaves (20). The increase in the VFI of animals fed supplemental diets conforms to the findings of other authors which reported that supplementation increased consumption of basal diets of low nutritional value (22, 23). This is because browsing of multipurpose trees provides rumen microbes with a source of nitrogen thereby boosting the rate of passage and digestibility of grasses with a corresponding increase in intake even at relatively small levels of supplementation (24). The significantly lower intake in the *G. sepium* supplementation compared to that of *M. oleifera* could be due to palatability problem encountered by small ruminants when fed with *G. sepium* (25).

The higher VFI in *M. oleifera* and *G. sepium* fodder combination supplement than the single supplement of *M. oleifera* and *G. sepium* is in line with the reports of other authors who asserted that there were better production outcomes in browse mixtures than single browses (9, 26). This could be as a result of improved palatability due to the combination of the two fodders.

The apparent lowest dry matter (DM) digestibility in the control diet could be due to the higher fibre content of the control diet than others. Dry matter digestibility is dependent on the cell wall constituents of diets with feedstuffs having higher fibre being less digestible than those with lower fibre (27). Crude protein and CF digestibility followed the same trend with the DM digestibility. However, the CP and CF digestibility in this study (58.00-60.62% CP; 60.22-68.65% CF) were lower than the values of CP (75.56%) and CF (89.35%) obtained by Asaolu *et al*, when sole *M. oleifera* and its mixtures with equal proportions of *L. leucocephala* and *G. sepium* were fed to WAD goats (28). The difference could be due to higher nutrient concentration in the browses than grasses which revealed higher digestibility for multipurpose tree fodders than grasses (29).

Nitrogen intake, nitrogen balance and nitrogen retention increased with supplementation of the basal diet in this study. This is in consonant with the observation by McDonald *et al*. which reported that dietary nitrogen intake in animal is directly related to the proportion of nitrogen in the diet (30). The higher nitrogen retention values obtained for the supplemented diets indicate a higher efficiency of protein utilization as a result of the supplementation.

5. Conclusion

Supplementation of *Panicum maximum* and cassava peels-based diet with *Moringa oleifera* or *Gliricidia sepium* fodders improved the intake of basal diet and enhanced better nutrient utilization of WAD sheep. Mixtures of the two fodders in the same ratio were superior to either of the supplement fed alone.

References

[1] Paterson, R. T., Karanja, G. M., Roothaert, R. L., Nyaata, O. Z., and Kariuki, I. W. (1998). A review of tree fodder production and utilization within smallholder agroforestry systems in Kenya. *Agroforestry Systems,* 41(2): 181-199.

[2] Fasae, O. S., and Alokan, J. A. (2006). Growth performance of weaner Yankassa sheep fed varying levels of *Leucaena leucocephala* leaf residues. *ASSES SERIES* A6 (2): 323-328.

[3] Agishi, E. C. (1985). Forage resources of Nigerian rangelands. In: Small Ruminants Production in Nigeria. *Proceedings of the National Conference on Small Ruminant Production*, Zaria Nigeria, 6-10 October 1985. National Animal Production Research Institute, Shika, Zaria, Nigeria. 115-140.

[4] Devendra, C. (1992). Nutritional potential of fodder trees and shrubs as protein sources in ruminant nutrition. Legume trees and other fodder trees as protein source for livestock. *FAO Animal Production and Health Paper*, (102): 95-113.

[5] Lowry, J. B. (1990). Toxic factors and problems: methods of alleviating them in animals. In: Shrubs and Tree fodders for Farm Animals. C. Devendra (ed.) IDRC, Ottawa, Canada, 76-88.

[6] Akinbamijo, O.O., Adediran, S. A., Nouala, S., and Seacker, J. (2006). *Moringa* fodder in ruminant nutrition in the Gambia – Trees for life journal – a forum on beneficial trees and plants. Publication of International Trypanotolerance Centre, Banjul T retrieved on 6/4/2010 at http: // www.tfjournal.org.

[7] Asaolu, V. O., Odeyinka, S. M., Akinbamijo, O. O., and Sodeinde, F. G. (2010). Effects of moringa and bamboo leaves on groundnut hay utilization by West African Dwarf goats. *Livestock Research for Rural Development, 22* (1).Retrieved on May 28 2013, from http://www.lrrd.org/lrrd22/1/asao2212.htm.

[8] Fuglie, L. J. (1999). The Miracle Tree: *Moringa oleifera*: Natural Nutrition for the Tropics. Church World Service, Dakar.66 pp.Revised in 2001 and published as The Miracle Tree: The Multiple Attributes of Moringa, 172 pp.

[9] Bosman, H. G., Versteegden, C., Odeyinka, S. M., Tolkamp, B. J. (1995). Effect of amount offered on intake, digestibility and value of *Gliricidia sepium* and *Leucaena leucocephala* for West African Dwarf goats. *Small Ruminant Research*, 15: 247-256.

[10] Phiri, D. M., Coulman, B., Steppler, H. A., Kamara, C. S., and Kwesiga, F. (1992). The effect of browse supplementation on maize husk utilization by goats. *Agroforestry Systems,* 17(2): 153-158.

[11] Adebowale, E. A., and Ademosun, A. A. (1981). Carcass characteristics and chemical composition of organs and muscles of sheep and goats fed brewers' dried grain-based ration. *Tropical Animal Production*, 6: 133-137.

[12] Van Soest, P.J. (1994). Nutritional ecology of the ruminant. 2nd edition Cornell Univ. Press, Yuthaca, NY.

[13] Allen, M. S. (1996). Physical constraints on voluntary intake of forages by ruminants. *Journal of Animal Science*, 74(12): 3063-3075.

[14] NAPRI (1984) Highlights of research achievements on animal production. Science and Technology briefing Lagos, December 1984. 3-17.

[15] AOAC (2005). Official Method of Analysis, 18th Ed. Association of Official Analytical Chemists, Washington DC.

[16] Pearson D (1976) Chemical Analysis of Foods. 7th Ed. Livingstone, London Church. 11: 7-11.

[17] Adegbola, A.A., and Asaolu, O. (1986). Preparation of cassava peels for use in small ruminant production in Western Nigeria. In Preston T.R. and Nuwanyakpa, M.Y. (Eds) Proceed. of a workshop held at the University of Alexandria, Egypt, October, 1985, ILCA, Addis Ababa, Ethiopia, 109-115.

[18] Asaolu, V., Binuomote, R., Akinlade, J., Aderinola, O., and Oyelami, O. (2012). Intake and Growth Performance of West African Dwarf Goats Fed *Moringa oleifera*, *Gliricidia sepium* and *Leucaena leucocephala* Dried Leaves as Supplements to Cassava Peels. *Journal of Biology, Agriculture and Healthcare*, *2*(10): 76-88.

[19] Brisibe, E. A., Umoren U. E., Brisibe F., Magalhaes P. M., Ferreira J. F., Luthria D., Wu X., and Prior R. L. (2009). Nutritional characterization and antioxidant capacity of different tissues of *Artemisia annua L. Food Chem.* 115(4): 1240-1246.

[20] Waldroup, W.P. and Smith, K. (2008). Fact sheet soyabean use – poultry soyabean meal information centre http://www.soymeal.org/pdf/poultry soybean use. pdf.Retrieved on 2/11/12.

[21] Minson, D. J. (1990). Forage in Ruminant Nutrition. Academic Press, San Diego, CA.

[22] Okoli, I. C., Anunobi, M. O., Obua, B. E., and Enemuo, V. (2003). Studies on selected browses of South Eastern Nigeria with particular reference to their proximate and some endogenous anti-nutritional constituents. *Livestock Research for Rural Development*, *15*(9): 3-7.

[23] Van Soest, P.J. (1994). Nutritional ecology of the ruminant. 2nd edition Cornell Univ. Press, Yuthaca, NY.

[24] Aye, P.A., and Adegun, M.K. (2010). Digestibility and growth in West African Dwarf (WAD) sheep fed Gliciridia-based multinutrient block supplements. *Agricultural and Biology Journal of North America* 1 (6): 1133 – 1139.

[25] Leng, R. A. (1990). Factors affecting the utilization of “poor-quality" forages by ruminants particularly under tropical conditions. *Nutrition Research Reviews*, *3*(1): 277-303.

[26] Shelton, H. M. (2000). Tropical forage tree legumes in agroforestry systems. *UNASYLVA-FAO-*, 25-32.

[27] Aregheore, E. M., and Perera, D. (2004). Effects of *Erythrina variegata, Gliricidia sepium* and *Leucaena leucocephala* on dry matter intake and nutrient digestibility of maize stover, before and after spraying with molasses. *Animal Feed Science and Technology*, *111*(1): 191-201.

[28] Bakshi, M. P. S., and Wadhwa, M. (2004). Evaluation of Forest Tree Leaves of Semi-hilly Arid Region as Livestock Feed. *Asian-Aust. J. Anim. Sci.*17(6): 777-783.

[29] Asaolu, V.O., Binuomote, R.T., Akinlade, J.A., Oyelami, O.S., and Kolapo K.O. (2011). Utilization of *Moringa oleifera* fodder combinations with *Leucaena leucocephala* and *Gliricidia sepium* fodders by West African dwarf goats. *Int. J. Agric. Res*, 6(8): 607-619.

[30] Ulyatt, M. J. (1981). The feeding value of herbage: can it be improved. *New Zealand Agricultural Science*, *15*(4): 200-205.

[31] McDonald, P., Edward, R.A., Greenhalgh, J.F.D.F., and Morgan, C.A. (1996). Animal Nutrition. Longman Scientific and Technical, Harlow, UK. 155pp low, UK. 155pp.

8

Yield and pomological characteristics of three pomegranate *(Punica granatum* L) cultivars: Wonderful, Acco and Herskovitz

Serhat Usanmaz[1], İbrahim Kahramanoğlu[1, 2, *], Nihat Yılmaz[3]

[1]European University of Lefke, Faculty of Agricultural Sciences and Technologies, Cyprus, Mersin 10 Turkey
[2]Alnar Pomegranates Ltd., Ataturk Ave. 109/1 Guzelyurt, Cyprus, Mersin 10 Turkey
[3]Erciyes Üniversitesi, Safiye Cıkrıkcıoglu Vocational College, Kayseri, Turkey

Email address:
ibrahimcy84@yahoo.com (İ. Kahramanoğlu)

Abstract: Pomegranate fruit has been associated with a high level of nutrients and many health benefits. This study evaluated the yield and pomological characteristics of three pomegranate cultivars: Wonderful, Acco and Herskovitz grown in Cyprus conditions. Researches performed by collecting data from 17 pomegranate orchards between 2011 (4 years old) and 2012 (5 years old). Yield (kg/tree) was determined according to total harvest. Ten samples were randomly selected from each orchard and each cultivar to determine average fruit weight, number of arils per fruit, fruit width and height, aril per cent, juice per cent and total soluble solids. Results indicated that the Wonderful cultivar had the highest fruit weight and second highest yield (481.12 g/fruit and 14.17 kg/tree). The second highest fruit weight was obtained from Herskovitz which had the highest yield (431.04 g/fruit and 15.44 kg/tree). The lowest fruit weight and yield obtained from Acco (350.31 g/fruit and 11.43 kg/tree). Acco had highest juice content followed by Wonderful and Herskovitz (40.22% > 35.60% > 29.42%). Wonderful had the highest juice content (5.05 l/tree). Juice content per tree for Acco and Herskovitz were determined 4.58 l and 4.53 l, respectively.

Keywords: Pomegranate, Yield, Juice Per Cent, Aril Per Cent, Fruit Weight, Fruit Diameter

1. Introduction

Pomegranate was one of the first five crops together with figs, dates, olives and grapes which are old known cultivated plants. Domestication of pomegranate started 3000 – 4000 BC in the North of Iran and Turkey [1]. Pomegranate (*Punica granatum* L.) belongs to the Punicaceae family with two species namely: *Punica granatum* L. and P. *protopunica* Balf. P. *protopunica* is reported to be endemic to the Socotra Island and is the only congeneric relative of P. *granatum* [2, 3]. Pomegranates are native to central Asia [4, 5] and are adaptable to a wide range of climate and soil conditions. Thus, it is grown in many different geographical regions including the Mediterranean basin, Asia and California.

Pomegranate fruit has been traditionally known to include high level of nutrients and many health benefits. Recent scientific findings confirm traditional usage of pomegranate for medical purposes [6] and reported that tissues of pomegranate fruit, flowers, bark and leaves contain bioactive phytochemicals that are antimicrobial, reduce blood pressure and act against serious diseases such as diabetes and cancer. It is also a good source of antioxidants, three times of wine or green tea, vitamins as it includes vitamin A, C and E and a good source of potassium, calcium, magnesium, iron and zinc. Many studies conducted about pomegranates [7, 8, 9, 10, 11] and revealed the benefits of pomegranate for human health. The findings of these studies increased the public awareness about pomegranate and thus consumption of pomegranate fruit increased. Total pomegranate import for the European Union countries was 171,993 tons in 2002 and it reached 269,093 tons in 2012. This means 56.5% increase in 10 years [12].

In Cyprus, it is known that pomegranate was grown for fresh consumption and exportation during early 1900 [13]. However, from 1960 to 2007, number of pomegranate trees was decreased and only remained around other fruit orchards as wind barriers. Since there is a huge increase in the demand for the pomegranates, new plantations started to be constituted since 2007 with the "alternative crops" projects of USAID in Cyprus. The project funded by USAID and implemented by the Economic Development and Growth for Enterprises (EDGE) Project. Seventeen voluntary farmers attended to this project and established totally around 22 ha of pomegranate orchards.

Determination of the pomological characteristics of fruits is an important aspect for producers, packers, juice producers, marketers and consumers. It is also important for the engineers to design equipment and processes for harvesting, handling, sorting, sizing, packing, storing and processing. Several researches [14-20] reported a great variation in pomological characteristics among the pomegranate cultivars. Moreover, no information is available for the pomegranate cultivars grown in Cyprus. Therefore, this study aimed to assess and compare some pomological and yield characteristics of the three cultivars of pomegranates: Wonderful, Acco and Herskovitz.

2. Materials and Methods

2.1. Materials

This study performed by collecting data from 17 pomegranate orchards between 2011 (4 years old) and 2012 (5 years old). Each orchard covers 1.3 ha with 80% coverage of Wonderful cultivar, 10% with Acco and 10% with Herskovitz. Wonderful cultivar (originated in Florida and first propagated in California) has sweet-tart taste, deep purple-red fruits with soft seeds and delicious vinous flavour. Herskovitz (Israel cultivar: 116) has sour taste fruits with red skin and hard seeds whereas Acco (Israel cultivar: 128) has very sweet, good quality fruits with pinky skin and notably softer seeds.

All fields are suitable for growing pomegranates with clay loam soil and moderate water salinity (Table 1.).

Table 1. *Soil and water characteristics of the studied orchards.*

Orchard Code	Soil Characteristics				Water salinity (ppm)
	pH	Lime%	Salinity %	Org. Matter %	
O-1	7.3	6	0.19	3	1420
O-2	7.5	42	0.09	1.85	3230
O-3	7.4	6	0.11	1.9	705
O-4	7.5	17	0.09	0.7	1600
O-5	7.5	20	0.05	1.3	1600
O-6	7.5	15	0.12	2.3	950
O-7	7.4	29	0.07	1.33	3700
O-8	7.5	9	0.09	2.2	1250
O-9	7.4	13	0.12	1.2	950
O-10	7.6	2	0.13	1.9	1050
O-11	7.4	3	0.09	1.7	1360
O-12	7.5	12	0.11	1.7	1410
O-13	7.4	10	0.08	1.3	1380
O-14	7.9	10	0.12	1.7	1500
O-15	7.4	10	0.13	1.7	1250
O-16	7.5	1	0.09	1.7	800
O-17	7.6	2	0.1	2.5	1800

All orchards were established by 5 x 3 m distance and pruned as globe shape with one trunk. Minimum temperature was 5.4 C (in February) and Maximum temperature was 33.7 C (in August). Average minimum and maximum temperatures were 11.6 C and 24.7 C respectively whereas total precipitation was 315 mm in 2012. Irrigation is performed by drip irrigation according to the basic needs and rainfall (Table 2.).

Table 2. *Pure nitrogen (N), phosphorus (P), potassium (K) and water requirements of pomegranate trees.*

Months	Pure nutrient (gr) / tree / month						Water (l) / tree / day	
	4 years old			5 years old			4 years old	5 years old
	N	P	K	N	P	K		
January	0,0	0,0	0,0	0,0	0,0	0,0	0	0
February	0,0	0,0	0,0	0,0	0,0	0,0	0	0
March	12,5	30,0	0,0	15,0	45,0	0,0	11	13
April	62,5	36,0	0,0	75,0	54,0	0,0	19	21
May	62,5	42,0	20,0	75,0	63,0	25,0	22	27
June	50,0	12,0	40,0	60,0	18,0	50,0	28	40
July	37,5	0,0	50,0	45,0	0,0	62,5	32	50
August	25,0	0,0	50,0	30,0	0,0	62,5	32	45
September	0,0	0,0	40,0	0,0	0,0	50,0	28	40
October	0,0	0,0	0,0	0,0	0,0	0,0	24	30
November	0,0	0,0	0,0	0,0	0,0	0,0	0	0
December	0,0	0,0	0,0	0,0	0,0	0,0	0	0
Yearly / tree	250	120	200	300	180	250	5880	7980

Cultivation and management practices of the 17 orchards are same. Studied orchards generally face with *Alternaria* sp., Aphids (*Aphis punica* and *Aphis pomi*), *Deudorix livia*, *Zeuzera pyrina*, *Planocochus citri* and *Ceratitis capitata*. Management program for these pests and diseases is given in Table 3.

Table 3. Management program of the pomegranate pests and diseases.

Months	Pests and diseases	Management
December – February	Alternaria sp.	Copper sulphate application is used
March – May	Aphids (Aphis *punica* and Aphis *pomi*)	Acetamiprid or Primicarb is used
March – May	Deudorix *livia*	Indoxacarb or Spinosad is used
May – August	Zeuzera *pyrina*	Chlorpyrifos Ethyl is used
June – September	Planocochus *citri*	Spirotetramat application and fruit thinning is performed
August – November	Ceratitis *capitata*	Attract-and-kill traps and Cypermethrine are used

2.2. Methods

Seventeen orchards with three pomegranate cultivars: Wonderful, Acco and Herskovitz were used for this study. Orchards were checked monthly and data collected from February 2011 to November 2012. Pomological and yield data of the selected seventeen orchards were collected regularly and evaluated. The data includes; fruit weight, yield (kg/tree), number of arils per fruit, fruit width and height, percentage of fruit bark, aril yield, fruit juice yield and total soluble solids (TSS). Harvested pomegranates were packed and stored at the packing house of Alnar Pomegranates Ltd. Total yield of each orchard divided to the number of trees to determine the yield (kg/tree). Thus, 10 fruits were selected randomly from each orchard, for each cultivar and physical properties were determined. Fruit weight was measured by using digital scale (sensitive to ±0.01 g). Number of arils per fruit was determined by extracting the arils carefully by hand. Fruit width and height were measured by a digital caliper with ±0.01 mm accuracy. Percentage of aril was determined by using digital scale (±0.01 g). Percentage of juice was calculated with simple formula as percentage after squeezing the arils and TSS in the juice were determined with a hand refractometer (°Brix). Data from the two year experiments were subjected to ANOVA to determine any statistical differences among species. Mean separations were done by using the Duncan's multiple range test at $P < 0.01$.

3. Results

Fruit characteristics and yield of three pomegranate cultivars are given in Table 4. The yield were recorded among cultivars ($P < 0.01$) with no significant differences. According to the results, it is clear that the highest fruit weight is obtained in the Wonderful (461.72 g) followed by Herskovitz (412.15 g) and Acco cultivars (336.23 g) in 2011 of 4 years old trees. The average fruit mass of pomegranate cultivars ranged from 350.31 g to 481.12 g for 5 years old trees. Results for the fruit width and height are in accordance with the results for fruit weight. When comparing the cultivars, the highest width and height were determined in the order of Wonderful > Herskovitz > Acco in both years. In 2012, when the trees are 5 years old, average fruit width and height were 107.12 mm and 93.90 mm for Wonderful, 97.03 mm and 83.81 mm for Herskovitz and 81.60 mm and 68.38 mm for Acco. No statistical significance was reported when comparing the yields of three cultivars. However, on the other hand, it is clear from the two years results that Herskovitz is having higher yield than the Wonderful and Acco.

Pomological characteristics of three pomegranate cultivars are presented in Table 5. Statistically significant differences were determined among cultivars ($P < 0.01$). The average number of arils per fruit for cultivars varied from 539.41 to 759.23 in 2011 and from 560.09 to 790.95 in 2012. These results are in accordance with the fruit weight, width, height and yield results where all characteristics are increased in 2012. Significant differences were obtained in fruit aril per cent and fruit juice per cent for the three cultivars. For the other characteristics, like fruit weight and number of arils, are recorded high for the Wonderful cultivar whereas the percentage of aril and juice are recorded high for Acco cultivar. The aril and juice per cent of three cultivars were slightly higher in 2012 than 2011 without significant differences. The aril per cent of Acco, Wonderful and Herskovitz is 68.32%, 60.55% and 58.36%, respectively, in 2012. Results for the juice per cent of three cultivars are in accordance with the results for aril per cents. The highest juice content is recorded for Acco (40.22%) and followed by Wonderful (35.57%) and Herskovitz (29.42%). Total soluble solid contents of the studied cultivars ranges from 16.00 to 21.45.

Table 4. Fruit characteristics and yield of pomegranate cultivars: Wonderful, Acco and Herskovitz

Ages and Cultivars		Average Fruit weight (g)	Average Fruit width (mm)	Average Fruit height (mm)	Average Yield (kg/tree)
4 years old	Wonderful	461.72 a	104.91 a	90.92 a	8.21±7.05 a
	Acco	336.23 c	80.04 c	66.82 c	8.35±6.73 a
	Herskovitz	412.15 b	95.59 b	82.37 b	13.71±8.63 a
5 years old	Wonderful	481.12 a	107.12 a	93.90 a	14.17±8.13 a
	Acco	350.31 c	81.60 c	68.38 c	11.43±7.64 a
	Herskovitz	431.04 b	97.03 b	83.81 b	15.44±10.13 a
Significance level		0.01	0.01	0.01	NS

Table 5. Some pomological characteristics of pomegranate cultivars: Wonderful, Acco and Herskovitz

Ages and Cultivars		Average number of arils per fruit	Percentage of aril	Percentage of juice	TSS (oBrix)
	Wonderful	759.23 a	60.47 b	35.57 b	21.42 a
4 years old	Acco	539.41 c	68.09 a	40.08 a	17.22 b
	Herskovitz	666.17 b	58.30 c	29.39 c	15.81 c
	Wonderful	790.95 a	60.55 b	35.60 b	21.45 a
5 years old	Acco	560.09 c	68.32 a	40.22 a	17.29 b
	Herskovitz	694.62 b	58.36 c	29.42 c	16.00 c
Significance level		0.01	0.01	0.01	0.01

4. Discussions

The fruit weight of five year old trees showed considerable variation with the fruit weight of 481.12 g for Wonderful, 431.04 g for Herskovitz and 350.31 g for Acco during 2012 due to early ripening of Acco and Herskovitz cultivars. Pantelidis, et al. [21] reported that the mean fruit weight of Wonderful and Acco cultivars are 474.2 g and 273.6 g, respectively, in Greece but the fruit weight of Acco in present study is significantly higher. On the other hand, Bartual, et al., [22] reported that the average fruit of Wonderful cultivar is 324.86 g in Spain, which is significantly lower than the results of present study. A wide variation among the fruit mass of different pomegranate cultivars ranging from 103 g to 505 g were reported [14-16, 18-20].

Fruit width and height results are in accordance with the results for fruit weight. Moreover, recent studies reported that fruit width is ranging from 60.6 to 106.99 mm and fruit height of 52.9 to 95.52 mm [16-18]. Similarly with fruit mass, fruit width and height of pomegranate cultivars under study are in accordance with the findings of previous works but the fruit weight of Wonderful (2012) is higher than the other pomegranate cultivars studied in other researches: Eksinar, Hicaznar, Ciparski, Konjski and Pastun. The Acco and Herskovitz are early cultivars with smaller dimensions and less ripening period than Wonderful. The yield of three cultivars of 4 and 5 years old is found to be superior over the fruit yield of Malas e Torsh e Saveh (Iran cultivars; 8.10 kg/tree) [19]. It is observed that some orchards have recorded significant difference in fruit yield as reported here: 31.6 kg/tree for Wonderful, 31.4 kg/tree for Acco and 39.8 kg/tree for Herskovitz.

The number of arils per fruit indicated that there is a significant difference among the cultivars. The number of arils per fruit for Wonderful is 790.9, 694.6 for Herskovitz and 560.9 for Acco. Results indicated that both three cultivars under study have superior aril content to that of the Eksinar which is having mean of 502 and maximum of 618 arils per fruit [16]. The aril per cent of Acco, Wonderful and Herskovitz are 68.3%, 60.6% and 58.3%, respectively. The significant edible percent for Wonderful and Acco were reported in Greece [21] for the same cultivars as 53.0% and 53.6%, respectively. Some previous studies reported a wide variation among the aril per cent for different pomegranate cultivars ranging from 46.30 to 68.31 [15, 16, 20].

The Acco cultivar have 40.22 per cent of juice content whereas 35.57 per cent for Wonderful and 29.42 per cent for Herskovitz. Similar kind of observations was reported with 34 to 35 per cent juice content in Wonderful cultivar [21, 22]. The juice content of Acco cultivar (two years of experimental results) is higher to that of the same cultivar grown in Greece (37.9%) [21]. Akbarpour, *et al.*, [15] reported that the juice content of 12 cultivars in Iran varies from 20.18% to 59.83% but of Spanish varieties with juice content of 50.26% to 64.17% [23]. Total soluble solid contents of the cultivars vary from 16.00 to 21.45. These findings are in accordance with the results of other works [15, 19, 20]. Among the studied cultivars, Wonderful has the TSS content of 21.45 followed by Acco (17.29) and Herskovitz (16.00). This observation is in agreement with the reports of research workers in Greece and Spain [21, 22]

5. Conclusions

Results of present study showed considerable variation in the pomological characteristics of widely grown pomegranate cultivars under Cyprus conditions. Wonderful cultivar is the best for fruit size production. But no statistical difference was determined among cultivars in spite of good performance of Herskovitz (15.4 kg/tree, Table 6).

Table 6. Summary results for the two years average of yield and fruit Juice for Wonderful, Acco and Herskovitz.

Cultivars	Average yield (kg/tree)	Fruit Juice (% of a fruit)	Fruit Juice (l / tree)
Wonderful	14.2	35.59	5.05
Acco	11.4	40.15	4.58
Herskovitz	15.4	29.40	4.52

Because of the sour taste, the Herskovitz is not having good marketing potential and thus the Wonderful cultivar, with its sweet-tart taste and high yield, seems to be the best cultivar for fresh consumption. The Acco cultivar's performance is poor with low pomological characteristics such as fruit weight, yield, width, height and number of arils but have good aril and juice per cents: 68.32% and 40.15%. Considering the yield (kg/tree) and juice content

(%) together, again the Wonderful cultivar is good with juice content of 5.051 per tree.

Acknowledgment

Special thanks to USAID and EDGE Project for the alternative crops project, which gave chance to the farmers to produce pomegranates and made this study possible. Authors gratefully acknowledge to Prof. Dr. Kazım Abak for his valuable advice and to Mr. Erdinc Bilgin, chairman of Cyprus Pomegranate Producers Union for his kind help.

References

[1] C. Lye, "Pomegranate: preliminary assessment of the potential for an Australian industry" Rural Industries Research and Development Corporation of Australian Government. RIRDC Publication No 08/153, 2008, 17p.

[2] D., Chatterjee and G. S. Randhawa, "Standardized names of cultivated plants in India. 1. Fruits". Indian J. Hort. Vol. 9, 1952, pp.24-36.

[3] C. Yılmaz, "Nar" Hasad Yayıncılık, 2007, 190p.

[4] J. Morton, "Pomegranate". In: Fruits of warm climates. Miami, F, L. 1987, pp. 352–355.

[5] D. Holland, K. Hatip and I. Bar-Ya'akov, "Pomegranate: Botany, Horticulture and Breeding", Horticultural Reviews, Volume, 35, Edited by Jules Janick, John Wiley & Sons Inc, 2009, pp.127-191.

[6] J. M. T. Jurenka, "Therapeutic Applications of Pomegranate (Punica granatum L.): A Review", Alternative Medicine Review Vol. 13, No. 2, 2008, pp.128-144.

[7] M. I. Gil, F. A. Tomas-Barberan, B. Hess-Pierce, D.M. Holcroft and A. A. Kader, "Antioxidant activity of pomegranate juice and its relationship with phenolic composition and processing", J. Agric. Food Chem. Vol. 48, 2000, pp.4581-4589.

[8] M. Aviram and L. Dornfeld, "Pomegranate juice consumption inhibits serum angiotensin converting enzyme activity and reduces systolic blood pressure", Atherosclerosis Vol. 158, 2001, pp.195-198.

[9] E. P. Lansky, W. Jiang, H. Mo, L. Bravo, P. Froom, W. Yu, N.M. Harris, I. Neeman and M.J. Campbell, "Possible synergistic prostate cancer suppression by anatomically discrete pomegranate fractions", Invest New Drug, Vol. 23, 2005, pp.11-20.

[10] M. Haidari, M. Ali, S.W. Casscells and M. Madjid, "Pomegranate (Punica granatum) purified polyphenol extract inhibits influenza virus and has a synergistic effect with oseltamivir", Phytomedicine, Vol. 16, No. 12, 2009, pp.1127-36.

[11] G. Turk, M. Sonmez, M. Aydin, A. Yuce, S. Gur, M. Yuksel, E.H. Aksu and H. Aksoy, "Effects of pomegranate juice consumption on sperm quality, spermatogenic cell density, antioxidant activity and testosterone level in male rats", Clinical Nutrition Vol. 27, 2008, pp.289-296.

[12] EC Market Access Database, European Commission, Market Access Database. http://madb.europa.eu/madb/statistical_form.htm, access date: 01 July 2013.

[13] W. Bevan, "Notes On Agriculture in Cyprus and its Products". London, Hazell, Watson & Viney, ld., printers, 1919.

[14] Y. Ozkan, "Determination of pomological characteristics of Niksar district pomegranates (Punica granatum L.) of the Tokat Province". Acta Hort. Vol. 598, 2003, pp.199-203.

[15] V. Akbarpour, K. Hemmati and M. Sharifani, "Physical and chemical properties of pomegranate (Punica granatum L.) fruit in maturation stage". American-Eurasian J. Agric. & Environ. Sci. Vol. 6, No. 4, 2009, pp.411-416.

[16] A. Celik and S. Ercisli, "Some physical proterties of pomegranate cv. Eksinar", Int. Agrophysics Vol. 23, 2009, pp.295-298.

[17] S. Gozlekci, S. Ercisli, F. Okturen, and S. Sonmez, "Physico-chemical characteristics at three development stages in pomegranate cv. Hicaznar". Not. Bot. Hort. Agrobot Cluj Vol. 39, No. 1, 2011, pp.241-245.

[18] J. Gazde, S. Voca, Z. Cmelik, I. Mustac, S. Ercisli and Radunic, "Physico-chemical characteristics of main pomegranate (Punica granatum L.) cultivars grown in Dalmatia region of Croatia", J. App. Bot. Food Quality, Vol. 85, 2012, pp.202-206.

[19] M. Hasani, Z. Zamani, G. Savaghebi and R. Fatahi, "Effects of zinc and manganese as foliar spray on pomegranate yield, fruit quality and leaf minerals", J. Soil Sci. Plant Nutr. Vol. 12, No. 3, 2012, pp.471-480.

[20] M. M. Mir, I. Umar, S.A. Mir, M.U. Rehman, G.H. Rather and S.A. Banday, "Quality Evaluation of Pomegranate Crop – A review", International Journal of Agriculture & Biology Vol. 14, 2012, pp.658-667.

[21] G. Pantelidis, P. Drogoudi and A. Manganaris, "Physico-chemical and antioxidant properties of pomegranate genotypes in Greece", Options Mediterraneenes Vol. 103, 2012, pp.335-337.

[22] J. Bartual, H. Valdes, J. Andreu, A. Lozoya, J. Garcia and M.L. Badenes, "Pomegranate improvement through clonal selection and hybridization in Elche", Options Mediterraneenes Vol. 103, 2012, pp.71-74.

[23] J. J. Martinez, P. Melgarejo, F. Hernandez, D.M. Salazar and R. Martinez, "Seed characterisation of five new pomegranate (Puica granatum L.) varieties", Scienta Horticulturae, Vol. 110, No. 3, 2006, pp.241-246.

9

Rock-phosphate solubilising bacteria and their effect on soybean (*Glycine max*) growth under pot grown conditions

Fankem Henri[1, *], Nguesseu Njanjouo Ghislain[1], Ngo Nkot Laurette[1], Tchuisseu Tchakounte Gylaine Vanessa[1], Nwaga Dieudonné[2], Etoa François-Xavier[2]

[1]Department of Plant Biology, Faculty of Science, University of Douala, Douala, Cameroon
[2]Department of Microbiology, Faculty of Science, University of Yaoundé I, Yaoundé, Cameroon

Email address:
fankemhenri@yahoo.fr (F. Henri)

Abstract: The modern agriculture is dependent on phosphorus (P) derived from phosphate rock. However, the direct application of low-grade rock phosphate as a P source in soils need an addition of inoculums of phosphate solubilising microorganisms to improve the rock phosphate efficiency as a phosphorus source. Phosphate solubilising bacteria (PSB) were screened for their phosphate solubilising ability on plates and in liquid cultures supplemented with Malian, Moroccan or Mexican rock phosphates. They were subsequently tested on soybean grown in pots filled with non sterile soil amended with Moroccan rock phosphate for their aptitude in promoting soybean growth. The activity of the different strains on plates indicates *Panthoea* sp. and *Bacillus* sp. as the most efficient strains able to show halo zone on plates supplemented all different rock phosphates, with a solubilisation index (SI) of 3.65, 4.10 and 5.42 (*Panthoea* sp.) and 2.93, 3.13 and 2.13 (*Bacillus* sp.) respectively for Malian, Moroccan and Mexican rock phosphates. *Panthoea* sp. remains the strains showing the highest concentration of the solved P with all rock phosphates: 1038.25, 996.67 and 1207.87 µgP/g for Malian, Moroccan and Mexican rock phosphates respectively. It is followed by *Klebsiella* sp. (862.57, 615.19 and 426.29 µgP/g respectively) and *Bacillus* sp. (810.86, 270.92 and 180.95 µgP/g). In general, *Panthoea* sp. and *Bacillus* sp. better contribute to the soybean growth with the effect of 35% and 34% respectively compare to non inoculated control supplied with non soluble Moroccan rock phosphate. The activity of *Klebsiella* sp. (13%) that is low in general seems to be stimulated when associated with the two other strains (33%). This suggests that the use of rock phosphate combined with the co-inoculation with those strains would ensure soybean production in economically profitable and environmentally friendly conditions.

Keywords: Phosphate Solubilising Bacteria, Rock Phosphate, Rock Phosphate Solubilisation, Soybean Growth

1. Introduction

After nitrogen, phosphorus (P) is an essential plant nutrient whose deficiency restricts crop yields severely [1]. It plays an important role in virtually all major metabolic processes in plant including photosynthesis, energy transfer, signal transduction, macromolecular biosynthesis and respiration [2]. Tropical and sub tropical soils are predominantly acidic, and often extremely phosphorus deficient [3] with high phosphorus fixation capacities. Phosphorus is added to soil in the form of phosphatic fertilizers, part of which is utilized by plants and the remainder converted into insoluble fixed forms. It is well documented that major fractions of soil phosphorus are usually present in the forms which are unavailable to plant [4]. This leads to the need of frequent application of phosphate fertilizers, but its use on a regular basis has become a costly affair and environmentally undesirable [5]. Therefore, the necessity to develop economical and eco-friendly technologies is steadily increasing [6, 7, 8]. Natural rock phosphates have been recognised as a valuable alternative for P fertilisers, the modern agriculture being dependent on phosphorus derived from phosphate rock [9]. However, the direct application of low-grade rock phosphate as a P source in neutral and alkaline soils was of little importance [10, 11], but addition of an inoculums of phosphate solubilising microorganisms to soil has also been found to improve the rock phosphate efficiency as a phosphorus source [12]. Many soil microorganisms, including bacteria and fungi, are able to mobilize phosphorus from sparingly soluble rock phosphates, and they have an enormous potential in providing soil phosphates for plant growth [13, 14]. These organisms are ubiquitous but vary in density and mineral/rock phosphate solubilising ability from soil to soil or from one production system to another. They are generally isolated from rhizosphere and nonrhizosphere soils, rhizoplane, phyllosphere, and rock P deposit area soil and even from stressed soils using serial plate dilution method or by enrichment culture technique [4]. Once a potential isolate is identified, it must be further tested for direct contribution to P plant nutrition and not necessarily to general growth promotion, as commonly done because promotion of growth, even by phosphate solubilising bacteria (PSB), can be the outcome of other mechanisms [15] and ability to solubilise P is not necessarily correlated with the ability to promote plant growth [16]. Therefore, this study is conducted to characterize three bacterial strains in solubilising rock phosphates as well as their impact in promoting soybean growth under pot grown conditions.

2. Materials and Methods

2.1. Microorganisms

All the strains used in this study are from the strain bank of the Laboratory of Biotechnology, Faculty of Science, University of Douala. Strains were isolated from soil collected in three agro ecological zones of Cameroon; *Klebsiella* sp. in zone I, *Panthoea* sp. in zone II, and *Bacillus* sp. in zone IV. They have been previously screened for their ability in solubilising sparingly soluble phosphates including $Ca_3(PO_4)_2$, $AlPO_4$, $FePO_4$ and sodium-phytate, and would be recognized as inorganic/organic phosphate solubilisers.

2.2. Rock Phosphates

Rock phosphates of different origins and analyzed for their chemical contents (Table 1) were used: the Tilemsi rock phosphate from Mali, Gafsa rock phosphate from Mexico and Moroccan rock phosphate.

To get rid of their soluble fractions, the different rock phosphates were washed 4 times with warm water following the cycle: 1 hour - 24 hours - 1 hour - 24 hours. They were then dried at 60 °C until complete evaporation of water and homogenized before use.

Table 1. *Mineralogical composition of the different rock phosphates used (Magallón-Servín, unpublished).*

Origin rock P	Mineral elements (%)								mg/kg		
	Total P_2O_5	Available P	K	Ca	Mg	Na	Fe	Al	Mn	Zn	Cu
Mali	30	12.98	0.056	28.19	0.131	0.232	3.844	0.80	8360	87	51
Mexico	28	8.87	0.219	25.94	0.222	0.358	0.442	0.58	788	103	18
Morocco	13	9.33	0.093	28.83	1.93	0.552	0.267	0.42	96	219	38

2.3. Preparation and Evaluation of the Concentration of the Inoculums

To prepare inoculums from each bacterial strain, pure bacterial colony was individually suspended into 50 ml Nutrient Broth (NB) (5 g Peptone, 1 g Beef extract, 2 g Yeast extract, 5 g Sodium chloride, 1000 ml Distilled water, pH 7.0) and incubated at 28 °C, 150 rpm, for 3 days. Cultures were then spin at 10.000g, 10 minutes at 4 °C, followed by three times washing with 0.85% sterile NaCl at the same conditions. Bacterial cells were resuspended into 0.85% sterile NaCl and the OD adjusted to 0.2 at 620 nm. To assess the number of bacterial cells per milliliter, 1 mL of bacterial suspension, OD 0.2 was serial diluted until 10^{-7}. 200 μL of dilutions 10^{-7}, 10^{-6}, 10^{-5} were used to inoculate Nutrient Agar (NA) (5 g Peptone, 1 g Beef extract, 2 g Yeast extract, 5 g Sodium chloride, 15 g Agar, 1000 ml Distilled water, pH 7.0) plates in duplicate. After incubation at 28 °C, for 4 days, bacterial colonies were counted and the number of Colony Forming Unit (CFU) per mL recorded. Counting colonies allowed the determination of the concentration of the inoculum evaluated to $1\text{-}1.7\times10^{9}$ CFU / mL. Consortia of strains were freshly prepared by mixing equal volumes of the different strain suspensions.

2.4. Bacterial Rock Phosphate Solubilising Capacity on Agar Plate

The characterisation of strains for their rock phosphate solubilising ability was assessed on plates filled with the National Botanical Research Institute's Phosphate growth medium (NBRIP) [17] with some modifications and containing per litre of distilled water: 20 g glucose, 5 g $MgCl_2{\cdot}6H_2O$, 0.25 g $MgSO_4{\cdot}7H_2O$, 0.2 g KCl, 0.1 g $(NH_4)_2SO_4$ and one rock phosphate type at 5 $g.L^{-1}$ (Malian RP, Moroccan RP or Mexican RP), plus 0.5% bromocresol green [18], pH 7.5. A stock solution of 0.5% dye was prepared by dissolving a corresponding weight of bromocresol green into 70% ethanol and the final pH adjusted to 6.5 with 1M KOH. Five microliters of each bacterial suspension obtained as described above were transferred onto a single point of compartmented Petri dish. The plates were sealed and incubated at 28 °C for 5 days and the phosphate solubilisation recorded through the halo/yellow zone surrounding the bacterial colony. The index of solubilisation (IS) as defined by Qureshi et al. [19]

was used as an indicator for the isolate efficiency: IS = (Colony diameter + diameter of halo zone) / Colony diameter.

2.5. Quantitative Estimation of Phosphate Solubilisation by Bacteria in Liquid Media

Bacteria were tested in liquid media to assess their capability in releasing phosphorus from insoluble rock phosphate sources. 50 ml NBRIP medium were distributed into 250 mL Erlenmeyer flasks, individual rock phosphate types (Malian RP, Moroccan RP or Mexican RP) were added to the medium at the concentration of 5 $g.L^{-1}$ and the pH adjusted to 7.5. After sterilization and cooling, 200 µl bacterial suspensions of $1\text{-}1.7\times10^{9}$ CFU / mL were used to inoculate flasks containing the different rock phosphates. Each treatment was performed in triplicate and non-inoculated flasks supplemented with the different rock phosphates supplied with 200 µl 0.85% sterile NaCl served as controls. Incubation was made at 28 °C, 150 rpm for 7 days. At end of the incubation time and in all cases, the cultures were transferred into sterile falcon tubes, centrifuged at 10.000g for 10 minutes at 4 °C, a part of the supernatant taken for pH measurement and another part for P determination following the method described by Murphy & Riley [20].

2.6. Pot Experiment

The pot experiment was conducted to assess the effect of a single strain or consortia of strains on soybean (*Glycine max*) growth. The experiment was conducted in 3 liters pots containing homogenized non-sterile soil with the following characteristics: pH H_2O, 5.5; pH KCl, 4.4; nitrogen, 0.06%; available phosphorus, 7.54 ppm; organic matter, 0.59%; organic carbon, 0.34%; iron, 1.75 ppm; aluminium, 0.29 méq/g; calcium, 1.04 méq/g; magnesium, 1.60 méq/g; potassium, 0,.15 méq/g; sodium, 0.07 méq/g. The experiment consisted of 12 treatments including 10 microbial treatments and two controls. The microbial treatments consisted of three single inoculations respectively labeled A (*Panthoea* sp.), B (*Klebsiella* sp.), C (*Bacillus* sp.) and two consortia AC (*Panthoea* sp. + *Bacillus* sp.) and ABC (*Panthoea* sp. + *Klebsiella* sp. + *Bacillus* sp.). All the pots, except the positive control were amended with Moroccan rock phosphate at the rate of 80 $kg.ha^{-1}$ (0.3625g for 0.8836 dm^3) to increase the amount of phosphorus in soil. All the pots were equally supplied with potassium at the rate of 80 kg $K_2O.ha^{-1}$. In all inoculated pots, one pre-germinated seed was soaked with 1 mL of bacterial suspension at sawing. The control treatments consisted of a positive control (C^+) supplemented with soluble KH_2PO_4 at the concentration of 350 mgP/g soil, and a negative control (C^-) supplemented with Moroccan rock phosphate, both without bacterial inoculation.

The experimental design is a completely randomized block system with 12 treatments, 1 host plant and 4 replications, resulting in a total of 48 experimental units. Plant growth was followed during 6 weeks within which, each pot received 500 ml water three times per week. Growth parameters (number of leaves, plant high, stem base diameter) were taken the third and the sixth week. At the end of the growth, plants were harvested, the aerial part separated from the root part, and then dried at 60 °C until the dry mass of materials became stable to determine shoot, root and total dry weight.

2.7. Statistical Analysis

Statistical analyses were performed with Sigma plot 12.0. The analysis of variance (ANOVA) was run to find difference between factors and the HSD Turkey test to compare the different treatments.

3. Results

3.1. Strains Characterization on Plates Supplemented with Rock Phosphates of Different Origins

3.1.1. Solubilisation Index of Strains

The efficacy of the different strains in solubilising sparingly soluble rock phosphates was first assessed on plates using solubilisation index (SI) an indicator for the strain efficiency, higher the SI, greater the solubilisation ability. *Panthoea* sp. and *Bacillus* sp. were able to solubilise the three rock phosphate types (SI>1), while *Klebsiella* sp. could only solubilised the Malian rock phosphates (Figure 1). *Panthoea* sp. with the solubilisation index of 3.65, 4.10 and 5.42 respectively for Malian, Moroccan and Mexican rock phosphates is the strain showing the greatest solubilisation index regarding the different rock phosphates. Significant differences can be clearly observed within *Panthoea* sp. and *Bacillus* sp. (2.93, 3.13 and 2.13 respectively) and within *Bacillus* sp. and *Klebsiella* sp. (1.81, 1 and 1 respectively). Regarding the activity of the different strains on plates, *Panthoea* sp. can be considered as the most efficient strains, followed by *Bacillus* sp. and *Klebsiella* sp. (Figure 2).

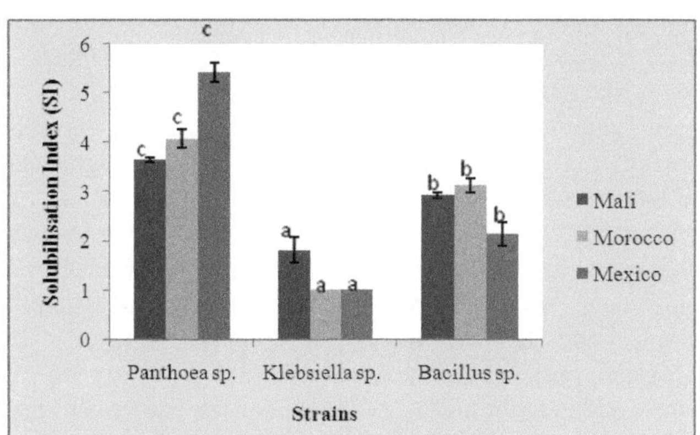

Figure 1. Solubilisation of different types of phosphate by each strain on plates. The different letters within the same phosphate type are significantly different (p<0.05).

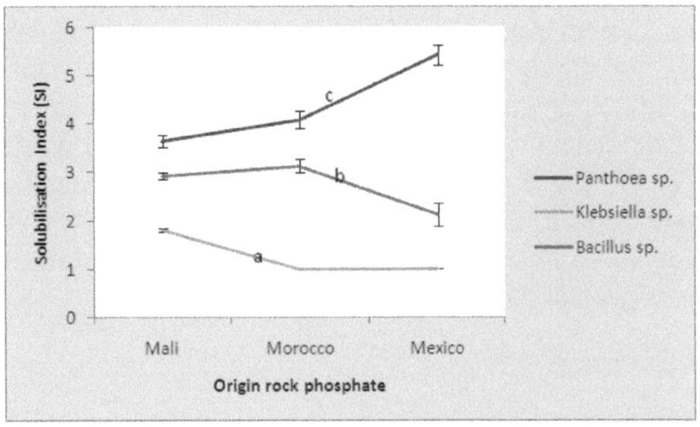

Figure 2. Relative efficiency of strains in solubilising rock phosphates of different origins on plates. The different letters indicate a significant difference (p<0.05) between strains regardless the phosphate origin.

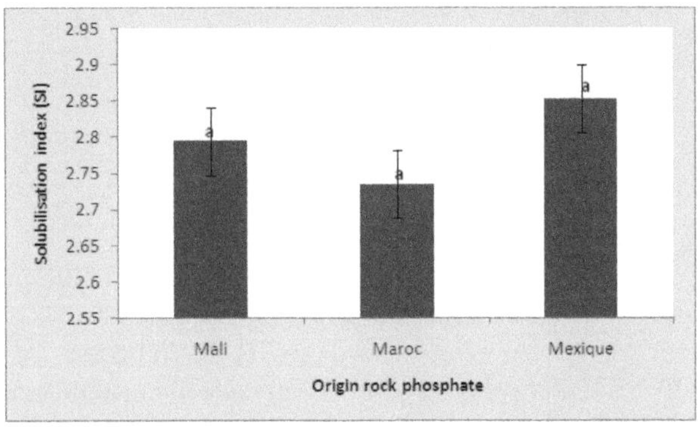

Figure 3. Fluency of the different rock phosphates to be solved by the strains on plates. The different letters indicate a significant difference between rock phosphates (p < 0.05).

3.1.2. Fluency of the Different Rock Phosphate to be Solved on Plates

There is no significant difference between the different rock phosphate solubilisation on plates (Figure 3). The different rock phosphates with an average solubilisation index of 2.794 (Mali), 2.734 (Morocco) and 2.852 (Mexico) show the same capacity to be solved.

3.2. Strains Characterization in Liquid Cultures Supplemented with Rock Phosphates of Different Origins

3.2.1. Concentration of the Solved-P in Liquid Cultures and pH of the Corresponding Media

In general, all the strains were able to mobilise phosphorus from insoluble sources, with significant difference within strains and control (Figure 4). Same like on plates, *Panthoea* sp. is the strain showing the highest efficiency in mobilising phosphorus from insoluble rock phosphates sources with an average amount of 1038.25, 996.67 and 1207.87 µgP/g for Malian, Moroccan and Mexican rock phosphates respectively. Contrary to the result obtained on plates, *Klebsiella* sp. appears much more efficient in liquid culture with average solved P of 862.57, 615.19 and 426.29 µgP/g respectively, compared to *Bacillus* sp. with average solved P of 810.86, 270.92 and 180.95 µgP/g respectively. Furthermore, *Klebsiella* sp. that was not able to show any visible activity on plates amended with Moroccan and Mexican rock phosphates could conversely mobilised significant amount in broth containing the same rock phosphate types. In general, the rock phosphate solubilisation is accompanied with drop of pH of the media, but no strong correlation could be established between the amount of the solved P and the pH of the corresponding media.

3.2.2. Efficiency of Strains According to the Concentration of the Solved P

There is a significant difference (p<0.05) between ability of strains in solubilising the different rock phosphates (Figure 5). *Panthoea* sp. presents the highest efficiency in solubilising all rock phosphates in liquid cultures followed by *Klebsiella* sp. and finally by *Bacillus* sp. Great differences can be easily observed within *Panthoea* sp. and *Klebsiella* sp. and within *Klebsiella* sp. and *Bacillus* sp.

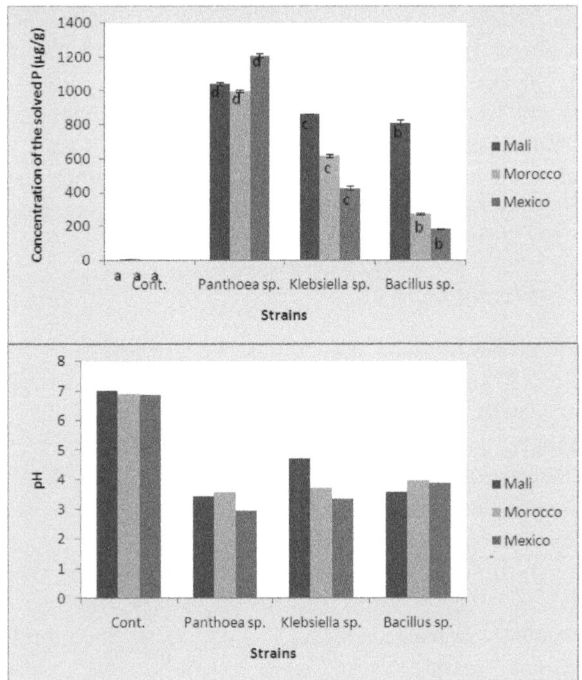

Figure 4. Concentration of the solved P in liquid cultures amended with rock phosphates of different origins and the pH of the corresponding media. The different letters within the same phosphate type are significantly different (p < 0.05).

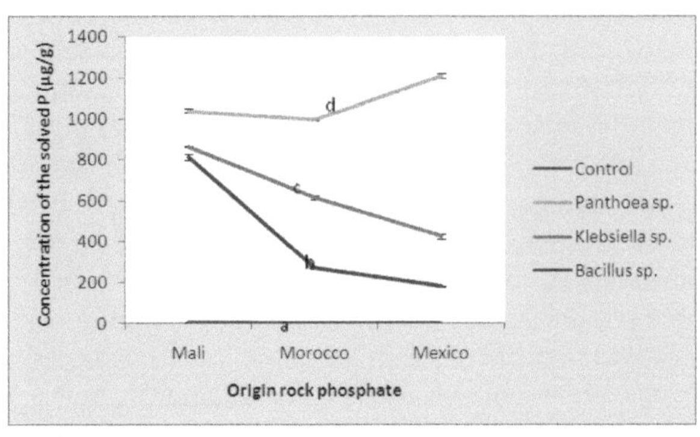

Figure 5. Aptitude of strains in solubilising rock phosphates of different origins in liquid media. The different letters indicate a significant difference between strains regardless the phosphate type (p<0.05).

3.2.3. Fluency of the Different Rock Phosphate to be Solved in Liquid Media

Contrary to the result obtained on plates with no significant difference on the ability of the different rock phosphates to be solved, there is a significant difference between the different rock phosphates in liquid cultures (Figure 6). The Malian rock phosphate (679.59 µgP/g) is the easiest phosphates to be solved by the different strains, followed by Moroccan (470.69 µgP/g) and Mexican rock phosphate (453.79 µgP/g).

3.3. Effect of Inoculation by Strains on the Soybean Growth

3.3.1. Growth Parameters Three Weeks After Planting

The number of leaves varies between 5 and 6 while the stem base diameter varies between 2.48 and 3.40 cm with no significant difference between treatments in both cases at three weeks after planting (Table 2). However, there is a significant difference ($p<0.05$) between treatments regarding the plant high. While the negative control (C-) shows the lowest score (12.40 cm), the best one is obtained with *Bacillus* sp. (15.68 cm) followed by *Panthoea* sp. (15.00 cm) and the consortium of the three strains ABC (14.58 cm). This shows the real aptitude of strains, either in single or in consortia to promote the plant growth regarding the plant high parameter.

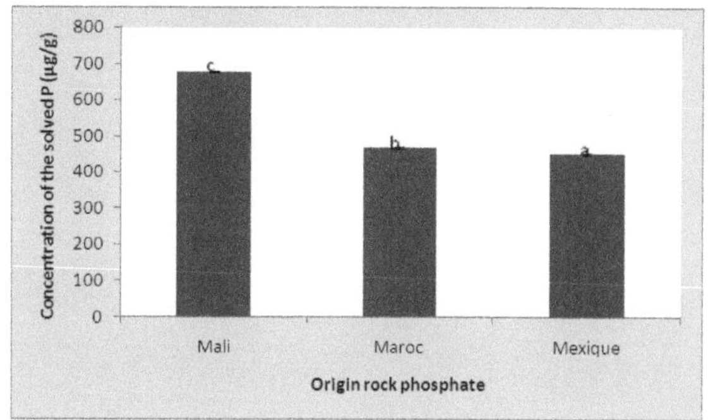

Figure 6. *Facility of the different rock phosphates to be solved by the strains in liquid media. The different letters indicate a significant difference between rock phosphates (p<0.05).*

Table 2. *Number of leaves, plant high and stem base diameter of soybean plants the third week after planting.*

Treatment	(a) Number of leaves	Standard deviation	(b) Plant high (cm)	Standard deviation	(c) Stem base diameter (cm)	Standard deviation
C-	5 a	0.50	12.40 a	0.43	2.70 a	0.48
C+	6 a	0.58	13.78 bc	0.33	3.40 a	0.43
A	5 a	0.00	15.00 cd	0.71	2.78 a	0.39
B	5 a	0.00	13.08 ab	0.82	2.48 a	0.36
C	5 a	0.50	15.68 d	0.40	3.27 a	0.21
AC	5 a	0.50	14.08 bc	0.15	2.75 a	0.35
ABC	5 a	0.50	14.58 cd	0.66	2.63 a	0.21

(a) Source of Variation	DF	SS	MS	F	P
Between Groups	6	1.857	0.310	1.625	0.190
Residual	21	4.000	0.190		
Total	27	5.857			
(b) Source of Variation	DF	SS	MS	F	P
Between Groups	6	30.244	5.041	16.970	<0.001
Residual	21	6.238	0.297		
Total	27	36.481			
(c) Source of Variation	DF	SS	MS	F	P
Between Groups	6	2.810	0.468	3.645	0.012
Residual	21	2.698	0.128		
Total	27	5.508			

The different letters within the same column indicate a significant difference ($p<0.05$) between treatments. A (*Panthoea* sp.), B (*Klebsiella* sp.), C (*Bacillus* sp.), AC (*Panthoea* sp. + *Bacillus* sp.), ABC (*Panthoea* sp. + *Klebsiella* sp. + *Bacillus* sp.).

3.3.2. Growth Parameters Six Weeks After Planting

Six weeks after planting, no real differences can be observed between treatments regarding the number of leaves. However, significant differences are noted between treatments as far as concern the plant high and the stem base diameter. Same like three weeks after planting, the best record is obtained with *Bacillus* sp. in both plant high (40.50 cm) and stem base diameter (4.93 cm) among the inoculation treatments. In general, the stem base diameter varies with treatments, the highest performance obtained by the positive control (5.20 cm), followed by *Bacillus* sp., *Panthoea* sp. (4.37 cm) and the consortium (AC) of both *Bacillus* sp. and *Panthoea* sp. (4.58 cm). This shows the real aptitude of strains, either in single or in consortia to promote the plant growth regarding the different parameters.

Table 3. *Number of leaves, plant high and stem base diameter of soybean plants six weeks after planting.*

Treatment	(a) Number of leaves	Standard deviation	(b) Plant high (cm)	Standard deviation	(c) Stem base diameter (cm)	Standard deviation
C-	9 ab	0.50	32.33 a	3.09	3.69 ab	0.21
C+	10 ab	0.50	39.75 bc	2.75	5.20 e	0.28
A	10 ab	0.58	35.33 a	0.94	4.37 cd	0.45
B	9 ab	0.50	34.93 a	0.90	3.30 a	0.22
C	10 ab	1.00	40.50 c	0.35	4.93 d	0.25
AC	9 ab	0.50	36.21 ab	0.85	4.58 cd	0.19
ABC	8 a	0.50	35.61 ab	1.76	4.22 bc	0.14

(a) Source of Variation	DF	SS	MS	F	P
Between Groups	6	7.214	1.202	3.258	0.020
Residual	21	7.750	0.369		
Total	27	14.964			
(b) Source of Variation	DF	SS	MS	F	P
Between Groups	6	194.355	32.392	9.960	<0.001
Residual	21	68.299	3.252		
Total	27	262.653			
(c) Source of Variation	DF	SS	MS	F	P
Between Groups	6	10.658	1.776	25.403	<0.001
Residual	21	1.469	0.0699		
Total	27	12.127			

The different letters within the same column indicate a significant difference ($p<0.05$) between treatments. A (*Panthoea* sp.), B (*Klebsiella* sp.), C (*Bacillus* sp.), AC (*Panthoea* sp. + *Bacillus* sp.), ABC (*Panthoea* sp. + *Klebsiella* sp. + *Bacillus* sp.)

3.3.3. Root, Shoot and Total Plant Dry Weight Six Weeks After Planting

Significant differences ($p<0.05$) are observed within strains regarding the root, shoot and total plant dry matter. Concerning the root dry mass, only one inoculated treatment *Panthoea* sp. (2.89 g) is significantly different from the negative control C- (1.95 g), but the best score is performed by the positive control C+ (3.92 g) supplied with soluble phosphate (Table 4). Regarding the root dry matter, all inoculated treatments are significantly different from the negative control. The best record is obtained by the consortium of the three strains ABC (2.75 g) followed by *Bacillus* sp. (2.30 g), *Panthoea* sp. (2.24 g) and then *Klebsiella* sp. (1.79 g). As far as concern the total plant dry matter, the best record is still obtained with the positive control C+ (6.21 g). Among the inoculated treatments, *Panthoea* sp. (5.14 g) and the consortium of the three strains ABC (4.97 g) show the best performance, followed by *Bacillus* sp. (4.60 g) and *Klebsiella* sp. (4.00 g). This shows the real aptitude of strains, either in single or in consortia to promote the plant growth regarding the shoot, root and total plant dry masses.

Table 4. *Shoot dry weight, root dry weight and total dry weight of soybean plants six weeks after planting.*

Treatment	(a) Shoot dry weight (g)	Standard deviation	(b) Root dry weight (g)	Standard deviation	(c) Total plant dry weight (g)	Standard deviation
C-	1.95 a	0.04	1.32 a	0.24	3.27 a	0.27
C+	3.92 c	0.17	2.29 cd	0.11	6.21 e	0.19
A	2.89 b	0.11	2.24 cd	0.09	5.14 d	0.04
B	2.21 a	0.16	1.79 b	0.12	4.00 bc	0.22
C	2.30 a	0.18	2.30 d	0.11	4.60 cd	0.16
AC	2.25 a	0.28	1.43 ab	0.25	3.69 ab	0.52
ABC	2.23 a	0.19	2.75 e	0.21	4.97 d	0.30

(a) Source of Variation	DF	SS	MS	F	P
Between Groups	6	10.884	1.814	60.890	<0.001
Residual	21	0.626	0.0298		
Total	27	11.509			
(b) Source of Variation	DF	SS	MS	F	P
Between Groups	6	6.471	1.079	36.136	<0.001
Residual	21	0.627	0.0298		
Total	27	7.098			
(c) Source of Variation	DF	SS	MS	F	P
Between Groups	6	23.857	3.976	51.568	<0.001
Residual	21	1.619	0.0771		
Total	27	25.476			

The different letters within the same column indicate a significant difference ($p<0.05$) between treatments. A (*Panthoea* sp.), B (*Klebsiella* sp.), C (*Bacillus* sp.), AC (*Panthoea* sp. + *Bacillus* sp.), ABC (*Panthoea* sp. + *Klebsiella* sp. + *Bacillus* sp.)

3.3.4. *Effect (%) of the Different Treatments on Growth, Root, Shoot and Total Dry Weight of Soybean Six Weeks After Planting*

Based on the effect of inoculation on the different measured plant parameters compared to the negative treatment without inoculation, the different treatments can be classified as follow:

C+ > A = C = ABC > B = AC. In general, *Panthoea* sp. (35%) and *Bacillus* sp. (34%) better contribute to the soybean growth compare to non inoculated control supplied with non soluble Moroccan rock phosphate. The activity of *Klebsiella* sp. (13%) that is low in general seems to be stimulated when associated with the two other strains (33%).

Table 5. *Effect (%) of the treatments on growth, root, shoot and total dry weight of soybean plants six weeks after planting under pot grown conditions.*

Treatment	Number of leaves	Plant height	Stem base diameter	Root dry weight	Shoot dry weight	Total dry weight	Average
C+	11	23	41	101	73	90	57
A	9	9	18	48	70	57	35
B	0	8	0	13	35	22	13
C	9	25	34	18	74	41	34
AC	0	12	24	15	9	13	12
ABC	0	10	14	14	108	52	33

The effect (%) is calculated according to the formula: $Effect = \frac{(T-C-)}{C-}$ x 100. Where, T is a given treatment and C- the negative control. A (*Panthoea* sp.), B (*Klebsiella* sp.), C (*Bacillus* sp.), AC (*Panthoea* sp. + *Bacillus* sp.), ABC (*Panthoea* sp. + *Klebsiella* sp. + *Bacillus* sp.)

4. Discussion

Several strains of bacterial and fungal species have been described and investigated in detail for their rock phosphate solubilising capabilities [21, 22]. Bacterial strains exhibiting P solubilising activity were detected on plates by the formation of halo/yellow zone around their colonies. This reaction is used to assess the P solubilisation activity of strains with the SI value as an indicator for the strain efficiency [23]. Although *Klebsiella* sp. did not showed any visible halo zone on plates amended with Moroccan and Mexican rock phosphates, it could conversely mobilized important amounts of phosphorus from these rock phosphate sources and could be considered as rock phosphate solubilisers. These contradictory results between plate halo detection and phosphate solubilisation in liquid cultures found earlier by Deubel and Merbach [24] and Fankem et al. [25] indicates that liquid culture should be associated to halo zone for rock phosphate solubilisers characterization. However, the plate method is still a feasible way to pre-screen the isolates that possess phosphate solubilising ability. While clear correlation was found between solubilisation ability on plate and in liquid culture with *Panthoea* sp., no correlation was observed with *Klebsiella* sp. and *Bacillus* sp. between solubilisation ability on plate and in liquid culture as mentioned in the previous investigations by Azziz et al. [26] and Fernández et al. [27]. Although all the strains showed good solubilisation in liquid cultures amended with the different rock phosphates, the amount of the solved P varied with the strain as well as the rock phosphate origin. Solubilisation of phosphate is commonly accompanied by a remarkable drop in pH [22, 28, 29], though no strong correlation was observed within the drop of pH and the amount of the solved P. Regarding the efficiency of strains on plates and in liquid cultures, *Panthoea* sp. shows the highest performance, followed by *Bacillus* sp. This performance was maintained on soybean growth regarding the different parameters measured, with a significant increase of 35% compared to the negative control. Although the activity of *Klebsiella* sp. (13%) was not strong enough in this study, that strain altogether with *Pseudomonas*, *Enterobacter* and *Microbacterium* have been reported in several papers [30] as plant promoting strains. This value is therefore in agreement with Tandon [31] who reported that inoculation with phosphate solubilising microorganisms generally induces a growth increase of approximately 10-15%. However, values greater than that range have been reported by Abou-el-Seoud and Abdel-Megeed [32] and Fankem et al. [33] who respectively found an increase of 26% and 36%, with corn, and in this case 35% with soybean. Increasing the bioavailability of P in the soils with combined inoculation and rock materials has been reported by many researchers [34, 35, 36], which may lead to increased P uptake and plant growth [37, 38, 39]. In general, while *Bacillus* and *Pseudomonas* [40], *Aspergillus* and *Penicillium* [41] are considered as important genera of mineral phosphate solubilising bacteria and fungi respectively, only few reports indicated both *Panthoea sp.* and *Klebsiella* sp. as mineral phosphate solubilising bacteria and rare reports indicating their rock phosphate solubilising ability. The present study highlights the aptitude of *Panthoea sp.*, *Klebsiella* sp. and *Bacillus* sp. in single or in consortia in solubilising rock phosphates of different origins as well their aptitude in promoting the soybean growth. This came to confirm the aptitude of those strains in solubilising sparingly soluble inorganic/organic phosphates including $Ca_3(PO_4)_2$, $AlPO_4$, $FePO_4$ and sodium-phytate, and would be recognized as inorganic/organic/rock phosphate solubilisers. Growth enhancement by bacteria may relate to their ability to produce extensive root length [42] and can also improve root development and increase the rate of water and mineral uptake [43, 44]. This increase in growth may be attributed to auxin production [45], ACC-deaminase activity [46, 47] and production of organic acids [33] or phosphatases [48] to solubilise/mineralize P, thereby increasing phosphate nutrition of inoculated plants. Therefore, the positive effect on growth with non-soluble P may result from the synergic combination of both bacterial capacities for IAA production and P mobilisation. The use of phosphate-solubilising bacteria as inoculants simultaneously increase phosphorus uptake by the plant and crop yield [49]. Moreover, according to Amellal et al. [50] *Panthoea agglomerans* NAS206 can play an important role in the regulation of the water content (excess or deficit) of the rhizosphere of wheat by improving soil aggregation. This may hence justified its efficiency on the performed tests.

5. Conclusion

Natural rock phosphates have been recognised as a valuable alternative for P fertilisers, the modern agriculture being dependent on phosphorus derived from phosphate rock. However, the direct application of low-grade rock phosphate as a P source in soils is of little importance and need an addition of inoculums of phosphate solubilising microorganisms to improve the rock phosphate efficiency as a phosphorus source. The organisms possessing phosphate solubilising ability can convert the insoluble phosphatic compounds into soluble forms into soil and make it available to the crops. While *Bacillus* and *Pseudomonas* are generally considered as important genera of mineral phosphate solubilising bacteria, only few reports indicated both *Panthoea sp.* and *Klebsiella* sp. as mineral phosphate solubilising bacteria, and rare reports their rock phosphate solubilising ability. The present study highlights the aptitude of *Panthoea sp.*, *Klebsiella* sp. and *Bacillus* sp. in solubilising rock phosphates of different origins as well their aptitude in promoting the soybean growth. These bacterial strains can be therefore recognized as inorganic/organic/rock phosphate solubilisers.

Acknowledgments

The authors would like to thank the Rothamsted International, African Fellow Programme, London, UK for providing the fund and the James Hutton Institute, Dundee, Scotland where part of this work was done.

References

[1] M.S. Khan, A. Zaidi, P.A. Wani, "Role of phosphate solubilising microorganisms in sustainable agriculture". In Sustainable Agriculture. E. Lictfouse et al. (Eds) Springer, 2009, p 552, DOI: 10.1007/978-90-481-2666-8_34.

[2] M.S. Khan, A. Zaidi, M. Ahemad, M. Oves, P.A. Wani, Plant growth promotion by phosphate solubilizing fungi – current perspective, vol. 56. Arch Agron Soil Sci., 2010, pp 73–98.

[3] A. Gaume, "Low P tolerance of various maize cultivars: the contribution of the root exudation". Ph.D. dissertation, Swiss Federal institute of Technology, Zurich, Switzerland, 2000.

[4] A. Zaidi, M.S. Khan, M. Ahmad, M. Oves and P.A. Wani, "Recent advances in plant growth promotion by phosphate solubilizing microbes". In Microbial strategies for crop improvement. M.S. Khan, A. Zaidi and J. Mussarat (Eds.). Springer Verlag Berlin Heidelberg, 2009, pp. 23-132.

[5] W.I.A. Saber, K.M. Ghanem and M.S. EL-Hersh. Rock phosphate solubilisation by two isolates of *Aspergillus niger* and *Penicillium* sp. and their promotion to mung bean plants, vol. 4(7). Research Journal of Microbiology, 2009, pp. 235-250.

[6] M. Vassileva, N. Vassilev and R. Azcon, Rock phosphate solubilisation by *Aspergillus niger* on olive cake-based medium and its further application in a soil-plant system, vol. 14. World J. Microbio. Biotechnol., 1998, pp. 281-284.

[7] M.S. Reddy, S. Kumar, K. Babita and M.S. Reddy, Biosolubilisation of poorly soluble rock phosphates by *Aspergillus tubingensis* and *Aspergillus niger*, vol. 84. Bioresour. Technol., 2002, pp. 187-189.

[8] [8] C.C. Chuang, Y.L. Kuo, C.C. Chao and W.L. Chao, Solubilization of inorganic phosphates and plant growth promotion by *Aspergillus niger*, vol. 43. Biol. Fert. Soils, 2007, pp. 575-584.

[9] D. Cordell, J-O. Drangert, S. White, The story of phosphorus: Global food security and food for thought, vol. 19. Global Environmental Change, 2009, pp. 292–305.

[10] H.T. Rogers, R.W. Pearson, L.E. Ensminger, Comparative efficiency of various phosphate fertilizers, vol. 4. Agronomy, 1953, pp. 189-242.

[11] G.W. Cooke, The value of rock phosphate for direct application, vol. 24. Empire Journal of Experimental Agriculture, 1956, pp. 295-306.

[12] A.C. Gaur, & K.P. Ostwall, Influence of phosphate dissolving bacilli on yield and phosphate uptake of wheat crop, vol. 10. Indian Journal of Experimental Biology, 1972, pp. 393-394.

[13] A.H. Babana, H. Antoun, Effect of Tilemsi phosphate rock solubilizing microorganisms on phosphorus-uptake and yield of field grown wheat in Mali, vol. 287. Plant and Soil, 2006, pp. 51-58.

[14] H. Hamdali, K. Moursalou, G. Tchangbedji, Y. Ouhdouch, H. Mohamed, Isolation and characterization of rock phosphate solubilizing actinobacteria from a Togolese phosphate mine, vol. 11(2). Afr. J. Biotechnol., 2012, pp. 312-320.

[15] Y. Bashan, A.A. Kamnev, L.E. De Bashan, A proposal for isolating and testing phosphate-solubilizing bacteria that enhance plant growth, vol. 49. Biol Fertil Soils, 2013, pp. 1–2.

[16] M.M. Collavino, P.A. Sansberro, L.A. Mroginski, O.M. Aguilar, Comparison of in vitro solubilization activity of diverse phosphate-solubilizing bacteria native to acid soil and their ability to promote *Phaseolus vulgaris* growth, vol. 46. Biol Fertil Soils, 2010, pp. 727–738.

[17] C.S. Nautiyal, An efficient microbiological growth medium for screening phosphate solubilizing microorganisms, vol. 170. FEMS Microbiology Letters, 1999, pp. 265-270.

[18] R.S. Gadagi, & T. Sa, New Isolation Method for Microorganisms Solubilizing Iron and Aluminium Phosphates Using Dyes, vol. 48(4). Soil Science and Plant Nutrition, 2002, pp. 615-618.

[19] M.A. Qureshi, Z.A. Ahmad, N. Akhtar, A. Iqbal, F. Mujeeb, M.A. Shakir, Role of phosphate solubilizing bacteria (PSB) in enhancing P availability and promoting cotton growth, vol. 22. The journal of animal & plant sciences, 2012, pp. 204-210.

[20] J. Murphy, & J.P. Riley, A modified single solution method for the determination of phosphate in natural waters, vol. 27. Analytica Chimica Acta, 1962, pp. 31-36.

[21] A.H. Babana, A.H. Dicko, K. Maïga & D. Traoré, Characterization of rock phosphate-solubilizing microorganisms isolated from wheat (*Triticum aestivum* L.) rhizosphere in Mali, vol. 1(1). Journal of Microbiology and Microbial Research, 2013, pp. 1-6.

[22] H. Fankem, , L. Ngo Nkot, G. Nguesseu Njanjouo, G.V. Tchuisseu Tchakounte, A.V. Tchiazé Ifoué, D. Nwaga, Rock Phosphate Solubilisation by Strains of *Penicillium* spp. Isolated from Farm and Forest Soils of three Agro Ecological Zones of Cameroon, vol. 2(2). American Journal of Agriculture and Forestry, 2014, pp. 25-32.

[23] L.A. Fernandez, P. Zalba, M.A. Gomez, M.A. Sagardoy, Phosphate solubilization activity of bacterial strains in soil and their effect on soybean growth under greenhouse conditions, vol. 43. Biol. Fert. Soils., 2007, pp. 805-809.

[24] A. Deubel & W. Merbach, "Influence of microorganisms on phosphorus bioavailability in soils". In Microorganisms in soils: roles in genesis and functions. F. Buscot & A. Varma (Eds). Springer, Berlin Heidelberg, 2005, pp. 177–191.

[25] H. Fankem, M. Abba, L. Ngo Nkot, A. Deubel, W. Merbach, F-X. Etoa & D. Nwaga, "Selecting indigenous P solubilizing bacteria for cowpea and millet improvement in nutrient deficient acidic soils of southern Cameroon". In Innovations for a green revolution in Africa: Exploring the Scientific Facts. A. Bationo, B. Waswa, J. Okeyo, F. Maina (Eds), Springer Science+Business Media B.V. 2011, pp. 391-398.

[26] G. Azziz, N. Bajsa, T. Haghjou, C. Taulé, Á. Valverde, J.M. Igual, A. Arias, Abundance, diversity and prospecting of culturable phosphate solubilizing bacteria on soils under crop–pasture rotations in a no-tillage regime in Uruguay, vol. 61. Appl. Soil Ecol., 2012, pp. 320–6.

[27] L. Fernández, B. Agaras, P. Zalba, L.G. Wall, C. Valverde, *Pseudomonas* spp. isolates with high phosphate-mobilizing potential and root colonization properties from agricultural bulk soils under no-till management, vol. 48(7). Biol. Fert. Soils, 2012, pp. 763–73.

[28] J. Kloeppe, R. Rodriguez-Kabana, A. Zehnder, J. Murphy, E. Sikora, C. Fernandez, Plant root-bacterial interactions in biological control of soil borne diseases and potential extension to systemic and foliar diseases, vol. 28(1). Australas Plant Pathol., 1999, pp. 21–6.

[29] N. Gupta, S. Jyotasanmayee, P. Reena and Dipikakerkatta, Solubilization of tricalcium phosphate and rock phosphate by microbes isolated from chromite, iron and manganese mines, vol. 66(2). Acta Bot. Croat., 2007, pp. 197–204.

[30] H. Rodriguez & R. Fraga, Phosphate solubilizing bacteria and their role in plant growth promotion, vol. 17. Biotechnol. Adv., 1999, pp. 319–39.

[31] H.L. Tandon, "Phosphorus Research and Production in India". Fertilizer Development and Consultation Organisation New Delhi, 1987, p. 160.

[32] Abou-el-Seoud and A. Abdel-Megeed, Impact of rock materials and biofertilizations on P and K availability for maize (*Zea Mays*) under calcareous soil conditions, vol. 19. Saudi Journal of Biological Sciences, 2012, pp. 55–63.

[33] H. Fankem, L. Ngo Nkot, A. Deubel, J. Quinn, W. Merbach, F-X. Etoa and D. Nwaga, Solubilization of inorganic phosphates and plant growth promotion by strains of *Pseudomonas fluorescens* isolated from acidic soils of Cameroon, vol. 2. African Journal of Microbiology Research, 2008, pp. 171-178.

[34] Q.M. Lin, Z.H. Rao, Y.X. Sun, J. Yao, L.J. Xing, Identification and practical application of silicate-dissolving bacteria, vol. 1. Agric. Sci. China, 2002, pp. 81–85.

[35] H.S. Han & K.D. Lee, Phosphate and potassium solubilizing bacteria effect on mineral uptake, soil availability and growth of eggplant, vol. 1(2). Res. J. Agric. Biol. Sci., 2005, pp. 176–180.

[36] P. Marschner, "The role of rhizosphere microorganisms in relation to P uptake by plants". In The Ecophysiology of Plant–Phosphorus Interactions, Plant Ecophysio. Series. P.J. White & J.P. Hammond (Eds). Springer, Heidelberg, 2009, pp. 165–176.

[37] H.S. Han, E. Supanjani, K.D. Lee, Effect of co-inoculation with phosphate and potassium solubilizing bacteria on mineral uptake and growth of pepper and cucumber, vol. 52(3). Plant Soil Environ., 2006, pp. 130–136.

[38] M.A. Jorquera, M.T. Hernandez, Z. Rengel, P. Marschner, M.L. Mora, Isolation of culturable phosphobacteria with both phytate mineralization and phosphate-solubilization activity from the rhizosphere of plants grown in a volcanic soil, vol. 44. Biol. Fertil. Soils, 2008, pp. 1025–1034.

[39] S.J. Sabannavar & H.C. Lakshman, Effect of rock phosphate solubilization using mycorrhizal fungi and phosphobacteria on two high yielding varieties of *Sesamum indicum* L., vol. 5(4). World J. Agric. Sci., 2009, pp. 470–479.

[40] P. Illmer, and F. Schinner, Solubilization of inorganic phosphates by microorganisms isolated from forest soil, vol. 24. Soil Biol Biochem., 1992, pp. 389–95.

[41] M.R. Motsara, P.B. Bhattacharyya, B. Srivastava, "Biofertilizers: their description and characteristics". In Biofertilizer Technology, Marketing and Usage. A Sourcebook-cum- Glossary, Fertilizer Development and Consultation Organisation, New Delhi, India, 1995, pp. 9–18.

[42] X.F. Sheng & W.Y. Huang, Physiological characteristics of strain NBT of silicate bacterium, vol. 38. Acta Pedol. Sin, 2001, pp. 569–574.

[43] M. Alexander, "Introduction to Soil Microbiology". Second ed. John Wiley & Sons Inc., New York. 1997.

[44] K.M. Saghir, A. Zaidi, P.A. Wani, Role of phosphate solubilizing microorganisms in sustainable agriculture: a review, vol. 27. Agron. Sustain. Dev., 2007, pp. 29–43.

[45] P. Gyaneshwar, G. Naresh Kumar, L.J. Parekh, P.S. Poole, Role of soil microorganisms in improving P nutrition of plants, vol. 245. Plant and Soil, 2002, pp. 83-93.

[46] P.R. Naik, G. Raman, K.B. Narayanan & N. Sakthivel, Assessment of genetic and functional diversity of phosphate solubilizing fluorescent pseudomonads strain from rhizospheric soil, vol. 8. BMC Microbiology, 2008, p. 230.

[47] M. Jayadi, B. Baharuddin & Ibrahim, In vitro selection of rock phosphate solubility by microorganism from Ultisols in South Sulawesi, Indonesia, vol. 1(4). American Journal of Agriculture and Forestry, 2013, pp. 68-73.

[48] R. Chabot, H. Antoun, M.P. Cescas, Growth promotion of maize and lettuce by phosphate-solubilizing *Rhizobium leguminosarum* biovar *phaseoli*, vol. 184. Plant and Soil, 1996, pp. 311-321.

[49] S. Mehta & S.C. Nautiyal, An efficient method for qualitative screening of phosphate solubilizing bacteria, vol. 43. Current Microbiol., 2001, pp. 51-56.

[50] N. Amellal, G. Burtin, F. Bartoli, and T. Heulin, Colonization of Wheat Roots by an Exopolysaccharide-producing Pantoea agglomerans strain and its effect on rhizosphere soil aggregation. Applied and Environmental Microbiology, 1998, pp. 3740–3747.

Distribution and morphometric characters of the Mediterranean brown ray, *Raja miraletus* (Chondrichthyes: Rajidae) in the Gulf of Gabes (Tunisia, Central Mediterranean)

Hasna Kadri[1,2*], Sondes Marouani[1,2], Mohamed Nejmeddine Bradai[1], Abderrahmen Bouaïn[2], Eric Morize[3]

[1]Institut National des Sciences et Technologies de la Mer, Centre de Sfax, B.P. 1035, Sfax 3018, Tunisie
[2]Faculté des Sciences de Sfax, B.P. 802, Sfax 3028, Tunisie
[3]Lemar/ IRD 230 Broennou, 29870 Landéda, France

Email address:
hasnakadri@yahoo.fr (H. KADRI)

Abstract: To determine some morphometric characters of the Mediterranean brown ray Raja miraletus Linnaeus, 1758 in the Gulf of Gabes (Tunisia), samples were analysed and described on data from the commercial trawl at monthly intervals between December 2006 and January 2007. A total of 501 male and 629 female specimens were collected during the sampling period. The largest (in terms of Total length: TL) female and male were 58 cm and 56 cm, respectively. Some various morphmetric characters were measured. Relationships between the characters were defined separately for both sexe. Morphometric characters were strongly correlated with total length but the nature of allometry differed according to the characters.

Keywords: *Raja Miraletus*, Morphometric and Numeric Characters, Gulf of Gabes, Mediterranean Sea

1. Introduction

Skates (order Rajiformes) are one of the most speciose elasmobranch orders and include at least 27 genera and more than 245 species (Ebert and Compagno 2007). Skates are important elements of the marine biodiversity, but they are highly vulnerable to commercial Exploitation (Barbara 2010). The brown ray, *Raja miraletus* Linnaeus, 1758 is a medium-sized skate inhabiting the Mediterranean Sea and the western African coast (Séret and Opic 1990; Serena et al. 2010). Despite the increasing fishing pressure, there is a paucity of information on the life history of *R. miraletus* resulting in an assessment of 'minor preoccupation' on IUCN Red List assessments for chondrichthyans in the Mediterranean Sea (Cavanagh and Gibson 2007).

Information on distribution species is reported for some Mediterranean areas (Capapé and Quignard 1974a; 1974b; 1975; Capapé and Azouz 1975; Abdel-Aziz 1992; and Relini et al. 1999) including the Central Adriatic Sea (Zupanovic 1961; Jardas 1973; Jardas 1981), but data from Southern Tunisia are scanty. Therefore, morphometric conversions are particularly helpful when, for example, a specimen is damaged, or when dealing with commercially preprocessed specimens, in which not all morphometric traits can be measured. As conversion factors differ between species, they may also serve as a tool for species identification, or for accurate identification of problematic specimens. This is particularly important at landing ports where fish need to be identified on site.

In the present paper some features of the brown ray biology in the Southern Adriatic are reported, in order to add information referring to the just mentioned basin. These data were compared with results from other Mediterranean marine areas, in order to highlight possible intraspecific variations of life parameters among different populations. The present work focuses on some morphometric characters.

The present study aimed to (i) estimate relationships between different body measurements to increase the information available for the brown ray in Tunisian waters, and (ii) investigate the ability of these measurements to discriminate between species to provide additional tools for assisting in the identification of *R.miraletus* on the Gulf of Gabes.

This study reported the first description of biometric relationships for *R. miraletus*, which would be useful for the sustainable conservation of this rear fishery in the Gulf of Gabes and also other countries.

2. Materials and Methods

2.1. Study Area

The Gabes region is located in southern Tunisia and in southern Mediterranean Sea. It extends along 750 km representing 58% of the Tunisian coast (Bradai et al. 1992). It ex- tends from the city of Chebba (35.3°N) in the north of the region to the Island of Djerba (33.8°N) in the south, a distance of 200 km. In breadth, it extends from 10°E to just under 11.2°E. The annual cycle of water temperature is very pronounced (13°C to 29°C) and resembles that of a lagoon. In summer, the particularly high temperatures and salinities (38 to 39) lead to the appearance of red algae. It was recorded with high concentrations in summer (47- 48‰) and often in winter (40- 42‰). The obtained data set was registered and maps of population distribution were drawn. The second important feature of the Gulf of Gabes is the amplitude of the high tide which attains 1.8 m, the highest in the Mediterranean.

2.2. Sex Ratio and biological sample

A total of 1130 *R. miraletus* specimens were collected monthly in the Gulf of Gabes between December 2006 and January 2007 from commercial bottom-trawl landings in the Gulf of Gabes and using a beam trawl lasted for 2 h, during both day and night. Specimens were captured using a 22-mm stretched-mesh size cod-end, operating at over 30m depth (Figure 1).

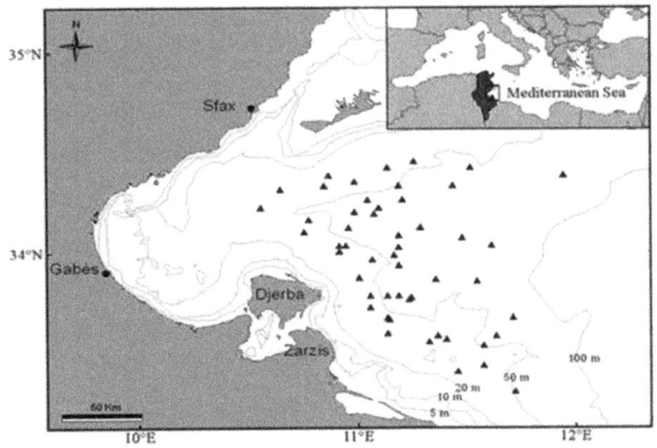

Figure 1. *Map of the study area, indicating the sampling location (Gulf of Gabes, Southern Tunisia, Central Mediterranean).*

The authority responsible for a national park gave us permission to collect these samples and for this work studies did not involve endangered or protected species.

The sex ratio according to fish size provide a useful tool to examine the biological characteristics of the fish species, such as sexual inversion, longevity in relation to sex, vulnerability to fishing gear and the spatial, seasonal and even daily distribution of species. With the numerical abundance by sex, the following ratio was done by computing (Anato 1999).

The sex ratio was expressed as the number of males/number of both sexes combined. A Chi- square test ($\chi 2$) was used to examine the homogeneity of the sex ratio. On the other hand, the size frequency was analyzed at a 5 cm interval standard length class using a histogram to determine the type of distribution, which characterizes the fish population. Length-frequency distributions of females and males were compared using the Kolmogorov–Smirnov two-sample test.

2.3. Morphometric characters measurement

The (TL) was measured to the nearest centimeter. The total body mass (TM) of each specimen was weighted to the nearest 10 and 0.01g respectively for large and small Specimens. Numeric characters tooth were also counted directly on specimens by making incisions at the jaw angles to expose the teeth, the tooth shape was noted (the number of the Tooth rows lower, the number of the tooth rows upper jaw, number of pectoral fin rays (NP), number of nictitating lamellae (NL), number of pseudo-branchial lamellae (NPL), the number of the trunchal vertebrae. The parameters of distribution (average) and parameters of dispersion (minimum, maximum, ecartype, interval of confidence) of this morphometric characters were calculated.

3. Results and Discussion

This study provides morphometric information for the Mediterranean brown ray *Raja miraletus* in southern Tunisian waters (Gulf of Gabes). A total of 1130 *Raja miraletus* collected 629 were females (13.5-56 cm TL; [8.5 to 35 cm DW] and 7-980g TM) whereas 501 were males (13.5-58cm TL; [9 to 37 cm DW]; 8-850g TM). Although males and females covered similar length ranges, the distribution of length within these ranges was significantly different (Kolmogorov-Smirnov test, D=0.088, n=1130, p=0.02) (Figure 2). The variation in the proportion of both sexes, depending on the size, showed a highly significant difference (χ^2 cal = 137.37, $P < 0.001$). There was no significant difference in the regression slopes between sexes for the TL and DW relationship (ANCOVA, F1, 1167=3.13, P□ 0.05) (Figure 3). The combined sex TL and DW (cm) relationship is described by (Figure 3). The TM-TL relationships was significantly different between sexes (ANCOVA, F1, 1151= 29.17, $P<0.05$) (Figure 4). Therefore, the nonlinear relationship of TL-TM is presented separately for each sex.

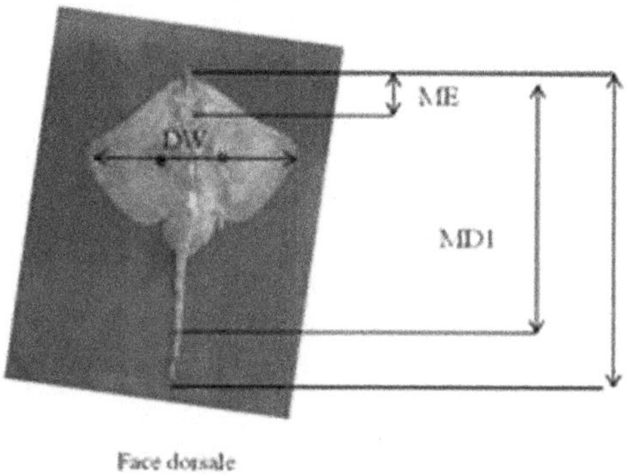

Figure 2. Measurements made on Raja miraletus.

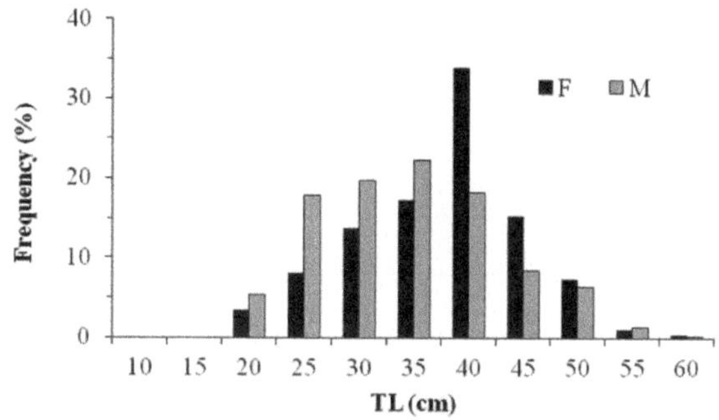

Figure 3. *Length-frequency distributions (5 cm length classes) of female and male Raja miraletus sampled in Gulf of Gabes (Tunisia).*

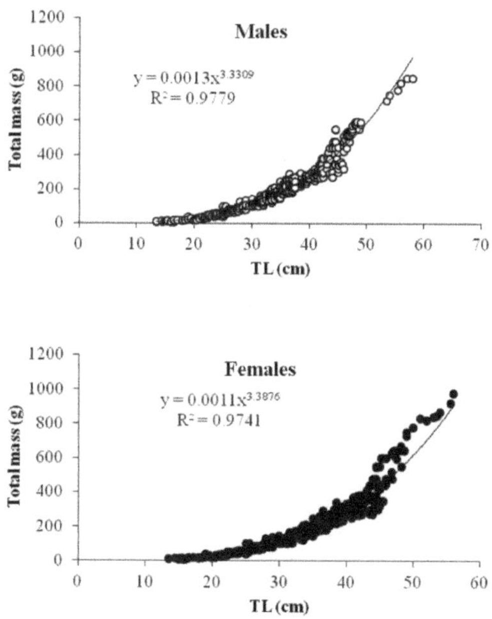

Figure 4. *Relationship between Total length (TL) and Total mass (TM) of Raja miraletus in the Gulf of Gabes (Tunisia). r^2 = Coefficient of determination.*

A study led by Ungaro in 2004 showed that in the Southern Adriatic Sea (Mediterranean basin), there are not significant differences in length-weight between sex. The same indication came from Stergiou and Moutopoulos (2001), while other authors report two different curves (male and female) (Jardas 1973; Capapé and Quignard 1974a; Relini et al. 1999). The length/width-weight relationships have many benefits for indicators of the condition and can be used to calculate biomass and to estimate the recovery of edible meat from crabs of various sizes (Lagler 1968). Previously, we demonstrated that differences in growth between sexes are a common feature in Raja species. Females typically attain larger size than males, but they grow more slowly (Kadri et al. 2012; 2013 a; 2013 b). It is important to mention that weight-size relationships can provide useful information about the increase in weight of a population and this parameter could also be important for comparative studies between populations (Mori et al. 1990).

This study provides morphometric information for the Mediterranean brown ray *Raja miraletus* in southern Tunisian waters (Gulf of Gabes). The number of females in all seasons was higher than males, there is statistical significance (Table 1). Our results are in accordance with Capapé et al. 2010 who reported a 1:0.95 ratio of male and female of *R. miraletus* on the coast of Senegal (eastern tropical Atlantic). Similarly, (Ezzat et al. 1987) shown higher numbers of females sampled in Mediterranean waters of Alexandria in each month. The sex ratio in Southern Adriatic basin is an expected result of 1:1.39 in favor of females (ungaro 2004).

Table 1. *Variation in the proportion of male and female Raja miraletus according to season in the Gulf of Gabes (Tunisia), and comparison through a t-Student test ($p<0.05$; bold values: statistically different). NS: not significant, SS: significant).*

	Summer	Autumn	Winter	Spring	Annual
Males	148	111	122	120	501
Females	154	158	173	144	629
Total	302	269	295	264	1130
%Males	27.25	35.35	48.72	46.08	37.02
%Females	50.99	58.74	58.64	54.55	55.66
X^2cal	0.12	8.21	8.82	2.18	11.91
P	0.73	0.004	0.003	0.14	<0.001
significance	NS	NS	NS	NS	SS

In our study, these individuals were measured 13.5 to 58 cm in Total length, 8.5 to 37 cm in disk width and 7 to 980 g in weight. In the Gulf of Tunis (Capapé and Quignard 1974a), *Raja miraletus* was measured 32 to 33 cm in the maximum disk width, 54 to 57 cm in maximum total length of males and females respectively, and 151 to 1338g in weight. The (TL) range was 165-510 mm and 150-495 mm TL for female and male individuals respectively, and the weight ranged between 20-750g in the Southern Adriatic basin (Ungaro 2004), but in Senegalese coast, the weight for *R.miraletus* was 1205 to 1336 g (Capapé et al. 2007).

The b value for the weight/ total length relationship for female *Raja miraletus* sampled in this study (3.38), and were higher than that for males (3.33). Ungaro 2004 reported a positive allometric growth pattern of *R. miraletus* (146 individuals) collected from the Southern Adriatic basin.

The b values vary according to species, sex, age, seasons

and feeding. Furthermore, many parameters were described to affect the b growth exponent such changes in physiological conditions or different amounts of available food life span or growth increment (Le Cren 1951; Ricker 1975).

Weight-size relationships can provide useful information about the increase in weight of a population and is also important for comparative studies between populations (Mori et al. 1990).

The maximum size for males and females in this work (65 and 80 cm TL, respectively) is greater than the previously recorded maximum size in other studies (Capapé and Quignard 1974a; Consalvo et al. 2010).

Morphometric characters were strongly positively correlated except for TL-LD and LT-MO and show negative allometric growth (figures 4 and 8). Other numeric counts characters are also presented (Table 2).

In this study, the counts of numeric caracters for *R.miraletus* were similar with results observed in the Gulf of Tunis (Capapé and Quignard 1974 a) (Table 3).

Table 2. *Morphometric characters of the Mediterranean brown ray (Raja miraletus) in the Gul of Gabes (Tunisia). number of nictitating lamellae (NL), number of pectoral fin (NP), number of pseudobranchial lamellae (NPL), number of the Tooth rows (NT) (upper jaw, lower jaw), minimum (Min), maximum (Max) and average Mean±SE.*

Parameters	NV	(NL)	(NP)	(NPL)	NT	
					upper jaw	lower jaw
Min	22	11	77	10	28	32
Max	26	13	84	15	44	44
Mean±SE	23.24±0.44	12.14± 0.14	78.96 ±0.34	12.28± 0.25	36.48± 0.95	37.47± 0.68

Table 3. *Comparison of numeric counts various of other studies in the Mediterranean.*

	Capapé and Quignard (1974) (Gulf of Tunis)	Quignard (1965) (Gulf of Tunis)	Present study (Gulf of Gabes)
Number of the trunchal vertebrae	24-26	23-25	22-26
Number of the nictitating lamellae (NL)	11-13		11-13
Number of the pectoral fin (NP)	76-80	76-79	77-84
Number of pseudobranchial lamellae (NPL)	12-14	10-12	10-15
Number of the upper jaw	38-48	41-49	28-44
Number of the lower jaw	40-50	40-43	32-44

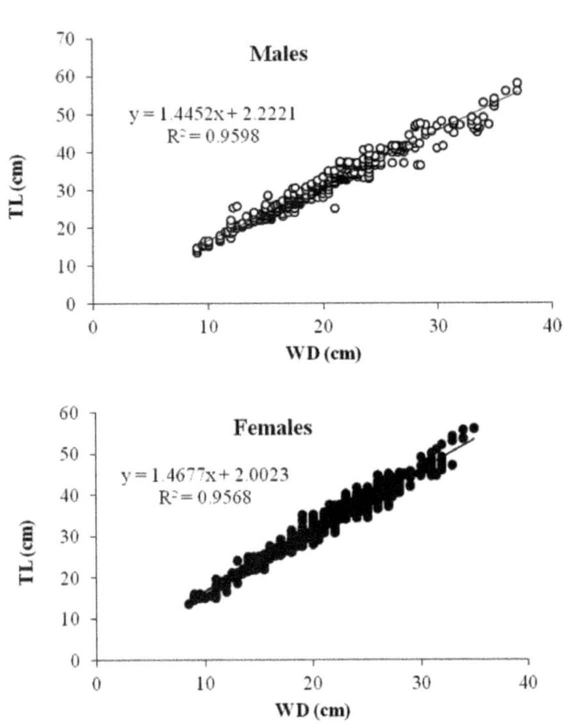

Figure 5. *Relationship between Total length (TL) and Disc width (DW) of Raja miraletus in the Gulf of Gabes (Tunisia). r^2 = Coefficient of determination.*

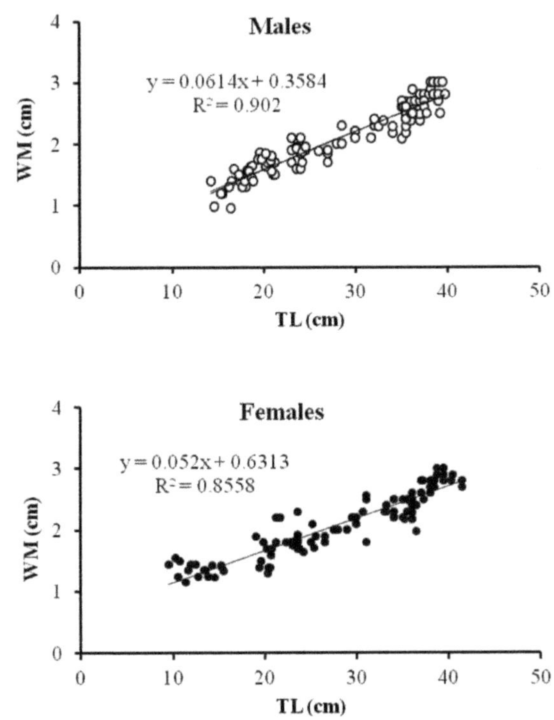

Figure 6. *Relationship between Total length (TL) and snout tip to eye (SE) of Raja miraletus in the Gulf of Gabes (Tunisia). r^2 = Coefficient of determination.*

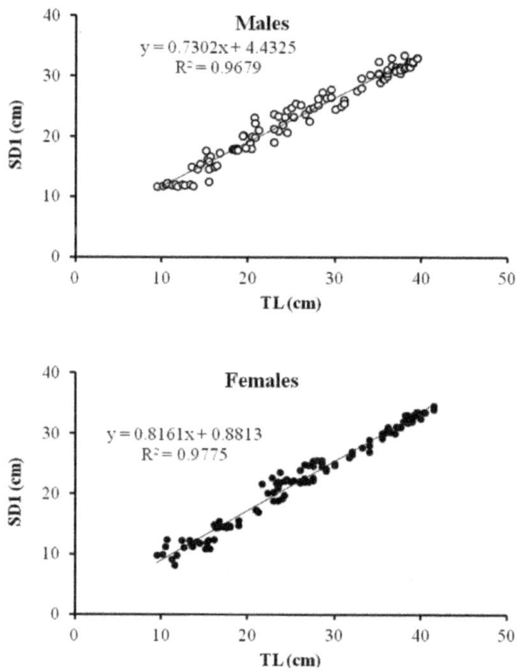

Figure 7. *Relationship between Total length (TL) and width of the mouth (WM) of Raja miraletus in the Gulf of Gabes (Tunisia). r^2 = Coefficient of determination.*

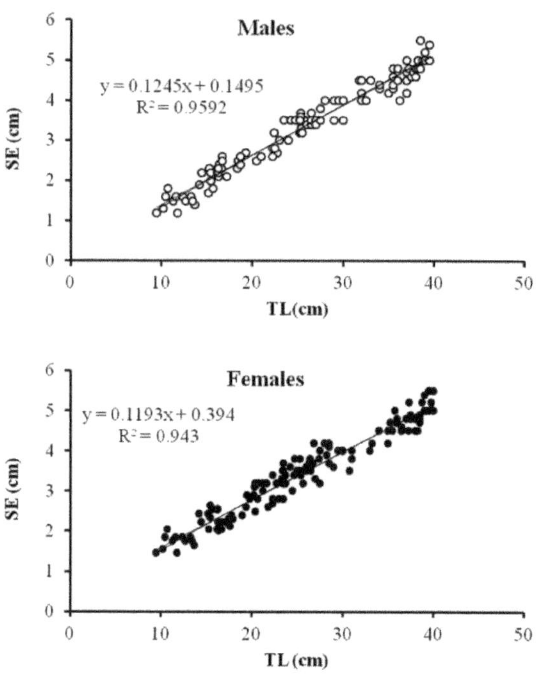

Figure 8. *Relationship between Total length (TL) and snout tip to first dorsal (SD1) of Raja miraletus in the Gulf of Gabes (Tunisia). r^2 = Coefficient of determination.*

In conclusion, this study shows that there was a strong relationship between DW, SE, WM, SD1, H and TL for the *R. miraletus* population in the Gulf of Gabes in southern Tunisia.

This study would be an effective tool for fishery biologists, managers and conservationists to initiate management strategies and regulations for the sustainable conservation of the remaining stocks of this species in the Gulf of Gabes (Southern Tunisia) ecosystem. In addition, results of this study provide invaluable information for the online FishBase database, as well as an important baseline for further studies.

Acknowledgements

This work was undertaken with the aid of funds provided by National Institute of Sciences and Technologies of the sea (INSTM, Sfax Center). The authors are also grateful to the anonymous referees whose suggestions and comments improved the submitted manuscript.

References

[1] Abdel-Aziz SH (1992). The use of vertebral rings of the brown ray Raja miraletus (Linnaeus 1758) off the Egyptian Mediterranean coast for estimation of age and growth. Cybium. 16: 121–132.

[2] Anato CB (1999). Les Sparidae des côtes béninoises: milieu de vie, pêche, presentation des espèces et biologie de Dentex angolensis Poll et Maul, 1953, Thèse de Doctorat d'Etat Sciences, Fac. Sci. 1060 Tunis, 277 pp.

[3] Babara Serra-Pereira, Ines Farias, Teresa Moura, Leonel Serrano Gordo, Miguel Santos and Ivone Figueiredo (2010). Morphometric ratios of six commercially landed species of skate from the Portuguese continental shelf, and their utility for identification ICES J. Mar.Sc. 67: doi:10.1093/icesjms/fsq056.

[4] Bradai MN, Ghorbel M, Bouain A and Abdelmouleh A (1992). Observations ichtyologiques effectuées dans la région du golfe de Gabès (Tunisie). Bull. Inst. Natl. Sci. Tech. Oceanogr. Peche Salammbo. 19: 57-65.

[5] Capapé and Azouz (1975). Etude du régime alimentaire de deux raies communes dans le golfe de Tunis: Raja miraletus Linné, 1758 et R. radula, Delaroche, 1809. Arch. Inst. Pasteur, Tunis. 52: 233–250.

[6] Capapé C and Quignard (1974a). Contribution à la biologie des Rajidae des côtes tunisiennes. I. Raja miraletus Linné, 1758: Répartition géographique et bathymétrique, sexualité, reproduction, fécondité. Arch. Inst. Pasteur, Tunis. 51(1-2): 39-60.

[7] Capapé C and Quignard JP (1974b). Contribution à la biologie des Rajidae des ôtes tunisiennes. XVI. Raja miraletus Linné, 1758: relations taille-poids du foie, poids du corps, poids des gonades, coefficient de condition, rapports hépato- et gonadosomatique. Ann. Inst. Michel Pacha. 10: 19-46.

[8] Capapé C and Quignard JP (1975). Essai d'évaluation de la fécondité chez les Sélaciens ovipares: cas de Raja miraletus Linné, 1758 et R. radula Delaroche, 1809 des côtes tunisiennes. Arch. Inst. Pasteur, Tunis. 52: 263-276.

[9] Capapé C, Diatta Y, Ndiaye P, Reynaud C and Oddone MC (2010). New biological data on the brown ray, Raja miraletus (Chondrichthyes: Rajidae), off the coast of Senegal (eastern tropical Atlantic).Cienc. Mar. 36(3): 301–309.

[10] Capapé C, Guélorget O, Siau Y, Vergne Y and Quignard JP (2007). Reproductive biology of the thornback ray Raja clavata L., 1758 (Chondrichthyes: Rajidae) from the coast of Languedoc (southern France, northern Mediterranean). Vie Milieu 57: 83–90.

[11] Cavangh RD and Gibson C (2007). Overview of the Conservation Status of Cartilaginous Fishes (Chondrichthyans) in the Mediterranean Sea. IUCN, Gland (Switzerland) and Malaga, 50p.

[12] Consalvo I, Iraci Sareri D, Bottaro M, Tudisco A, Cantone G and Vacchi M (2010). Diet composition of juveniles of rough ray Raja radula (Chondrichthyes: Rajidae) from the Ionian Sea. Ital. J. Zool. 77: 438–442.

[13] Ebert DA and Compagno LJV (2007). Biodiversity and systematic of skates (Chondrichthyes: Rajiformes: Rajoidei). Environ. Biol. Fish. 80: 111–124.

[14] Ezzat A, Abdel-Aziz SH, El-Charabawy MM and Hussein MO (1987). The food of Raja miraletus Linnaeus, 1758 in Mediterranean waters off Alexandria. Bull. Inst. Oceanogr. Fish. Cairo, 13: 59–74.

[15] Jardas I (1973). A contribution to our knowledge of the biology and ecology of thornback ray (Raja clavata L.) and brown ray (Raja miraletus L.) in the Adriatic. Acta Adriat. 15(3): 1-42.

[16] Jardas I (1981). Morphometric characteristics of the brown ray, Raja (Raja) miraletus L. (Pisces, Chondrichthyes) in the Adriatic Sea. Biosistematika. 7(2): 159-171.

[17] Kadri H, Marouani S, Saïdi B, Bradai MN, Ghorbel M, Bouaïn A and Eric M (2012). Age, growth and reproduction of Raja miraletus (Linnaeus, 1758) (Chondrichthyes: Rajidae) of the Gulf of Gabès (Tunisia, Central Mediterranean Sea). Mar. Biol. Res. 8: 309-317.

[18] Kadri H, Marouani S, Saïdi B, Bradai MN, Ghorbel M, Bouaïn A and Eric M (2013a). Age growth and reproductive biology of the rough skate, Raja radula (Chondrichthyes: Rajidae), off the Gulf of Gabes (southern Tunisia, central Mediterranean). Mar. Fres. Res http://dx.doi.org/10.1071/MF12218.

[19] Kadri H, Marouani S, Saïdi B, Bradai MN, Ghorbel M, Bouaïn A and Eric M (2013b). Age, growth and reproductive biology of the rough skate, Raja clavata (Chondrichthyes: Rajidae), off the Gulf of Gabes (southern Tunisia, central Mediterranean). Mar. Biol. Res. http://dx.doi.org/10.1080/17451000.2013.797584.

[20] Lagler KF (1968). Capture, Sampling and Examination of Fishes, in Ricker, eds, Methods for Assessment of Fish Production in Fresh Waters, IBP Handbook No:3, Blackwell Scientific Publication. Oxford.

[21] Le Cren ED (1951). The Length-Weight Relationships and Seasonal Cycle in Gonad Weight and Condition in Perch (Perca fluviatilis). J. Anim. Ecol. 20: 210-219.

[22] Mori M, Manconi R and Fanciulli G (1990). Notes on the reproductive biology of Carcinus aestuarii Nardo (Crustacea, Decapoda) from the lagoon of San Teodoro (Island of Sardinia, Italy). Riv. Idrobiol. 29: 763–774.

[23] Relini G, Bertrand J and ZAMBONI A (1999). Synthesis of the knowledge on bottom fishery resources in Central Mediterranean (Italy and Corsica). Bio l. Mar. Medit. 6(suppl. 1): 868.

[24] Ricker WE (1975). Computation and interpretation of biological statistics of fish populations. Bull. Fish. Res. Board. Can. 191: 1-382.

[25] Serena F, Mancusi C and Barone M (2010). Field identifcation guide to the skates (Rajidae) of the Mediterranean Sea. Guidelines for data collection and analysis. Biol. Mar.Medit. 17(Suppl. 2): 204 pages.

[26] Séret B and Opic, P (1990). Poissons de mer de l'ouest africain tropical. Init.-Doc., 49. Paris: ORSTOM, 416 pp.

[27] Stergiou KI and Moutopoulos DK (2001). A review of length-weight relationships of fishes from Greek marine waters. Naga, ICLARM Quat. 24(1-2): 23-39.

[28] Ungaro N (2004). Biological parameters of the brown ray Raja miraletus in the southern Adriatic Basin. Cybium. 28: 174–176.

[29] Zupanovic S. 1961. Contribution à la connaissance des poisons de l'Adriatique. Acta Adriatica 9:1-84.

11

Climate change and green technology for enduring agriculture

Mihir Kanti Sarkar[1], Md. Rajib Rahman[2], G. M. Faysal Ahmed[3], Ringko Kabiraj[4], ATM Jashim Uddin[1], Ananya Biswas[5], Md. Humayan Kabir[3]

[1]Bangladesh Agricultural Research Council (BARC), Farmgate, Dhaka- 1215, Bangladesh
[2]Soil Resource Development Institute (SRDI), Farmgate, Dhaka- 1215, Bangladesh
[3]Ministry of Agriculture, Dhaka- 1000, Bangladesh
[4]Bangladesh Computer Council (BCC), Sher-E-Bangla Nagar, Agargaon, Dhaka-1207, Bangladesh
[5]Faculty of Agriculture, Patuakhali Science and Technology University (PSTU), Dumki, Patuakhali- 8602, Bangladesh

Email address:

mihircse@gmail.com (M. K. Sarkar), rahman.rajib_pt@yahoo.com (R. Rahman), faysal_5408@yahoo.com (G. M. F. Ahmed), rinku.bcc@gmail.com (R. Kabiraj), atm_uddin@yahoo.co.uk (A. J. Uddin), ananya.pstu@gmail.com (A. Biswas), mh.kabir.moa01@gmail.com (H. Kabir)

Abstract: Climate is the primary determinant of agricultural productivity. Given the fundamental role of agriculture in human welfare, concern has been expressed by federal agencies and others regarding the potential effects of climate change on agricultural productivity. Climate change is expected to influence crop and livestock production, hydrologic balances, input supplies and other components of agricultural systems. However, the nature of these biophysical effects and the human responses to them are complex and uncertain. Technology can be termed as the application of knowledge for practical purposes. In this context, technology allows people to become more efficient or to do things that were not possible before. To benefit from technology, it needs to be successfully linked with country's overall development objectives and applied to solving socio-economic problems. In this paper we propose some practical applications of green technology in agricultural production to regain the loss due to climate change.

Keywords: Agriculture, Climate Change, Green Technology, Knowledge

1. Introduction

Environmental technology is the key to conservation ecology, a science of protecting biological diversity. Conservation ecology also termed as conservation biology refers to the application of science to the conservation of genes, populations, species, and ecosystems.

The wastewater and sewage disposal has been major threat to human health in developing Asia. The liquid waste discharged by domestic residences, commercial properties, industry or agriculture generates potential contaminants and concentrations that to some extent is minimized or recycled in the developed world. It necessitates the adoption of available and affordable technology for renewable energy including sunlight, wind, rain, and geothermal heat, which are naturally replenished. The technologies that are available are solar power, wind power, hydroelectricity/micro hydro, biomass and bio fuels for transport. The importance for considering primary energy use (includes both renewable and non-renewable energy contained in raw fuels) is because about 13 per cent of world's primary energy comes from renewable sources, most of which comes from traditional biomass like wood-burning. The understanding of these energies helps us devise policies for sustainable development, which includes environmental sustainability, economic sustainability and socio-political sustainability.

2. Friendly Technologies for Agriculture

Widespread environmental degradation, severe poverty around the globe and the burning concerns about achieving and maintaining good quality of life were the principal factors for taking interest in intergenerational equity, in relation to access to natural resources. As most good agricultural land has already been farmed and the regions have exceeded the safe limit, primarily in Asia, the natural resources availability for further farming expansion is practically exhausted.

Furthermore, the pressure from worldwide urbanization, manufacturing and population growth necessitates a renewed commitment to clean energy and environment solutions. The need is a balanced mix of alternative energies and the development of new technologies. The advances in solar, wind, bio-fuels and energy efficiency design has speeded up the development on technology-driven energy and cost efficiency mechanism to justify economic growth. The strength and quality of the execution of public policy determines the proper use of renewable energy by protecting environment. There is no controversy in developing agriculture to obtain higher yield and increased income of the farmer without affecting the environment. This approach, in fact, is sustainable agriculture. Sustainability has been given due consideration, because in Asian countries, land use has been intensified and problems such as, unplanned exploitation of the natural soil and water resources, have been realized. The skyrocketing costs of energy and agricultural inputs have reduced profitability, which has severely damaged the environment. This necessitates the proper assessment of the constraints and potentials of natural resources by examining policies of respective governments, and appropriateness of agro-technologies.

3. Feasible Technologies

3.1. Solar Photovoltaic Technology

Solar photovoltaic technology for instance, converts sunlight into electricity using semi conductor modules. Used generally for meeting lighting requirements, they can also be used for pumping water, refrigeration, communication, and charging batteries. Solar photovoltaic has application as green agricultural energy source for pumping water, street lighting in villages, lighting in rural houses and pest management.

3.2. Wind Energy

Wind energy is in a boom cycle. Overall, wind energy contributes only 1% of global electricity generation, but some countries and regions are already producing up to 20%. Its importance is increasing since it produces less air pollutants and greenhouse gases compared to other conventional energy sources.

3.3. Bio-Fuel

Bio fuel, as bio-ethanol and bio diesel, has the potential to assume an important portfolio in future energy platter. Food security concerns and risks to environment and biodiversity are parameters that necessarily need to be assessed, while analyzing sustainability linkage of agriculture with bio fuel use. Also, conversion of wasteland to farmland with some crop options can be viewed as positive impacts. This area is going to be the hot cake for future research.

Biogas is the product of anaerobic digestion of organic matter by methanogenic bacteria. Biogas qualifies on the merits that this technology utilizes organic agricultural waste and converts it to fuel and fertilizer. Direct impacts of biogas are fuel-wood savings agriculture residue management, livestock manure, and kerosene savings. Increases in soil fertility and crop production have also been observed.

3.4. Hydropower

Hydropower plants ranging from 500 kW to 25 MW capacities are conceived renewable. Generally, used in rural electrification, hydropower plants can take an equally important role in facilitating irrigation and value addition at source of agricultural products.

Some advanced technologies may also be used that are very much environment friendly for agriculture such as, Biomass, Improved Water Mills (an intermediate technology based on principle of traditional water mills), Geothermal technology, Bio Transgenics(BT) etc.

4. Agriculture Technology and Poverty Reduction

Agriculture plays a fundamental role in the economy of the least developed countries (LDCs), both in terms of size of an economic sector and element of a development strategy (Sadoulet and de Janvry 1995). It accounts for a large share of gross domestic product (GDP) (ranging from 30 to 60 per cent in about two thirds of them), employs a large proportion of the labor force (from 40 per cent to as much as 90 per cent in most cases), represents a major source of foreign exchange (from 25 per cent to as much as 95 percent in three quarters of the countries), supplies the bulk of basic food and provides subsistence and other income to more than half of the LDCs' population (FAO 2001). Since agriculture sector creates both forward and backward linkage through the income linkage, any shock in agriculture creates impact on the whole economy. Despite the importance of agriculture sector, the productivity of this sector is declining in underdeveloped countries. The per capita food production decreased by 0.8 per cent in 1980's and 0.1 per cent in 1990's (FAO, 2001). In the context of MDG to halve poverty by 2015, it is imperative to increase benefit from agriculture sector, as there are volumes of evidences that increased agriculture productivity is the key to reduce poverty.

Continued agricultural growth is a necessity, not an option, for most developing countries, and this growth must be achieved on a sustainable basis so as not to jeopardize the underlying natural resource base or to impose costly externalities on others (Hagos, 2003). Gallup et al. (1997) found that every 1% increase in per capita agricultural output led to a 1.61% increase in the incomes of the poorest 20% of the population. Thirtle et al. (2001) concluded from a major cross-country analysis that, on average, every 1% increase in agricultural yields reduced the number of people living on less than US$ 1 a day by 0.83%. Hence, development of agriculture sector, defined in terms of increased production with decreased average cost, becomes prerequisite for the overall development of an underdeveloped economy.

World Bank in its World Development Report 2007 has laid emphasis to the agriculture development for poverty alleviation and has advised to put this sector in the centre of development agenda, if world wants to halve poverty by 2015. In the mean time economists have taken it as challenge to sustain and expand agriculture sector's unique poverty-reducing power. It is logical to assume that agriculture should itself be sustainable to keep and expand this power.

Sustainable agriculture integrates three main goals-environmental health, economic profitability, and social and economic equity. Sustainability of agriculture can be maintained in various ways. The decision, about which particular method or combination of methods to choose, varies from country to country and region to region. Some of the common ways towards sustainable agriculture are:

a) Integrated Pest Management (IPM)
b) Rotational Grazing
c) Soil conservation
d) Water quality/wetlands
e) Cover crops
f) Crop/ landscape diversity
g) Nutrient management
h) Agro-forestry
i) Marketing

The demand for cereals in the developing world is expected to increase by 59% in next 25 years. The data shows, the growth rate in cereal yields has declined from an annual rate of 2.9% in 1967-82 to 1.8% in 1982-1994 (Janvry, Graff and et al., 2000). The need for alternative agriculture technology is thus realized to increase the yield and meet the demand of increasing population.

In spite of major institutional changes and dramatic achievements in food security, the developing countries in Asia still house bulk of the world's poor. The new challenge is to improve food and nutrition security at the household level, poverty reduction and sustainable agriculture. The concern is the failure of technological innovation such as Green Revolution to prevent a steady decline in the growth of yields.

It is important to understand the relationship between technological change and well being of smallholder farm households. Literature has shown a robust and positive effect of agricultural technology adoption on farm household well-being. Technological change can make both the direct and indirect impact on reducing poverty. Increased production, higher gross revenues from higher volumes of sales, reduced production cost, lower yield risks, and improved natural resource management means an increased welfare to the farmers, which is the direct impact of innovation. The indirect impact is the reduced food price for net buyers, employment and wage effects in agriculture and other sectors of the economy and agriculture's contribution from foreign exchange earnings to overall economic growth.

5. Selection of Technology

In general, the seven criteria have been proposed to judge the appropriateness of technology by Robert C. Wicklein in his paper entitled, "Design Criteria for Sustainable Development in Appropriate Technology: Technology as if People Matter". The issues have been analyzed in the present study by linking the strength or weaknesses of the stated technological applications to find out if they help in meeting the stated objectives. In other words, the current study attempts to use most of these criteria while assessing the technology for sustainable development.

a) System Independence: It is the ability of the technological device to stand alone for doing the required job. Whether the technology will require relatively more capital or labor will be analyzed to check system independence of the technology. Since, all three countries covered in the study are developing countries with high population pressure and unemployment, labor intensive technology will be system independence on the ground of cost. It is also kept in mind that required input for the technology is available or not.

b) Image of Modernity: People should perceive themselves as modern by adopting the technology. The message is people's realization that technological device can elevate the user's social status as well as need a basic human need. Image of modernity requires that the social status of people who adopt it either increases or remains unchanged. Although social status is contextual and is quite difficult to measure, it is common in all three countries that people who are member of social organizations have high social status. It has also been found that the people who are engaged in formal organization are more likely to adopt new technology. A study carried out by Yokoyama and Ali (2006) found that Malaysian rice farmers who are members of formal organization produce more rice. Similarly, Adhikari (2006) found that farmers with membership in social organization have more probability to adopt new technology. Similar evidences are available for India as well. Hence to check image of modernity, effort is made to analyze whether there is any probability that people with membership in formal organization refuse to adopt the technology.

c) Individual Technology vs. Collective Technology: It is the criteria to look into the societal/cultural standards in which the technology operates. In other words, it is the careful assessment of the technology that is based on group approach and becomes more system dependent. A society geared towards individual or single family unit will need more system independent technology. Collective technologies are more easily adopted as collective action reduces transaction cost. The success story of community forest in Nepal is an evidence to prove that collective technology relatively becomes more successful. Similar situation has been experienced in Malaysia and India.

d) Cost of Technology: Affordability of the technology is an important indicator for their wider use, since cost is the major factor in encouraging or discouraging the application of appropriate technology in developing economies. Although, the level of cost (high or low) is a relative concept, in all three countries labor is relatively cheaper than capital, and therefore, labor-intensive technologies are less costly.

e) Risk Factor: It is an important factor to find out how smoothly technology works in the local production system and the supportive system that explains to what degree is the technology system dependent or system independent. This indicates the need for understanding two types of risk- both the internal and external risk. Although analysis of risk is necessary before applying new technology, it is almost impossible to remove all risks. In this analysis risk, is considered on the basis of change in income level of people. If there is possibility of decreasing the income level of people then that technology will be more risky.

f) Evolutionary Capacity of Technology: If the chosen device is static, it will relatively reflect the short-lived solutions to a much larger problem. The technology, which supports the continuation of development by enhancing capability to expand, can be expected to compete at the regional, national and international level.

g) Single-Purpose and Multi-Purpose Technology: In contrast to single purpose technology, multipurpose technologies are the ones that furnish a variety of applications (e.g. a tiller that can be used for tilling the land, powering water pump, and drying rice).

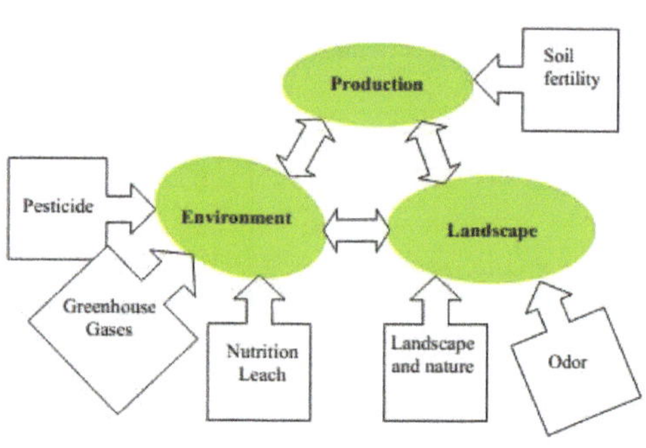

Production, Environment, Landscape and Nature Values – Constitute A Unity When Green Technologies Are Selected For Future Agricultural Systems

Figure 1. *Three Bottom Lines*

According to Borch et al. (2004) "The future environmentally friendly farming technologies are assessed and chosen in accordance with their contribution on three areas, where operating economical plans, green accounts and nature plans are treated equally." Components of Green Technology can be explained as shown in figure1.

6. Situation Analysis

The situation analyses of selected technologies in selected countries are made to find out how they are working, what are the costs and benefits and how the system can be sustained. In our analysis following factors are considered;

a) Current Performance: Performance of the technology is explained on the basis of their success and failure stories. The current performance has also been assessed on the basis of percentage share of population adopting particular technology.

b) Policy: Current government policy towards such technology has been identified. The review of selected regulatory measures in the chosen countries is made to find out their strengths and weaknesses to facilitate GT initiative and recommendations are made to develop favorable regulatory frameworks for the further development of GT.

c) Cost-Benefit: Cost-Benefit analysis of the technology has been carried out. A detailed cost-benefit analysis with valuation in quantitative technique is not possible due to time constraint. So components of costs and benefits along with risk factors and externalities have been identified.

d) Performance Indicator: The performance and sustainability of the technology has been reviewed. Therefore, the operational, technical, economical and environmental aspects of the selected technology have been considered.

7. Structure of Energy Consumption in Bangladesh

Dhaka has made the estimation of greenhouse emissions from the fuel combustion in Bangladesh. The United States Country Studies Programme (USCSP) also states that GHG inventory was constructed, the mitigation options of GHG emissions in energy sector were identified and vulnerability assessment for water resources and agriculture were carried out. The result shows that GHG emissions and per capita GHG emissions for Bangladesh is relatively very small in the region. One of the reasons for low GHG emissions is negligible percentage share of thermal (diesel and coal) power generation, the major contributor of GHG emissions, in Bangladesh. Thousand tons of oil equivalent.

Table 1. Structure of Energy consumption, Bangladesh

Energy Source	2005-06	2006-07	2007-08	2008-09	2009-10	2010-11	2011-12
Traditional	7,397	7,544	6,377	5,989	5,674	5,213	4,678
Fuel Wood	6,956	7,153	6,578	6,143	5,768	5,463	4,978
Agricultural Waste	465	480	435	413	345	375	351
Animal Dung	576	614	598	534	456	414	365
Commercial	1,654	1,567	1453	1378	1234	1134	1098
Coal	203	197	201	187	154	143	133
Petroleum	867	813	734	712	684	657	634
Electricity	185	197	187	175	163	152	123
Others	68	65	68	61	63	57	54
Total	18,371	18,630	16,631	15,592	14,541	13,608	12,414

8. The use of Information and Communication Technology (ICT) for Facilitating Green Technology

While information and communication technologies could add little direct value to agricultural yield, possibilities also are that they can be the centre of paradigm shifts.

Since, GT has emerged as a strong force to influence climate change and build globally based, environmentally sustainable solutions, organizations are using ICT to reduce carbon emissions and develop and support business models with a green focus. This relationship is considered while ICT is recommended to serve as a green technology.

It is believed that the use of ICT improves energy efficiency in the economy, starting with buildings, lighting and the power grid. ICT enables economy a green behavior. For example, the most advanced computer servers consume the same amount of energy as a standard light bulb; if widely used they can offer potential energy savings up to 70%. The European Commission presumes that real gains from green ICT will come from developing energy efficient ICT solutions that impact the other 98% of global emissions (http://europa.eu/rapid/pressReleasesAction.do?reference=IP/08/733). The intention of elaborating ICT in the present Green Technology is in this context.

Precision agriculture uses information and communication technologies (ICT) to cover the three aspects of production namely, a) for data collection of information input through options as Global Positioning System (GPS) satellite data, grid soil sampling, yield monitoring, remote sensing, etc; b) for data analysis or processing through Geographic Information System (GIS) and decision technologies as process models, artificial intelligence systems, and expert systems; c) and for application of information by farmers. Adjustments in volume and timing of fertilizer and pesticide inputs and limited input leakage to environment are expected in precision agriculture (Hrubovcak, 1999). ICT can also be useful in knowledge intensive farm management as IPM. Global Plant and Pest Information Services of FAO http://www.ecoport.org, United States' Department of Agriculture's National IPM Network at http://www.ipmcenters.org/index.cfm, North American Plant Protection Organization's Phytosanitary Alert System http://www.pestalert.org/index.cfm? NAPPO Language Pref='English' are some information portals for IPM. Opportunities to link niche products and producers to markets can also give impetus to low input indigenous agricultural production.

The Government of Nepal's Information and Technology Act 2004 supports ICT intervention for agricultural development and environmental management (ITP, 2004). Ninth Malaysian Plan commits to use of ICT for providing access to market and trading information. Supply and Demand Virtual Information (SDVI) system, e-trade, and Agribazaar portals, Medan InfoDesa facilities are expected to support agro marketing. (NMP, 2006).

9. Case Studies

9.1. NEPAL: Biogas

In the simple words biogas, a combustible gaseous mixture primarily of methane and carbon dioxide, is a product of anaerobic digestion of organic matter by methanogenic bacteria. When produced under controlled conditions inside a biogas plant, it can serve to meet energy requirements for cooking, lighting, refrigeration, electricity generation, and internal combustion engines. Slurry or digested effluent from biogas plant has high fertilizer value. Both energy and agriculture products of biogas technology and its minimal adverse environmental implications make it an appropriate green agricultural technology. Revived biogas sector of Bangladesh is the result of a consolidated effort of public and private stakeholders; and donor agencies. Establishment and institutionalization of the elements of biogas industry in Nepal further suggests an appropriate and inclusive scheme for renewable energy promotion to all strata of the society. Biogas Support Programme (Phase I-IV) has been instrumental in promotion of biogas in Bangladesh.

Biogas has links with agriculture, forest, environment and overall livelihood of the people. These links can be shown with the aid of following figure 2. Biogas can be deployed to increase incomes and save the environment and has great impact on the health and time saving for women and children.

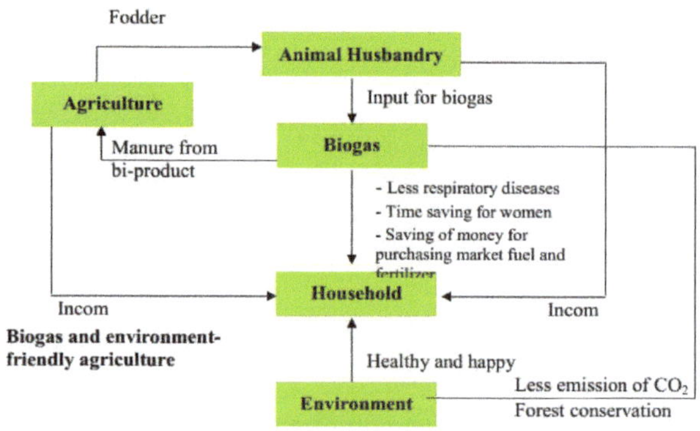

Figure2. *Sustainability and Biogas*

9.2. Agro-Forestry and Environment-Friendly Agriculture

Adoption of agro-forestry technology improves the various environmental components of agriculture. Some of them are resource conservation, carbon sequestration, biodiversity conservation, improvement in soil fertility and structure etc. Agro-forestry helps to improve the soil fertility through (Rai and Handa, NA):

a) More efficient nutrient cycling within the system and consequently more efficient utilization of nutrient that are either inherently present in the soil or externally applied;

b) Biological nitrogen fixation and solubilization of relatively unavailable nutrients;

c) Increase in the plant-cycling fraction of nutrients, with their resultant reduction beyond the nutrient-absorbing zone of the soil;

d) Complementary interaction between the component species of the system, resulting in a more efficient sharing of nutrient resources among the components;

e) Enhanced nutrient economy, because of different nutrient-absorbing zones of the root system of the component species;

f) Moderating effect of additional soil-organic matter on extreme soil reactions and consequently improved patterns of nutrient- release ability.

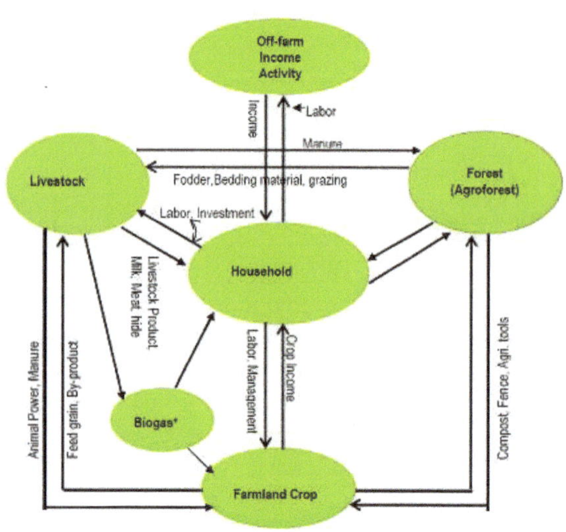

Figure 3. *Inter-Linkages of Agro forest, Agriculture, Household and Livestock*

There is indirect benefit of agro-forestry for environment friendly agriculture through the route of agro-forestry=> livestock=> ┌ biogas=> environment friendly agriculture.

10. Impact of Green Technology

There is an international call for all types and sizes of enterprises- government, non-government to not-for-profits to promote sustainable and environmentally responsible business practices. These enterprises can collectively jump on the "green" bandwagon. The message for increasing market opportunities for their products can be disseminated through environmental-awareness initiatives. The advantage of reducing the impact of climate change through the reduction of carbon footprint can be explained by creating awareness about new technologies. The awareness is helpful in understanding market trends, environmental threats and opportunities.

Recently, a new Green Technology Initiative (GTI) has been launched in London to help IT stakeholders to limit environmental impact of their IT infrastructure. In other words, it means limiting the carbon emissions associated with IT equipment. Reducing the carbon footprint is in line with British Governments' target of a 20% reduction in greenhouse gas emissions by 2020. There is an overwhelming interest in making the system greener, but majority lack information on how to do it?

Non-sustainable land-use practices and highly erosive monsoon rains are the contributing factors for low agricultural productivity and hence high incidence and severity of poverty in many parts of Asia. The discussion on sustainable agriculture becomes incomplete if we do not consider environmental degradation in the context of development. Sustainability should also be resource-saving, not one-sided resource usages. Rapid growth of urbanization and increasing demand has made the poor a victim of natural resource degradation largely in the form of shortage of fuel, fodder, and drinking water.

Since, energy improves productivity in major sectors of the economy that contributes to GDP, it can create employment and increase income. With backward agriculture and high poverty levels, the policy for making choices of the appropriate alternative energy technology is very important. The per capita energy consumption in developing countries stands at 400 units as against a minimum of 8000 units in developed world. The possibility of improving per capita energy consumption levels is through renewable energy technologies.

11. Discussion

An overview of the challenges and relevant options are briefly elaborated in this section. There is a consensus that although there is a strong business case for "sustainability", it is one of the most difficult and complex tasks to balance environmental and business concerns. In other words, there is a problem in balancing between the societal benefits of "green" practices and regulations with their costs. This situation is elaborated by Professor Jim Sweeny by saying "....in at least two crucial areas — energy management and materials management — a gap remains between societal benefits and private business benefits".

Storms, floods, droughts and wildfires are the results of climate change. We know the earth's average temperature has risen by 0.7 degree Celsius since pre-industrial times. As CO_2 emissions are the major factor for causing climate change, the major task is to adopt energy demand reduction measures and significant growth in renewable energy for reducing CO_2 emissions.

The use of technologies at the *status quo* level will create health hazards to the people and other life that inhabits the planet. The Inter-governmental Panel on Climate Change (IPCC) in its Fourth Assessment Report outlines human activity for creating scary effects of climate change to damage the world environment. This scenario necessitates the early success in deploying renewable carbon-free technologies by moving away from the coal and oil based economy to solar, wind, nuclear, bio-fuels, hydroelectricity, batteries, hybrid cars, etc. Efforts in recent years are found in building giant space mirrors to reflect solar radiation back into space for commercializing renewable energy.

As the adverse impact from climate change in ecosystem, agriculture, fresh water and human health is going to be the major threat, APCAEM countries need to immediately prepare themselves for need-specific alternative technologies. It is advisable to seeking technical and financial support from relevant UN agencies in capacity building programs for the mitigation of unforeseen eventualities.

Although rich countries can relatively take more benefits from renewable sources to meet their energy demands, it seems extremely difficult to derive adequate electricity or liquid fuels to sustain the current high per capita rates of consumption from renewable sources. This suggests the fact that sustainable future may be possible only when we get success in significantly reducing current material "living standards" and in gross economic activity. The limitations of conventional energy sources are as such that resources for power generation i.e., coal, gas, etc. are limited. Solar photovoltaic (PV) technology facilitates in converting the solar energy to electrical energy and is being used for applications such as lighting, water pumping, communication systems, remote area village's electrification, telecom applications, traffic signaling and railway signaling systems. This system has proved to be effective in rural electrification projects around the world. The fundamental problem of PV is the inadequate information on the potential and limitations of its application. Therefore, unless the potential contribution of PV to rural development and poverty alleviation is not properly examined, it will be difficult for the developing countries to make financial as well as political commitment.

12. Conclusion

The review shows when used correctly agro-environment-friendly technology has promoted sustainable agriculture growth and reduced widening rural-urban income disparities. The application of green technology is the answer for sustainable development but poor countries have not been able to use applicable technology largely because of their inability to afford to the available alternatives. Efforts are needed to align economic development policies with the goal of increasing the realization of human capabilities. GT in the reviewed countries has mostly minimized environmental hazards and made farmers more productive through enhanced efficiency. The lesson that has been drawn from this study is since distributive structure and property system especially in India and Nepal is not very satisfactory, care should be given to access land and credit; the access in health, the access in clean water and education; and above all the access in employment.

In assessing the applied technologies, it was realized that sustainable agriculture as defined and elaborated by FAO was resource conserving; environmentally non-degrading; technically appropriate; and economically and socially acceptable. However, as it is generally agreed in case we fail to precisely define the scope of work, the term "sustainability" becomes both ambitious and ambiguous.

The greater is the chance of the failure of macroeconomic policies; the worse is the case in income inequality. If the policies on food security and income generation fail to address poverty and malnutrition, the objective of elevating economic status of majority of the poor at the individual and household level through the application of green technology remains incomplete. The major concern of the United Nations through the use of agricultural technology is therefore, to guarantee the distribution of the benefits of increased agricultural productivity rather equitably.

Information technologies and their inter-linkages with agro-food sector are vital to understand the strength of pro-poor and environment-friendly agricultural technology. An open debate needs to be organized and facilitated at regular intervals on the need for investment in the technology, reason for their acceptance, strength and weaknesses for the adoption, and investigating the reason for their rejection.

In terms of the use of electronic equipment, the weak regulatory measures (regulation, directives, and standards) especially on the purchase, operation and disposal, has attracted interest on the global political agenda about the environmental impact of the electronics industry. Although the adoption of technology and regulatory compliance is on the average painful in the developing world, there are many valuable business benefits to be gained by opting for "greener" power. Examples from selected countries have revealed clearly that GT reduces operational cost.

Electronic communication networks as a powerful information infrastructure connects enterprises and knowledge bases. As the new information system has been developed as a parallel traditional physical world, the development of its infrastructure to connect indigenous practices into the modern system should be given top priority.

The present study shows adoption of Green Technologies have increased agricultural output without depleting presently available resources beyond the point of recovery. Though selection of technology is by default condition dependent one or the other renewable energy technology and green agriculture strategy can in all circumstances be pragmatic. Inter linkages of technologies with society should be deliberated in advance to access their roles in achieving the expected outcomes. The review on alternates shows an assessment should not occur in isolation. Instead, implications on income and opportunity creation; output, input, and ecological balance, gender equity, etc. should be considered with due attention. Green technologies assure potency for sustainable agricultural growth but significant effort will be required to substitute conventional practices.

Several studies have shown food production among small landholders is positively correlated to the increased use of industrial inputs and marketing opportunities for food crops. This demands a favorable macro environment that allows farmers to enhance the use of inputs. A concerted effort is, therefore, needed to create environment that helps farmers connect safely with market, making them able to avoid externality.

Acknowledgements

First of all we consign our limitless thanks to the Almighty God for giving us strength, endurance and ability to complete this research.

We would like to express our deepest gratitude to our colleagues for their direct and indirect help for completion the research on "Climate change and Green Technology for enduring Agriculture".

References

[1] J.W. McKinsey- Jr, R.E. Evenson, "Technology–Climate interactions in the green revolution in India" center discussion paper no.805, Economic growth center, Yale University.

[2] D. Schüller, "On the Optimal Allocation of Green Technology under Climate Change Agreements", FNI Report 8/2009, Fridtjof Nansen Institute, N-1326 Lysaker, Norway.

[3] P.F. Moore, B.J. Parai, "THE GREEN REVOLUTION", Technology in Contemporary Society.

[4] K. P. O. Brien, S. Franjevic, J. Jones, "Green Chemistry and Sustainable Agriculture -The Role of Bio pesticides" Advancing Green Chemistry September, 2009.

[5] Aikens, M. T., A. E. Havens et al. (1975). The Adoption of Innovations: the Neglected Role of Institutional Constraints. Department of Rural Sociology, Ohio State University.

[6] Amatya, S.M. and Newmann, S.M.(1993). Agroforestry in Nepal: research and practice. Agro forestry Systems, 21: 215--222, 1993.

[7] Bartlett, A. (2005). Farmer Field School to Promote Integrated Pest Management in Aisa: The FAO Experience available at http://www.communityipm.org/downloads.html.

[8] Wang, X. and Z. Feng. (1994. Atmospheric Carbon Sequestration through Agro forestry in China. Energy: 20(2).

[9] Regmi, B. (2003). Contribution of agroforestry for rural livelihoods: A case of Dhading District, Nepal. Paper Presented at The International Conference on Rural Livelihoods, Forests and Biodiversity 19-23 May 2003, Bonn, Germany.

[10] Pehu, E. and C. Ragasa, (2007).Agricultural Biotechnology Transgenics in Agriculture and their Implications for Developing Countries. Background Paper for World Development Report 2008, The World Bank.

[11] Neill, S. P. and D. R. Lee. (2001). Explaining the Adoption and Disadoption of Sustainable Agriculture: The Case of Cover Crops in Northern Hounduras. Economic Development and Cultural Change 49(4): 793-820.

12

Field study for the effectiveness of some plants leaf extracts against insect *Eutectona macheralis* or Teak skeletonizer in forest nursery of Indore, Madhya Pradesh

Hemant Pathak[1], Saurabh Maru[2], Satish Chandra Silawat[1]

[1]Forest Research and Extension Circle, Indore, Madhya Pradesh, India
[2]Laboratory of Pharmacology, Department of Pharmacy, Shri G S Institute of Technology and Science, Indore, Madhya Pradesh, India

Email address:
Hemantpathak777@gmail.com (H. Pathak)

Abstract: The naturally occurring pesticides thus appear to have a prominent role in the development of future economical pesticides not only for agricultural and forestry crop productivity but also for the safety of the environment and public health. The harmful environmental implications of the synthetic pesticides have compelled to search for some alternative methods. This lead to increased development of compounds based on the models of naturally occurring toxins of biological origin, having various biological activities. This includes plant extract, which are now known because they are environmentally harmless and host specific. These different concentrations of plant extracts *viz. Annona sqamosa* leaf extract, *Lantana camara* leaf extract, *Albizzia lebbeck* leaf extract were tested for feeding inhibition properties against *Eutectona macheralis* insect larvae on host plant *Tectona grandis.* Prepared leaf extracts were tested in field condition and the most effective concentration has been worked out.

Keywords: Eutectona Macheralis, Tectona Grandis and Leaf Extracts

---

1. Introduction

Unintentional poisonings kill an estimated 355 000 people globally each year (The world health report 2003). In developing countries – where two thirds of these deaths occur – such poisonings are associated strongly with excessive exposure to, and inappropriate use of, toxic chemicals. In many such settings, toxic chemicals may be emitted directly into soil, air, and water – from industrial processes, pulp and paper plants, tanning operations, mining, and unsustainable forms of agriculture – at levels or rates well in excess of those tolerable to human health.

Acute exposure to pesticides can lead to death or serious illness. Chronic pesticide exposure is most often a problem in the occupational setting, particularly among poor rural populations where men, women, and children all work and live in close proximity to fields and orchards where chemicals are applied and stored. Long-term exposure to pesticides can increase the risk of developmental and reproductive disorders, immune-system disruption, endocrine disruption, impaired nervous-system function, and development of certain cancers. Children are at higher risk from exposure than are adults (Human development report 1998).

The production of pesticides started in India in 1952 with the establishment of a plant for the production of BHC near Calcutta, and India is now the second largest manufacturer of pesticides in Asia after China and ranks twelfth globally [1]. There has been a steady growth in the production of technical grade pesticides in India, from 5,000 metric tons in 1958 to 102,240 metric tons in 1998. In 1996–97 the demand for pesticides in terms of value was estimated to be around Rs. 22 billion (USD 0.5 billion), which is about 2% of the total world market.

The pattern of pesticide usage in India is different from that for the world in general. As can be seen in Figure, in India 76% of the pesticide used is insecticide, as against 44% globally. The use of herbicides and fungicides is correspondingly less heavy. The main use of pesticides in India is for cotton crops (45%), followed by paddy and wheat [2].

Thus there is increasing need to find pesticides with selective toxicity to pest and safe to humans and animals.

The use of botanical insecticides/Biopesticides may serve as suitable alternative as they are relatively safe, degradable and economical (abundantly available in nature). Biopesticides are derived from natural materials such as animals, plants, bacteria, and certain minerals widely used for controlling insects and disease causing pathogens. Though several plants from different families have been reported for their pesticidal activity, only a few botanicals have moved from the laboratory to field use. In present study we are going to observe the effect of biopesticide formulations on feeding inhibition properties against *Eutectona macheralis* insect larvae on host plant *Tectona grandis.*

2. Methodology

2.1. Host Plant

Tectona grandis L.f. is a large deciduous tree with a rounded crown and, under favorable conditions, a tall clean cylindrical bole of more than 25 m. Leaves are broad, elliptical or obovate and usually 30 to 60 cm long. Over most of its range, teak occurs in moist and dry deciduous forests. Teak (*Tectona grandis* L.f.) is one of the world's premier hardwood timbers, rightly famous for its mellow color, fine grain and durability. It occurs naturally only in India, Myanmar and Thailand, and it is naturalized in Java, Indonesia. Although relatively unimportant in terms of the volume of world timber production, because of its strength and aesthetic qualities teak is the tropical hardwood most in demand for a specific market of "luxury" applications including furniture, shipbuilding and decorative building components. It is thus of major importance in the forestry economies of its main producing countries [3].

The leaf of Tectona grandis L.f. was observed to have damage of mesophyll tissue on surface of leaf and Skelton of leaf observed. The insect caterpillars of leaf skeltonizer or Eutectona Macheralis were responsible for damage of leaf mesophyll tissue. Due to damage of mesophyll tissue plant leaves were unable to photosynthesis process. Further leaves were dry and discarded.

2.2. Collection of Test Plants Material

Sample of test plants viz. *Annona squamosa L., Lantana camara* (L.) Moldenke., and *Albizzia lebbeck* (L.) Benth., were collected from different places of Indore district and identified. *Annona squamosa L.* plant contains aporphine alkaloids, carvone, linalool,limonene squamosin [4]. The samples containing leaves of the selected plant materials were air-dried for 6-7 days. After complete drying the plant parts were pulverized into powder with the help of mortar and pestle. The plant material was extracted by Soxhlet extraction method.

2.3. Soxhlet Extraction

The ordinary method of extraction was not efficient to yield good amount of active principle of the plant material. To extract more active principle from all the plant materials, Soxhlet extraction was used. Known amount (50g) of plant material of each species was filled into the Soxhlet apparatus. A cotton plug was used at the place of thimble to stop the entry of the crude material into the siphoning tube. The required solvent (ethanol/methanol) was filled up five times more than total amount of the sample material into the flask of the apparatus. The apparatus was then connected with the water supply to the condenser. The temperature of the heating mantle was maintained. The process was carried out for 5 to 6 hours for each sample. The extract was transferred to Petri plates and solvent was allowed to evaporate [5].

The evaporated material was weighed and mixed with neem oil based emulsion. Neem oil based emulsion was a mix. Emulsion of neem seed oil (8ml), liquid soap (4ml) and water (2L). 2.5 gm of plant part extract of each species in 10% different solvent mixed with 250ml neem oil based emulsion. Different types of formulation were fill up in different spray pump and sprayed on host plant species.

2.4. Treatment (Process of Dose Administration)

The dose were prepared in glass jars and filled in spring bottles. The formulation was sprayed on selected plants in a month of (August – September). This is the period, usually atmospheric temperature is normal but during the present study the Moisture was also present in air. During our trials test plants were exposed to ample sunlight regularly. The weather was rainy. These are optimum condition for growth of insect and pest.

In field trial, treatments of formulation on host plants were given in regular interval of 7and 15 days from first day of treatment. The trials were started with leaf Soxhelt Ethanol/ Methanol extract of *Annona, Lantana* and *Albizzia* plants on host *Tectona grandis* L.f..

3. Result and Discussion

The results of the effect of ethanol/methanol extracts of three plants with neem oil to control teak skletonizer or *Eutectona macheralis* presented in Table 1. Effects of treatments on *Eutectona macheralis* evaluated from the reduction in numbers of infected leaves after application of treatments. The analysis of variance for the effects of treatments generally indicated the plant extracts treatments significantly reduced infected leaves than the untreated during the trial seasons. Table1 reveals that Ethanolic leaf extract of *Annona squamosa* L., *Lantana camara* (L.) and *Albizzia lebbeck* (L.) Benth. with Neem oil based solution found to be effective on infected plant and Methanolic leaf extract of *Annona squamosa* L., *Lantana camara* (L.) and *Albizzia lebbeck* (L.) Benth. with Neem oil based solution found to be more effective compare to ethanol extract, on infected leaves. Only neem oil based emulsion also found to be effective. Neem oil contains azadirachtin, nimbidin, limonoids, terpenoids, coumarins, sulphur compounds, phenolics etc., [6] which have proven value against many insect larvae, Leaf folder, aphids, Jassids, fruit borer and stem borer insect [7].

This is the first time when the effect of *Annona squamosa L.* leaf extract against teak skletonizer larvae was proved in the field study. Earlier it has been studied in laboratory and reported that both topical and leaf applications of cold alcohol extract of *Annona squamosa L.* seeds are highly effective in controlling the lepidopteran pest, *S. litura.* [8]. This is also the first time when *Annona squamosa L.* leaf extract was used in combination with neem oil in the field studies on economically important trees. The present study indicated that the leaves of *Annona squamosa L.* reported to contain annonins which is toxic or repellent for insect larvae. The insecticidal activity of the seed extracts of *A. squamosa* is attributable to annonins (*i.e.*, annonin I = squamocin), adjacent bis-tetrahydrofuran (THF) ring acetogenins studied [9]. Table 1 revealed that Methanol extract of *Annona squamosa L.* leaves found more effective than Ethanol as Methanol has been inhibited by 1.8% infected leaf compare to 1.4% increase by Ethanol extract. Insecticidal activity of ethanol extract of *Annona squamosa L.* leaves also reported [10]. *A. squamosa* leaf extract with petroleum ether effected on *Eutectona macheralis* larvae [11].

Table 1. *Treatment on Tectona grandis with Soxhlet leaf Ethanol/Methanol Extract of selected plant species*

S. No. of block	Treatment Type	Total No. of leaves in 10 plants of each block			Total No. of Infected leaves in 10 plants of each block (%)			Effect of treatment on infected leaves in %
		A	B	C	A	B	C	
1.	1.Control	83	84	70	4 (4.8)	8 (9.5)	16 (22.8)	18 increase
2.	2.Base sol.	65	67	73	9 (13.8)	4 (5.9)	22 (30.1)	16.3 increase
3.	3.Base sol. with *Annona* leaf extract in Ethanol	51	56	69	6 (11.7)	4 (7.14)	9 (13.1)	1.4Increase
4.	4.Base sol. with *Annona* leaf extract in Methanol	60	61	61	7 (11.6)	6 (9.8)	6 (9.8)	1.8Decrease
5.	5.Base sol. with *Lantana* leaf extract in Ethanol	72	78	73	9 (12.5)	6 (7.6)	10 (13.6)	1.1Increase
6.	6.Base sol. with *Lantana* leaf extract in Methanol	70	72	73	9 (12.3)	8 (11.1)	7 (9.5)	2.8Decrease
7.	7.Base sol. with *Albizzia* leaf extract in Ethanol	69	72	75	10 (14.4)	11 (15.2)	11 (14.6)	0.2Increase
8.	8.Base sol. with *Albizzia* leaf extract in Methanol	72	60	56	22 (30.5)	18 (30)	17 (30.3)	0.2Decraese

Table 1 revealed that Methanol extract of *Lantana camara* (L.) leaves found more effective than Ethanol as Methanol has been inhibited by 2.8% infected leaf compare to 1.1% increase by Ethanol extract. *Lantana camara* (L.) leaf extract found effective, [12] mentioned Antifungal property of *L. camara* and *O. sanctum* against *Drechslera sorokiniana* were reported. The above discussed effects of crude aqueous extract of *L. camara* leaves may be due to the active toxic group like Lantadene present in the leaves. The possibility of toxic effects of some other chemical groups such as lantoniside, linaroside and carmarinic acid cannot be ruled out and has to be elucidated in future along with some other bioactivities [13]. Leaf applications of crude aqueous extract of *L. camara* leaves are highly effective in controlling the lepidopteran pest *S. litura* reported [14]. Adulticidal activity of essential oil of *Lantana camara* (L.) leaves against mosquitoes also reported [15].

Catterpilar of *Eutectona macheralis*.

Effect of *Eutectona macheralis* on *Tectona grandis* leaf.

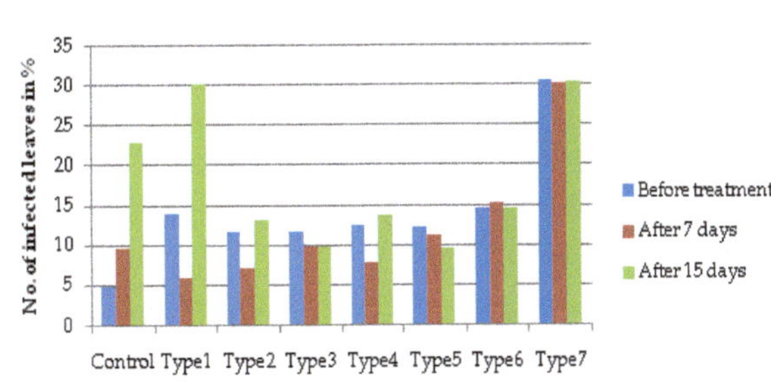

Graph 1. *Treatment on Tectona grandis with soxhlet leaf Ethanol/Methanol Extract*

The Methanol extract of *Albizzia lebbeck* (L.) Benth. leaves found more effective than Ethanol as Methanol has been inhibited by 0.2% infected leaf compare to 0.2% increase by Ethanol extract. *Albizzia lebbeck* (L.) Benth. leaf extract found least effective which resulted that the plant species have some insecticidal properties. It was also reported [5] Methanolic extract of *A. lebbeck* seeds, leaves and bark contain alkaloids and triterpens that may be affect 20-hydroxy ecdyson which is triterpen. Also phytoecdysteroids are class of triterpenoides that composed of saponine, phytosterols and phytoecdysteroids[16]. The plant leaf powders of *W. somnifera, O. sanctum* and *A. lebbeck* possess the property to safeguard the stored chickpea grain [17]. Seeds of *Albizzia lebbeck* (L.) Benth. have saponins, oleic and linoleic acid[18].

4. Conclusion

This plant product is also eco-friendly, easily available and economically viable. Biopesticides considered being safe to natural enemies and free from any residue problem on the crop and in the environment [19]. The use of plant extracts with insecticidal properties has the potential of reducing the effects of insect pests of forest tree and agricultural crops. These can be of importance to the resource-poor farmers in many areas of the developing world. Now we were concluded that the Leaf Extract of test plants found effective against teak skletonizer insect. If these biopesticides produced commercially and farmers are trained for their use there is no doubt that these eco-safe products can replace the hazardous chemicals of the field in coming days.

Acknowledgement

Authors are thankful to Forest Research and Extension Circle, Indore for conducting experiments.

References

[1] SC.Mathur, Future of Indian pesticides industry in next millennium. Pesticide Information, 24(4):9–23. (1999).

[2] Md., Aktar wasim, Dwaipayan Sengupta, and Ashim Chowdhury, Impact of pesticides use in agriculture: their benefits and hazards. *Interdiscip. Toxicol.* 2(1): 1–12. (2009).

[3] D. Pandey and C. Brown Teak: a global overview, Unasylva 201, Vol. 51. (2000).

[4] O. Ekundayo A review of the volatiles of the *Annonaceae. J Essent Oil Res.*1: 223. (1989)

[5] Maru et. Al. Anticancer property of Bryophyllum pinnata (Lam.) Oken. leaf on human cervical cancer cells *BMC Complementary and Alternative Medicine* 12:152012,

[6] Subbalakshmi Lokanadhan, P. Muthukrishnan and S.Jeyaraman, Neem products and their agricultural applications, *J. Biopest,* 5 (Supplementary): 72-76 (2012)

[7] Shiv Narayan Sharma, Zenu Jha, Mahendra Sharan Tiwari, Dinesh Baghel and D.K. Sharma Standardization and Quality Evaluation of Herbal Pesticide *African Journal of Basic & Applied Sciences* 2 (5-6): 184-187, (2010).

[8] Pratibha V. Deshmukhe, Ashok A. Hooli and S. N. Holihosur Bioefficacy of cold ethyl alcohol extract of *Annona squamosa* against *Spodoptera litura Fabricius. Journal of Biopesticides* 3(1 Special Issue) 271 - 274. (2010).

[9] M. Sahai, S. Singh, Y.K. Gupta, S. Akashi, and R. Yuji, Annonaceous acetogenins from the seeds of *Annona squamosa* adjacent *bis*-tetrahydrofuranic acetogenins. *Chem. Pharm. Bull. (Tokyo)* 42:1163-1174. (1994).

[10] N. Begum, B. Sharma, and R.S. Pandey Evaluation of Insecticidal efficacy of *Calotropis procera* and *Annona squamosa* ethanol extracts against *Musca domestica. J. Biofertil. Biopestici.*; 1:1,.doi.org/10.4172/2155-6202.1000101. (2010).

[11] P. B. Meshram, Role of some biopesticides in management of some forest insect pests, Journal of Biopesticides 3 (1 Special Issue) 250 – 252 (2010).

[12] K.P. Varma, R. Hedge,and S. Kulkarni, *In vitro* evaluation of phytoextracts and biocontrol agents against *Drechslera sorokiniana,* In: *Asian Congress* of *Mycology, Plant Patholology, Indian Society of Mycology, Plant pathology,* University of Mysore (Abstract), October 1-4, 241. 2002.

[13] N. Kulkarni, K.C. Joshi, and B.N. Gupta, Antifeedant property of *Lantana camara* var. *aculeata* and *Aloe vera* leaves against teak skeletonizer, *Eutectona machaeralis* Walk. (Lepidoptera: Pyralidae) Entomon, 22: 61-65. (1997).

[14] Pratibha V.Deshmukhe, Hooli A. Ashok and S. N. Holihosur Effect of *Lantana camara* (L.) on growth, development and survival of tobacco caterpillar. *Karnataka J. Agric. Sci.,* 24 (2): (137-139) (2011).

[15] V.K. Dua, A.C. Pandey and A.P. Dash Adulticidal activity of essential oil of *Lantana camara* leaves against mosquitoes, *Indian J Med Res* 131, , pp 434-439 (2010),

[16] A.S. Al-Dahmin, Evaluation of the field efficacy of crude extracts of *Albizzia lebbeck* (L.)Benth. and Actara insecticide on biological performance of *Ommatissus lybicus* (Homoptera : Tropiduchidae). Ph.D. Thesis. College of Science, University of Baghdad. (2008).

[17] Sarwar Muhammad some possibilities on the effectiveness of plant powders as grain protectants against cowpea weevil, *Callosobruchus maculatus* (Fabricius) Walp (Coleoptera: Bruchidae) infestation in chickpea, *International journal of Agronomy and Plant Production.* Vol., 1 (2), 45-50. (2010)

[18] faisal Mohammed, P.P. Singh and R. Irchhaya review of albizzia lebbek a potent herbal drug, *IRJP,* 3 (5). (2012)

[19] U. Mukherjee, and H. N. Singh, Eco-friendly approaches to manage Diamond-Back moth, *Plutella xylostella* (L.) in cauliflower. *Journal of Appllied Zoological Reasearch,* 17(1): 57 – 60. (2006).

Performance evaluation of New Zealand white rabbits fed *Alchornea cordifolia* leaf meal as replacement for soya bean meal

Philip Cheriose Nzien Alikwe[1], Elijah Ige Ohimain[2, *], Amos Ebiowei Kester[1]

[1]Biochemistry/Biotechnology Research Unit, Animal Science Department, Niger Delta University, Wilberforce Island, Bayelsa State
[2]Agricultural and Veterinary Microbiology Research Unit, Biological Sciences Department, Niger Delta University, Wilberforce Island, Bayelsa State

Email address:
eohimain@yahoo.com (E. I. Ohimain)

Abstract: The Twenty-four weaned rabbits were fed graded levels of *Alchornea cordifolia* leaf meal (ACLM) at different inclusion levels in the diet. A control (0%ACLM) and three other diets containing different levels of ACLM (5%, 10% and 15%) as partial replacement for soya bean meal in the diet were fed for six weeks with the aim of evaluating the effects of ACLM on the growth performance and carcass characteristics of the rabbits. The study revealed a decrease in the growth rate with the highest level of inclusion i.e treatment D (15%ACLM). The highest mean weight gain of 1.36±0.04kg and feed intake of 131.94±7g were recorded in treatment C (10% ACLM inclusion in the diet) and the lowest mean weight gain of 1.14±0.04kg was recorded in the treatment A (0% ACLM) and the lowest feed intake of 94.44±10g was recorded in the treatment D. There was no significant difference (P>0.05) in feed conversion ratio (FCR) across the treatment groups. Carcass characteristics showed that live weight of rabbits fed 15% ACLM diet was lower (P<0.05) than other experimental diets with lower percentage of ACLM which implies that 15% ACLM diet possibly depressed the final weight of the experimental rabbits . This could be attributed to the higher percentage of the anti-nutritional factors in ACLM in treatment D which has been implicated for reduced weight gain. This study concluded that the use of ACLM may be recommended as protein source for rabbits but at 5-10% inclusion levels.

Keywords: *Alchornea Cordifolia*, Leafmeal, Soyabean, Rabbits, Supplements

1. Introduction

Like many developing countries Nigeria has for long been plagued with the problem of worsening situation of inadequate consumption of animal protein. Small scale rabbit production has been advocated as a viable income generating activity [1], which emerged as low cost answer to food insecurity among rural communities within the Country.

A panacea for the protein shortage is to target prolific livestock species with short reproductive cycle, high fecundity but also with high (21%) crude protein and low (5.40%) crude fat content in its meat [1]. The crude fat in rabbit meat is lower (8%) when compared with beef (35%), mutton (31%), pork (29.5%) and chicken (12%) [3]. When compared with meat from other species, the fat in rabbit meat contains less stearic and oleic acids but a high proportion of the essential poly-unsaturated fatty acids such as linoleic[5]. Dandu *et. al* [6] reported that non-conventional feedstuff (NCF) offer the best alternative for the reduction in feed cost which ultimately leads to reduction in the prize of meat and other animal products. However, for this non-conventional feed-stuff to meet the need of reducing cost of production they must be available all year round and should be easy to procure and process when needed. Non-conventional feed-stuff may be of animals, plants or mineral origin. *Alchornea cordifolia* is one of the recently discovered plants used as performance enhancers in livestock and poultry industry [7]. The root and bark is used as stimulating intoxicant and aphrodisiac by the people of Congo in Africa Oliver [8]. After drying it is ground to a fine powder, and is mixed either with food or macerated for several days in palm wine and consumed to produce energy for tribal festivities and in warfare. It is said to produce a state of intense excitement followed by a deep and sometimes fatal depression [9]. In Nigeria, goats and sheep browse on its leaves. Its root and bark have 0.03-0.26% of total alkaloids while the leaves contain 10% of tannins (alkaloids) [9]. Wekle and Njoku [7] reported the effect of *Alchornea cordifolia* on weight gain and organ size in broilers and observed no significant effect on the overall weight gain but interestingly discovered its hypo tropic effects on the gonads which may lead to increase in testosterone production. They also discovered atrophy of the ovary of the females which may lead to the suppression of oestrogen production and related hormones and thus precipitate masculinity and reproductive impotency. The present study was therefore designed to determine the performance and carcass characteristics of rabbits fed *Alchornea cordifolia* leaf meal as partial replacement for soya bean meal.

2. Materials and Methods

Processing of *Alchornea* Leaf:

Fresh leaves of *A. cordifolia* were harvested from within the Campus of Niger Delta University, Wilberforce Island, Bayelsa State, Nigeria. The leaves were chopped and air-dried for two weeks and milled into powder with grinding machine, then sieved and bagged in polythene bags and stored until needed. Samples of the powder (ACLM) were sent to the Biochemistry Laboratory of the Institute of Agriculture, Research and Training, Ibadan, Oyo State Nigeria for further analysis

Experimental Diets:

Four experimental diets were formulated. All the experimental diets were fed as a meal, and all major ingredients were ground in a hammer mill and mixed manually. In the experimental diets: 5, 10 and 15% ACLM were added in partial replacement of soya bean meal (Table 1) as a protein feed-stuff.

Table 1. *Ingredients composition of the Diets*

Ingredients	Dietary levels of ACLM%			
	0.0	5.0	10.0	15.0
Maize	43.0	43.0	43.0	43.0
Soya bean meal	17.0	12.0	7.0	2.0
ACLM	0.0	5.0	10.0	15.0
Palm kernel meal	12.0	12.0	12.0	12.0
Wheat offal	15.0	15.0	15.0	15.0
Crayfish meal	3.0	3.0	3.0	3.0
Bone meal	7.0	7.0	7.0	7.0
Oyster shell	2.0	2.0	2.0	2.0
Salt	0.50	0.50	0.50	0.50
Premix	0.50	0.50	0.50	0.50
Total	100	100	100	100
Calculated crude protein	18.0	16.7	15.4	14.1

Experimental Design

Twenty-four 5-6 weeks old rabbits of mixed breeds comprising twelve (12) males and twelve (12) females with an initial average weight of 960g were used for this study. The rabbits were housed in hutches with wire-mesh cages. Two hutches were used for this study with each hutch having six compartments (cages). The twelve cages were made to sit in a square wire-mesh base for easy cleaning. The rabbits were housed as two rabbits per cage and tagged according to the experimental groups/replicate for proper identification. There were six animals per group representing one control (A) and three treatment groups (B, C and D) in a complete randomized design (CRD). The rabbits were weighed at the initiation of the experiment and subsequently at weekly intervals. Feed intakes were recorded daily. Experiment lasted for 42 days (six weeks). On the 43rd day, two animals from each group were randomly selected and slaughtered and the following measurements were taken and analyses done; Live weight, carcass weight, meat, skin, head, thigh, limb, liver, spleen and stomach weight and length of intestine at the Animal Science Laboratory of Niger Delta University, Bayelsa State, Nigeria.

Data Collection and Analysis

The initial body weight of all replicates were noted and thereafter weekly body weight and feed consumption were recorded throughout the experiment. The rabbit's performance data on weight gain, feed intake, Feed Conversion Ratio (FCR) and survivability were calculated based on the records of body weight, feed intake and mortality respectively. Data on performance and carcass characteristics were subjected to Analysis of Variance (ANOVA) in SPSS Program Version 16 and were calculated as Standard Errors of the Mean (SEM), while Duncan's Multiple Range Test (DMRT) was used in assessing the significant differences among the treatment means. Significant was accepted at 0.5% level of probability.

3. Results

Proximate, Anti-nutritional and Essential Amino-acid Characteristics;

The proximate composition of *Alchornea cordifolia* leaf meal is presented in (Table 2) while the quantitative phytochemical analysis is presented in Table 3

Table 2. *Proximate Composition of Alchornea cordifolia Leaf Meal**

Component	Composition (%)
Dry matter	90.04
Moisture content	9.96
Crude protein	17.94
Crude fibre	16.84
Ash	11.38
Energy	3.37

Alchornea leaf meal contain 17.94% Crude protein, 16.84% Crude fibre, 11.38% Ash and 3.37% kcal/kg Energy.

The phytochemical composition (Table 3) shows the presence of Phytate, Oxalate, Saponins, Phenols, Cardiac glycoside and Hydrocyanic acid.

Table 3. *Phytochemical composition of Alchornea leaf*

Phytochemicals	Powder (%)
Phytate	1.21
Oxalate	0.86
Saponins	2.04
Phenols	1.16
Cardiac glycoside	0.11
Hydrocyanic acid	22.30

Table 4. *Essentials Amino acids Compositions of Alchornea Leaf*

Essential Amino acid	%
Histidine	1.35
Isoleucine	2.48
Leucine	4.56
Lysine	3.84
Methionine	1.16
Phenyl alanine	2.04
Threonine	0.87
Tryptophan	1.03

Growth Feed Intake and Feed Conversion

Addition of *Alchornea cordifolia* leaf meal to rabbit diet at 5, 10 and 15% caused no significant difference (P>0.05) between treatments A to C in feed intake, but there was significant difference (P<0.05) between treatment D and the treatments A, B and C. Treatment C recorded the highest feed intake of (131.94± 7g) and treatment D had the lowest feed intake of (94.44±10g). Results on weight gain showed no significant difference (P>0.05) across the treatment groups. Treatment C had the highest weight gain of (1.36±04kg) and treatment A had the lowest weight gain of (1.14±0.04kg). There was also no significance difference (P>0.05) in Feed Conversion Ratio (FCR) among the treatment group.

Table 5. *Performance of Rabbits Fed Alchornea cordifolia Leaf Meal (ACLM)*

Parameter	A (0%l)	B (5%)	C (10%)	D (15%)
Initial Live weight (kg)	1.02±0.02	1.12±0.03	1.03±0.01	1.03±0.03
Final Live weight (kg)	1.48±0.03	1.43±0.05	1.49±0.02	1.25± 0.04
Weight gain (kg)	0.46±0.04	0.31±0.07	0.46±0.04	0.22±0.05
Initial Feed intake (kg)	0.11±0.01	0.11±0.04	0.08±0.05	0.06±0.03
Final feed intake (kg)	0.119±0.03	0.124±0.01	0.155±0.02	0.113±0.02
Feed intake (kg)	0.09±10	0.014±10	0.075±7	0.053±10
FCR	0.196	0.452	0.163	0.241
Mortality (%)	0.1666	0.1666	-	0.3333

Carcass Characteristics

Live weight is taken as weight before bleeding/slaughter. Carcass weight is taken as weight after bleeding/slaughter, meat weight is taken as the edible part of the rabbits i.e., after removal of the viscerals and skin along with the furs. Weight and measurement were collected from other parts/cuts of the rabbits as shown below in Table 6.

Table 6. *Carcass Characteristics of Rabbits Fed Alchornea Leaf Meal*

Parameter	0.0	5.0	10.0	15.0
Live weight (kg)	1.50±0.00	1.65±0.07	1.23±0.04	0.95± 0.07
Carcass weight (kg)	0.70±0.35	0.85±0.21	0.68±0.04	0.55± 0.00
Meat (kg)	0.65±0.00	0.70±0.25	0.62±0.04	0.40±0.00
Skin (g)	138.0±7.07	120.0±41.01	87.5±13.44	77.0±5.66
Head (g)	144.5±10.61	125.5±21.92	111.0±1.41	101.0±0.00
Thigh (g)	92.0±9.89	89.0±18.39	87.0±0.00	66.5±2.12
Limb (g)	54.5±9.19	49.0±12.73	46.5±0.71	34.5±2.12
Back(g)	0.30±0.07	0.23±0.04	0.21±0.01	0.20±0.01
Intestine (g)	4.27±0.18	4.27±0.67	3.70±0.42	2.55±0.78
Liver (g)	42.50±2.12	34.00±4.24	34.00±0.00	24.00±4.24
Spleen (g)	0.50±0.00	0.25±0.07	0.30±0.00	0.25±0.07
Stomach (g)	96.50±23.34	92.00±5.66	66.50±10.61	53.00±4.24

4. Discussion

The proximate analysis of the treatment diet fed to the rabbits was as follows: Crude Protein (CP) and Metabolizable Energy (ME) of 18.0% and 2.59%kcal/kg. It offered the normal nutrition required for maximum performance. Little is known about the proximate composition and anti-nutritional factors in *Alchornea cordifolia* leaf or the effects of processing procedures such as air and sundrying on the compounds. This remains for future research. But based on chemical analysis ACLM is a valuable source of nutrient for monogastrics such as pig, poultry and rabbit. However no systematic research is known to have been done in the nutritional quality and phytochemical properties of *Alchornea cordifolia* leaves for any monogastric species. The significant difference (P<0.05) in feed intake of the rabbits between 15% ACLM and the rest of the treatments is an indication that inclusion levels of 15% and above of ACLM in rabbit diet may decrease feed intake while inclusion levels between 5 – 10% had better performance in feed intake and this is probably due to the lower inclusion levels compared to treatment D, also the caecum of the rabbits can digest highly fibrous plant materials. This agrees with the findings of Aduku and Olukosi, [11], who stated that rabbits are pseudo-ruminants and as such are able to utilize forages as well as concentrates. There was no significant difference (P>0.05) in weight gain across the treatment groups. Treatment with 10% ACLM recorded the highest weight gain and there was also no significant difference (P>0.05) in feed conversion ratio among the Treatment groups. There was significant difference (P<0.05) observed in the values of parameters obtained in live weight and head weigh whereas skin, thigh, carcass, meat, limb, and back weights were similar across the treatment diets. No treatment effect (P<0.05) was observed in some carcass parameters such as shoulder, tail, kidney, heart, lungs and caecum, while on the other hand significant difference ($P<0.05$) were noticed on other carcass parameters; stomach, spleen, liver and intestine decline with increasing levels of ACLM in the diet which may be as a result of some anti-nutritional factors in *Alchornea cordifolia* which affects the rabbits in terms of final product (carcass). This agrees with the findings of Bangbose *et. al*, [12] who stated that high inclusion of Tigernut Meal (TNM) in the diet of weaner rabbits decrease protein percentage in carcass.

5. Conclusion

This study concludes that the use of ACLM may be recommended as protein source for rabbits at 5 – 10% inclusion levels in the diet. Additional work to investigate the economic significance of using ACLM as rabbit non-conventional feed supplement/feed additive compared to other feedstuff is on- going in our research laboratory. Further studies will be carried out to see if any health effects can be observed due to ACLM feeding to detect quality of rabbit meat fed this ACLM.

Acknowledgement

The author wishes to thank Sylvester C. Izah and Tariwari C.N Angaye for the editorial work.

References

[1] S. A. Majgandi, and I. F. Ngang, Performance of rabbit fed varying levels of calabash seed cake. Journal of Agriculture and Environment 2002, pp. 201 – 208.

[2] S. K. Das, and K. M. Bujarbaru, Carcass Trait of Rabbits, Organoloeptic properties and consumption pattern of rabbit meat in the North Eastern Hill Region of India. Pan-American Rabbit Sci. Newsletter 2005, pp. 39-43.

[3] F. Lebas, P. Coudert, R. Rouvier and H. de Rochambeau, The Rabbit-Husbandary, Health and Production. FAO, Rome, 1986.

[4] J. Adrian, G. Legrand and R. Frangne, Dictionaire de biochimie elmimentaire et de nutrition. Paris, France. Technique et Documentation, 1981.

[5] O.M. Daudu M,Orunmuyi and G.E. Joktham, Use of non conventional feeds: potentials and constraints for rabbit production in Nigeria. Proceedings of the 34th Annual conference of the Nigerian Society for Animal Production (NSAP).March 15th -18th , University of Uyo, Uyo,. Nigeria, (2009) page 198-201

[6] S.N. Wekhe, and C. O Njoku, Preliminary investigations of the effect of *Alchoria cordijolia* on weight gain and organ size of broilers. Proc. 5th annual conf. of ASAN, Sept. 19-23, 2000. Port Harcourt. pp. 56-58.

[7] B. Oliver, Medical Plant in Nigeria. Nigerian College of Arts. Science and Technology Vol. 1, 1960, pp. 138.

[8] B. Oliver – Bever. Medical Plants in Tropical West Africa. Cambridge University Press, London. 1986, pp. 115.

[9] P. C. N. Alikwe, and O. J. Owen, Evaluation Of The Chemical And Phytochemical Constituents Of *Alchornea Cordifolia* Leaf Meal As Potential Feed For Monogastric Livestock African Journal of Food Agriculture Nutrition and Development 2013, in press.

[10] A. O. Aduku and J. O. Olukosi,Rabbit Management in the Tropics Living Books Series. GU Publications, Abuja, Nigeria. Vol. 1, 1990, pp. 33-34.

[11] A. M. Bamgbose, M. Abimbola, W. A Olayemi, A. O. Osofowora, A. O. Oso, and O. T. Ojo, Performance of weaner rabbits fed supplemented *Tridax procumbens*.. Proc. 7th Anun. Conf. Anim. Sci. Ass. of Nig. (ASAN). *In:* Fanimo, OA (eds). September 16-19, 2002, University of Agriculture, Abeokuta. Nigeria.

Rock phosphate solubilisation by strains of *Penicillium* spp. isolated from farm and forest soils of three agro ecological zones of Cameroon

Fankem Henri[1, *], Ngo Nkot Laurette[1], Nguesseu Njanjouo Ghislain[1], Tchuisseu Tchakounte Gylaine Vanessa[1], Tchiazé Ifoué Alice Virginie[1], Nwaga Dieudonné[2]

[1]Department of Plant Biology, Faculty of Science, University of Douala, Douala, Cameroon
[2]Department of Plant Biology and Physiology, Faculty of Science, University of Yaoundé I, Yaoundé, Cameroon

Email address:
fankemhenri@yahoo.fr (H. Fankem)

Abstract: Rock Phosphate (RP) is theoretically the cheapest fertilizer and most abundant, but its direct application in the soils is not always agronomically effective due to its low phosphorus (P) availability. An environmentally friendly and economically alternative to traditional processing of these deposits is the use of Phosphate Solubilising Microorganisms (PSM). Three *Penicillium* strains were screened for their aptitude in solubilising sparingly soluble phosphates as well as phosphate rock on plates and in broth supplemented with the different phosphate types. The three strains show the ability in solubilising at least two from the sparingly soluble phosphates types, with the solubilisation index (SI) varying from 1.44 to 4.74. All the strains were able to show halo zone on plates supplemented with Malian and Algerian rock phosphates with solubilisation index ranging from 1.2 to 1.8. No halo zone was detected on plates with Moroccan rock phosphate as sole phosphate source. However, significant amount of phosphorus were mobilised by all strains in broth with the different rock phosphates, including the Moroccan rock phosphate. The amount of solved P varied from 257.46 mgP.L^{-1} (*Penicillium* sp. 43) to 326.65 mgP.L^{-1} (*Penicillium* sp. 109) for Malian rock phosphate, from 152.61 mgP.L^{-1} (*Penicillium* sp. 43) to 337.56 mgP.L^{-1} (*Penicillium* sp. 27) for Algerian rock phosphate, and from 283.79 mgP.L^{-1} (*Penicillium* sp. 43) to 347.16 mgP.L^{-1} (*Penicillium* sp. 109) for Moroccan rock phosphate. The rock phosphate solubilisation was associated with pH media drop and both parameters were strongly correlated.

Keywords: *Penicillium* Spp, Rock Phosphates, Fixed Phosphates, Phosphate Solubilisation

1. Introduction

The optimal development of crops demands a high, often costly input of P fertilizers. Current concepts in sustainability involve application of alternative strategies based on the use of less expensive natural sources of plant nutrients like rock phosphate. The beneficial effect of rock phosphate has made this material an attractive component for management in agriculture [1]. However, the direct application of low-grade rock phosphate as a P source in neutral and alkaline soils was of little importance [2,3]. In Mali, the use of Tilemsi Rock Phosphate (TPR) as phosphorus source was limited by its inefficiency in many agricultural soils [4]. However, Addition of an inoculum of phosphate solubilising microorganisms to soil has also been found to improve the rock phosphate efficiency as a phosphorus source [5]. Many soil microorganisms, including bacteria and fungi, are able to mobilize phosphorus from sparingly soluble rock phosphates, and they have an enormous potential in providing soil phosphates for plant growth [6,7]. Microbial solubilisation of rock phosphate, especially low grade and its use in agriculture is receiving greater attention. This process not only compensates for higher cost of manufacturing fertilizers in industry but also mobilises the fertilizers added to soil [8]. The present study aims at characterizing the rock phosphate solubilising ability of three *Penicillium* strains isolated from different agro ecological zones of Cameroon.

2. Materials and Methods

2.1. Sampling Zones Description

The three fungi strains were obtained from soils collected in three out of the five agro ecological zones of Cameroon: The High Guinea Savanna Zone II, the southern plateau raises northward to the grassy, rugged Adamawa Plateau. This feature stretches from the western mountain area and forms a barrier between the country's north and south. Its average elevation is 1,100 meters and its temperature ranges from 22 °C to 25 °C with high rainfall. The Western Highlands Zone III, an irregular chain of mountains, hills, and plateaus known as the Cameroon range extends from Mount Cameroon on the coast (Cameroon's highest point at 4,095 meters) almost to Lake Chad at Cameroon's northern tip. This region has a mild climate, particularly on the Western High Plateau, although rainfall is high. Its soils are among Cameroon's most fertile, especially around volcanic Mount Cameroon. This area has been delineated by the World Wildlife Fund as the Cameroonian Highlands forests ecoregion. The Bimodal Humid Forest Zone V, the South Cameroon Plateau rises from the coastal plain to an average elevation of 650 meters. Equatorial rainforest dominates this region, although its alternation between wet and dry seasons makes it less humid than the coast. This area is part of the Atlantic Equatorial coastal forests ecoregion. In general, the climate varies with terrain, from tropical along the coast to semiarid and hot in the north. Exceedingly hot and humid, the coastal belt includes some of the wettest places on earth. Top soils (0-25 cm depth) were collected from forest and farmed locations of these agro ecological zones for strains isolation and characterization.

2.2. Isolation and Characterization of Strains on Plates Supplemented with Sparingly Soluble Phosphates

Rock-phosphate-solubilising fungi were isolated from farm and forest soils collected in three agro ecological zones of Cameroon using the dilution plate method. Ten grams (10) soil sample were mixed with 90 ml sterile distilled water, vigorously shacked and left to stand for 5 minutes. Series of the homogeneous soil was diluted and the 10^{-5}-10^{-6} dilutions plated on non selective LB medium (10 g Peptone, 5 g Yeast, 10 g NaCl, 15 g Agar, 1000 ml Distilled water, pH 7.5) used for broad spectrum microorganisms. Fungi were purified by several transfers of fungal mycelia into new media until obtaining the pure fungal culture. Fungi strains were maintained on LB agar slants until use.

To assess the phosphate solubilisation activity of microorganisms, pure fungal colonies were individually suspended into 10 ml LB broth and incubated at 28 °C, 150 rpm, for 5 days. Ten microliters of each fungal suspension were transferred onto a single point of Petri dish containing P-free minimal medium (MM) containing per liter distilled water; 0.4 g NH_4Cl, 0.78 g KNO_3, 0.1 g NaCl, 0.25 g $MgSO_4$, 0.1 g $CaCl_2{\cdot}2H_2O$, 0.27 mg $FeSO_4$, 1.56 mg $MnSO_4{\cdot}H_2O$, 1.40 mg ZnSO4·7H2O, 10 g Glucose, 20 g Agar, and one of the four insoluble phosphate types (either $Ca_3(PO_4)_2$ (Ca-P) or $AlPO_4$ (Al-P), or $FePO_4$ (Fe-P), or Na-Phytate (Na-phy)) at 30 mM P, plus 0.5 % BCG (Bromocresol green) dye [9] for better observation. A stock solution of 0.5 % dye was prepared by dissolving a corresponding weight of Bromocresol green into 70 % ethanol and the final pH adjusted to 6.5 with KOH 1M. The plates were sealed and incubated at 28 °C for 5 days and the phosphate solubilisation recorded through the halo/yellow zone surrounding the fungal colony. The index of solubilisation (IS) as defined by Qureshi et al. [10] was used as an indicator for the isolate efficiency:

IS = (Colony diameter + diameter of halo zone) / Colony diameter.

2.3. Characterization of Strains on Plates Supplemented with Rock Phosphate from Different Origins

Three rock phosphates of different origins with characteristics indicated in Table 1 were used. To remove the soluble fraction, the different rock phosphates were washed four times with worm distilled water according to the cycle: 1 hour – 24 hours - 1 hour – 24 hours. They were then completely dried in oven at 60 °C and homogenised before use.

Table 1. *Mineral composition of the different rock phosphates used. (Magallón-Servín, unpublished).*

Origin	Mineral elements (%)								mg/kg		
Rock P	Total P_2O_5	Available P	K	Ca	Mg	Na	Fe	Al	Mn	Zn	Cu
Mali	30	12.98	0.056	28.19	0.131	0.232	3.844	0.80	8360	87	51
Morocco	13	9.33	0.093	28.83	1.93	0.552	0.267	0.42	96	219	38
Algeria	29	Nd	Nd	Nd	Nd	Nd	Nd	Nd	Nd	Nd	Nd

Nd : Not determined.

The characterisation of strains for their rock phosphate solubilising ability was assessed as previously described but the medium used was the National Botanical Research Institute's Phosphate growth medium (NBRIP) [11] with some modifications and containing per liter of distilled water: glucose, 20 g; $MgCl_2{\cdot}6H_2O$, 5 g; $MgSO_4{\cdot}7H_2O$, 0.25 g; KCl, 0.2 g; $(NH_4)_2SO_4$, 0.1 g and one rock phosphate at 5 g L^{-1}, plus 0.5 % BCG dye as previously described, pH 7.5. After 5 days incubation at 28 °C, the solubilisation index of each fungal strain was determined.

2.4. Ability of Fungi to Solubilise Rock Phosphates in Liquid Culture

Microorganisms were tested in liquid media to assess their capability in releasing phosphorus from insoluble sources; 50 ml NBRIP medium were distributed into 250 ml

Erlenmeyer flasks, individual rock phosphate types (Malian RP, Moroccan RP, Algerian RP, or Mexican RP) was added to the medium at the concentration of 5gL^{-1} and the pH adjusted to 7.5. After sterilisation and cooling, 200 µl mycelia suspensions were used to inoculate flasks containing the different rock phosphates in triplicate. Triplicate non-inoculated flasks supplemented with the different rock phosphates served as controls. Incubation was made at 28 °C, 150 rpm for 7 days. Cultures were then transferred into sterile falcon tubes, centrifuged at 10000 rpm for 10 minutes and aliquots taken for both P and pH measurement.

2.5. Identification of Fungal Strains

Identification was made from pure isolates diluted and pre heated at 95 °C for 10 minutes. Amplification of 18S rRNA gene was performed from the genomic DNA of strains. Fungal PCR cocktails (24µl) contained Water (18.1 µl), Buffer (2.6 µl), 817F (0.5 µl), 1536R (0.5 µl), $MgSO_4$ (1 µl), dNTPs (0.5 µl), BSA (0.5 µl), Platinum Taq (0.1 µl) and HhaI (0.2 µl). Possible contamination with microbial template was minimised through pre digestion of components with hha I. One microliter of the preheated fungal cells was added to the predigested Master mix and the amplification was performed in a DNA thermo cycler at 94 °C for 4.5 minutes followed by 30 cycles of 30 seconds at 94 °C denaturation, 30 seconds at 55 °C annealing and 1.5 minute at 68 °C with final extension of 68 °C for 10 minutes. A 3 µl aliquot of each amplification product was electrophoresed on a 1% agarose gel in 1× TAE buffer at 50 V for 30 min and the PCR products were visualised with a UV transilluminator. PCR products adjusted to 25-30 ng/µl were sequenced with automated DNA sequencer with specific primers, either 817F or 1536R. The sequences were edited and compared with 18S sequences on the GenBank, using the algorithm BLASTN-NCBI in order to find the homologous sequences.

2.6. Statistical Analysis

Statistical analyses were performed with Sigma plot 12.0. The analysis of variance (ANOVA) was run to find difference between factors and the HSD Turkey test to compare the different treatments. The correlation within the solved P and pH of the corresponding media were assessed using Statistica 9.

3. Results

3.1. Identification of Strains

The three *Penicillium* species were respectively labelled Ber-fm F_1, isolated on farm field in agro ecological zone V, with 27 as strain number; Wak-fm F_2 isolated on farm field in agro ecological zone II, with 43 as strain number and Dsc-ft F_5 isolated on forest field in agro ecological zone III, with 109 as strain number. The sequences of the different strains are indicated in Table 2. .

3.2. Strains Characterization on Plates Supplemented with Sparingly Soluble Inorganic and Organic Phosphates

The solubilisation index which equals to 1 indicates the non solubilisation ability by the fungal strains while the rating indicates the phosphate type that was solubilised by each strain (Table 3). P.27 was the strain able to show halo zone on plates amended with the different P types, while the strain P.109 was able to solubilise Ca-P and Al-P only. The highest SI was performed by P.27 on plates supplemented with Ca-P (2.43) and Na-phy (1.45), that strain being the only one able to show halo zone on plate amended with Fe-P. However, the strain P. 109 shows the highest SI on plate with Al-P (4.74). An example of halo zone displayed by P.27 on plates amended with the different phosphate types is shown in Figure 1.

Table 2. *Sequence of the different strains.*

Strains	Sequences
Penicillium sp. 27	AGGGATAGTCGGGGGCGTCAGTATTCAGCTGTCAGAGGTGAAATTCTTGG ATTTGCTGAAGACTAACTACTGCGAAAGCATTCGCCAAGGATGTTTTCAT TAATCAGGGAACGAAAGTTAGGGGATCGAAGACGATCAGATACCGTCGTA GTCTTAACCATAAACTATGCCGACTAGGGATCGGACGGGCTTCTATGATG ACCCGTTCGGCACCTTACGAGAAATCAAAGTTTTTGGGTTCTGGGGGGAG TATGGTCGCAAGGCTGAAACTTAAAGAAATTGACGGAAGGGCACCACAAG GCGTGGAGCCTGCGGCTTAATTTGACTCAACACGGGGAAACTCACCAGGT CCAGACAAAATAAGGATTGACAGATTGAGAGCTCTTTCTTGATCTTTTGG ATGGTGGTGCATGGCCGTTCTTAGTTGGTGGAGTGATTTGTCTGCTTAAT TGCGATAACGAACGAGACCTCGGCCCTTAAATAGCCCGACCCGCTTTTGC GGGCCGCTGGCTTCTTAGGGGGACTATCGGCTCAAGCCGATGGAAGTGCG CGGCAATAACAGGTCTGTGATGCCCTTAGATGTTCTGGGCCGCACGCGCG CTACACTGACAGGGTCAGCGAGTACATCACCTTGGCCGAGAGGTCTG.
Penicillium sp. 43	GACGCCGTATGATAATAGGGATAGTCGGGGGCGTCAGTATTCAGCTGTCA GAGGTGAAATTCTTGGATTTGCTGAAGACTAACTACTGCGAAAGCATTCG CCAAGGATGTTTTCATTAATCAGGGAACGAAAGTTAGGGGATCGAAGACG ATCAGATACCGTCGTAGTCTTAACCATAAACTATGCCGACTAGGGATCGG

Strains	Sequences
	GCGGGGTTTCTATGATGACCCGCTCGGCACCTTACGAGAAATCAAAGTTT TTGGGTTCTGGGGGGAGTATGGTCGCAAGGCTGAAACTTAAAGAAATTGA CGGAAGGGCACCACAAGGCGTGGAGCCTGCGGCTTAATTTGACTCAACAC GGGGAAACTCACCAGGTCCAGACAAAATAAGGATTGACAGATTGAGAGCT CTTTCTTGATCTTTTGGATGGTGGTGCATGGCCGTTCTTAGTTGGTGGAG TGATTTGTCTGCTTAATTGCGATAACGAACGAGACCTCGGCCCTTAAATA GCCCGGTCCGCGTTTGCGGGCCGCTGGCTTCTTAGGGGGACTATCGGCTC AAGCCGATGGAAGTACGTGGCAATAACAGGTCTGTGATGCCCTTAGATGT TCTGGGCCGCACGCGCGCTACACTGACAGGGTCAGCGAGTACATCACCTT GGCCGAGAGGTCTGGGTAATCTTGTTACACCCTGTCGTGCTGGG.
Penicillium sp. 109	TAGGGATAGTCGGGGGCGTCAGTATTCAGCTGTCAGAGGTGAAATTCTTG GATTTGCTGAAGACTAACTACTGCGAAAGCATTCGCCAAGGATGTTTTCA TTAATCAGGGAACGAAAGTTAGGGGATCGAAGACGATCAGATACCGTCGT AGTCTTAACCATAAACTATGCCGACTAGGGATCGGACGGGCTTCTATGAT GACCCGTTCGGCACCTTACGAGAAATCAAAGTTTTTGGGTTCTGGGGGGA GTATGGTCGCAAGGCTGAAACTTAAAGAAATTGACGGAAGGGCACCACAA GGCGTGGAGCCTGCGGCTTAATTTGACTCGACACGGGGAAACTCACCAGG TCCAGACAACATAAGGATTGACAGATTGAGAGCTCTTTCTTGATCTTTTG GATGGAGGTGCATGGCCGTTCTTAGATGGAGGAGTGATTTGTCTGCTTAA TTGCGATAACGAACGAGACCTCGGCCCTTAAATAGACCGACCCGCTTTTG CGGGCCGCTGGCTTCTTATGGGGACTATCGGCTCAAGCCGATGGAAGTGC GCGGCAATAACAGGTCTGTGATGCCCTTATATGTTCTGGGCCGCACGCGC GCTACACT.

Table 3. *Solubilisation Index of strains on plates supplemented with sparingly soluble inorganic and organic phosphates.*

Strains	Solubilisation Index (SI)				
	(a) MM with $Ca_3(PO_4)$	(b) MM with $AlPO_4$	(c) MM with $FePO_4$	(d) MM with Na-Phytate	Rating
Penicillium sp. 27	2.43	3.92	1.43	1.45	a,b,c,d
Penicillium sp. 43	2.29	3.89	1.00	1.44	a,b,d
Penicillium sp. 109	1.64	4.74	1.00	1.00	a,b

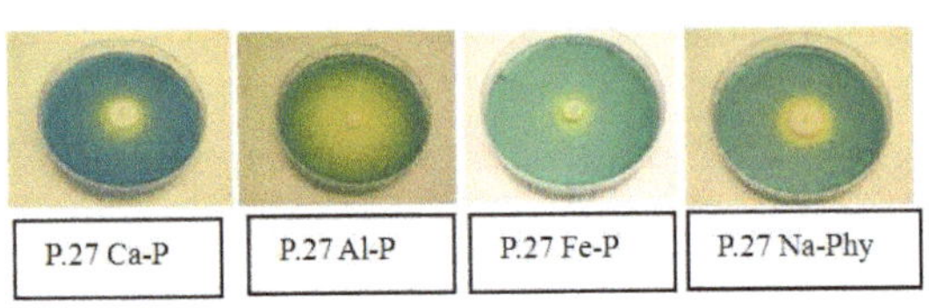

Figure 1. *Halo (yellow) zone displayed by P.27 on plates amended with the different sparingly soluble phosphates. The yellow colour indicates the drop of pH of the medium associated to the process of P solubilisation..*

3.3. Strains Characterization on Plates Supplemented with Rock Phosphates of Different Origins

3.3.1. Solubilisation Index of Strains

The solubilisation index was equally used here to evaluate the efficiency of strains in solubilising rock phosphate of different origins on plates. All the strains were able to solubilise Malian and Algerian rock phosphates (Figure 2), with the highest index displayed by strain P.109 on both Malian and Algerian rock phosphates. However, none of the strains was able to show any visible halo zone on plate amended with Moroccan rock phosphate (SI = 1). An example of halo zone (Figure 3) displayed by the different strains on plate supplemented with Algerian rock phosphate can be observed.

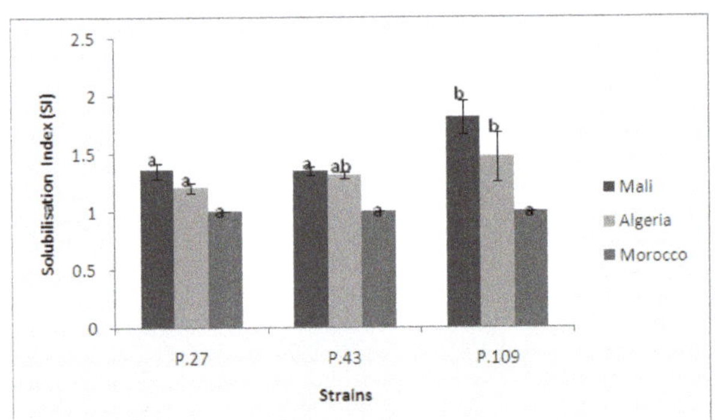

Figure 2. *Solubilisation index of strains on plates supplemented with rock phosphates of different origins. The different letters within the same phosphate type are significantly different ($p < 0.05$).*

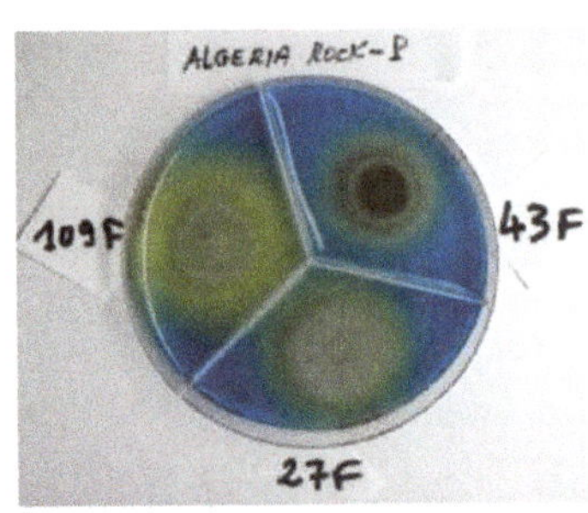

Figure 3. *Halo (yellow) zone displayed by the different strains on plate supplemented with Algeria rock phosphate. The yellow colour indicates the drop of pH associated to the process of phosphate solubilisation.*

3.3.2. *Efficiency of Strains according to their SI*

According to figure 4, the strain P.109 appears to be the most efficient on plate, independently to phosphate type, while strains P.27 and P.43 displayed the equal aptitude.

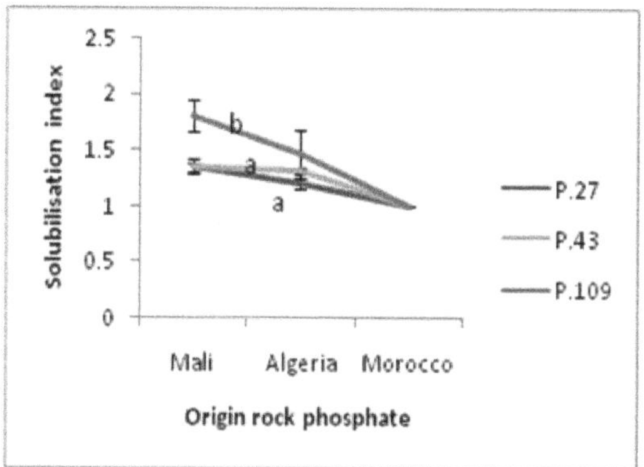

Figure 4. *Aptitude of strains in solubilising rock phosphates of different origins on plates. The different letters indicate a significant difference between strains regardless the phosphate type.*

3.3.3. *Fluency of the Different Rock Phosphate to be Solubilised on Plates*

The Malian rock phosphate appears the easiest to be solubilise on plate (Figure 5), followed by Algerian rock phosphate, while Moroccan rock phosphate seems to be the most recalcitrant on plate.

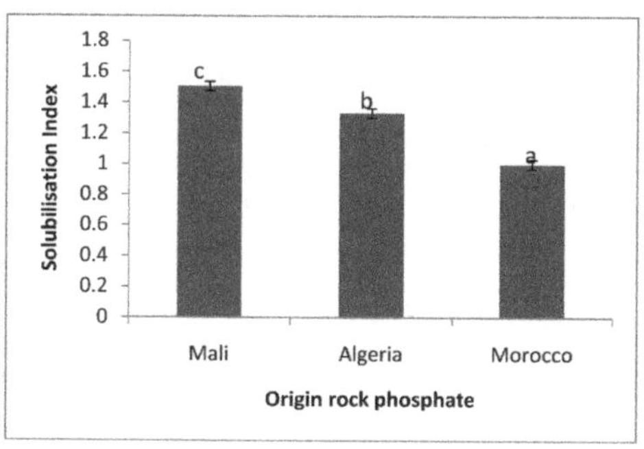

Figure 5. *Fluency of the different rock phosphates to be solubilised by the strains on plates. The different letters indicate a significant difference between rock phosphates ($p < 0.05$).*

3.4. *Strains Characterization in Liquid Cultures Supplemented with Rock Phosphates of Different Origins*

3.4.1. *Solved Phosphorus and pH in Liquid Cultures*

Important amount of phosphorus was mobilised from the insoluble rock phosphate sources (Figure 6a) by all the strains compared to the non-inoculated control. Significant differences ($p < 0.05$) are observed within control and strains regarding all the phosphate types. The solubilisation of rock phosphate is associated to the pH decrease of the media amended with the different P types (Figure 6b). While the pH of control media is maintained at neutral, that of the inoculated media is almost acidic, and the acidity seems to be linked to the amount of the solved P for strains P.109 and P.27.

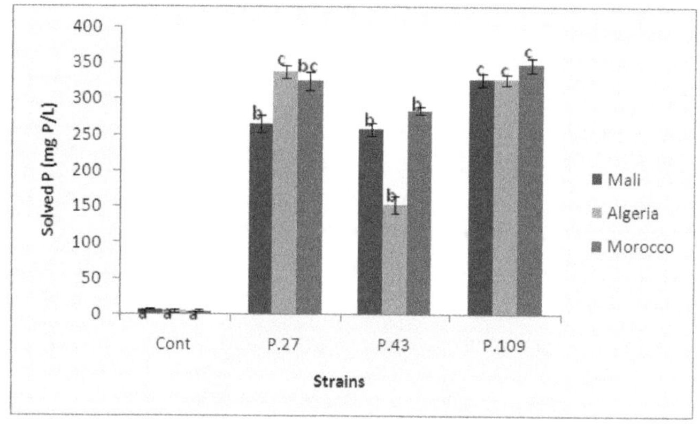

Figure 6a. *Concentration of the solved P in liquid cultures amended with rock phosphates of different origins. The different letters within the same phosphate type are significantly different ($p < 0.05$).*

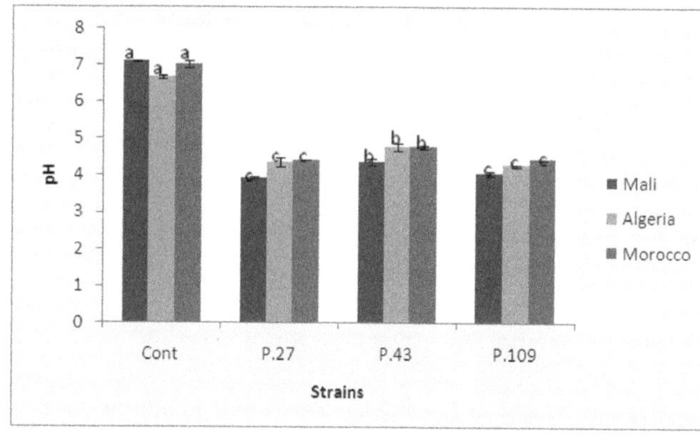

Figure 6b. *pH of the corresponding media after seven days of incubation. The different letters within the same phosphate type are significantly different ($p < 0.05$).*

3.4.2. *Efficiency of Strains according to the concentration of the Solved P*

While significant amount of P are mobilised by all the strains compared to control, the highest amount is mobilised by strains P.109 and P.27 (Figure 7). Nevertheless, P.43 shows high performance regarding Malian and Moroccan rock phosphates, but with low activity on Algerian rock phosphate.

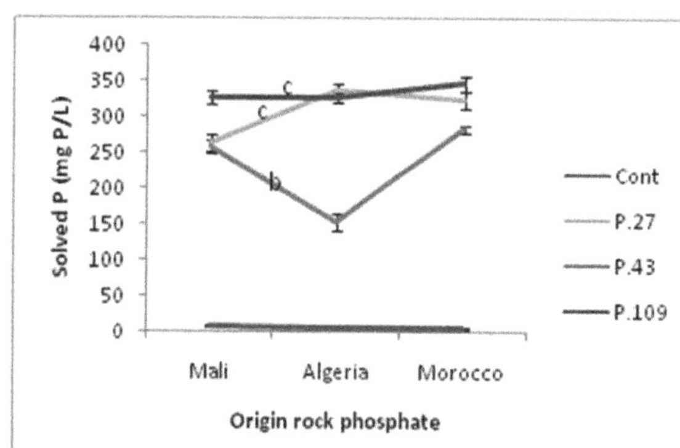

Figure 7. *Aptitude of strains in solubilising rock phosphates of different origins in liquid media. The different letters indicate a significant difference between strains regardless the phosphate type ($p < 0.05$).*

3.4.3. Fluency of the Different Rock Phosphate to be Solubilised in Liquid Media

Contrary to the results obtained on plates, the Moroccan rock phosphate seems to be easily mobilised by strains in liquid medium, compared to Malian and Algerian rock phosphates (Figure 8). So, all the strains that did not show any halo zone on plates amended with the Moroccan rock phosphate were able to mobilise significant amounts in liquid media containing the same rock phosphate type.

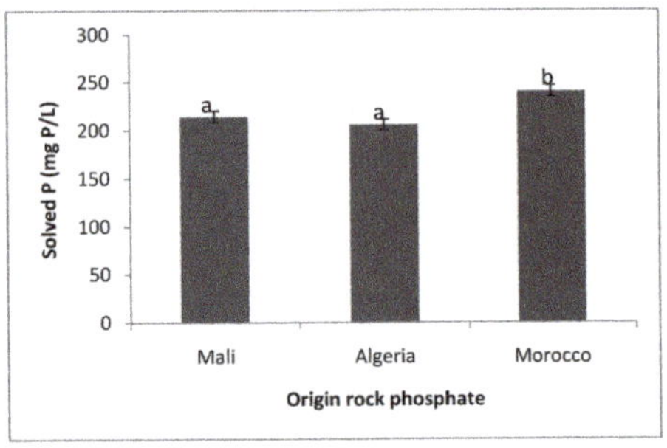

Figure 8. Facility of the different rock phosphates to be solubilised by the strains in liquid media. The different letters indicate a significant difference between rock phosphates ($p < 0.05$).

3.5. Correlation between Concentration of the Solved P and the pH of the Corresponding Media

Strong correlation was observed within the solved P and the corresponding media. Here, higher the amount of the solved P, lower is the pH of the corresponding media.

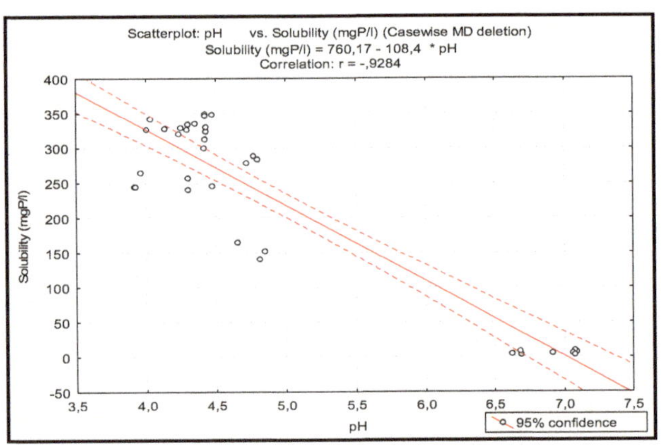

Figure 8. Correlation between pH and the solved P at the end of the incubation time ($p < 0.05$).

4. Discussion

The first phosphate solubilising activity of the strains was assessed on agar plates containing sparingly soluble inorganic phosphates (Ca-P, Al-P, Fe-P), one organic phosphate (Na-phy) as well as rock phosphate plus dye for better observation [9,12,13]. This reaction, shown as a halo or clear or yellow (in the present case) zone on the plate is used to assess the P solubilisation activity of strains [14], with the SI value as an indicator for the strain efficiency. Thus, the higher the SI value, the greater is the activity of the tested strain. The yellow colour associated to the strain activity indicates the drop of pH at this place, related to the change of the dye (Bromocresol green), moving from green (pH greater than 6) to yellow (pH ranging from 5.4 to 3.8). The strains were considered as phosphate solubilisers (PS) or rock phosphate solubilisers while solubilising at least one of the inorganic/organic phosphate types and one rock phosphate. Some *Penicillium* species have been characterized for their inorganic and organic phosphate solubilisation as well as rock phosphate solubilisation ability. These include *Penicillum rugulosum* [15,16]; *P. oxalicum* [9], *P. Bilaiae*, *P.simplicissimum* and *P. minioluteum* [17]. Although P.109 was not able to solubilise sparingly soluble Fe-P and Na-phy on plate, it appears the best rock phosphate solubiliser on plate, independently to rock phosphate origin. While comparing the relative efficiency of strains using the different RP, it was observed that RP solubilisation depended on the nature of RP and the organism, and the relative efficiency of strains for RP solubilisation may have been due to the nature and quantity of organic acids secreted in the medium [8]. Overall, fungi showed better solubilisation of all the rock phosphates. However, the degree of phosphate solubilisation varied with the type of organisms involved as well as the rock phosphate origin. None of the strains showed any halo zone on plates amended with Moroccan rock phosphate. They conversely mobilised an important amount of P in liquid culture supplemented with the same rock phosphate. These contradictory results between plate halo detection and phosphate solubilization in liquid cultures were found by Fankem et al. [13] and Deubel and Merbach [18]. This indicates that liquid culture should be associated to halo zone for rock phosphate solubilisers characterization [4]. However, while screening a large number of microorganisms, the halo detection method can be regarded as generally reliable for isolation and preliminary characterization of rock phosphate-solubilising microorganisms. In agreement with the present findings, two fungal strains, *Penicillium bilaii* and *P*. cf. *fuscum* were found to solubilise relative amounts of rock phosphates in liquid culture [19], and the inorganic P as well as the rock phosphate solubilisation were directly related to the pH drop generated by each strains [20]. In the present study, a strong correlation could be established between the quantity of phosphate solubilised and drop in pH. This is not in agreement with the findings of Narsian and Patel [8], but could be explained by the significant positive correlation observed by Babana et al. [4] between the solved P and organic acids produced by PSM, suggesting that organic acids may play an important role although it is not the only possible mechanism for P solubilisation. As rock phosphates are theoretically the cheapest fertilizers but with low P availability, the assessment of the three *Penicillium* strains as rock phosphate solubilisers, is important as rock phosphate can be directly applied to soil and their solubilisation optimized by these strains. This agrees with view of Yu et al. [21], Silva and Vidor [22] who investigated the solubilization of rock phosphate in liquid culture by *Aspergillus niger* and *Penicillium oxalicum*. However, more work is required regarding the nature of rock phosphates and their interaction with efficient PSM.

5. Conclusion

The results obtained on agar plates and in liquid media amended with either sparingly soluble phosphates or rock phosphates showed the real aptitude of the three *Penicillium* strains in mobilising important amounts of phosphorus from insoluble phosphate sources. The efficiency of strains was assessed on plates through the solubilisation index that is a quick way to characterize and select strains for their phosphate solubilising activity. Liquid culture experiment involved the P measurement as well as the pH of the corresponding media. All the three strains were capable to mobilise important amount of phosphorus from the insoluble rock phosphate sources, and the pH of the corresponding media were correlated to the amount of the solved P. These strains can be used for the solubilisation of fixed phosphates into the agricultural soil systems. However, the selection of phosphate solubilising *Penicillium* strains as possible inoculation tools for phosphate-deficient soils should focus on the integral interpretation of laboratory assays, greenhouse experiments as well as field trials.

Acknowledgments

The authors would like to thank the Rothamsted International, African Fellow Programme, London, UK for providing the fund and the James Hutton Institute, Dundee, Scotland where part of this work was done.

References

[1] S.S.S. Rajan, J.H. Watkinson, A.G. Sinclair, Phosphate rocks for direct application to soils, vol. 57. Adv. Agron. 1996, pp. 77-159.

[2] H.T. Rogers, R.W. Pearson, L.E. Ensminger, Comparative efficiency of various phosphate fertilizers, vol. 4. Agronomy, 1953, pp. 189-242.

[3] G.W. Cooke, The value of rock phosphate for direct application, vol. 24. Empire Journal of Experimental Agriculture 1956, pp. 295-306.

[4] A.H. Babana, A. H. Dicko, K. Maïga and D. Traoré, Characterization of rock phosphate-solubilizing microorganisms isolated from wheat (*Triticum aestivum* L.) rhizosphere in Mali, vol. 1(1). Journal of Mıcrobıology and Mıcrobıal Research 2013, pp. 1-6.

[5] A.C. Gaur, & K.P. Ostwall, Infuence of phosphate dissolving bacilli on yield and phosphate uptake of wheat crop, vol. 10. Indian Journal of Experimental Biology 1972, pp. 393-394.

[6] A.H. Babana, H. Antoun, Effect of Tilemsi phosphate rock solubilizing microorganisms on phosphorus-uptake and yield of field grown wheat in Mali, vol. 287. Plant and Soil, 2006, pp. 51-58.

[7] H. Hamdali, K. Moursalou, G. Tchangbedji, Y. Ouhdouch, H. Mohamed, Isolation and characterization of rock phosphate solubilizing actinobacteria from a Togolese phosphate mine, vol. 11(2). Afr. J. Biotechnol. 2012, pp. 312-320.

[8] Varsha Narsian, H.H. Patel, *Aspergillus aculeatus* as a rock phosphate solubilizer, vol. 32. Soil Biology & Biochemistry, 2000, pp. 559-565.

[9] R.S. Gadagi, T. Sa, New Isolation Method for Microorganisms Solubilizing Iron and Aluminium Phosphates Using Dyes, vol. 48(4). Soil Science and Plant Nutrition, 2002, pp. 615-618.

[10] M.A. Qureshi, Z.A. Ahmad, N. Akhtar, A. Iqbal, F. Mujeeb, M.A. Shakir, Role of phosphate solubilizing bacteria (PSB) in enhancing P availability and promoting cotton growth, vol. 22. The journal of animal & plant sciences, 2012, pp. 204-210.

[11] C.S. Nautiyal, An efficient microbiological growth medium for screening phosphate solubilizing microorganisms, vol. 170. FEMS Microbiology Letters, 1999, pp. 265-270.

[12] S. Mehta, S.C. Nautiyal, An efficient method for qualitative screening of phosphate-solubilizing bacteria, vol. 43. Current Microbiol., 2001, pp. 51-56.

[13] H. Fankem, M. Abba, L. Ngo Nkot, A. Deubel, W. Merbach, F.X. Etoa and D. Nwaga, "Selecting indigenous P solubilizing bacteria for cowpea and millet improvement in nutrient deficient acidic soils of southern Cameroon". In Innovations for a green revolution in Africa: Exploring the Scientific Facts. A. Bationo, B. Waswa, J. Okeyo, F. Maina (Eds.). Springer Science+Business Media B.V. 2011 (DOI 10.1007/978-90-481-2543-2_40), pp. 391-398.

[14] L.A. Fernandez, P. Zalba, M.A. Gomez, M.A. Sagardoy, Phosphate solubilization activity of bacterial strains in soil and their effect on soybean growth under greenhouse conditions, vol. 43. Biol. Fert. Soils, 2007, pp. 805-809.

[15] Reyes, R. Baziramakenga, L. Bernier, and H. Antoun, Solubilization of phosphate rocks and minerals by a wild-type strain and two UV-induced mutants of *Penicillium rugulosum*, vol. 33. Soil Biology and Biochemistry, 2001, pp. 1741-1747.

[16] Reyes, L. Bernier, H. Antoun, Rock phosphate solubilization and colonization of maize rhizosphere by wild and genetically modified strains of *Penicillum rigulosum*, vol. 44. Microbial Ecology, 2002, pp. 39-48.

[17] S.A. Wakelin, R.A. Warren, P.R. Harvey, M.H. Ryder, Phosphate solubilization by *Penicillium* spp. closely associated with wheat roots, vol. 40. Biology and Fertility of Soils, 2004, pp. 36-43.

[18] A. Deubel, W. Merbach, "Influence of microorganisms on phosphorus bioavailability in soils". In Microorganisms in soils: roles in genesis and functions. F. Buscot, A. Varma (eds). Springer, Berlin Heidelberg, 2005, pp. 177–191.

[19] P.E.A. Asea, R.M.N. Kucey, J.W.B. Stewart, Inorganic phosphate solubilization by two *Penicillium* species in solution culture and in soil, vol. 20. Soil Biology & Biochemistry, 1988, pp. 459-464.

[20] S.A. Omar, The role of rock-phosphate-solubilizing fungi and vesicular-arbusular-mycorrhiza (VAM) in growth of wheat plants fertilized with rock phosphate vol. 14. World Journal of Microbiology & Biotechnology, 1988, pp. 211-218.

[21] S.L. Yu, Y.N. Liu, G.L. Jing, B.J. Zhao, S.Y. Guo, Analysis of phosphate-accumulating organisms cultivated under different carbon sources with polymerase reaction-denaturing gradient gel electrophoresis assay, vol. 17. J. Environ. Sci., 2005, pp. 611–614.

[22] G.N. Silva, C. Vidor, Phosphate solubilising activity of microorganisms in the presence of nitrogen, iron, calcium and potassium, vol. 36. Pestic Agro Bras., 2001, pp. 1495–1508

15

Technical efficiency of vegetable farmers in Peri-Urban Ghana influence and effects of resource inequalities

Samuel Twumasi Amoah[1, *], Irene Akobour Debrah[2], Razak Abubakari[3]

[1]Department of Environment and Resource Studies, University for Development Studies, WA, Ghana
[2]Faculty of Education, University for Development Studies, WA, Ghana
[3]School of Business and Law, University for Development Studies, WA, Ghana

Email address:
atwumasi@uds.edu.gh (S. T. Amoah)

Abstract: In Ghana, statistics indicate that women account for about 70% of total food production and are the most important actors in the food chain which begins from the farm production, market and intra household distribution of food. They play a lead role in post-harvest activities such as shelling of grains, storage, processing and marketing. They are also becoming increasingly visible in farm tasks which traditionally have been designated as male preserve. Despite the increasing central role of women in food production, they have much more limited access to resources than their male counterparts especially in the areas of education, land, agricultural extension services and access to credit, all of which combine to restrain their ability to increase productivity. This study was therefore undertaken toexamine the technical efficiency of male and female vegetable farmers in the Kumasi Metropolis using the stochastic production frontier model. Female vegetable farmers were found to be producing at high levels of inefficiency. The predicted efficiencies differed substantially from between 2 and 85 percent, with mean efficiency of 24 percent. The low mean efficiency index is an indication of inefficiencies in resource use. Also, female headed farms recorded a mean technical efficiency of 16.5 percent with a range between 2 and 66 percent. The male headed farms, on the other hand, showed a mean technical efficiency of 30.8 percent, and a range between 2 and 85 percent. The results imply that on the average, female vegetable farmers were relatively technically inefficient than their male counterparts. The paper concludes that since women farmers contribute immensely of domestic food supply in Ghana, it is important that efforts be made to build their capacity to produce efficiently.

Keywords: Inefficiency, Stochastic Frontier Model,Peri-Urban, Vegetable, Farmers

1. Introduction

In Ghana, agriculture is probably, the most important national asset since the economy and majority of the population is dependent on it. This sector is the largest in the economy in terms of its contribution to the Gross Domestic Product (GDP) and employment. The sector accounted for 34.5% of GDP in 2009. With increased demand for agricultural inputs such as land and water resources, in a framework of increasing environmental concerns, it has become important that measures are taken not only to improve productivity levels of farmers but also to do so in a manner that can sustain the environment. Increasing the productivity of farmers will require improvements in production technology through research and development, as well as a strong extension base that can offer a robust platform for technology transfer and monitoring. This is to ensure adoption and feedback to research stations for continuous evaluation and improvements [16].

Agriculture's contribution to the overall economic growth and development of the country has been, and continues to be, very significant, notwithstanding Ghana's oil find [16]. It is therefore important that for Ghana to achieve Millennium Development Goal One (MDG1) which aims at eradicating poverty and hunger, there is the need to put pragmatic measures in place to implement appropriate policies and projects. These should aim at promoting and substantially increasing agricultural productivity so that the country can attain food self-sufficiency. This requires that all sub-sectors in the agricultural sector are given special attention. This is because any changes or developments made in the any of the sub-sectors can have a prominent knock-on-effect on the entire agriculture sector and the economy in general. Unfortunately, there seem to be over concentration on the traditional export crops such as cocoa, and coffee to the neglect of the other non- traditional crops banana, pineapple among others. However, low producer prices for the traditional crops on the international market, persistent threats from climate change and issues of food insecurity have brought to the fore the need to pay much attention to the less developed sub-sectors of the agricultural sector. One important sub-sector that is worthy of mention in this direction is the vegetable sub-sector.

The worldwide production of vegetables has doubled over the past quarter century and the value of global trade in vegetables now exceeds that of cereals.The vegetable sector is the source of livelihood for millions of growers and thousands of traders and processors. As of 2007, a total volume of over 880 million tonnes of vegetables were being produced globally and it was expected to reach the 1.2 billion tonnes by 2015 [16]. Vegetable is an important cash crop, as well as staple food consumed by majority of Ghanaians in both rural and urban communities across the country. The vegetable sector has seen lots of fluctuations at the production levels. For instance, in 2010, 410,794 tonnes of vegetables were produced but it reduced to 329,019 tonnes in 2011, representing a drop of 2,417 tonnes. This correspondingly affected total revenue accruing to the sub-sector. In 2010, total revenue from the sub-sector was GH□102,091,000 (40,371,477.71 USD) but in 2011 total revenue fell to GH□101,617,000 (40,184,036.21 USD)[16]. This was due to the liberalization of the domestic market in 1992 in the course of the structural adjustment programmes which started in 1983 and led to the invasion of the local vegetables market by foreign imports from neighbouring countries like Burkina Faso, Mali and Niger [14]. The importation of vegetables is ousting the local vegetable producers from the domestic market and that is making many local vegetables farmers lose their source of livelihood. Besides the huge importation of vegetables, local vegetable farmers are not technically efficient enough to meet the demand of the market. It is therefore, not surprising that the 800,000 Ghanaian vegetable farmers in the country could not satisfy the vegetable demand in 2011 [25]. The increasing volume of imported vegetables implies increasing expenditure on the commodity.

Factors such as climate change and poorly organized markets for local vegetables have been implicated for the low output which may account for the significantly higher imports experienced in recent years. The current widespread practice of traditional vegetable cultivation, which is dominated by women, is perceived to be the key reason for the deficit. The vegetable industry in Ghana can be regarded as having three distinct components: Commercial/Market gardening areas sited around the principal cities such as Accra, Kumasi, Takoradi and Tamale; a form of truck farming in which vegetables are produced in rural areas from where they are purchased by contractors or middlemen and transported by road to the cities and small domestic or backyard gardening [7]. The importance of vegetables cannot be over emphasized. Almost all vegetables are low in fat and calories, none has cholesterol, and many of them are great sources of fiber. The high levels of fiber in vegetables keep the digestive system healthier, allowing the individual to avoid issues with constipation. Since vegetables are low in calories, it enables individuals to eat lots of vegetables without consuming excess energy [18]. The presence of many vitamins and other substance in vegetables provide nutrients to the body. Vegetables provide essential amino acids that the body needs to survive.

Vegetable farming activities in tropical Africa has always been dominated by women who grow a substantial amount of the staple foods, yet, they have much less access to knowledge, technology, credit and land than their male counterpart [24]. Generally, resource constrained women farmers on the African continent are believed to produce up to 70-80% of the domestic food supply in most sub-Saharan African societies [17]. On average, they also provide 46% of the agricultural labour. This notwithstanding, women's yields are too low by Green Revolution standards (3 to 4 tonnes per hectare). Again, women's yields are much lower than men's yields in societies where a comparison can be made (ibid). According to Quisumbing [31]., gender differences in productivity goes beyond management styles to include differences in the intensity of use of productive inputs such as fertilizer, manure, labour, credit, extension training, and education. It has also been established by research that The Structural Adjustment Programme (SAP) embarked upon by Ghana further worsened the plight of women in agriculture who are already disproportionately burdened. The SAP was accompanied by the removal of subsidies from agricultural inputs and rather introduced trade liberalization measures which eventually resulted in the importation of vegetables that were hitherto being produced locally [29]. This situation had a telling effect on the local economy particularly local vegetable producers, especially women farmers [22]. It is for this reason that estimates by Udry [36]opines that if productive inputs such as fertilizer, manure and credit could only be reallocated within the African household, food crop production could increase in the range of 10-20%. For anyone interested in women's welfare, the cultivation of African traditional vegetables offer an important entry point. Vegetable cultivation provides an important economic pillar upon which women's livelihood is supported. This is because the production, handling and marketing are mostly done by women [27].

Vegetable farming which is mostly done in urban and peri-urban areas in Ghana has unfortunately not received the appropriate public and institutional support despite its significant contributions to urban food security, poverty alleviation, women empowerment and improved human nutrition through the provision of balanced diets. Ghana has a high potential and positive comparative advantage for vegetable production but considerable evidence suggests that serious bottlenecks exist in the functioning of the production system. Anecdotal evidence and inquiry suggest that, a number of factors are responsible for the low vegetable production at the household level. The question then arises as to how efficient farmers are using or combining the available scarce resources at their disposal to produce the maximum desired output.

The vegetable production sector in Ghana is largely not properly institutionalized, not properly organized and inefficient [25]. The sector is plagued with Post-harvest problems. It is also fragile to harsh weather conditions at the retail level and this often results in high losses, which sometimes serves as disincentive and discourages producers, marketers and consumers [25]. The problems are acute for dry season vegetable crops. The depth of research and knowledge on the level of production efficiency and on ways to improve the efficiency of vegetable production in Ghana is limited. The literature search on the use of stochastic frontier approach to determine the technical efficiency of vegetable farmers yielded very scanty results. While it is obvious that the vegetable production sector in Ghana is not efficient, knowledge about the exact level of is quite blurred. It is also not clear what the impediments, particularly the extent of their impact to efficient vegetable production are. In order to adopt measures in solving the problem of inefficiency in the vegetable production sector, there is need to obtain more specific evidence as to the magnitude of inefficiency. These are key issues central to this study and whose investigation can be useful for the formulation of policies to strengthen and improve the efficiency of vegetable production system

The study is organized in four main parts. After the introduction, the next section outlines the methodology adopted for the study, which draws on the stochastic frontier production function model and the last two sections are devoted to the results and policy implication of the study.

2. Materials and Methods

2.1. Study Area

Kumasi is located in the transitional forest zone and is about 270km north of the national capital, Accra. It is between latitude 6.35° – 6.40° and longitude 1.30°C – 1.35°C, an elevation which ranges between 250 – 300 metres above sea level with an area of about 254 square kilometres. The unique centrality of the city as a traversing point from all parts of the country makes it a special place for many to migrate to. The Metropolis falls within the wet sub-equatorial type. The average minimum temperature is about 21.5°C and a maximum average temperature of 30.70c. The average humidity is about 84.16 per cent at 0900 GMT and 60 per cent at 1500 GMT. The moderate temperature and humidity and the double maxima rainfall regime (214.3mm in June and 165.2mm in September) have a direct effect on agricultural activity.

Figure 1. *Vegetable Production Sites in Kumasi (Adams Abdulai, [2])*

The major soil type of the metropolis is the Forest Ochrosol. The detailed soil associations are the following: Kumasi - Offin Compound Association; Bomso – Offin Compound Association; Nhyanao - Tinkong Association; Bomso – Suko Simple Association; Bekwai – Oda Compound Association and Bekwai – Akumadan – Oda Compound Association. It is a very rich type of soil that has made it possible for a lot of foodstuff (vegetables, plantain, cassava etc) to be grown in the periphery.Agriculture in the metropolis has seen a dramatic change in the last two decades due to rapid urbanization. The demand for residential, industrial and commercial land uses has become much greater than that of agricultural land use. Following this, it has been estimated that about 80% of the arable lands have been displaced by the construction of houses and other physical infrastructure.It has been estimated that the metropolis has 12,000 hectares of irrigable lands consisting of swampy and marshy areas (Metro.Agriculture Directorate). Agricultural land use in the metropolis has been consigned to crop farming in the peri-urban communities (eg. Dichemso, Takyiman, Parkoso, Apeadu, Kokoben etc) and along the banks and valleys of rivers/streams. Some of the common vegetables that are commonly cultivated by the farmers include,Cabbage, Carrots, Green pepper, Lettuce, Garden eggs, Pepper, Okra and Shallots among others. The main locations for vegetable cultivation are Gyinyasi, KNUST, Manhyia, Kakaro, Georgia and Asokore Mampong. Besides vegetable production, the cultivation of food crops like yam, cocoyam, plantain, cereals, legumes and fruits take place concurrently.

2.2. Stochastic Frontier Production Model

The Stochastic Production Frontier (SPF) is used to study technical efficiency of production. The model was originally developed by Aigner, Lovell, and Schmidt [4] The production and cost analysis in the stochastic frontier framework involves two steps. With the first step, the frontier model is estimated, usually by maximum likelihood while in the second, the model so estimated is used to determine measures of inefficiency or efficiency. The technical relationship between inputs and outputs of a production process is described by a production function which establishes the maximum level of output attainable from a given measure of input when using the stochastic frontier approach. This is the reason why it is called the production frontier.

The SPF has been more popular because it can handle statistical noise, resulting in more accurate specification. A more complete specification is essential for accurate efficiency measures since the estimated frontier is conditional on the functional form. In the writings of Bravo-Ureta and Reiger [8], modelling production functions using the stochastic frontier analysis is in consonance with production theory. According to Kumbhakar and Lovell [23], there are two main objectives in stochastic frontier analysis. First, it serves as a benchmark against which to estimate technical efficiency of producers. Second, it incorporates exogenous variables that are neither input nor output in the production process, but which nonetheless affect producer performance. Although, the stochastic frontier method has the foregoing strengths; it has been criticized on the grounds that there is no a priori justification for the selection of any particular distributional form for the technical inefficiency term, *ui.* The stochastic frontier analysis concerns the estimation of frontiers, which envelop data, rather than with functions which intersect data [23]. The frontier production shows the maximum output obtainable from given quantities of inputs representing maximum efficiency. Technical inefficiency is measured from the frontier level. The stochastic frontier production function specification enables the separation of output shortfalls due to technical inefficiency from those caused by random disturbances.

The stochastic frontier production function model used by Parikh and Shah [30] begins by considering a stochastic production function with a multiplicative disturbance term of the form:

$$y = f(x;\beta)e^{\varepsilon} \quad (1)$$

where:

y is the quantity of agricultural output

x is a vector of input quantities

β is a vector of parameters and

e is a constant/exponent

ε is a stochastic disturbance term consisting of two independent elements u and v, with

$$\varepsilon = v - u \quad (2)$$

Equation (1) can thus be re-written as:

$$y = f(x;\beta) \cdot exp\ (v - u) \quad (3)$$

The measure of technical efficiency for each farm can be calculated using the formula:

$$TE = exp \cdot [E\ \{u/\varepsilon\}] \quad (4)$$

And u in equation (4) is defined as:

$$u = f\ (Z_b; \delta) \quad (5)$$

Where Z_b is a vector of farmer-specific factors and δ is a vector of parameters.

The symmetric component, v, accounts for random variation in output due to factors outside the farmer's control, such as weather and diseases. It is assumed to be independently and identically distributed as $N{\sim}(0, \sigma_v^2)$. A one-sided component $u \geq 0$ reflects technical inefficiency relative to the stochastic frontier. The advantage of this model is that it allows for simultaneous estimation of the production part of the frontier model and the individual technical efficiency/inefficiency component.

2.3. Data Collection

This cross-sectional survey utilized data from vegetable producers in the Kumasi Metropolitan Area (KMA). Five peri-urban communities were purposively selected based on the concentration of vegetable producers. These villages are Ayigya, Gyenyaase, Karikari, Georgia and D-Line. At least 50 farmers were randomly selected from each of the vegetable producing communities. Respondents in the survey included farmers who grow a particular type of vegetable on a particular piece of land and farmers who grow two types of vegetables on the same piece of land at different times of the year. Farmers were selected on the basis of gender, input and output per head. A total of 250 farmers were interviewed. At least, 6 out of 10 respondents were men. Additionally, five different focus group discussions were held, one in each of the five communities selected for the study. Each focus group had seven participants of which four were males and three were females with average age of 33 and 30 respectively. With respect to educational qualification, on the average highest level was Junior High School. In all, more males took part in the study than females and this is explained by the land tenure arrangement in the study area which favored males. Some of the issues that were probed included access to land and its security, availability of fertilizers, loans, farm inputs and improved seeds among others. The discussions were done in the local language and later transcribed. The purpose of the FGD was to stamp or otherwise, the responses obtained from the questionnaire interview. This approach is deemed appropriate when the object of the research is to explore altitudes or reactions of a group or community in response to some commonly experienced aspects of their environment [37] Through such interactive discourse, participants are able to offer insights on the perspective of the community, revealing clues to the social contexts that shape their opinions ([33]; [34]).

3. Results and Discussions

3.1. Results

Our hypotheses were stated in null terms. The first null hypothesis stipulates that there is no technical inefficiency effect in the production function, see Table 1. This null hypothesis is rejected for vegetable producers in the study area. Thus, it can be concluded that the explanatory variables in the efficiency model do contribute significantly to the explanation of the technical inefficiency effects for the vegetable producers. As pointed out by Coelli and Battese [11], if a null hypothesis includes $\gamma = 0$, then the statistic has asymptotically a mixed Chi-squared distribution, since by its definition γ has to be non-negative.In the second null hypothesis, we state that the variables included in the inefficiency effect model have no effect on the level of technical inefficiency. This null hypothesis is also rejected for vegetable producers, showing that the joint effect of these variables on technical inefficiency is statistically significant. A standard *t*-test for independent samples was used to test for statistical significance of the difference in gross revenue of male vegetable farmers and their female counterparts in the KMA. The mean difference for the gross revenue of male vegetable farmers and their female counterparts is significantly different from zero. The null hypothesis of equality of means is rejected under variance assumption ($Pr(T < t) = 0.009$). The gender index is negatively associated with the level of technical inefficiency at the 1% level of significance. This implies that men are more productive than the women in the area so far as vegetable production is concerned.

Table 1. *Hypothesis tests*

Null Hypotheses	Likelihood	χ^2statistic	$\chi^2_{0.95}$
1. H_o: $\gamma = 0$	-162.5	70.3	11.07
Decision: Reject the null; no technical inefficiency in the model			
2.$H_0: \delta_1 + \ldots + \delta_6 = 0$	-365.13	113.4	13.4
Decision: Reject the null; variables included in the inefficiency effect model have no effect on the level of technical inefficiency			

*The critical values for the test of null hypothesis was obtained from table 1 of Kodde and Palm [20]

3.1.1. Stochastic Production Frontier Model Approximation

The utmost likelihood estimates of the model parameters are computed using the frontier models routine of the Statistical package STATA with a Cobb-Douglas functional form. The real investigations for the occurrence of inefficiency were calculated by estimating the stochastic frontier production function and conducting a likelihood-ratio test assuming the null hypothesis of no technical inefficiency. This test statistic is computed automatically when the frontier model is estimated using STATA. Table 2 is the presentation of the parameters and related statistical test obtained from the stochastic frontier production function analysis. The estimated coefficient for farm size, labour, fertilizer and seeds are positive and statistically significant at least at the 5% level. A 1% increase in expenditure on fertilizer, and seed will lead to about 0.95% and 1.87% increase in gross revenue respectively (all things being equal).

Table 2. *Maximum Likelihood Estimates of The StochasticProduction Function*

Variable	Coefficient	Standard Error	T-statistic
Production part			
Constant	14.360*	7.66	0.061
Farm Size	0.667**	0.303	0.026
Labour	0.148**	0.069	0.012
Herbicides (cedis)	0.881	0.634	0.164
Fertilizer (cedis)	0.950**	0.473	0.037
Seed (cedis)	1.875***	0.536	0.008
Inefficiency effect (μ)			
Constant	-0.654	0.982	0.666
Gender of the farmer (male=1, 0 otherwise)	-0.011***	0.003	3.667
Age (years)	0.413***	0.015	27.533
Education (years)	-0.517	1.223	0.131
Farming experience (years)	-0.114***	0.012	-9.500
Extension contacts	-0.948	0.612	-1.549
Credit (Access=1, 0 otherwise)	-2.400***	0.166	-14.458
Variety of vegetables seed (improved= 1, 0 otherwise)	-1.056***	0.152	-6.947
Model diagnostics			
Likelihood ratio (1)	-162.50		
Sigma square	30.32		
Gamma	0.34		
Prob> Chi 2	0.008		
Wald Chi2 (6)	70		
Wald Chi2 (4)	113		
Likelihood ratio (2)	-365.5		

******and * represent statistical significance at the 0.01, 0.05and 0.1 levels, respectively.Source: computed from survey data, 2012

Similar studies by Chinwuba and Odjuvwuederhie [10] found that expenditure on labour, fertilizer and seed yams, which constituted the material input variables shown by the frontier production function, positively affected revenue of farmers in Southeastern Nigeria [10]. Fertilizer is a major land augmenting input that increases yield per hectare by improving fertility of the soil. Our result also confirms the work of Weir [38] who found fertilizer to have a positive and significant impact on output. Abdulai and Huffman [3] found a negative relationship between the use of fertilizer and the level of profit inefficiency in the Northern region of Ghana. The estimated coefficient of herbicide though positive is not statistically significant even at the 10% level of significance. This implies that expenditure on herbicide will not lead to any significant increase in the gross revenue of vegetable producers in the area (Table 2).

3.1.2. Efficiency Levels

The sources of efficiency/inefficiency were examined using the estimated δ coefficients associated with the inefficiency effect. The inefficiency effects examined are those relating to age, gender, farming experience, education, access to credit, extension and seed variety. A negative sign on a parameter explaining inefficiencies means that the variable is improving technical efficiency, while for a positive sign, the reverse is true. As indicated in Table 2, the estimated coefficient on age was positive and statistically significant at the 1% level. This implies that farmers who had advanced in age were relatively less efficient in vegetable production. Since labour productivity decreases with age, younger farmers tend to be more productive than their older counterpart. The estimated coefficient on gender of the farmer (male = 1, female = 0) is negative and in consonance with *a priori* expectation. It is statistically significant at the 1% level. This implies that inefficiency is less on male plots than their female counterparts. The result is consistent with the findings of Due and Gladwin [13]. Also in relation to the findings of Akinwuni and Djato [5], they emphasized that the relative inefficiency of women vegetable farmers is not just because they are female but rather due to constraints that disproportionately affect them.

The estimated coefficient on years of experience is negative, conforming to *a priori* expectation, and it is statistically significant at the 1% level. The implication is that farmers with more experience in vegetable production were more efficient than the inexperienced ones in the area. The result is consistent with the work of Bravo-Ureta [9] who observed positive relationship between economic efficiency and experience in a study of dairy farms in New England. Vegetable farmers in the study area tend to use the knowledge acquired through experience on soil, and crop management in their farm operations.

The seed variety dummy was found to be negative and statistically significant at the 1% level. This implies that those farmers who adopted new vegetable technology in the form of improved vegetable seed were more efficient than those who used the traditional vegetable technology, a result consistent with that of Seyoum *et al.* [35] in Ethiopia. The estimated coefficient on education is not statistically significant but has the expected negative effect on inefficiency. The result, however, implies that the years of education tend to have no significant effect on the inefficiency levels of the vegetable farmers in the study area. The result is consistent with findings by Bravo-Ureta and Evenson [9] on peasant farmers in Paraguay. Seyoum *et al.*, [35] demonstrated that education does not significantly affect the efficiency of farmers using traditional methods. Weir [38] also found a threshold effect that implies that at least four years of schooling are required to lead to significant effects on farm level technical efficiency, irrespective of gender.

The estimated coefficient on access to credit was negative, agreeing with *a priori* expectation, and statistically significant at the 1% level. This suggests that farmers who have access to credit tend to be more efficient in vegetable production. This finding is consistent with the work of Abdulai and Eberlin [1], which shows a positive association between credit and input use and farm productivity in Nicaragua. The availability of credit helps to finance the procurement of material inputs which have a positive effect on vegetable production.

The estimated coefficient on number of contact with extension officers was negative according to *a priori* expectation but is not statistically significant even at the 10% level. The result implies that more contact with extension officers tend to have no significant effect on the inefficiency levels of farmers in the study area. The result is not consistent with findings obtained by other researchers ([6]; [28]) because it is expected that more contact with extension workers will increase the farmer's likelihood of adopting improved vegetable technologies which will eventually increase the efficiency level of the vegetable farmer. The result is not surprising because the rate of adoption of land augmenting technologies such as improved seed and fertilizer is very low in the KMA. About 70% of the respondents use the traditional input. Thus, knowledge disseminated by extension officers to the vegetable farmers may not affect their output significantly, as the knowledge is on improved vegetable variety.

The inefficiency component of the disturbance term (μ) of the estimated frontier model is significantly different from zero ($p = 0.008$) indicating the presence of statistically significant inefficiency in the data. The likelihood ratio, sigma square and gamma parameters present results on the behavior of the error term outlined in the inefficiency model. The gamma measures the percentage variations in plot output due to technical inefficiency. If gamma is significantly different from zero it implies that there is technical inefficiency in vegetable production. The percentage variation in plot output due to technical inefficiency (gamma) is 34%. The maximum likelihood estimates of the inputs parameters show the expected signs for all production inputs.

3.1.3. *Efficiency Indices among Vegetable Producers in the KMA*

Table 3 indicates that female vegetable farmers recorded a mean technical efficiency of 16.5 percent with a range between 2 and 66 percent and a standard deviation of 21.7 percent. The male vegetable farmers, on the other hand, showed a mean technical efficiency of 30.8 percent, a range between 2 and 85 percent and a standard deviation of 26.5 percent. The results imply that on the average, male vegetables producers are relatively technically efficient than their female counterparts, a result that is statistically significant at the 1% level.

Table 3. *Distribution of Efficiency Indices amongVegetables Producers*

Efficiency Index	Male %	Female%	Percent
0- 10	12	46	29
11- 20	34	26	30
21- 30	24	18	21
31- 40	6	4	5
41- 50	2	2	2
51- 60	10	2	6
61- 70	6	2	4
> 70	6	0	3
N	100	100	100
Min	2	2	2
Max	85	66	85
Mean	30.8	16.5	24.5
STD	26.5	21.7	23.3

Given the specification of the Cobb-Douglas production function, the predicted efficiencies differ substantially among farmers, ranging between 2 and 85 percent, with mean efficiency of 24.5% and a standard deviation of 23.3%. This low mean technical efficiency is an indication of high inefficiency in resource use. The distribution of the technical efficiency level in Table 3 shows that vegetable production on most plots is technically inefficient. For instance, the modal efficiency score group is 11-20 accounting for 30percent of the total sample of vegetable farms, followed by 0-10 accounting for 29percent. Only 3percent of vegetable farms have technical efficiency scores of more than 70 percent. The mean technical efficiency for the area is lower than that for most farmers in African countries whose mean technical efficiency range between 55 and 79 percent.

This finding is similar to the studies of Heshmati *et al.* [19] who obtained mean technical efficiency of 25 percent for the cereal farmers in rural Uganda and Weir *et al.* [39] who recorded a mean technical efficiency of 35percent for cereal crop farmers in rural Ethiopia. The mean technical efficiency observed in the study area is also close to that of Mochebelele and Winter-Nelson [26] who obtained mean technical efficiency of 36 percent for rural households that send migrant workers from Lesotho to South African mines and 24 percent for households that do not send migrant labour.

In figure 2, the modal efficiency class for the women vegetable producers was 1-10 and about 72 percent of the women's farms have technical efficiency levels below 21 percent. Their male counterparts on the other hand have their modal class as 11-20 and about 46 percent of the men's farms have technical efficiency levels below 21 percent. The result implies that technical efficiency is relatively higher on the men's vegetables farms than their female counterparts.

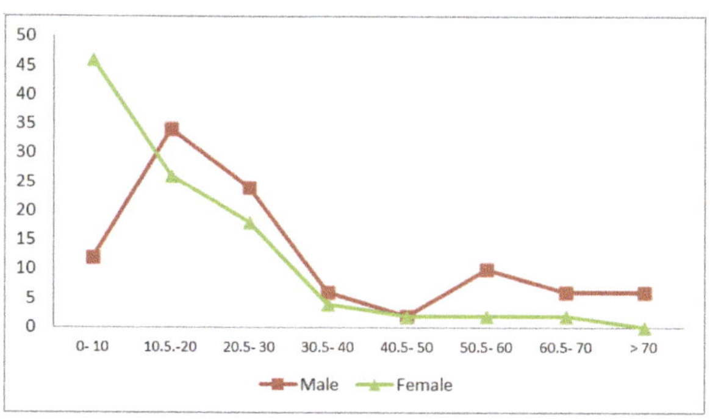

Figure 2. *Distributions of Farm-specific Efficiencies in the Stochastic Production Frontier*

3.2. Discussions

Vegetables are reputed to enrich household diets [21]. Adding to the debate, Dittoh [12] opines that vegetables add flavor to the food and also provide considerable protein, vitamins and minerals. He further asserts that most vegetables are low in starch content and are a good source of phyto- nutrients. They serve as roughage, which promotes digestion, and prevent constipation. We contend that if scarce production resources are efficiently allocated and utilized by vegetable farmers, not only will it provide employment, it will also improve the food self-sufficiency ratio of the country and eventually, Ghana will be a net exporter of vegetables. As intimated previously, the maximum likelihood approximation of the factor of the stochastic production function and inefficiency effects were calculated using the STATA software. This package estimates the two models simultaneously. To test for the functional form, the respective likelihood ratio tests were computed and their results compared with the critical values obtained from Kodde and Palm [20].

Seven factors were identified as sources of inefficiency/efficiency of the vegetable farmers. These were years of formal education, gender, access to credit, contact with extension officers, age, the farmer's experience in vegetables production and the seed variety planted by the farmer. The gender index was found to be a significant factor in reducing inefficiency in the area.

The seed variety index was found to be negative and statistically significant. This implies that those farmers who adopted new variety of vegetable seed were more efficient than those who used the traditional vegetables technology. The hypotheses tested show that first, all farmers are not operating at technically efficient level; secondly, expenditure on material inputs such as fertilizer and vegetable seed positively affect gross revenue; third, the variables included in the inefficiency model explain the observed variations; and fourth, there is relatively less observed inefficiency on the farms of male vegetable farmers compared to their female counterparts.

4. Conclusion and Policy Direction

The main purpose of this study was to examine the differences in technical efficiency among male and female vegetable farmers in the peri-urban areas of the Kumasi Metropolitan Area. Technical efficiency estimates were obtained using the Stochastic Efficiency Frontier Model. The results show that efficiency levels were significantly different across all production units. While some production units were efficient, there was some inefficiency in the production system. This suggests that a significant proportion of the error term in the production is explained by inefficiency effects. The frontier model used in this study was a static model. It must be emphasised at this point that our results only encapsulate the current levels of efficiency of the production units, which could change with time.

The results of the research show that there is more than enough opportunity for the vegetable farmers to increase production based on the existence of wide variations in the current level of productive efficiency. One of the variables that were significant in influencing production efficiency was age. Old age for example was associated with higher levels of inefficiency. It is therefore important that specific policies on vegetable production should be youth oriented.

Similarly, increasing and improving access to credit and extension services should also be a priority of policy makers as financial support and extension services will enable farmers acquire the much needed input and new knowledge in the industry and this will eventually increase efficiency and productivity of farmers. Finally, it is recommended that gender-sensitive safety net programs that recognize women as vegetable producers will be worth an effort to reduce the inefficiency levels of vegetable producers in the KMA of Ghana. Vegetable farmers, especially the women, need to be better supported, if Ghana is to ever experience an increase in local vegetables production.

References

[1] Abdulai, A. and Eberlin E"Technical Efficiency During Economic Reform in Nicaragua: Evidence From Farm Household Survey Data". Economic Systems,2001(25):113-125.

[2] Abdulai, R. T, "Is land title registration the answer to insecure and uncertain property rights in sub-Saharan Africa?" *RICS Research Paper Series,* 2006, Volume 6(6), pp. 1–28

[3] Abdulai, A., and Huffman, W,"Structural Adjustment and Economic Efficiency of Vegetables Farmers in Northern Ghana. Economic Development and Cultural Change, 2000(504-519).

[4] Aigner, D., Lovell, C.A., Knox, and Schmidt, P,"Formulation and estimation of Stochastic Frontier Production Function Models". Journal of econometrics, 1977(6) 21-37.

[5] Akinwumi, A., and Djato, K. K,"Relative Efficiency of Women as Farm Managers: Profit Function Analysis in Cote d'Ivoire": Agricultural Economics, 1997(16), 47-53.

[6] Ali, M., and Byerlee, D,"Economic Efficiency of Small Farmers in a Changing World: A survey of recent evidence. Journal of International Development, 1991, 13 (1): 1-27.

[7] Ayamba, I,"Backyard Crop Production in KMA". In: Kumasi Natural Resources Management Research Project (KNRMP)-Kumasi Urban Natural Resources Studies. R67-99. 1999.

[8] Bravo-Ureta B.E. and Reiger L,"Dairy Farm Efficiency Measurement Using Stochastic Frontiers and Neoclassical Duality", American Journal of Agricultural Economics, 1991, 73 (2): 421-428.

[9] Bravo-Ureta, B. E. and Evenson, R. E,"Efficiency in Agricultural Production, the Case of Peasant Farmers in Eastern Paraguay". Agricultural Economics, 1994(10) 27-37.

[10] Chinwuba P. I. and Odjuvwuederhie E. I,"Determinants of Yam Production and Economic Efficiency of Smallholder Farmers in Southern Nigeria". Journal of Central European Agriculture, 2006, 7(2): 337-342.

[11] Coelli T. J and G. E .Battese,"Identification of Factors which Influence the Technical Efficiency of Indian Farmers"; Australian Journal of Agricultural Economics, 1996, Vol. 40, pp103-128.

[12] Dittoh, S,"The Economics of Dry Season Vegetable Production in Nigeria". Acta Horticulture, 1992, No. 296, pp. 257-264.

[13] Due, J.M. and Gladwin, C. H,"Impacts of Structural Adjustment Programme African Women Farmers and Female-headed Households". American Journal of Agricultural Economics 73 (1991): 1431-1439.

[14] FAO,Statistical Year Book, Food and Agriculture Organization (FAO), Rome. 2005

[15] FAO, Agricultural data FAOSTAT. Food and Agriculture Organization of the United Nations. Rome, Italy. 2006.

[16] FAO, Country Statistics: Ghana. Last accessed in 2011. Available at http://countrystat.org/gha/cont/pages/page/indicators/en

[17] Gladwin, C.H., McMillan, D,"Is a turnaround in Africa Possible Without Helping African women to Farm?" Economic Development and Cultural Change, 1989(37) 345-369.

[18] Hart, F.L. And Fisher, H.J,"Modern Food Analysis". Springer, Berlin,Germany. 1971, pp:371-373

[19] Heshmati, A. and Mulugeta, Y. Technical Efficiency of the Ugandan Matoke Farms, Applied Economic Letters, 1996(3): 491-494.

[20] Kodde, D.A, and Palm A.C. Wald,"Criteria for Jointly Testing Equality and Inequality Restrictions". Ecometrica, 1986 (54): 1243-1248.

[21] Komolafe, M.F., A.A. Adegbola, L.A. Are and T.I. Ashaye,Agric. Sci. West Africa Schools, Coll. Univ. Press Ltd. Ibadan, 1980 pp: 176-183.

[22] Kraus, J,"The struggle over structural adjustment in Ghana". Africa Today, 1991, 38(4): 19-37.

[23] Kumbhakar, S. C. and Lovell, C. A. K,"Stochastic Frontier Analysis". Cambridge University Press, Cambridge. 2000.

[24] Minia, Z,"Climate Scenarios Developed for Climate Change Impact Assessment inGhana", Report prepared for the Environmental Protection agency under the Netherland. Climate Change Studies Assistance Programme (NCCSAP) Phase 2, Part 1, Accra. 2004.

[25] MoFA,"Agriculture in Ghana: Facts and Figures". Ministry of Food and Agriculture, Accra. 2011.

[26] Mochebelele, M.T. and Winter-Nelson, A"Migrant Labour and Farm Technical Efficiency in Lesotho", World Development, 2000, 28 (1): 143-153.

[27] Nekesa P, &Meso B,"Traditional African vegetables in Kenya: production, marketing and utilization" In: Traditional African Vegetables. Promoting the conservation and use of underutilized and neglected crops. 16. Guarino L editor. Proceedings of the IPGRI International workshop on genetic Resources of Traditional Vegetables in Africa: Conservation and Use, 29-31 August 1995, ICRAF-HQ, Nairobi, Kenya. Institute of Plant Genetic and Crop Plant Research, Gatersleben International Plant Genetic Resources Institute, Rome, Italy, 1997, pp.98-103.

[28] Owens, T., Hoddinott, J. and Kinsey, B,"The Impact of Agricultural Extension on Farm Production in Resettlement Areas of Zimbabwe". CSAE WPS/2001-6, Centre for the study African Economics, University of Oxford. 2001.

[29] Owusu-Sekyere et al.,"Gender, Resource Use and Technical Efficiency Among Rice Farmers In The Ashanti Region Ghana". Journal of Agricultural Economics and Development, 2013, Vol. 2(3), pp. 102-110,

[30] Parikh, A., Ali, F., and Shah, M.K "Measurement of Economic Efficiencyin Pakistani Agriculture, American Journal Agricultural Economics, 1995, 77 (8): 675-685.

[31] Quisumbing, A. R,"Male-Female Differences in Agricultural Productivity: Methodological Issues and Empirical Evidence". Economic Development and Cultural Change, 1996, (24): 1579-96.

[32] Rahman S,"Profit Efficiency among Blanglandesh Vegetables Farmers". Food Policy, 2003(2)8: 483-503.

[33] Scammell, M.L. & Dearry, A,"Advancing the Community-Driven Research Agenda in N.Lo.E.H, Research Triangle Park", North Carolina. 1997.

[34] Scammell, K.M., Senier, L., Darrah-Okike, J., Brown, P. & Santos, S,"Tangible Evidence, Trust And Power: Public Perceptions Of Community Environmental Health Studies". Social Science & Medicine, 2009, 68(1):143-153, 6 Nov

[35] Seyoum, E. T., Battese, G.E. and Fleming, E. M,"Technical Efficiency and Productivity of Maize Producers in Eastern Ethiopia: A study of farmers within and outside the Sasakawa- Global 2000 Project. Agricultural Economics", 1998, 19: 341-348

[36] Udry, C. Gender, "Agricultural Production, and the Theory of the Household". Journal of Political Economy, 1996, (104): 1010-1046.

[37] Ulin, P. R., Robinson, E. T., & Tolley, E. E,"Qualitative methods in public health: A field guide for applied research". San Francisco, CA: Jossey-Bass, 2005: Pp. 139-174.

[38] Weir, S,"The effects of Education on Farmer Productivity in Rural Ethiopia", Working Paper CSAE WPS99-7, Centre for the study of African Economics, University of Oxford, 1999.

[39] Weir, S. and Knight, J,"Education Externalities in Rural Ethiopia: Evidence from Average and Stochastic Frontier Production Functions", Working Paper CSAE WPS99-7, Centre for the study of African Economics, University of Oxford. 2000.

16

Effect of short-term feed restriction on production traits of brown and black plumage commercial layer strains at late phase of egg production

Simeon Olutoye Olawumi

Department of Animal Production and Health Sciences, Ekiti State University, Ado-Ekiti, Nigeria

Email address:
olawumisimeon@yahoo.com

Abstract: The effect of strain and different levels of quantitative feed restriction for short time (28d) on egg production, feed consumption and efficiency, mortality and egg quality traits were evaluated during late phase of egg laying. A total number of 90 layers aged 54 weeks old, comprising 45 birds each of Isa Brown (brown plumage) and Bovan Black (black plumage) were randomly distributed into three treatment groups having three replicates each. They were placed under three feeding regimens namely, T_1 (*ad libitum*- control), T_2 (90% *ad libitum*) and T_3 (80% *ad libitum*). Data collected include egg production, mortality, feed variables, external and internal egg quality characteristics. The results revealed that strain has insignificant (P>0.05) effect on egg numbers during the 28 days trial regardless of treatments imposed. This means that feed restriction on short-term has no negative impact on both the brown and black feathered layer birds productivity. However, feed restriction has significant (P<0.01) effect on both egg numbers and hen-day egg production regardless of strain of birds. Regarding the two traits, *ad libitum* (control) and 90% *ad libitum* had similar mean values, but higher than 80% *ad libitum*. As per feed efficiency, brown strain has higher (P<0.01) mean values than the black strain. The two strains subjected to feed restriction recorded similar mean values and performed equally well in egg weight, egg length and other external and internal egg qualities. However, in albumen height and Haugh unit (HU), brown birds had higher (P<0.01) mean values than black hens. Haugh unit is a measure of internal egg quality. The highest net revenue/hen and economic efficiency was recorded in 90% *ad libitum* ($1.19, 112.26), intermediate ($1.06, 100) in *ad libitum* and lowest (1.10, 95.28) in 80% *ad libitum*. Brown strain achieved higher (P<0.01) economic efficiency (100 vs 88.54) and net revenue ($0.96 vs $0.85) per hen than black strain at the end of the experimental period. There was significant strain x treatment interaction effect on egg number, percent egg production and feed variables. Brown feathered birds and adoption of milder feed restriction seem to hold prospect for increased production, profit and sustainability of egg producing industry.

Keywords: Strain, Feed Restriction, Haugh Unit, Egg Quality

1. Introduction

In livestock industry, productivity and profitability depend to a larger extent on the genetic make-up or constitution of an animal being reared for either meat or egg production. Other important factors influencing productivity include prevailing environmental conditions, management practices such as feed and feeding management and the technical knowledge of the farmers, [1]. Feed restriction in a layer flock entails limiting the number of hours birds have access to feeds or actual reduction of the amount of feeds given per day. The main goal of controlled feeding in commercial laying flock is to ensure that the birds do not build up excessive fat during production so as not to interfere with their performance. Restricted feeding according to some researchers helped in maintaining correct body weight, prevented over eating and from getting overly heavy [2,3]. However, there are different approaches to this concept and individual farmers are left with options best suited for his research study. Other investigators asserted that feed restriction includes: (a) limiting the bird's time to diet, (b) quantitative feed restriction, (c) use of low energy diet, and (d) dietary protein restriction [4].

In growing pullets and breeder chicks, quantitative feed restriction is usually employed to control growth by feeding a predetermined amount of balanced ration in order to achieve a good production during laying period [5]. The major problem associated with quantitative feed restriction programme is that of feed distribution in such a way that all the birds have equal opportunity and access to feeds [6]. Closely connected to that assertion is the fact that birds are not uniform in body weight and body weight gain, and the rate of egg production differs too. These aforementioned factors play significant effects on the amount of feeds consumed by individual bird even if they are of the same age.

In previous study, a significant effect of feed restriction on egg production in Bovan Near layer strain was documented [7]. The researchers however, reported insignificant influence of controlled feeding on egg quality traits. In contrast, other investigators reported that feed deprivation has no significant effect on egg number, hen-day egg production, egg weight and egg quality [8]. The researchers however, found significant effect of feed deprivation on yolk index and Haugh unit between Hisex Brown and Bovan White strains of layers. Feed restriction in growing pullets of egg-type and broiler breeders strains had long been advocated by previous researchers [9]. In another study, the authors observed that feed restriction during rearing period has no negative effect on egg production except when carried out during laying phase [10]. Some workers have reported that feed restriction delayed sexual maturity and reduced body weight at laying [9,11].

The ever increasing cost of feed ingredients constitutes a clog in the wheel of progress in livestock industry, and this now calls for drastic cost reducing measures without compromising on the wellbeing of the birds and product quality. One of such measures is controlled feeding in laying flock, commercial and breeder pullets. It is however, important to ascertain the inherent abilities of these birds before the application of this method so as not to impede their genetic potentials in the process. Birds have different genetic constitution and production potentials, application of controlled feeding therefore, must be done with caution in order not to sacrifice product quality on the altar of saving cost and feeding management practices.

In view of arguments for and against controlled feeding in layers, this study was conceived and undertaken to evaluate and compare the response to feed restriction of brown and black feathered layer strains reared under uniform management, housing and nutritional conditions. The objective was to assess the degree of tolerance to feed restriction between the two strains and how it affects their performance in terms of egg production, egg quality traits, mortality, feed consumption, feed conversion ratio and feed efficiency. The study also aimed at evaluating cost of production and net returns from the different feed restriction levels imposed on the layer strains.

2. Materials and Methods

2.1. Study Location

The study was carried out at the Animal Breeding Unit, Teaching and Research Farm, Ekiti State University, Ado-Ekiti and lasted for four weeks during 1st – 29th September, 2012. The city is situated along latitude $7^0 31^1$ and $7^0 49^1$ North of the Equator and longitude $5^0 71^1$and $5^0 27^1$East of the Greenwich meridian. The rainfall pattern in the agro-climatic zone had been reported [12].

2.2. Experimental Birds

A total number of ninety layers comprising 45 birds each of brown (Isa Brown) and black (Bovan Black) plumage layers were reared and housed in cages for the purpose of this experiment. The birds were of the same age at the time of the study and given uniform management and quality feed. During the growing period, the birds were vaccinated against Newcastle and Fowl Pox diseases, while antibiotics were administered against bacterial diseases to boost the immunity level of the birds. The birds were dewormed at three months interval and multivitamins given regularly to ameliorate the effect of vaccination and heat stress. Some other bio-security measures were taken to prevent disease transfer from external sources.

The forty-five experimental birds were on strain basis randomly divided into three treatments of 15 birds per treatment, and these were later subdivided into three replicates of five birds each. That is, we have T_1 (*ad libitum*), T_2 (90% *ad libitum*) and T_3 (80% *ad libitum*) representing the treatment groups and under these we have three replicates such as *ad libitum* (T_1R_1, T_1R_2, T_1R_3), 90% *ad libitum* (T_2R_1, T_2R_2, T_2R_3) and 80% *ad libitum* (T_3R_1, T_3R_2, T_3R_3). They were given two weeks adaptation period and fed *ad libitum* before the commencement of application of feed restriction. The layer diet contained 16.5%CP (crude protein) and 2500Kcal/kg ME (metabolizable energy). Water was supplied *ad libitum* to the birds throughout the observed period.

2.3. Data Collection

Hen-day egg production (eggs laid in a day by living birds) and mortality were recorded on daily basis. In addition, feed consumption, feed efficiency, feed conversion ratio and economic returns were recorded for both strains of commercial layers during the experimental period (from 54-58 weeks of age)

Egg quality was measured twice: 55 and 57 weeks of age. A total number of 27 fresh eggs per strain, that is, three eggs per replicate which amounted to nine eggs per treatment were randomly selected, numbered and then weighed on treatment and breed basis on a sensitive scale. These eggs were thereafter, broken on a flat plate at equatorial region for the determination of internal egg contents. Egg quality traits evaluated included egg weight, length, width, yolk weight and height, albumen height, shell weight and shell thickness.

Yolk and shell weight were measured using sensitive scale, while egg length, egg width, yolk height and albumen height were measured with vernier callipers.

Albumen weight was calculated by the formula:

Albumen weight= egg weight - (yolk weight + shell weight)

Haugh unit (Hu) = 100log ($H+7.57-1.7W^{0.37}$) (Haugh, 1937)

H= albumen height in mm

W= weight of egg in grammes

2.4. Statistical Analysis

The data from this study were subjected to a two-way analysis of variance for the effect of breed and treatment and their interactions. Means were separated by the use of Duncan New Multiple Range Test. Data were analyzed using the General Linear Model procedure of SAS software [13].

The appropriate statistical model used was:

$Yijk = \mu + Gi + Rj + \varepsilon ijk$

Yijk = observation on kth population, of ith strain and jth feed restriction

μ = common mean

Gi = fixed effect of strain (i=2)

Rj =fixed effect of feed restriction (j=3)

εijk = error term

3. Results and Discussion

3.1. Cumulative Egg Number and Hen-Day Egg Production

Averages of cumulative egg number as affected by strain and feed restriction were shown in Table 1. Strain has insignificant ($P>0.05$) effect on egg number during the 28 days trial regardless of treatments imposed. This means that feed restriction on short-term has no negative impact on both the brown (Isa Brown) and black (Bovan Nera) feathered layer birds productivity, and could be practiced to save cost and to obtain higher profit from the laying stocks. With regard to hen-day egg production, brown strain recorded higher ($P<0.01$) percentage than black strain. The difference in hen-day egg production could be attributed to their genetic variations. These results are in agreement with those of other authors [8,13], who have reported that brown strains had higher hen-day production than white strains. Therefore, it can be concluded that there exist differences in egg production between brown and black strains of commercial layers when subjected to feed restriction.

Feed restriction has significant ($P<0.01$) effect on both egg number and hen-day egg production regardless of strain of birds. Pertaining to egg number, *ad libitum* and 90% *ad libitum* had similar mean values, but higher than 80% *ad libitum*. In the same manner, *ad libitum* and 90% *ad libitum* recorded similar mean values in hen-day production, but superior to 80% *ad libitum*. This implies that milder feed restriction produces satisfactory result and can be applied to reduce cost and increase profit margin. On the other hand, severe feed restriction should be discouraged as this appears to impact negatively on birds' performance. In partial agreement to the results in this study, previous data indicated found that feed restriction has no adverse effect on hen-day egg production [8,14].

3.2. Cumulative Feed Intake, Feed Efficiency and Feed Conversion Ratio

The results presented in Table 1 indicated that both strains consumed equal amount of feed regardless of treatments imposed during the experimental period. As per feed efficiency, brown strain has higher ($P<0.01$) mean values than the black strain. The inference is that the former utilized the feeds given efficiently than the latter, and this might be responsible for its higher hen-day production observed in this study. Similarly in line with the results in this study, [8 and 15] have shown that brown egg strain had better feed efficiency than white egg strain. However, feed conversion ratio as influenced by strain was different from what was reported for feed efficiency; as brown strain has lower values than the black strain. The results show the inverse relationship between feed efficiency and feed conversion ratio, and that good and profitable egg production is achieved when feed efficiency is higher coupled with lower feed conversion ratio.

Feed restriction has significant ($P<0.01$) effect on feed intake, feed efficiency and feed conversion ratio as shown in Table 1. The birds fed *ad libitum* consumed more feeds than those fed 90% and 80% *ad libitum*. It was also shown that those fed 90% *ad libitum* has higher feed intake than 80% *ad libitum*. The reduced feed intake by birds in 90% and 80% *ad libitum* was due to the imposition of feed restriction on them. The birds in the two groups had limited and restricted access to daily ration when compared to *ad libitum* birds. Similar results have also been reported by other workers in the literature [16]. The authors indicated that feed restriction decreased average daily feed consumption in broiler female breeder hens. With regard to feed efficiency, restricted birds in 90% *ad libitum* has higher and superior ($P<0.01$) mean values to *ad libitum* and 80% *ad libitum*. In feed conversion ratio, 90% *ad libitum* recorded lower and better mean values than the other two treatments. The results indicate that quantitative feed restriction improves feed utilization during late production period, and that birds fed 90% *ad libitum* utilized their feeds efficiently than *ad libitum* and 80% *ad libitum*. The former therefore, could be recommended to farmers judging from higher mean values obtained on hen-day production and feed efficiency. Adoption of milder feed restriction seems to hold good prospect for increased production, profit and sustainability of egg producing industry. In partial agreement with these results, previous researcher found significant improvement in feed conversion ratio in birds fed 80% *ad libitum* [17]. On the contrary, some authors have reported that at late laying period, feed restriction has no significant effect on birds' production [8].

3.3. Egg Quality Traits

3.3.1. Effect of Strain

There was no significant (P>0.01) effect of strain on all the egg quality traits measured at 55 weeks of age with the exception of albumen height and Haugh unit (Table 2). Also, at 57 weeks (Table 3), no significant (P>0.05) effect of strain was observed in all the egg quality traits measured with the exception of shell thickness. The two strains subjected to feed restriction recorded similar mean values and performed equally well in egg weight, egg length and other external and internal egg qualities. But in albumen height and Haugh unit, brown birds had higher (P<0.01) mean values than black hens. Haugh unit is a determinant of internal egg quality, the higher the value, the higher the nutritive value of an egg. In this study, Isa Brown eggs with higher HU were adjudged superior in egg quality and could be recommended for human consumption and egg processing industry. In partial agreement with these results, reference [8] reported that Brown strain had higher HU values than white strain only at early and mid-age production periods. It is interesting to note that there was progressive increase in egg weight and other egg quality traits including Haugh unit as the birds advanced in age across strains.

The results of the present study also reveal that brown hens produced eggs with stronger shell than black hens. This is an important and another distinguishing genetic difference between the two layer strains. It is desirable that eggs meant for home use and hatchery have stronger shell in order to prevent breakages and microbial spoilage. Therefore, in places where storage facility is non-existent, it is expected that brown strain eggs will remain healthy and good after some length of time due to its stronger shell if not exposed to extremely hot temperature.

Table 1. *Least squares mean showing the effect of strain and feed restriction of production traits of commercial layers.*

Factors	Traits				
	Egg number	Egg percentage	Feed intake	Feed efficiency	Feed conversion ratio
Strains					
IB	11.54±0.18	77.35±1.19^{a}	1433.33±13.69	0.0081±0.0001^{a}	0.126±0.0029^{b}
BN	11.05±0.23	73.57±1.53^{b}	1433.33±13.69	0.0076±0.0001^{b}	0.134±0.0021^{a}
Feed restriction					
100%	11.98±0.21^{a}	79.76±1.42^{a}	1600±0.00^{a}	0.0075±0.0001^{b}	0.137±0.0028^{a}
90%	11.68±0.23^{a}	77.86±1.50^{a}	1400±0.00^{b}	0.0084±0.0001^{a}	0.122±0.0025^{b}
80%	10.21±0.26^{b}	68.77±1.78^{b}	1300±0.00^{c}	0.0077±0.0002^{b}	0.132±0.0036^{a}

abc: means along columns with different superscripts are significantly different (P<0.01)
IB- Isa Brown BN- Bovan Nera

Table 2. *Effect of feed restriction on egg quality traits of layer strains at 55 weeks of age.*

Factors	Traits (55 weeks)									
	EWT	EL	EW	YWT	YH	AWT	AH	HU	SWT	ST
Strains										
IB	62.97±1.19	6.07±0.09	4.87±0.11	15.97±0.27	2.33 0.08	38.54 0.57	1.32±0.06^{a}	110.90±1.89^{a}	8.26±0.11	0.56±0.02
BN	61.10±1.59	6.19±0.08	4.69±0.04	15.93±0.30	2.22±0.07	37.66±1.25	1.14±0.06^{b}	104.77±2.11^{b}	7.78±0.24	0.57±0.02
Feed restriction										
100%	61.48±1.20	5.98±0.08	4.82±0.11	15.68±0.26	2.23±0.08	37.85±0.78	1.17±0.11	105.11±3.75	7.85±0.31	0.58±0.02^{a}
90%	63.37±2.09	6.17±0.14	4.88±0.13	16.15±0.31	2.28±0.11	39.18±1.66	1.23±0.04	107.96±1.46	8.18±0.23	0.53±0.02^{b}
80%	61.25±1.92	6.23±0.08	4.63±0.06	16.02±0.45	2.31±0.10	37.27±0.96	1.30±0.08	110.43±2.43	8.01±0.20	0.57±0.02^{ab}

EWT-egg weight EL-egg length EW-egg width YWT-yolk weight YH-yolk height AWT-albumen weight
AH-albumen height HU-haugh unit SWT- shell weight ST-shell thickness
ab means along the columns with different superscripts are significantly different (no superscripts implies the means are similar)
IB: Isa Brown BN: Bovan Nera

Table 3. *Effect of feed restriction on egg quality traits of layer strains at 57 weeks of age.*

actors	Traits (57 weeks)									
	EWT	EL	EW	YWT	YH	AWT	AH	HU	SWT	ST
Strains										
IB	63.14±0.80	6.030.08	4.83±0.08	14.89±0.28	2.26 0.06	40.18 0.72	1.33±0.056	111.34±1.59	8.38±0.20	0.43±0.02^{a}
BN	63.26±4.47	6.24±0.07	4.86±0.08	14.93±0.31	2.31±0.05	40.31±1.32	1.28±0.04	109.45±1.57	8.32±0.28	0.38±0.02^{b}
Feed restriction										
100%	64.67±0.99	6.08±0.10	4.92±0.11	15.12±0.27	2.33±0.03	41.47±1.23	1.18±0.30^{b}	105.92±1.23^{b}	8.52±0.29	0.40±0.03
90%	63.30±1.46	6.17±0.11	4.85±0.08	14.48±0.45	2.27±0.09	40.68±1.12	1.38±0.06^{a}	112.98±1.88^{a}	8.08±0.28	0.40±0.03
80%	61.63±1.67	6.17±0.09	4.77±0.08	15.13±0.31	2.25±0.06	38.58±1.33	1.35±0.03^{a}	112.28±1.18^{a}	8.45±0.33	0.42±0.03

ab means along the columns with different superscripts are significantly different (no superscripts implies the means are similar).
EWT-egg weight EL-egg length EW-egg width YWT-yolk weight YH-yolk height AWT-albumen weight
AH_albumen height HU-haugh unit SWT- shell weight ST-shell thickness
IB: Isa Brown BN: Bovan Nera

3.4. Economic Efficiency and Mortality Rate

Feed restriction has highly significant (P<0.01) effect on total feed intake, total feed cost, net revenue and economic efficiency (Table 4). Total feed cost ($/hen) was highest (P<0.01) for control birds but decreased by feed restriction during the experimental period. The reduction in feed cost represented about 3.7% and 7.41% for 90% *ad libitum* and 80% *ad libitum,* respectively compared to those feed *ad libitum*. The reduction in total feed cost as a result of feed restriction could be attributed to the negative effect of quantitative feed restriction on total feed consumption. The highest net revenue and economic efficiency was recorded in 90% *ad libitum*, intermediate and lowest in *ad libitum* and 80% *ad libitum*, respectively. The superiority of 90% *ad libitum* in net revenue and economic efficiency over other two groups could be attributed to the better feed conversion and higher feed efficiency (Table 1), total egg number per hen and to the total egg price recorded by the former. The result agrees with the previous study [18]. The researchers opined that net egg income over feed cost increased significantly by quantitative feed restriction. On the contrary, reference [8] reported no significant differences in net revenue between the different feeding regimens imposed on the birds.

Mortality rate in this study was none for control and feed restricted birds. This might be due to the short duration of the experiment. In partial agreement with this study, [16] stated that conventional feed restriction decreased mortality rate in broiler breeder females compared to *ad libitum* feeding after peak of lay. Similarly, [17] reported that mortality rate was insignificant between layers fed *ad libitum* and 80% *ad libitum* restricted feeding regimen.

In this study strain has significant (P>0.5) effect on egg production per hen, net revenue and economic efficiency, but the reverse was the case for feed consumption and feed cost per hen (Table 4). Isa Brown strain achieved higher (P<0.01) net revenue per hen than Bovan Nera at the end of the experimental period. This superiority (6.08%) could be attributed to the better feed conversion and higher feed efficiency (Table 1), total eggs number per hen and to the total egg price produced by Isa Brown strain compared to Bovan Nera (Table 4). Total feed cost on strain basis in the present study did not influence net returns, but total eggs per hen and market price of eggs did. The obtained result was similar to the findings of [8] who reported significant differences between strains in feed conversion, total egg number per hen and total egg price per hen. The researchers also found significant strain differences in net return per hen; which is also corroborated this study .. Isa Brown appeared good, feed and economic efficient and suitable for profitable rearing under feed restriction regimen.

Table 4. *Cost implication and economic efficiency of feed restricted laying hens during 4 weeks.*

Classification	FeedIntake (kg)		Feed Cost @ $0.625/kg		Total Eggs/Hen	Total Egg Price/Hen @ $0.125/egg	Net RevenuePer Hen($)	Relative EconomicEfficiency(%)	Mortality (%)
	Total	Per Hen	Total	Per Hen					
A. Treatments									
ad libitum	92.40	3.08	57.75	1.93	23.93	2.99	1.06	100	-
90%	83.16	2.77	51.98	1.73	23.36	2.92	1.19	112.26	-
80%	73.92	2.46	46.20	1.54	20.43	2.55	1.01	95.28	-
B. Strains									
IB	124.74	2.77	77.96	1.73	21.53	2.69	0.96	100	-
BN	124.74	2.77	77.96	1.73	20.60	2.58	0.85	88.54	-

IB: Isa Brown BN: Bovan Nera
$= ₦160 (Nigerian currency)

4. Conclusion

From these results, it can be concluded that strain differ in their response to feed restriction. Brown strain appeared tolerant to short-term feed restriction than black strain. The former recorded higher and significant values in percent egg production, feed efficiency, net revenue and economic efficiency than the latter. In addition, brown strain obtained better and higher Haugh unit than black strain. With regard to feed restriction levels, 90% *ad libitum* was better and feed efficient than *ad libitum and 80% ad libitum*. The former also recorded higher net returns and economic efficiency. Brown birds and 90% *ad libitum* show promise and could be recommended for improved productivity and profit maximization.

Acknowledgments

The author is indebted to the Staff of Teaching and Research Farm, Ekiti State University, Ado-Ekiti for their support and assistance during the period of data collection.

References

[1] D.D. Bell and W.D. Weaver. “Commercial chicken meat and egg production”. 5th edition, Cambridge, Massachussets. Kluwer Academic Publisher, 2002.

[2] L. Kostal, C. J. Savory and B. O. Hughes. "Diurnal and individual variation in behaviour of restricted-fed broiler breeders". Applied Animal Behaviour, 32, pp. 361-374, 1992.

[3] C. J. Savory, E. Seawright and A. Watson. "Stereotyped behaviour in broiler breeders in relation to husbandry and opioid receptor blockade". Applied Animal Behaviour, 32, pp. 349-360, 1992.

[4] P. J. W. Lee, A. L. Gulliver and T. R. Morris. "A quantitative analysis of the literature concerning the restricted feeding of growing pullets". Br. Poultry Sci., 12, pp. 413-437, 1971.

[5] S. N. Ukachukwu and U. O. Akpan. "Effect of quantitative feed restriction on the performance and laying characteristics of pullets". Tropical and Subtropical Agroecosystems, 7, pp. 15-20, 2007.

[6] S. Leeson and J. D. Summers. "Feeding programmes for broilers: Commercial Poultry Nutrition. University Books, Guelph, pp. 207-254, 1997.

[7] A. O. Fasuyi and O. G. Ojo. "Quantitative feed restriction threshold in Bovan Nera laying pullets in the warm wet tropics". J. Animal Sci. Advances, 2(1), pp. 158-165, 2012.

[8] A. M. A. Osman, M. A. Toson, S. A. Abdel-Latif, H. H. M. Hassanien and T. M. A. Marwan. "Effect of fasting on reproductive performance of laying hens". Egypt Poultry Sci., 30(4), pp. 1031-1057, 2010.

[9] R. S. Gowe, A. S. Johnson, F. D. Crawford, J. H. Downs, A. T. Hill, M. F. Mountain, J. R. Pelletier and J. H. Strain. "Restricted versus full feeding during the growing period for egg production stock". Br Poultry Sci., 1, pp. 37-56, 1960.

[10] R. Blair, M. M. McCowan and W. Bolton. "Effect of food regulation during the growing and laying stages on the productivity of broiler breeders". Br. Poultry Sci., 39, pp. 205-212, 1976.

[11] D. Balnave. " The effect of feeding low protein diets on pullets from hatch to point of lay, the quantitative restriction of food during the subsequent laying period". Br. Poultry Sci., 15, pp. 395-403, 1974.

[12] S. O. Olawumi and I. Dudusola. "Assessment of long-term production traits of three breeds of exotic commercial layers in the Derived savannah zone of Nigeria". J. Applied and Natural Sci., 3(1), pp. 20-24, 2011.

[13] Statistical Analysis System, SAS Users Guide. Statistics, 8th edition, SAS Institute Cary, NC, USA, 2001 .

[14] F. G. Silversides, D. R. Korver and K. L. Budgell. "Effect of strain of layer and age at photo-stimulation on egg production, egg quality and bone strength". Poultry Sci., 85, pp. 1136-1144, 2006.

[15] J. V. Felts. "Dietary self-selection and feed restriction studies with growing and breeding turkeys". Ph.D Dissertation. Virginia Polytechnic Institute and State University, Blacksburg, Virginia, USA, 1993.

[16] A.Al-Nasser, A. Al-Saffar, M. Mashaly, H. Al-Khalaifa, F. Khalil, M. Albaho, and A. Al-Haddad. "A comparative study on production efficiency of white and brown pullets". World's Poultry Sci., 62, pp. 296-307, 2006.

[17] P. M. Hocking, R. Bernard and G. W. Robertson. "Effects of low dietary protein and different allocations of food during rearing and restricted feeding after peak rate of lay on egg production, fertility and hatchability in female broiler breeders". Br. Poultry Sci., 43, pp. 94-103, 2002.

[18] R. Hasnath. "Effect of feeding systems on the egg production of Fayoumi hens of model breeding units under PLDP programme in Bangladesh". M.Sc. Thesis, department of Animal Science and Animal Health and Network for Smallholder Poultry Development, Bangladesh, 2002.

[19] P. N. Mbugua and D. L. Cunningham. "Effects of feed restriction on production performance of replacement pullets". Poultry Sci., 62, pp. 1169-1176, 1983.

Evaluation of river sand as a medium for raising cocoa (*Theobroma cacao* L.) seedlings

Sampson Konlan*, **Michael Obour Opoku-Agyeman, Kofi Acheampong, Kwabena Opoku-Ameyaw, Gilbert John Anim-Kwapong, Godwin Addo**

Cocoa Research Institute of Ghana, P. O. Box 8, New Tafo-Akim

Email address:
Sampson.konlan@gmail.com (S. Konlan)

Abstract: A nursery experiment was conducted between November, 2011 and May, 2012 at the Cocoa Research Institute of Ghana to investigate the effect of using river sand and river sand-topsoil mixture on cocoa seedlings. Treatments tested were sole topsoil, sole river sand, river sand + foliar fertilizer (Sidalco liquid fertilizer, NPK-10:10:10), river sand + topsoil (1:1) and river sand + topsoil mixture (1:1) + foliar fertilizer. The results showed that sole topsoil retained more moisture and supported significantly taller seedlings. The river sand-topsoil mixture + foliar fertilizer seedlings had higher chlorophyll content and was more photosynthetically efficient thus resulting in significantly ($P<0.05$) larger stem diameter than all other treatments. Stem volume of seedlings in the river sand + topsoil +foliar fertilizer were also not different from those found in the sole topsoil medium. There were positive correlations between moisture retention and chlorophyll content of the seedlings. Chlorophyll content was found to be positively correlated to stem diameter and height gain by the seedlings. It was therefore concluded that cocoa seedlings can be raised in polybags filled with a mixture of river sand and topsoil (1:1) for a period of six months provided foliar fertilizer is applied at the rate of 10 mls NPK (10:10:10) in 15 litres of water at bi-weekly intervals.

Keywords: *Theobroma Cacao*, River Sand, Chlorophyll Fluorescence, Photosynthesis, Stem Volume, Foliar Fertilizer

1. Introduction

Raising of cocoa seedlings in nurseries using polybags is the recommended practice because the management of individual seedlings result in much viguor and improved capability of successful establishment [3, 13, 14]. This is attributed to minimized transplanting shock caused by the absence or reduced damage to the roots as compared to bare-root seedlings [12, 18]. The polybags also allow the nursery period to be prolonged until the environment is conducive for transplanting to the field [12]. These polybags are filled currently using topsoil. The major problem associated with the use of topsoil as seedling medium is scarcity [2]. This scarcity has led to use of various alternatives and mixtures of other media with topsoil for nursing seedlings of other crops such as coconut [7, 15]. In raising coconut seedlings, topsoil mixed with river sand in the ratio of 3:1 is a satisfactory mixture for filling the polybags [7]. Even if such a ratio works very well in the raising of cocoa seedlings, the requirement of topsoil could still be large. Related nursery studies [10] concluded that cocoa pod husk-based compost was suitable in combination with topsoil at ratios of 1:2 and 1:3. Earlier work [15] indicated that river sand did not adversely affect germination when substituted for topsoil. However, the rate of germination was slightly delayed and early seedling viguor was reduced in the sand substituted mixtures. Whereas potting media has not be shown to affect percent germination [8], nutrient composition and availability potting media significantly affects growth parameters such as leaf number, plant height, root and stem diameters, and subsequently dry matter accumulation [1, 5, 8, 9]. The addition of mineral nitrogen is also reported to [8] to improve chlorophyll concentration in the leaves. It is therefore possible to overcome these problems experienced with low nutrient media which also have poor water holding capacity by increasing the moisture and nutrient supply to the seedlings established in such media through frequent watering and application of fertilizers [8, 13].

The recommended polybag size for raising cocoa seedlings (17.5 x 25 cm) usually takes in an average of 6 kg of top soil per bag. The current directive from Ghana COCOBOD for the Seed Production Unit (SPU) to raise 20 million seedlings each year would amount to a topsoil requirement of 120,000 MT per annum, excluding that of private nurseries. Excessive scraping of the topsoil to meet this demand will eventually result in serious degradation of productive farmlands. It is therefore necessary to investigate possible options to replace or reduce the topsoil requirement as the sole medium for filling polybags, with particular interest in the use of river sand found abundantly in close proximity to most farm settlements. The sole objective of this work therefore was to determine the suitability of river sand in appropriate proportions with topsoil as alternatives to sole topsoil in raising cocoa seedlings.

2. Materials and Methods

2.1. Experimental Design and Treatments

The experiment was conducted within the nursery premises of the Cocoa Research Institute of Ghana (lat. 6° 13' N, long. 0° 22' W, altitude 222 m above sea level), from November 2011 to May 2012. The experiment was laid out in a randomized complete block design with four replicates and five treatments (Table 1). Nursery bags were first perforated at the bottom with a perforator to create a 4 cm^2 hole to facilitate drainage of excess water before filling with the respective media.

***Table 1.** Treatment combinations.*

Treatment	Description
i	Topsoil only (control)
ii	River sand only
iii	River sand + foliar fertilizer
iv	River sand (50 %) + topsoil (50 %)
v	River sand (50 %) + topsoil (50 %) + foliar fertilizer

For each treatment forty polybags were filled with topsoil and 2 seeds sown per polybag. The polybags were then arranged on thick polythene sheet to prevent the growth of the roots into the ground, to avoid possible uptake of nutrients. The experiment was carried out under a shade net allowing 40 % sunlight through. Two weeks after germination an approved liquid foliar fertilizer (Sidalco NPK - 10:10:10) was applied to the seedlings at a rate of 10 ml per 15 litres of water at bi-weekly intervals using a knapsack sprayer. One insecticide treatment (60 ml Confidor 200 SL in 4.5 l of water) was carried out once each month using a motorized sprayer to prevent seedling damage by insect pests.

2.2. Data Collected

The number of seedlings that emerged per treatment was counted two weeks after sowing and expressed as a percentage of the 80 seeds initially sown. Data on seedling survival was obtained through monthly census of living plants which was then expressed as percentage of the number of seedlings that emerged. Moisture contents of the polybag media were recorded monthly using soil moisture meter (HH2 by Delta Devices Ltd) after 48 hrs of no watering. Twenty core plants were randomly selected and tagged for physiological and morphological data measurements. A leaf count was conducted once every month on individual plants. Data on leaf chlorophyll content was measured with a chlorophyll content meter (CL-01 by Hansatech Instruments Ltd) while leaf chlorophyll fluorescence was measured with a leaf chlorophyll fluorometer (FP 100 by Photon System Instruments Ltd). The rates of stomatal conductance, transpiration and photosynthesis were measured at the leaf level on ten of the twenty test plants at monthly intervals with a portable infra-red gas analyzer (LcPro, ADC Bioscientific, UK). To provide estimates for growth, stem diameter and plant height measurements were taken at monthly intervals from the twenty test plants using digital calipers and a standard metre rule respectively. The data on stem diameter and height was then used to derive their respective stem volumes for treatments and months, following the procedure described in earlier report [11]

$$Vol_{st} = \frac{\pi}{2}(d^2 * h)$$

Where:
Vol_{st} - stem volume (cm^3)
d - stem diameter
h - seedling height

Total leaf area per plant (cm^{-2}), root length, below and above ground plant dry matter were taken in the sixth month after sowing by destructively harvesting all the twenty test plants from each treatment. The harvested plant materials were first washed with tap water and oven dried at 80 °C for three days until a consistent weight was obtained. The weights of the dried samples were then taken as the dry matter accumulated by the seedlings after 6 months of growth.

2.3. Data Analysis

Data on seedling growth and physiology were analysed using ANOVA and treatments separated by least significant difference. The data on final emergence, plant survival and bag moisture content were subjected to arcsine transformations before analysis.

3. Results

3.1. Percent Soil Moisture, Final Emergence and Number of Plants Surviving

The sole river sand and river sand + fertilizer recorded significantly ($P<0.05$) lower percent moisture content compared to other treatments throughout the study period (Fig. 1). Also, river sand-topsoil mixtures improved moisture retention compared to the sole river sand even though its moisture retention capacity was significantly lower ($P<0.05$) than that of the sole topsoil. The higher moisture content of the sole topsoil medium did not translate into significant differences ($P<0.05$) in percent final emergence and subsequently, survival of cocoa seedlings afterwards (Fig. 2).

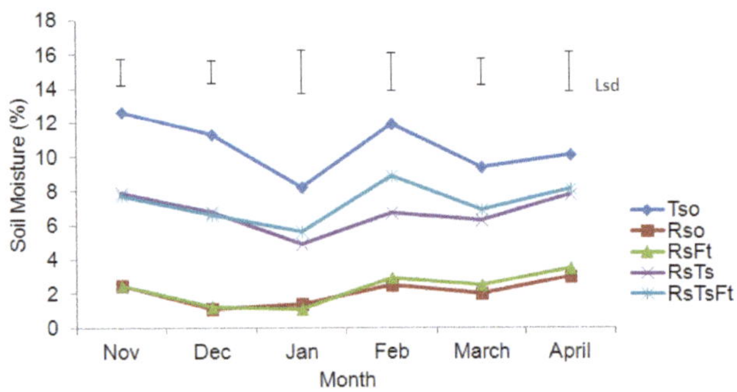

Note: Tso-topsoil only, Rso-river sand only, RsFt-river sand + fertilizer, RsTs-river sand + topsoil, RsTsFt-river sand + topsoil + fertilizer.

Figure 1. *Effects of polybag medium and foliar fertilizer on available soil moisture at various months after sowing.*

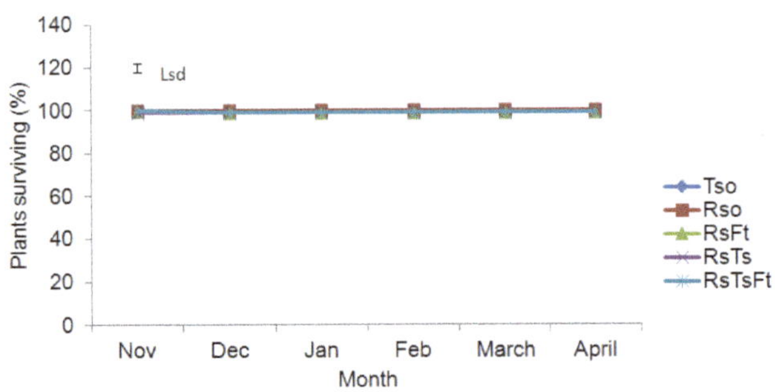

Note: Tso-topsoil only, Rso-river sand only, RsFt-river sand + fertilizer, RsTs-river sand + topsoil, RsTsFt-river sand + topsoil + fertilizer.

Figure 2. *Effects of polybag medium and foliar fertilizer on the proportion of plants surviving in the nursery at various months after sowing.*

3.2 Number of Leaves Per Seedling and Leaf Chlorophyll Content

The use of river sand as polybag medium with or without fertilizer significantly ($P<0.05$) reduced the number of leaves per cocoa seedling (Fig. 3). Leaf numbers of seedlings in the river sand + topsoil, river sand + topsoil + foliar fertilizer and sole topsoil treatments were similar throughout the six months of growth. That of the river sand + foliar fertilizer and sole river sand treatments were also similar during the first three months of growth, and in the 6th month, but significantly different ($P<0.05$) in the 4th and 5th months. Both treatments however, recorded leaf numbers that were significantly lower ($P<0.05$) than those of the remaining treatments throughout the six months (Fig. 3). The leaves of seedlings in the sole river sand and the river sand + foliar fertilizer treatments recorded similar chlorophyll values during the first two months of growth but was significantly different ($P<0.05$) from each other from the 3rd to the 6th month (Fig. 4). Leaf chlorophyll values of the sole topsoil and river sand + topsoil treatments which started very high showed a steady decline throughout the six months of seedling growth. From the 3rd month, leaf chlorophyll content of seedlings in the river sand + foliar fertilizer and river sand + topsoil + foliar fertilizer treatments continued to increase, recording significantly higher values from 4 - 6 months (Fig. 4). Leaves of seedlings grown in the sole river sand medium recorded the least chlorophyll content during the 6 months of monitoring.

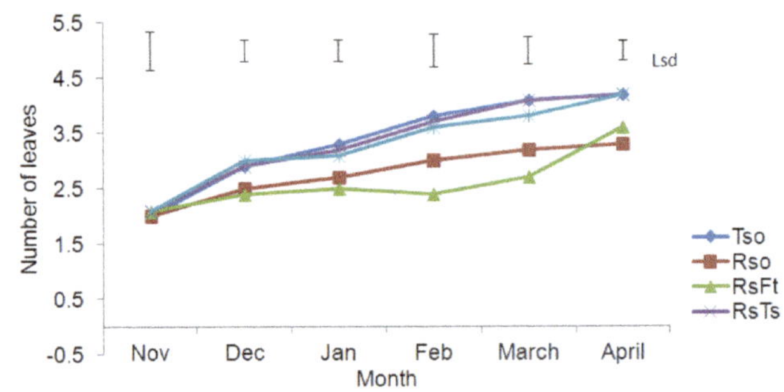

Note: Values are square root transformations. Note: Tso-topsoil only, Rso-river sand only, RsFt-river sand + fertilizer, RsTs-river sand + topsoil, RsTsFt-river sand + topsoil + fertilizer.

Figure 3. *Effects of polybag medium and foliar fertilizer on leaf number per plant of cocoa seedlings at various months after sowing in the nursery.*

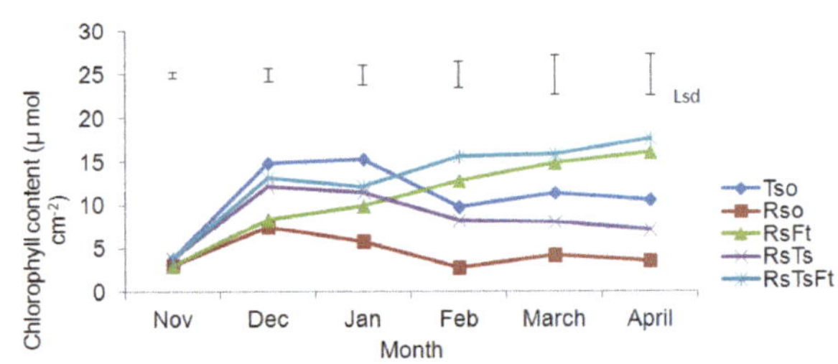

Note: Tso-topsoil only, Rso-river sand only, RsFt-river sand + fertilizer, RsTs-river sand + topsoil, RsTsFt-river sand + topsoil + fertilizer.

Figure 4. *Effects of polybag medium and foliar fertilizer on leaf chlorophyll content of cocoa seedlings in the nursery at various months after sowing.*

3.3 Stomatal Conductance, Transpiration Rates, Leaf Chlorophyll Fluorescence and Photosynthetic Rate

There were no significant differences ($P<0.05$) in stomatal conductance during the first three months of seedling growth (Fig. 5). Four months after sowing, seedlings in the sole river sand and river sand + foliar fertilizer media recorded significantly lower ($P<0.05$) stomatal conductance compared to those of seedlings in the remaining treatments. In the 5th and 6th months after sowing, stomatal conductance of seedlings grown in the sole topsoil medium was significantly higher ($P<0.05$) than that of the remaining treatments (Fig. 5).

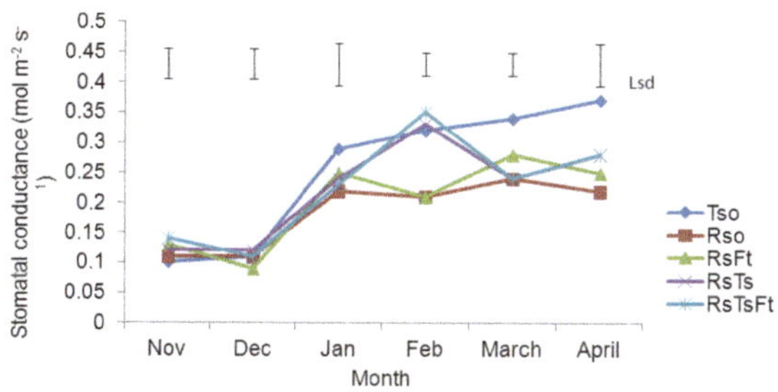

Note: Tso-topsoil only, Rso-river sand only, RsFt-river sand + fertilizer, RsTs-river sand + topsoil, RsTsFt-river sand + topsoil + fertilizer.

Figure 5. *Effects of polybag medium and foliar fertilizer on stomatal conductance of cocoa seedlings at various months after sowing in the nursery.*

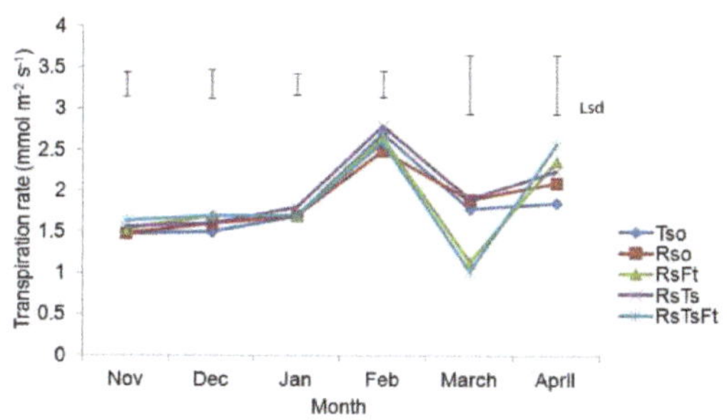

Note: Tso-topsoil only, Rso-river sand only, RsFt-river sand + fertilizer, RsTs-river sand + topsoil, RsTsFt-river sand + topsoil + fertilizer.

Figure 6. *Effects of polybag medium and foliar ferilizer on transpiration rates of cocoa seedlings at various months after sowing in the nursery.*

There were no significant treatment influences (P<0.05) on the transpiration rates of cocoa seedlings during the first four months, and at six months after sowing (Fig. 6). At five months, the rate of transpiration recorded by the river sand + foliar fertilizer and the river sand + topsoil + foliar fertilizer were similar and significantly lower (P<0.05) than that recorded by the remaining treatments (Fig. 6).

Chlorophyll fluorescence of cocoa seedlings was not significantly influenced (P<0.05) by treatments from 1 to 3, and at 5 months after sowing (Fig. 7). At four and six months however, leaf chlorophyll fluorescence of the river sand + foliar fertilizer was significantly higher (P<0.05) than that of the sole river sand treatment (Fig. 7).

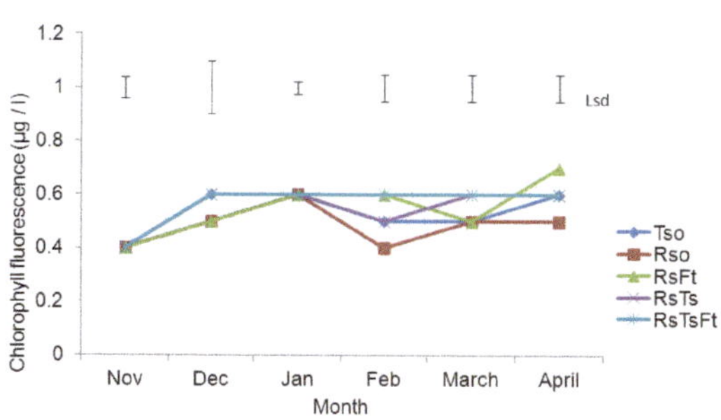

Note: Tso-topsoil only, Rso-river sand only, RsFt-river sand + fertilizer, RsTs-river sand + topsoil, RsTsFt-river sand + topsoil + fertilizer.

Figure 7. *Effects of polybag medium and foliar fertilizer on leaf chlorophyll fluorescence of cocoa seedlings at various months after sowing in the nursery.*

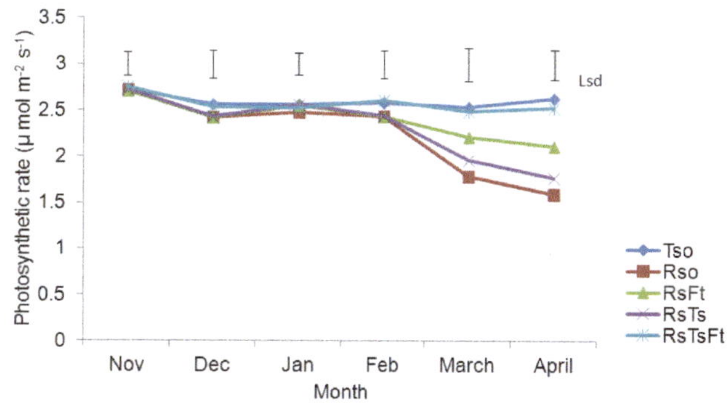

Note: Tso-topsoil only, Rso-river sand only, RsFt-river sand + fertilizer, RsTs-river sand + topsoil, RsTsFt-river sand + topsoil + fertilizer.

Figure 8. *Effects of polybag medium and foliar fertilizer on the rate of photosynthesis of cocoa seedlings at the leaf level at various months after sowing.*

Initially, all treatments recorded identical rate of photosynthesis at the leaf level up to the fourth month after sowing (Fig. 8). At five and six months however, the rate of photosynthesis of the sole topsoil and river sand + topsoil + foliar fertilizer were significantly higher (P<0.05) than the rates in the leaves of seedlings in the sole river sand and the river sand + topsoil media (Fig. 8).

3.4. Seedling Growth and Dry Matter Accumulation

There were no significant differences (P<0.05) in stem diameter of cocoa seedlings up to three months after sowing (Fig. 9). From four-six months after sowing, seedlings grown in the river sand + topsoil media and supplied with foliar fertilizer recorded significantly bigger (P<0.05) stems compared to the remaining treatments. Also, the effect of the sole topsoil treatment on stem diameter was significantly higher (P<0.05) than the effects of sole river sand, and river sand + topsoil treatments at five and six months after sowing.

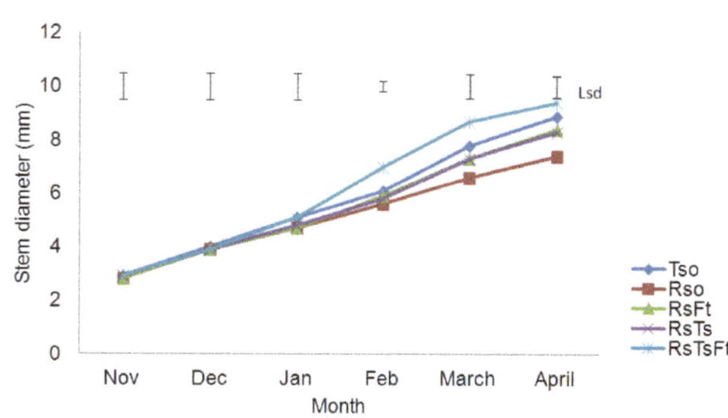

Note: Tso-topsoil only, Rso-river sand only, RsFt-river sand + fertilizer, RsTs-river sand + topsoil, RsTsFt-river sand + topsoil + fertilizer.

Figure 9. *Effects of polybag medium and foliar fertilizer on stem diameter increase in cocoa seedlings at various months after sowing.*

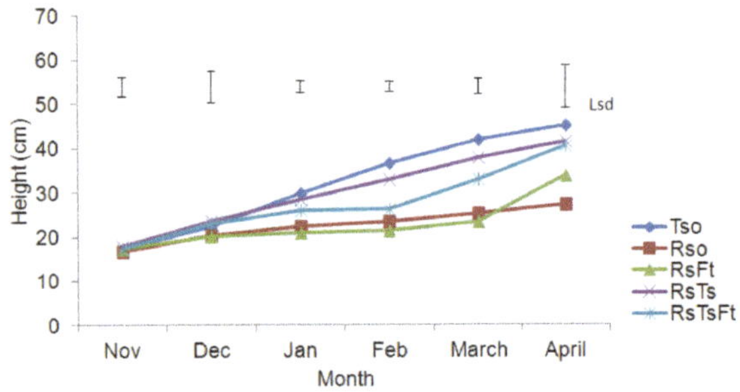

Note: Tso-topsoil only, Rso-river sand only, RsFt-river sand + fertilizer, RsTs-river sand + topsoil, RsTsFt-river sand + topsoil + fertilizer.

Figure 10. *Effects of polybag medium and foliar fertilizer on height gain in cocoa seedlings at various months after sowing.*

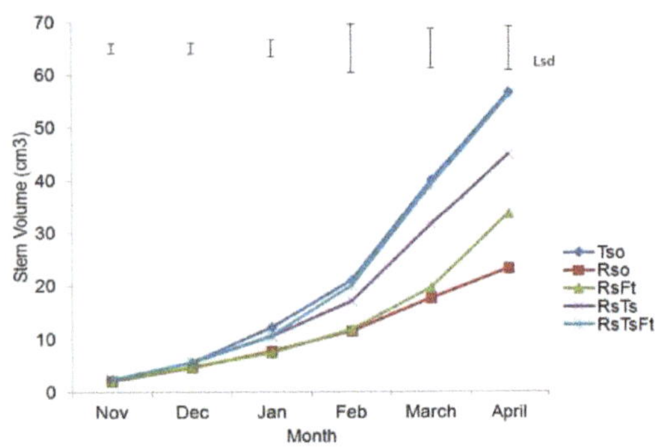

Note: Tso-topsoil only, Rso-river sand only, RsFt-river sand + fertilizer, RsTs-river sand + topsoil, RsTsFt-river sand + topsoil + fertilizer.

Figure 11. *Effect of polybag medium and foliar fertilizer on stem volume of cocoa seedlings at various months after sowing.*

Plant height was not significantly influenced ($P<0.05$) by treatments during the first two months of growth (Fig. 10). From three-six months, height of seedlings grown in the sole topsoil medium was significantly greater ($P<0.05$) than that of seedlings in the sole river sand and river sand + foliar fertilizer treatments. The river sand + topsoil + foliar fertilizer treatment also supported seedlings that were significantly ($P<0.05$)) taller than seedlings found in the sole river sand medium. Total leaf area of the sole river sand medium seedlings was significantly reduced ($P<0.05$) when compared to those of the sole topsoil and river sand + topsoil + foliar fertilizer treatments (Table 2). Apart from the reduced leaf area, seedlings in the sole river sand medium also recorded significantly shorter ($P<0.05$) roots and lower leaf, stem and root dry matter weights (Table 2). The sole topsoil and river sand + topsoil + foliar fertilizer media produced similar but significantly larger ($P<0.05$) leaf area, longer roots and higher leaf, stem, root and total dry matter compared to those of the remaining treatments (Table 2). Consequently, stem volume values recorded by seedlings grown in sole topsoil and river sand + topsoil + foliar fertilizer were similar and significantly larger than that of seedlings grown in sole river with or without foliar fertilizer application at five and six months after sowing (Fig. 11).

Table 2. *Leaf area, root length and dry matter accumulation of cocoa seedlings as affected by polybag medium and foliar fertilizer six months after sowing.*

Treatment	Leaf area(cm^{-2})	Root length (cm)	Dry matter accumulation (g) at 6 months			
			Leaf (dm)	Stem (dm)	Root (dm)	Total (dm)
Topsoil only	45.97^{a}	25.6^{a}	4.21^{a}	3.09^{a}	4.78^{a}	12.08^{a}
River sand only	39.01^{b}	20.6^{b}	1.88^{d}	1.58^{c}	1.82^{c}	5.38^{c}
River sand + Foliar	41.10^{ab}	22.8^{ab}	2.59^{c}	1.70^{c}	2.26^{c}	6.54^{c}
Topsoil +River sand	42.56^{ab}	26.1^{a}	3.30^{b}	2.45^{b}	3.41^{b}	9.16^{b}
Topsoil +River sand + Foliar	46.56^{a}	26.9^{a}	4.51^{a}	2.66^{ab}	4.20^{a}	11.37^{a}
$Lsd_{0.05}$	5.84	3.6	0.61	0.56	0.73	1.75
CV (%)	8.8	21.1	26.1	34.6	31.2	12.8

Note: cm (centimeters, g (grams), dm (dry matter). Differences between means in the same column followed by the same superscripted letter are not significant at 5 % probability level.

3.5. Correlations between Some Physiological and Growth Parameters

Correlations between some soil and growth parameters are presented in Table 3. Soil moisture was positive and highly correlated to seedling stem diameter ($r=0.69$, $P=0.01$) and height ($r=0.65$, $P=0.01$). Leaf area was also found to be positively and highly correlated to dry matter of leaf ($r=0.72$, $P=0.01$), stem ($r=0.73$, $p=0.01$), root ($r=0.74$, $P=0.01$) and consequently total dry matter ($r=0.77$, $P=0.01$). There was also positive and high correlation between leaf dry matter and root length ($r=0.66$, $P=0.01$). Seedling stem diameter was found to be positive and highly correlated to stem dry matter ($r=0.68$, $P=0.01$) and total dry matter ($r=0.63$, $P=0.05$). There was also positive and significant correlation between root length and stem dry matter ($r=0.55$, $P=0.05$). Root dry matter correlated positively and strongly with stem dry matter ($r=0.84$, $P=0.01$) while stem dry matter correlated positively and strongly with total dry matter production ($r=0.97$, $P=0.01$).

Table 3. *Correlations among some physiological and growth parameters of six month old cocoa seedlings grown in the nursery as affected by polybag medium and foliar fertilizer.*

1	%smc													
2	LA	0.45												
3	Ldm	0.68**	0.72**											
4	Lchlor	0.17	0.36	0.46										
5	Fv/Fm	0.06	0.17	0.30	0.62*									
6	Trans	-0.17	-0.09	0.11	0.14	0.36								
7	Stoma	-0.24	-0.26	-0.14	0.23	0.38	0.64**							
8	Prate	0.13	0.18	0.33	-0.17	0.01	-0.13	-0.18						
9	StemD	0.69**	0.38	0.57*	0.43	0.30	0.05	-0.08	0.06					
10	Height	0.65**	0.54*	0.60*	0.41	0.03	-0.36	-0.31	0.19	0.46				
11	Rlength	0.50	0.30	0.66**	0.23	0.28	0.38	0.11	0.43	0.55*	0.50			
12	Rdm	0.56*	0.74**	0.76**	0.19	-0.02	-0.19	-0.28	0.27	0.50	0.57*	0.37		
13	Sdm	0.82**	0.73**	0.88**	0.30	0.09	-0.11	-0.24	0.24	0.68**	0.74**	0.63*	0.84**	
14	Tdm	0.74**	0.77**	0.95**	0.35	0.14	-0.05	-0.23	0.30	0.63*	0.68**	0.60*	0.90**	0.97**
		1	2	3	4	5	6	7	8	9	10	11	12	13

Note:*significant at 0.05; **significant at 0.01. %smc – Percent soil moisture content; LA – Leaf area (cm^{-2}); Ldm – Leaf dry matter; Lchlor – Leaf chlorophyll content; Fv/Fm – Leaf chlorophyll fluorescence; Trans – Transpiration; Stoma – Stomata conductance; Prate – Photosynthetic rate; Rlength – Root length; Rdm – Root dry matter; StemD-Stem diameter; Sdm – Stem dry matter; Tdm – Total dry matter.

4. Discussion

The pattern exhibited by the polybag media contents with regard to moisture retention was expected as sand is known to have very low water holding capacity. The relatively lower moisture content of the sole sand and sand-topsoil mixtures primarily resulted from the presence of the sand component [16]. The wide variations in percent moisture content and potential availability of nutrients between the topsoil on one hand and the sole sand and sand-topsoil mixtures on the other hand did not translate into differences in percent final germination and plant survival afterwards. This was probably because the moisture and nutrient levels of the river sand treatments were sufficient for seed germination and continuous survival of the seedlings for up to six months. The relatively lower moisture and nutrient content of these treatments however, adversely affected leaf formation during the six months of growth in the nursery. With depleting nutrient levels in the polybags at the later stages, it became clear that leaf chlorophyll formation responded positively to the foliar fertilizer, with the highest leaf chlorophyll found in the river sand + topsoil + foliar fertilizer followed by the river sand + foliar fertilizer and the topsoil treatments. This higher leaf chlorophyll values did not lead to the expected improved photochemistry as seedlings recorded higher leaf chlorophyll fluorescence (fv/fm), indicating re-radiation of high amount of the photosynthetically active radiation (PAR) in long wave form [6]. This was probably due to low moisture levels and lack of complete complement of nutrients required for efficient photosynthesis since seedling growth was improved by the addition of foliar fertilizer to the river sand + topsoil medium over the sole topsoil medium [8]. The retardation of seedling growth by the sole river sand and river sand + foliar fertilizer treatments attributable to low moisture retention and nutrient availability support earlier findings [2] since the foliar fertilizer supplied only NPK and a few trace elements. The addition of the NPK to sole river sand only prevented the incidence of severe chlorosis compared to the sole river sand only treatment [8]. The increase in photosynthetic rate with increasing moisture and chlorophyll content was therefore translated into plant growth [2, 16]. This was probably the reason why the sole topsoil and river sand + topsoil + foliar fertilizer treatments with higher photosynthetic efficiency resulted in seedlings that had larger stem volumes, indicating better dry matter production and accumulation [1, 5, 8, 9]. Such seedlings will have a higher chance of establishment and survival in the field [4, 13, 14].

5. Conclusion

Based on the results obtained, It was concluded that the use of sole river sand with and without NPK (10:10:10) at this rate does not sufficiently support the growth of cocoa seedlings in the nursery. In situations of topsoil scarcity, cocoa seedlings can be raised in a mixture of sand and topsoil (1:1) and nutrient shortfalls provided for by application of foliar fertilizer.

Acknowledgement

The authors are grateful to Ofori Ruth Agyeiwaa, Ofosu Williams and Bernard Quarshie for their technical support.

This paper (CRIG/05/2014/044/002) is published with the kind permission of the Executive Director of Cocoa Research Institute of Ghana.

References

[1] Afaf MH. Effect of NPK and Growing Media on Growth and Chemical Composition of Fishtail Palm (*Caryota mitis* Lour). Life Science Journal. 2012; 9(4): 3159-3168.

[2] Alvim P. DE T, Alvim R. Environmental requirements of cocoa with emphasis on response to shade and moisture: Proceeding in: *International conference on cocoa and coconuts. The incorporated Society of Planters*. Kuala Lumpur, Malaysia. 1980; pp 93-111.

[3] Amoah FM, Opoku-Ameyaw K, Osei-Bonsu K, Oppong FK. Evaluation of bare-rooted methods for transplanting cocoa seedlings. *Ghana Journal of Agricultural Science*. 1999; 32: 69-77.

[4] Amoah FM, Opoku-Ameyaw K, Osei-Bonsu K, Oppong FK. The effect of seedling age on the survival rate of transplanted bare-rooted cocoa seedlings. In: Proceedings, 13th *International Cocoa Research Conference,* Kota Kinabalu, Malaysia, 2001; pp 1215-1219.

[5] Atiyeh RM, Subler S, Edwards CA, Bachman G, Metzger JD, Shuster W. Effects of vermicomposts and composts on plant growth in horticultural container media and soil. Pedo biologia, 2000. 44: 579–590

[6] Gitelson AA, Buschmann C, Lichtenthaler HK. The Chlorophyll Fluorescence Ratio F735/F700 as an Accurate Measure of the Chlorophyll Content in Plants. *Remote Sensing of Environment*. 1999; 69 (3): 296.

[7] Ho DT, Loh HP, Chan E. Germination, nursery practices and field planting techniques for the MAW A hybrid coconut. Proceedings of the International *Conference on Cocoa and Coconut.Planter,* Kuala Lumpur. 1978; 59: 207-214.

[8] Kirk WP, Desmond RL, Snake CJ, Michael GK. Growth Enhancement of Containergrown Pawpaw Seedlings as Influenced by Media Type, Root-zone Temperature, and Fertilization Regime. *HORTSCIENCE.* 2002; 37(2):329–333

[9] Mehmood T, Ahmad W, Ahmad KS, Shafi J, Shehzad MA, Sarwar MA. Comparative Effect of Different Potting Media on Vegetative and Reproductive Growth of Floral Shower (*Antirrhinum majus* L.) Tahir. *Universal Journal of Plant Science*. 2013; 1(3): 104-111

[10] Ofori-Frimpong K, Afrifa AA, Akrofi AY, Sarfo JE. Development of cocoa pod husk-based compost for nursing cocoa seedlings. In: Proceedings, 15th *International Cocoa Research Conference*, San Jose, Costa Rica. 2007; pp 1635-1644.

[11] Oliet J, Planne,,re R, Segua ML, Artero F, Jacobs DF. Mineral nutrition and growth of containerized *Pinus halepensis* seedlings under controlled released fertilizer. *Sciencia Hortuculturae*. 2001; 103: 113-129.

[12] Opoku-Ameyaw K, Baah F, Gyedu-Akoto E, Anchirinah V, Dzahini-Obiatey HK, Kudjoe AR, Acquaye S, Opoku SY. Cocoa Manual. A Source Book for Sustainable Cocoa production, 1st Edition. *Cocoa Research Institute of Ghana.* 2010; 105pp.

[13] Oppong FK, Opoku-Ameyaw K, Ofori-Frimpong K, Owusu-Ansah F. Effects of polybag size and foliar application of urea on cocoa seedling growth. *Ghana Journal of Agricultural Science.* 2007; 40 (2): 207-213.

[14] Oppong FK, Opoku-Ameyaw K, Osei-Bonsu K, Amoah FM, Brew KM, Acheampong K. The effect of time of planting at stake on cocoaseedling survival. *Ghana Journal of Agricultural Science*. 1999; 32: 79-86.

[15] Peries RRA, Everard JMDT. River sand as an alternative to topsoil for raising coconut seedling in polybags. *COCOS.* 1993; 9: 40 - 46.

[16] Shinji S, Andrew DN, Shawaeng R, Narong C. Imrpovement in water-holding capacity and structural stability of a sandy soil in Northeast Thailand. *Arid land Research and Management.* 2007; 21: 37-19

[17] Smillie RM, Hetherington SE. Photoabatement by Authocyanin Shields Photosynthetic Systems from Light Stress. *Photosynthetica*. 1999; 36 (3): pp451-463.

[18] Wuidart W. Production of coconut planting material - The polybag nursery Oleagineux. 1981; 19 (36) : 367 - 373.

Seasonality induced marginality: Vulnerability of wage earners' food and nutrition security in southern Bangladesh

Mohammad Monirul Hasan

Department of Economic and Technological Change, Center for Development Research (ZEF), University of Bonn, Bonn, Germany

Email address:
monir1021@gmail.com

Abstract: The paper examines the impact of wage-earning occupation in the food and nutrition security of the rural households' which is partly rooted in the process of marginalization due to seasonality. Seasonality is obvious in the nature but it becomes a problem for those individuals who are heavily dependent on it and they don't have any other buffering system to mitigate this shock such as savings, credit and social security. The result depicts that for being a wage-earner in agriculture, the vulnerability of food and nutrition security increases by 9% to 12.4% which are statistically significant at 5% level of significance. Marginalized households face seasonality every year and they lose their valuable assets to mitigate the adverse effect of natural calamities and idiosyncratic shocks. As a result the instrument to mitigate this seasonality becomes scarce and ineffective which results malnutrition and food insecurity. Because whenever the households do not have any other coping strategy, they just skip meals and start starving for the extended periods.

Keywords: Seasonality, Marginality, Food and Nutrition Security, Propensity Score Matching, Bangladesh

1. Introduction

Seasonality in agriculture is a common phenomenon which is directly related to the consumption smoothening of the rural households. Seasonality arising from agricultural crop cycle is manifested in household consumption through seasonality of income [1][2][4]. Almost 75% of the annual income of Indian rural households comes in 3 month period [2]. Besides income households consumption level also varies in rural economies [3][4][5]. Seasonality in consumption is largely driven by seasonal variation in income and partly by the inaccessible to credit market [1]. However, non-credit factors such as preferences, labor effort, seasonal variation in prices and precautionary savings motives affect the consumption seasonality [2][4] identified in rural Thailand that the observed seasonality in consumption pattern occurred due to variation in prices which is more acute than the households' inability to use savings or borrowing. According to permanent income hypothesis any change in consumption caused by shocks to income (transitory income) could be smoothed sufficiently by perfect capital market borrowing as the household would try to maximize utility. Household will borrow from market when it has transitory low income and by saving when having transitory high income. Hence the consumption patterns of households are largely determined by the change in permanent income, rather than the change in temporary income [6]. Hence lack of credit could be a potential determinant of seasonal consumption for rural economies [7]. It is well established in the economic literature that credit constraint are more vulnerable to smoothening consumption [8][9][2] and microcredit can help to mitigate the seasonality in consumption by diversifying agricultural income and employment [10]. In an agrarian society, households manage seasonality primarily through consuming their produced goods, self-insurance (utilization of buffer stock) or mutual insurance (through interfamily transfer), relatives' grant or loans and other means which are part of their crop cycle [1]. Any failure of these means could contribute an increase in the severity of seasonal deprivation. Lack of food entitlement resulting from economic and non-economic forces prevent the poor from having access to employment and other form of economic and social security and eventually making them marginalized in the society [11]. The term marginality means "an involuntary position and condition of an individual or group at the margins of social, political, economic, ecological, and biophysical systems, that prevent them from access to resources, assets, services, restraining freedom of choice, preventing the development of capabilities, and eventually causing extreme poverty" [12]. These households become marginalized because they have less income to purchase food and hence that reduces their productivity and hence restricts the development of capabilities which eventually make them marginalized.

With regard to the multi-dimensional nature of marginality, concept of marginality can be framed in to two dimensions as societal and spatial marginalization. Earlier framework attributes on human dimensions such as religion, social structure, wealth, culture, political view or ideology in connection with access to resources by individuals and groups. The latter dimension defined based on physical location and distance from centers of development, lying at the edge of or poorly integrated into system. In contrast, the definition of marginality which is considered by this paper has clearly described the multi-dimensionality of the concept using five different dimensions as social, political, economic, ecological, and biophysical dimensions. So if any given individual or a group of society fails the optimum accesses from at least one of the following mainstreams would possibly considered as a marginal or categorized as victim of marginality.

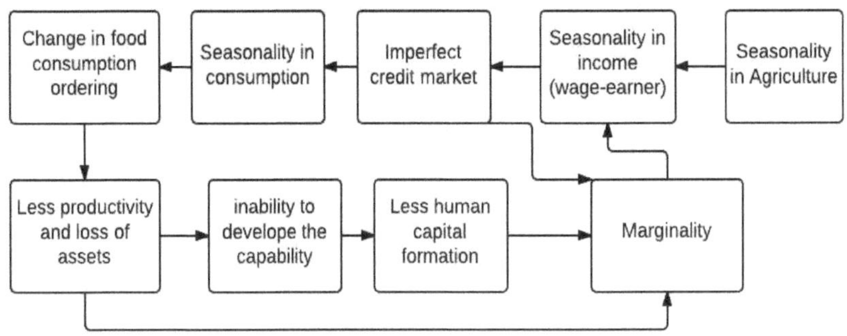

Source: Author's calibration

Figure 1. *The economics of marginalization*

Figure 1 shows the connection of seasonality and marginality; how one is generated from other one. When there is seasonality in agriculture, agricultural wage earners fall into the seasonality and their income also follow the same trend. But if the wage-earning agricultural worker does not have access to credit market, then seasonality in consumption arises and food consumption level falls for that reason. Less food makes them less productive and less shock absorbing capacity which depletes the assets of the households. The end results translate into marginality which works through the inability develop capability and human capital. Marginality translates into poverty because it restricts the translation of capability into functioning which is the main contribution of Sen, A. [11].

If variation in consumption are only transitory in nature and idiosyncratic across households, interventions such as cash transfer, food coupon and food-for-work can help mitigating seasonality in consumption but if it is because of other factors such as structural poverty arising from low income and productivity then those interventions would be ineffective and unsuccessful.

Household faces credit constraints and credit rationing due to distorted financial markets and principle-agent problems associated with it [13][14]. Hence households consumption is not completely smoothen with the exposer of imperfect financial market [15][16][17]. Households become credit constraints when they are unable to fill the temporary income gap by borrowing sufficiently [14]. Beside the credit constraints households precautionary behavior results the violation of the permanent income hypothesis [18][19][20]. Credit constraint households use personal savings, accumulated assets, external assistance and remittances or cash transfer to absorb the income shocks during lean period. Credit constraint becomes more persistent when they fail to generate sufficient savings for the future [21]. Adverse health shocks increase expenditure and also deplete the savings balance which has a long term impact on welfare [22].

Using the upper poverty line income the national poverty rate is 31.5% and in Barisal and Khulna region it is 39.4% and 32.1% respectively according to Household Income and Expenditure Survey (HIES), 2010. In southern Bangladesh, seasonality varies by rural occupation but there is a trend of seasonal income shock has a pattern to be mentioned. Households' average monthly income continues to fall from April to September and starts to escalate to the benchmark income level of BDT 3000 from September to November. Average income level remains above benchmark level from November to February [23][32].

Food and nutrition security is a major concern of Bangladesh especially in the southern part. According to Food and Agriculture Organization (FAO) food and nutrition Security (FNS) has four pillars- food availability, accessibility, utilization and stability. In this analysis, it is not possible to see the all the outcomes of FNS but focus on some very specific variables. This research is focused on food availability and the accessibility of the households. The short term household level FNS outcome for availability is frequency of meals eaten in a day and for accessibility is food expenditure share on households' total budget. About the drivers, the short term FNS drivers are for availability is household size and for accessibility is household dependency ratio, income, distance to nearest market etc. All these can be seen in a table 1.

Table 1. *FNS outcomes and drivers in household level*

	Availability	Accessibility
Outcomes (Short term)	Number of meals eaten in a day	food expenditure share on households total budget
Drivers (Short term)	Household size	Dependency ratio, Income, distance to nearest market
Drivers (Long term)	Farm land size	Access to savings and credit, income per capita.

Source: Pangaribowo, E. H. et al., 2013[24]

For this paper the outcome variable number of meal consumption eaten in a day will be the basis of analysis.

2. Data and Methods

2.1. Sample

The study is based on the data of marginalized households of southwestern Bangladesh. Palli Karma Sahayak Foundation (PKSF), Bangladesh introduced a program named PRIME (Programmed Initiatives for Monga Eradication) funded by DFID (Department for International Development), UKaid to eradicate poverty of the rural households in 2011. With the assistance of Partner Organizations (POs) PKSF initially stepped into six Upazilas (sub-district) of three districts in 2011. These districts are Khulna, Patuakhali and Satkhira. Subsequently, the program extended to five more Upazilas by 2012. PKSF conducted a census survey of 60,000 of households (Table 2) and sample for the baseline survey has been taken from this census.

Table 2. *Household Covered in the Original Census.*

District	Upazilla	Total household targeted for PRIME
Khulna	Dacope	7,588
	Koyra	13,632
Patuakhali	Golachipa	13,543
	Kolapara	5,745
Satkhira	Kaliganj	11,201
	Shaymnagar	8,344
Total		60,053

Source: [23] & PKSF Household Census for PRIME South.

Criteria for picking marginalized households were as follows- (1) Monthly income less than or equal to BDT 3,000 (Approx. EUR 30) per household during lean season; or (2) Main profession of the household head is daily wage earning (in farming, fishing, logging, honey assortment or other activities); or (3) occupying 50 decimal cultivable land or less.

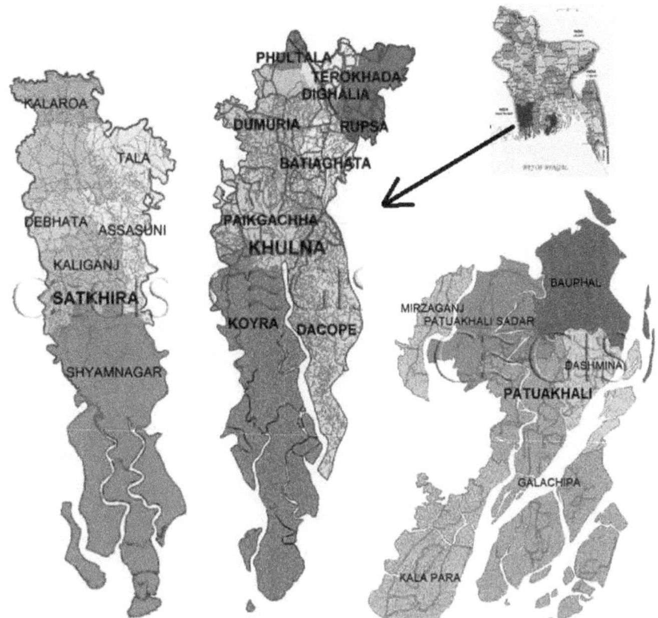

Districts: Satkhira Khulna Patuakhali

Source: Maps are generated by CEGIS. Map of Bangladesh (upper captioned) is from PKSF. [25]

Figure 2. *Study area in Southern Bangladesh.*

Institute of Microfinance (InM) conducted the baseline survey in 2011 and collected 4000 sample from which 3977 retains. The study area is shown in the figure 2.

2.2. Method

The Propensity Score Matching (PSM) is used for identifying the impact of being wage earner on food consumption vulnerability. The underlying assumption is that selection can be described solely in terms of observable characteristics. For every individual in the treatment group a matching individual is found to be identical individual in the non-treatment group on the basis of observable characteristics. Then average effect of treatment can be calculated as the average differences in the outcomes of two groups. PSM matches each participant with an identical nonparticipant and then measures the average difference in the outcome variable between the participants and the non-participants [27]. It tries to compare similar propensity scores to get the effect. If there is no match found, households are dropped.

The wage-earning households are considered as treated and the other occupation households are considered as non-treated or controlled.

$$y_i = wage\_earner = \begin{cases} 1 & \text{if households head's occupation is wage earning} \\ 0 & \text{if household head is not wage earner} \end{cases}$$

Wage earning household is defined as y_1 and the other occupation $(wage\_earning = 0)$ as y_0. The objective is to determine the average treatment effect on the treated (ATT). The average treatment effect on the treated is defined as

$$ATT = E(y_1 - y_o / wage\_earning = 1)$$

$$= E(y_1 / wage\_earning = 1) - E(y_0 / wage\_earning = 0)$$

The first term of the equation is observable whereas the second term is not observable as it is impossible to observe the same individuals as recipient as well as non-recipient simultaneously. The use of propensity score matching can eliminate this problem to estimate $E(y_0 / wage\_earning = 0)$

In the observational studies estimating ATT arises the problem of non-randomness of the selection of the treatment and control and therefore the estimation of ATT suffers from biasedness. PSM can fix this problem which encapsulates the pre-treatment characteristics of subject into a single index - the propensity score which is then used to generate the matching. PSM reduces the biasedness by comparing two groups based on observable characteristics. The validity of PSM depends on two conditions- (1) conditional independence (unobserved factors don't affect participation) and (2) sizable common support or overlap in propensity score across the participant and non-participants.

Any standard model such as Logit or Probit can be applied for estimating the propensity score. For the purpose of estimation Probit model takes the form as follows-

$$P(y_i = 1|x) = P(y_i^* > 0|x)$$

$$= p(x_i'\beta + \varepsilon_i > 0|x)$$

$$= P(\varepsilon_i > -x_i'\beta|x)$$

$$= 1 - F(-x_i'\beta)$$

Considering the error terms are independently and normally distributed:

$$P(y_i = 1|x) = 1 - \Phi\left(-\frac{x_i'\beta}{\sigma}\right), \sigma \equiv 1$$

$$= \Phi(x_i'\beta)$$

Here *Pr* represents probability, and Φ denotes the Cumulative Distribution Function (CDF) of the standard normal distribution. Parameters β are typically estimated by maximum likelihood methods. The usage of the standard normal distribution reasons no loss of generalization compared to using an arbitrary mean and standard deviation as adding a fixed value to the mean can be offset by deducting the same value from intercept and multiplying standard deviation (SD) with a fixed value can be compensated by multiplying the weights by the same value.

Matching participants and non-participants can happen in various ways such as (1) Nearest-neighbor matching, (2) Caliper or radius matching, (3) Stratification or interval matching, (4) Kernel and local linear matching, (5) Difference-in-difference matching etc.

In PSM, the first stage is to determine the propensity score and satisfy the balancing property. The propensity score that is within lowest and highest values for households in the treatment group is called area of common support. With the propensity score generated, the outcome of interest between treatment group and matched control group will be compared. This approach is also used by many authors such as [28][1][29][30][25][26].

3. Results

3.1. Descriptive Analysis

The descriptive analysis of the selected variables of the households is presented in table 3. The average age of the household head is 42 years and the years of schooling is 2. 53.33% of them are wage-earner and almost most of them are in rural agricultural labors. Almost 10% works in as self-employed in agriculture and 22% works as self-employed in non-agriculture. The rate of working outside of home is 13% who are mostly migrated workers. The average household size is 4 only. Average total land of the household is 13 decimals and of which 5 decimal is for agriculture. Household's yearly income is almost BDT 50,000 in which BDT 40,000 is spent on food and BDT 13,000 is spent on non-food. Households receive average amount of BDT 3,350 as social safety net program from the government. Majority of the households are from mainland but 25% of the households live in geographically inaccessible areas which is called char areas (river basins) in southern Bangladesh.

Table 3. *Summery statistics of the selected variables used in the model*

Characteristics	Mean	Standard Deviation
Age (Years)	42.6	13.8
Years of schooling	2.0	3.1
Wage earner	53.33%	49.90%
Self-employed in agriculture	9.88%	29.85%
Self-employed in non-agriculture	22.08%	41.48%
Live outside the household for work	12.75%	33.36%
Household size (Number)	4.0	1.46
Dependency ratio (female per male)	1.2	0.9
Total owned land (Decimal)	13.0	35.9
Total agricultural land (Decimal)	4.9	27.6
Number of cows	0.4	1.0
Number of goats	0.6	1.5
Number of poultry	4.0	5.5
Asset value including land (Taka)	58,940	123,666
Savings (Taka)	1,335	7,129
Total Income (Taka)	49,903	38,268
Expenditure on food (Taka)	39,409	15,110
Non-foods expenditure (Taka)	13,271	9,965
Unmet crisis in 2010-11 (Taka)	1,105.04	5,417
Distance from market place (km.)	3.9	2.5
Total formal loan (Taka)	1,234	4,475
Total informal loan (Taka)	1,643	12,216
Social Safety Net received (Taka)	3,351	7,121
Household in *char* areas	24.74%	43.15%

Source: Author's calculation.

In the southern part of Bangladesh the seasonality of occupations of the households can be observed in figure 3. The bold black line characterizes the average monthly wage income and the fade straight line characterizes the threshold level of household income at BDT 3,000.

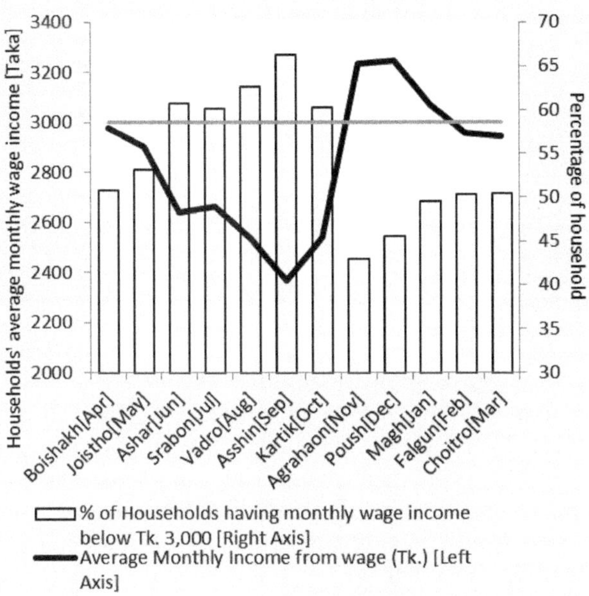

Source: Author's calculation.

Figure 3. *Seasonal Dynamics of Households' Monthly income from wage labor.*

The shaded bar-diagram characterizes the percentage of households possessing monthly wage income below threshold of BDT 3,000. Average monthly income jumps to fall from Bengali month Boishakh [April] and remains to fall until Asshin [September]. The bar chart of these months is higher than the other months which are the lean season reported by the households. The mean crisis season is 3.5 reported by the households [32]. This season is the monsoon in Bangladesh when the majority of the crop cultivation is hampered by flood.

Afterwards the month of Asshin [September], the wage income starts to rise again new cropping in the field in the beginning of autumn and winter seasons. Farmers start to cultivate and they employed agricultural labor in their fields. Henceforth the wage income flinches to rise until the end of Poush [December]. Once more the wage income starts to decline from Falgun [February] and the similar process repeats every year. In this season of September to February, households make good income (more than BDT 3,000) to consume food and non-food expenditure. This can be seen from the bar chart that the percentage of households having monthly wage less than BDT 3,000 is lowest in Agrahaon [November] and it is about 43% which is lower than the highest 67% in the month of Asshin [September]. Seasonality is not a problem if they could have savings enough to ensure food security in lean season. But in the study area, the households can't generate enough savings and loans to mark them better off in the lean season.

Households have reported the starting month and the ending month of their food consumption changing pattern. Figure 4 clearly depicts the diverse pattern of starting and ending month of the food insecurity of the households. Over 35% of the households reported that their deficiencies start on Ashar [June] and 25% reported it on Asshin [September]. Conversely almost 32% of the households reported that they end food deficiencies on Kartik [October]. The average length of food insecurity is 3.5 months. From the figure, it can be said that from October to March most of the households end their food insecurity and it starts again from April and continues until September. Albeit households have prior knowledge about this cycle, they can hardly do something against this shock.

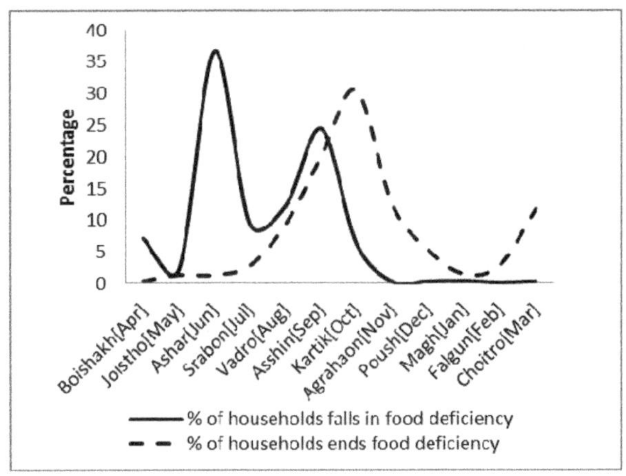

Source: Author's calculation.

Figure 4. *Starting and ending month of households' food insecurity.*

3.2. Households' Food and Nutrition Insecurity

The main purpose of this study is to examine the food and nutrition insecurity through the meal frequencies. Households described that their food consumption frequencies changes due to the seasonal income shock. Households find themselves capable to buy daily food in normal season but they have to ration their consumption in lean season. Table 4 characterizes the distribution of households having frequencies of food consumption in both normal and lean seasons. It illustrates that in normal time, about 0.28% of households suffer from occasional starvation, 19% experience consumption rationing and over 80% of the households enjoy full 3 meals in a day. But in lean season, around 9.46% of households fall in occasional starvation, 73% suffers from consumption rationing and merely 17.46% can consume full 3 meals in a day.

Table 4. Transition matrix of households' food and nutrition insecurity.

Consumption ordering in normal time	Consumption ordering in lean time			
	Occasional Starvation [<=1 meal]	Consumption rationing [2 meal]	Full 3 meals in a day	Total
Occasional Starvation	6 (54.55) (1.62)	3 (27.27) (0.1)	2 (18.18) (0.29)	11 (100) (0.28)
Consumption rationing	232 (31.02) (62.7)	508 (67.91) (17.77)	8 (1.07) (1.17)	748 (100) (19.12)
Full 3 meals in a day	132 (4.19) (35.68)	2,348 (74.47) (82.13)	673 (21.34) (98.54)	3,153 (100) (80.6)
Total	370 (9.46) (100)	2,859 (73.08) (100)	683 (17.46) (100)	3,912 (100) (100)
Pearson $\chi 2$	<0.01			

Source: Author's calculation, Note: Normal time characterizes when households earn more than BDT 3,000 and lean period characterizes when they earn less than this amount. Values in the parenthesis show row and column percentages.

In Table 4, the dark shaded box denotes the number of households fall in occasional starvation in lean season from normal season. This group of households is vulnerable of degree 2, as the benchmark is full 3 meals in a day. The less dark box embodies vulnerability of degree 1, because these households degrade one degree level from the benchmark and earlier level. The area from very light shaded box shows the number of households become better off in their declared lean season. They actually enjoy some benefit of seasonality. But the households who gain are very insignificant. The bulk of this group comes from the previous full 3 meals group and they continue it during their lean season indicating they ensure sufficient saving and income source to maintain a consistent consumption frequencies.

The households those were starving occasionally in normal season about 27% of them switched to consumption rationing and 18% to full 3 meals category in lean season. But the numbers of households are only 3 and 2 respectively. Conversely, households those who were experiencing rationing their consumption, about 31% of them fall in occasional starvation, 68% remain in the same group in lean season. Households who were enjoying full 3 meals in normal season, about 4.19% of them fall in occasional starvation, 74% fall in consumption rationing and only 21% could continue their regular full 3 meals in lean season.

3.3. Econometric Result

3.3.1. Impact of Wage Earning on Vulnerability

The adverse effect of seasonality in agriculture is higher for the wage-earning households as they are directly involved in this daily labor in agriculture. The paper tries to find the impact of being wage-labor in agriculture on FNS security of the rural households. An econometric technique such as PSM is applied to get the result. Table 5 represents the results of PSM which postulates that for being an agricultural wage-earner the vulnerability of FNS increases by 9 to 12.4 percentage point. There are three matching techniques applied here all of which shows the same trend and statistically significant results. For the nearest neighboring matching, 12.4% increase in vulnerability whereas in Stratification methods, almost 11.3% increase in the vulnerability in FNS. In Kernel Matching Methods, it is about 9.2%. So in every matching technique the result seems very robust and statistically significant at 95% level of confidence interval.

4. Discussion of Results

The study of seasonality not scant in the literature and they are also related to the consumption level of households [1][2][3][4][5]. But there are very few papers on how seasonality induces the marginality in the literature. Marginality is poorly addressed and there is affluent scope of linking seasonality in marginality. Authors such as [2][3][1] reported about the consumption in aggregate but there is limited number of paper describing the FNS in terms of meal consumption frequencies. The present paper strives to fill the gap by addressing seasonality through marginality and meal consumption frequencies which is one of the outcome variables of FNS. The absence of credit market is highlighted in many papers [8][9][2] as an important interventions advocates the access to credit can improves the situation. But it is not only the absence of credit that restricts them from consumption smoothening but also the right to access the credit. The marginalized households are incapable to accessing credit as they don't fulfill the requirement for getting the credit. So they have problem of availability, accessibility, utilization and stability. Besides, the lack of savings also makes the household food insecure for short term and marginalized in the long term.

Table 5. Estimation of Average Treatment Effect for the Treated (ATT): Impact of wage-earning occupation on vulnerability in FNS.

Matching Methods	Number of treated	Number of control	ATT	Standard Error	t-value
Nearest Neighbour method	2023	309	0.124**	0.035	2.249
Stratification method	2023	468	0.113**	0.048	2.354
Kernel Matching method	2023	468	0.092**	0.046	2.007

Source: Author's calculation.
Note: *** p<0.01, ** p<0.05, * p<0.1

Like [31] vulnerability in FNS is a transient poverty which is caused by unavailability of physical capital and adverse geographic conditions. Geographic location as well as natural disaster like cyclones can also make them marginalized [25]. Inadequate physical capital obstructs them developing their capabilities along with human capital formation and makes them marginalized. The contribution of this paper is to identify the impact of wage-earning occupation on food and nutrition insecurity of the marginalized households. The result depicts that for being a wage-earner in agriculture, the vulnerability of food and nutrition security increases by 9% to 12.4% which are statistically significant at 5% level of significance. The graphical presentations also support this result because the wide variation in income and food consumption due to seasonality is depicted in figure 3 and figure 4. It is true that there are other causes which are responsible for the variation in food consumption frequencies but the analysis claims that it is only 9% to 12.4% that is caused by the wage-earning occupation.

To identify the impact of occupation on food and nutrition security the problem of selection bias and confounding factor arises. The comparison of treated group and control group becomes questionable and the results they produce become inefficient and inconsistent. To solve this problem of selection bias and endogeneity, propensity score matching technique is used where the treatment group is compared with the control group based on some observable characteristics. The method is widely used to see the impact of any program [27][25] and the sophisticated software tool makes the calculation appropriately.

5. Conclusion

From the discussion throughout the paper, it is apparent that there is an impact of wage-earning occupation in the food and nutrition security of the rural households' which is partly rooted in the process of marginalization due to seasonality. Seasonality is obvious in the nature but it becomes a problem for those individuals who are heavily dependent on it and they don't have any other buffering system to mitigate this shock such as savings, credit and social security. Marginalized households face seasonality every year and they lose their valuable assets to mitigate the adverse effect of natural calamities and idiosyncratic shocks. As a result the instrument to mitigate this seasonality becomes scare and ineffective which results malnutrition and food insecurity. Because whenever the households don't have any other coping strategy, they just skip meals and start starving for the extended periods. It is needed to diversify their occupation or income source so that they can smoothen their consumption especially meal consumption and can secure their food and nutrition. The diversification of occupation may occur in various ways. Engagement in non-agricultural wage earning and non-agricultural self-employment can contribute smoothening the food consumption throughout the year. Besides engaging in non-agricultural activities household head can migrate to other places where the job availability is affluent. One possibility is to migrate to cities and do informal jobs. Government and non-government organization can play role to disseminate relevant information for migration and job placement. Various incentive programs such as subsidy for migration, easy loan, money transfer services etc. can motivate individual to migrate or finding new jobs during this seasonally lean period. Different kinds of financial services such as savings, crop insurance and employment guarantee can reduce the shocks arising from seasonality in agriculture.

Acknowledgement

The study is a part of a Master thesis chapter that was presented and submitted to the Faculty of Business Administration at Corvinus University of Budapest, Hungary & Faculty of Agriculture, University of Bonn, Germany as in partial fulfilment of the requirements for the degree of Masters of Science. The author is grateful to Professor Dr. Imre Fertő, the first supervisor and Professor Dr. Thomas Heckelei, the second supervisor for their valuable directions. The study was supported by European Union's Erasmus Mundus Program through AFEPA consortium. The author is immensely indebted to Institute of Microfinance (InM) and Palli Karma-Sahayak Foundation (PKSF) for providing data for the study.

References

[1] Khander, S.R., Khalily, B., and Samad, A., "Vulnerability to seasonal hunger and its mitigation in northwest Bangladesh", Working Paper No. 4, 2010, Institute of Microfinance (InM).

[2] Chaudhuri, S and Paxon, C., "Smoothing consumption under income seasonality: Buffer stock vs. credit." Colombia University, Department of Economics, Discussion paper series, 2001.

[3] Sahn, D. E., ed. Seasonal Variability in the Thirld World Agriculture: The Consequences for Food Security. Baltimore: Johns Hopkins University Press. 1989.

[4] Paxon, C., "Consumption and Income Seasonality in Thailand", Journal of Political Economy, vol. 101, no. 1:39-72.

[5] Dercon, S., & Krishnan, P., "Vulnerability, seasonality and poverty in Ethiopia". Journal of Development Studies, Vol. 366, No. 6:2000, pp 25–53.

[6] Friedman, M., A theory of the consumption functions. Princeton University Press. 1-6, 1957.

[7] Townsend, R. M., "Consumption insurance: An evaluation of risk-bearing systems in low income economies". Journal of Economic Perspectives, 9(3):1995,pp 83-102.

[8] Rosenzweig, M., "Risk, Implicit Contracts and the Family in Rural Areas of Low-income Countries", Economic Journal, Vol. 98, No. 393: 1988, pp1148-1170.

[9] Rosenzweig, M and Wolpin, K., "Credit Market Constraints, Consumption Smoothing and the Accumulation of Durable Production Assets in Low-Income Countries: Investment in Bullocks in India", Journal of Political Economy, vol. 101, No. 2: 1993, pp223-244.

[10] Pitt, M., & Khandker, S., "Credit programmes for the poor and seasonality in rural Bangladesh". Journal of Development Studies, Vol. 39, No. 2:2001, pp 1-24.

[11] Sen, A., Poverty and Famines: An essay on entitlement and deprivation. Oxford University Press, New York, NY. 1981.

[12] Von Braun, Joachim and Gatzweiler, Franz W., Marginality: Addressing the Nexus of Poverty, Exclusion and Ecology, Springer. 2014.

[13] Conning, J., & Udry, C., "Rural financial markets in developing countries". Handbook of Agricultural Economics, 3, 2007: pp2857-2908.

[14] Morduch, J., "Income smoothing and consumption smoothing." The Journal of Economic Perspectives, 9(3), 1995:pp103-114.

[15] Dercon, S., and Krishnan, P., "Vulnerability, seasonality and poverty in Ethiopia", Journal of Development Studies, Vol. 366, No. 6, 2000, pp. 25–53.

[16] Duflo, E., & Udry, C. Intra-household resource allocation in Cote d'Ivoire: Social norms, separate accounts and consumption choices (NBER Working Paper No. w10498).2004.

[17] Goldstein, M. Intrahousehold efficiency and individual insurance in Ghana. London School of Economics Working Paper no.DEDPS38. 2004.

[18] Deaton, A. "Saving and liquidity constraints". Econometrica, 59(5),1991:pp 1221-1248.

[19] Morduch, J., Risk, production and saving: Theory and evidence from Indian households. Harvard University, Manuscript. 1990.

[20] Paxson, C. "Using weather variability to estimate the response of savings to transitory income in Thailand". American Economic Review, 82(1),1992:pp 15-33.

[21] Armendariz, A., and Morduch, J., The economics of microfinance. Cambridge, Massachusetts: The MIT Press. 2005.

[22] Kochar, A. "Explaining household vulnerability to idiosyncratic income shocks". American Economic Review, 85(2), 1995: pp159-1964.

[23] InM, "PRIME intervention in the coastal areas of south-west Bangladesh", Institute of Microfinance (InM), 2011.

[24] Pangaribowo, E. H., Gerber, N. and Torero, M. Food and Nutrition Security Indicators: A Review, Working paper 108, Center for Development Research (ZEF) 2013, Germany.

[25] Hasan, Mohammad Monirul. Climate Change Induced Marginality: Households' Vulnerability in the Meal Consumption Frequencies. American Journal of Environmental Protection. Vol. 3, No. 3, 2014, pp. 103-112. doi: 10.11648/j.ajep.20140303.11

[26] Hasan, M Monirul, "Does credit improve the food consumption vulnerability of the extreme poor? - Empirical evidence from Bangladesh", MPRA Paper No. 28192. 2010. http://mpra.ub.uni-muenchen.de/28192/1/MPRA_paper_28192.pdf

[27] Khandker, S. R., Koolwal, G.B., and Samad, H. A., Handbook of impact evaluation- quantitative methods and practices. The World Bank. 2010.

[28] Godtland, E., Elisabeth, S., Alain, J., Rinku, M., and Oscar, O., "The impact of farmer- field-schools on knowledge and productivity: A study of potato farmers in the Peruvian Andes", Economic Development and Cultural Change, 52 (1), 2004, pp. 129 58.

[29] Heckman, J., Ichimura, H., and Petra T., "Matching as an econometric evaluation estimator", Review of Economic Studies, 65 (2), 1998, pp. 261–94.

[30] Caliendo, M., and Kopeinig, S., "Some practical guidance for the implementation of propensity score matching", Journal of Economic Surveys, 22 (1), 2008, pp. 31–72.

[31] Jalan, J., & Ravallion, M., "Transient poverty in post-reform rural China", Journal of Comparative, Vol. 26, No. 2:1998, pp338-387.

[32] Hasan, M. Monirul, "Determinants of Food Consumption Vulnerability of the Extreme Poor- Empirical Evidence from Southern Bangladesh", Erasmus Mundus AFEPA 2011-13, University Catholique de Louvain, Belgium. 2013.

19

Resource use efficiency and profitability of fluted pumpkin production in Ukwuani local government area of Delta State, Nigeria

Ogisi O'raye Dicta, Begho Toritseju*, Ewolor Scholastica Agatha

Department of Agricultural Economics and Extension,. Delta State University, Asaba Campus, Asaba. Nigeria

Email address:
tbegho@yahoo.com (B. Toritseju)

Abstract: The paper examined resource-use and profitability of fluted pumpkin (*Telfaria occidentalis* Hook F. Cucurbitaceae) production in Ukwuani local Government Area of Delta State, Nigeria. Primary data were obtained from questionnaire administered to 140 fluted pumpkin farmers drawn using systematic sampling techniques from the study area. Data were analyzed using descriptive statistics, marginal value productivity, multiple regression and a 5-point likert scale. Results showed that most of the respondents were females, illiterate and married. Farmers operated on subsistence level, about 95% practiced mixed cropping with farming experience not less than 5 years. An adjusted R^2 value of 0.664 indicated that the independent variables; farm size, labour and quantity of seed in the model explained 66% of variations in the dependent variable (output). The F value of 60.09 showed the overall statistical significance with a P value of zero. Net return per hectare was $644.90. The benefit-cost ratio was $1 to $1.23 which implies that for every $1 invested, 23cents was obtained thus depicting that the enterprise was profitable in the study area. Resources under-utilized were quantity of seed and farm size while family and hired labour were over-utilized. Major constraints faced by farmers were; high cost of labour, lack of adequate and quality seeds, pests and disease attack, storage and transportation cost. It is recommended that effective agricultural extension services, credit and loan schemes and infrastructure should be provided to farmers in the study area for increased profitability, yield and resource-use efficiency of fluted pumpkin production.

Keywords: Efficiency, Fluted Pumpkin, Profitability, Resource Use

1. Introduction

Fluted pumpkin (*Telfaria occidentalis* Hook F. Cucurbitaceae) is the first in the indigenous vegetable crops priority rating of south-eastern Nigeria. This is in accordance with the preference of farmers, researchers and extension agents. It is one of the most important vegetable in Nigeria referred to as "*Ugwu*" in Igbo, "*Iroko*" in Yoruba and "*Umee*" in Urhobo.

Telfaria occidentalis is a leafy vegetable (Enabulele & Ilavbarhe, 2001) an important component of the traditional farming systems and daily diet of Nigerians (Akoroda, 1990) mainly grown for its highly nutritive value. Consumers of fluted pumpkin have increased tremendously over the years, reason for this increase and acceptance is that the seeds, shoots, leaves and stem are consumed as food for man and fodder for animals.

Fluted pumpkin is perennial when grown on well-drained soils, slightly shaded and mulched but not so soggy soils (Idowu, Alimi, Tijani & Okobi, 2007). In the study area, the crop is grown on poor soils as an annual during the rainy season and also during the dry season around rivers or water sources. It is grown along side crops such as yam, melon, pepper and cassava.

Fluted pumpkin is very important in the diet of children, men, women, nursing mothers as well as livestock due to its high nutritive value. But in Nigeria, the output has not been able to meet the demand for human food not to mention that of livestock feed. As a result of the growing need, the task of producing enough fluted pumpkin poses an increasing challenge.

In Ukwuani Local Government area, fluted pumpkin plays an important role not only as the major source of vitamins in diets but also as a source of livelihood for the majority of farmers who grow and sell this vegetable.

Efficiency of resource use is a very fundamental strategy that leads to increased productivity. The inability of farmers to combine resources efficiently with other factors such as price and non-price factors affect real farm profit (Kwanashie, Garba & Ajilimar, 1997). This study was conducted to isolate the factors that affect profitability and determine their effect on the production of fluted pumpkin in the study area. The broad objective of the study therefore was to evaluate resource-use efficiency and profitability of fluted pumpkin (*Telfaria occidentalis* Hook F. Cucurbitaceae) production in Ukwuani Local Government Area of Delta State. The specific objectives were to: determine cost and benefit of fluted pumpkin in the study area, examine the resource-use efficiency in fluted pumpkin production, factors affecting production of fluted pumpkin and identify production constraints to fluted pumpkin.

Several studies in Nigeria have been carried out to analyse resource-use efficiency as well as profitability of crops and more specifically vegetables.

Udoh & Akpan, (2011) in their study estimated the efficiency of resource use among urban vegetable (Talinium triangulare) farmers in Akwa Ibom State. They reported showed that waterleaf farmers were inefficient in the use of these resources with land and manure was underutilized, while labour was over utilized. In terms of profitability, Udoh & Akpan, (2011) reported that farmers made profit of $1765.53 per hectare.

In studying resource use efficiency and profitability, several methods have been widely adopted. Ala (2013) examined of profitability and resource-use efficiency of yam production by women in Bosso Local Government Area of Niger State, Nigeria using farm budgeting, multiple regression and MVP/MFC ratio. The result of Ala (2013) showed that farmers obtained a net profit of $63.39/ha and the MVP/MFC ratio test revealed that farm size, fertilizers and farm labour were underutilized by the farmers. Similarly, Omonona *et al.,* (2011) in their study on profitability and resource-use efficiency among ofada rice farmers in southwest, Nigeria adopted the Marginal value product (MVP) approach and found that all identified resources were underutilize.

Jirgi, Ogundeji, Viljoen & Adiele, (2010) examined the profitability and resources-use efficiency of millet/cowpea mixed farmers production in Niger state Nigeria using the farm budgeting technique and exponential production function technique. The results showed that the estimated gross margin, net farm income, was $353.67 per hectare, while the results of allocative efficiency show that seeds, family labour and agrochemicals were under-utilized.

More specifically, Ayinde, Akerele, & Ojeniyi, (2007); Idowu, Alimi, Tijani & Okobi, (2007) examined economic factors affecting the production for fluted pumpkin using a modified cost-route approach and regression analysis. They reported fluted pumpkin production as being profitable under existing production systems with an average net income of $205.90/ha on a mean farm size of 0.301 ha. Cost of fertilizer and farm size were however underutilized.

Regarding constraints faced by fluted pumpkin farmers, Nwosu, Onyeneke & Okoli, (2012) reported that lack of credit facilities, lack of availability of inputs, pests and disease infestation, inadequate information about input and output prices, and poor road network respectively.

2. Methodology

2.1. The Study Area

The study was carried out in Ukwuani Local Government Area of Delta State, Nigeria. The study area was selected for this research because a considerable quantity of fluted pumpkin is produced and marketed in this area. Ukwuani Local Government Area shares boundaries with Ndokwa West in the East, Ethiope-East in the south and Orhionwhon Local Government area of Edo state in the North. It lies approximately between longitude 4.5^0 6.3^0 W and latitude 5.5^0 30^0 N of the equator. The climate is tropical, marked by two distinct seasons; wet and dry with an average temperature of 30^0 C. The inhabitants engage in farming and crops commonly grown in the area are fluted pumpkin, melon, sweet pepper, cassava, yam and maize.

The local Government Area consists of ten communities. Out of these, four contiguous communities were purposively selected for the study. Each of these communities was visited and a list of farmers that cultivated fluted pumpkin above subsistence level was made to form the working sample frame. Systematic sampling was applied to draw at least 35 farmers from each community. This resulted in a total sample size of 140. Primary data were collected using copies of well-structured questionnaires which were administered to respondents within a time frame of July 2013 to January 2014. Cost-benefit analysis was done to determine the cost and benefit of fluted pumpkin in the study area, the relationship between marginal value productivity (MPV) and marginal factor cost (MFC) was used to examine the resource-use efficiency in fluted pumpkin production while multiple regression analysis was used to determine factors affecting production of fluted pumpkin in the study area.

2.2. Model Specification

Cost Benefit analysis was calculated as:

BCR = TR/TC

2.3. Multiple Regression

Multiple regression was used to express causal relationship between the dependent variable (Yield output of fluted pumpkin) and the independent variables. Four functional forms were tried and the model with best fit was taken as the lead equation.

$Y = f(X_1, X_2, X_3, X_4, X_5, X_6, \mu)$

Linear Function

$Y = b_0 + b_1X_1 + b_2X_2 + b_3X_3 + b_4X_4 + b_5X_5 + b_6X_6 + \mu$

Exponential Function

$$LogY = b_0 + b_1X_1 + b_2X_2 + b_3X_3 + b_4X_4 + b_5X_5 + b_6X_6 + \mu$$

Semi Log Function

$$Y = b_0 + b_1LogX_1 + b_2LogX_2 + b_3LogX_3 + b_4LogX_4 + b_5LogX_5 + b_6LogX_6 + \mu$$

Double-Log Function

$$LogY = b_0 + b_1LogX_1 + b_2LogX_2 + b_3LogX_3 + b_4LogX_4 + b_5LogX_5 + b_6LogX_6 + \mu$$

Where:

Y = Quantity of fluted pumpkin produced in land area

X_1 = Farm size

X_2 = Family labour (Man days)

X_3 = Hired Labour (Man days)

X_4 = Quantity of seeds (kg)

X_5 = Quantity of fertilizer (kg)

X_6 = Quantity of Herbicides (kg)

μ = stochastic error term

Marginal Value Product and Marginal Factor cost

$$MVP = MPP \times P_y$$

Where:

MVP = Marginal value product

MPP = Marginal physical product

P_y = Unit price of output of fluted pumpkin

When

$\frac{MVP}{MFC} > 1$ The resource use is below Optimum (under-utilization)

$\frac{MVP}{MFC} < 1$ The resource use is above Optimum (over-utilization)

$\frac{MVP}{MFC} = 1$ The resource use is at Optimum (Optimum-utilization)

3. Results and Discussions

3.1. Socioeconomic Characteristic of Respondents

The socioeconomic characteristic of respondents are as presented in Table 3.1. The results showed that majority of respondents were female and within the age group of 41 – 50. About 78 percent of the farmers were married with an average household size of 5 persons. This implies that pumpkin farmers were within the ages regarded as economically productive in a population. About 53% had no education, 28% had primary education, 12% had secondary education while only 2% had tertiary education. This implies that a high percentage of pumpkin farmers are illiterate Majority (73%) of the farmers had above 5 years of farming experience and about 95% of them practiced mixed or intercropping. Crops intercropped with fluted pumpkin were maize, yam, cassava and pepper. The years of experience and wealth of knowledge of the farmers have some impact on the management ability and influences output. Most of the farmers which accounted for 80% cultivated 1.0ha and below thus output was small and fragmented. Olayide, *et al.*, (1980) classified small scale farmers as those having 0.1- 5.9ha farm size. Since land is a very important fixed asset, its method of acquisition is crucial. It was observed that inheritance (72%) was the common method of land ownership in the study area. Others were through rent, purchase and shared cropping.

Table 3.1. *Socioeconomic Characteristics of Respondents*

Characteristics	Frequency	Percentage
Age (Years)		
< 30	6.0	4.3
31-40	46.0	32.9
41-50	71.0	50.7
>50	17.0	12.1
Educational Level		
No formal education	81.0	57.9
Primary	39.0	27.9
Secondary	17.0	12.1
Tertiary	3.0	2.1
Gender		
Male	59.0	42.1
Female	81.0	57.9
Marital status		
Single	4.0	2.8
Married	109.0	77.9
Widowed	27.0	19.3
Farming Experience (Years)		
<1	0.0	0
1-5	28.0	20
6-10	66.0	47.2
>10	46.0	32.8
Cropping pattern		
Sole	7.0	5
Mixed	133.0	95
Farm size		
<0.5	36.0	25.7
0.5-1.0	77.0	55
1.1-2.0	27.0	19.3

3.2. Cost and Benefit Analysis

Labour cost: The unit cost of hired labour per man-day in the study area was $4.30 and the average number of man-days per hectare utilized in the study area was $0.82 as shown in Table 3.2. Labour cost accounted for 98.5% of the total variable cost.

Seed Cost: On an average, seeds from four pods of fluted pumpkin were used per hectare. The average cost of a pod was $1.54 thus average cost of seeds per hectare was $ 6.15. Seed cost accounted for 1.5% of the total variable cost.

Fertilizer and Herbicide cost: None of the respondents applied fertilizer and or herbicides on their farms therefore, cost of fertilizer was not included.

Cost of Land Rentage: The average cost of rent was $18.44per hectare. Majority of farmers used inherited land which was valued at the rent value in the study area. This accounted for 63% of the total fixed cost.

Table 3.2. Average Cost of Production for Fluted Pumpkin in the Study Area

Input cost ($)	Quantity	Unit cost	Total
Variable input			
Labour (Man-Days)	0.69	4.30	486.17
Seeds (pods)	0.02	1.54	6.15
Total Variable Cost (A)			492.32
Fixed Inputs			
Cutlasses			7.38
Hoes			3.69
Files			0.92
Land rent (ha)			18.44
Total Fixed cost (B)			30.42
Total cost (A+B)			522.74

3.3. Fixed Costs

The cost of each hectare was calculated using the straight line depreciation method. For the purpose of this study, a salvage value of zero (0) was assumed. The average life span of land was insurmountable only its nutrients can depreciate but for farming implements it was about three years.

Table 3.3. Relative cost of Variable Inputs

Variable input	Variable cost ($)	Percent distribution
Labour	486.17	98.75
Seed	6.15	1.25
Total	**492.32**	**100**

3.4. Gross Return

Gross return per hectare for a farmer on the average was $644.90 thus Net profit on the average amounted to $122.16.

The Benefit Cost Ratio (BCR) of 1.23 was obtained. By implication of this criterion for selection of enterprise, if BCR is greater than 1, the business enterprise is viable and profitable. As such, fluted pumpkin production is profitable in the study area.

Table 3.4. Relative Shares of Total Variable Cost and Total Fixed Cost

Cost	Shares ($)	Relative percent
Total Variable cost	492.32	94.2
Total fixed cost	30.42	5.8
Total	**522.74**	**100**

3.5. Percentage Gain Per Hectare

The percentage gain per hectare is 0.23 (23%). This implies that for every $1.00 spent in production, there is return of 23cents. This is similar to the findings of Idowu, Alimi, Tijani & Okobi, (2007) who reported that fluted pumpkin was profitable under existing production systems with an average net income of $205.90/ha on a mean farm size of 0.301ha.

Table 3.5. Cost and Return per hectare of fluted Pumpkin

Gross Return per ha	644.90
Total variable cost	(492.32)
Total fixed cost	(30.42)
Net Return/ha	**122.16**

3.6. Factors affecting Output of Fluted Pumpkin

The production function for this study was estimated using four functional forms. The linear form was chosen as the lead equation thus was further discussed. An Adjusted R^2 of 0.664 implies that about 66% of the variation in output was explained by the independent variables. Farm sizes, family labour and quantity of seeds were significant at $P<0.05$. It therefore implies that output can be increased to some extent by increasing these significant independent variables.

Table 3.6. Factors affecting Output of Fluted Pumpkin

Variable	Coefficient	T-statistics	P-value
Farm size	27.000	1.775	0.010*
Family Labour	3.595	0.974	0.028*
Hired Labour	-0.072	0.005	0.067
Quantity of seed	13.435	3.064	0.005*
R^2	0.673		
Adjusted R^2	0.664		
F-Cal	60.095		

* Significant at $P<0.05$

3.7. Resource-Use Efficiency

Marginal value productivity analysis was used to determine the efficiency with which each variable was used. Results as shown in Table 3.7 show that labour (family and hired) was over utilized while farm size and quantity of seeds were under-utilized. This suggests thus a wider scope for improvement by reallocating resources optimally in order to reduce waste and maximize output.

Table 3.7. Efficiency Analysis of Variable inputs

Variable	Coefficient	MFC	MVP	MVP/MFC	Efficiency status
Farm size	27.000	18.44	580.41	31.478	Under-utilized
Family Labour	3.595	4.30	2.88	0.681	Over-utilized
Hired Labour	-0.072	4.30	0.06	-0.041	Over-utilized
Quantity of seeds	13.435	1.54	866.42	563.867	Under-utilized

3.8. Constraints of Fluted Pumpkin in the Study Area

The mean ranking for each identified constraint was calculated and the most serious ones were determined by the mean score of 2.5. From the study, unpredictable market, high cost of labour, problem of storage and land acquisition, seed scarcity, lack of equipment and high cost of transportation were serious constraint to fluted pumpkin production in the study area.

Table 3.8. *Constraints of fluted Pumpkin in the Study Area*

Challenges	Ranking					Product of respondents	Mean ranking
	5	4	3	2	1		
High cost of Labour	10	48	2	35	2	323	3.23*
Labour scarcity	6	46	1	49	0	311	3.11*
Problem of Land acquisition	1	82	5	9	3	369	3.69*
Problem of storage	6	21	63	10	0	323	3.23*
High cost of Transportation	1	74	11	13	1	361	3.61*
Seed scarcity	0	77	5	18	0	359	3.59*
Inadequate Finance	0	3	1	86	10	197	1.97

4. Conclusion

We find that most of the respondents were females, illiterate and married. Farmers operated on subsistence level, about 95% practices mixed cropping with farming experience not less than 5 years. An adjusted R^2 value of 0.664 indicated that the independent variables; farm size, labour and quantity of seed in the model explained 66% of variations in the dependent variable (output). Net return per hectare was $644.90. The benefit-cost ratio was $1 to $1.23 thus it could be said that the enterprise was profitable in the study area. Resources under-utilized were quantity of seed and farm size while family and hired labour were over-utilized. Further, it was established that constraints faced by farmers were; high cost of labour, lack of adequate and quality seeds, pests and disease attack, storage and transportation cost.

From the results obtained, we make the following recommendation: provision of improved seeds and encourage fertilizer use, training of farmers on method to avoid waste in order that resources are put to optimum use. Also efforts should be made to provide credit and loan facilities to farmers to ease financial and capital constraints while improvement in infrastructural facilities will better the socioeconomic status of farmers.

Acknowledgements

The authors wish to acknowledge the enumerators and anonymous referees who read through the manuscript.

References

[1] Akoroda, M.O "Ethnobotany of *Telfaria occidentalis* Hook. F. (Cucurbitaceae) among Ibos of Nigeria". *Economic Botany* 44(1) 29-30. 1990.

[2] Ala, A. L. "Determination of Profitability and Resource-use Efficiency of Yam Production by Women in Bosso Local Government Area of Niger State, Nigeria". *European Scientific Journal, 9*(16). 2013.

[3] Ayinde, I. A., Akerele, D. and Ojeniyi, O. T. "Resource use ef ciency and profitability of fluted pumpkin production under tropical conditions". *International Journal of Vegetable Science, 17*(1), 75-82. 2010.

[4] Enabulele, H.N and Ilavbarhe K.O "A trend in the Analysis of the Production of some Horticultural Crops in Nigeria". *Nigerian Journal of Horticultural Sciences* (NJHS), May 2001, 5 (1): 1-9. 2001.

[5] Idowu, E.O, T. Alimi, A. Tijani and C.N. Okobi. "Profitability and Resource Use Efficiency of Fluted Pumpkin". *International Journal of Vegetable science*. Vol 13, Issue 1. 2007.

[6] Idowu, E. O., Alimi, T., Tijani, A. A., & Okobi, C. N. "Profitability and resource use efficiency in fluted pumpkin." *International Journal of Vegetable Science, 13*(1), 73-84. 2007.

[7] Jirgi, A. J., Ogundeji, A. A., Viljoen, M. F., & Adiele, M. A. "Resource Use Efficiency of Millet/Cowpea Intercropping in Niger State, Nigeria." In *Joint 3rd African Association of Agric. Econs.(AAAE) and 48th Agric. Econs. Asso. of South Africa (AEASA) conference at Cape Town South Africa Sept 19–23*. 2010.

[8] Kwanashie, M.A, Garba A. and Ajilimar A., "Policy Modelling in Agriculture. Testing the respose of Agriculture to Adjustment Policies in Nigeria". *African Economic Research-Consortium*, Research paper 57. 1997.

[9] Nwosu, C. S., Onyeneke, R. U., and Okoli, V. B. N. "Socio-economic determinants of fluted Pumpkin leaf (telferia occidentalis) production in Ezinihitte Mbaise local government area of Imo state, Nigeria". *Agricultural Science Research Journal, 2*(6), 355-361. 2012.

[10] Olayide, S.O., Eweke, J.A. and Bello-Osagie, V.E. (Eds). "Nigerian Small farmers. CARD University of Ibadan, for Benin-Owena River Basin Development Authority, Benin City". 1980.

[11] Omonona, B. T., Lawal, J. O., Oyebiyi, I. D., Tenywa, J. S., Taulya, G., Kawube and Santos, L. "Profitability and resource-use efficiency among Ofada rice farmers in southwest, Nigeria". In *10th African Crop Science Conference Proceedings, Maputo, Mozambique, 10-13 October 2011*. (pp. 69-71). African Crop Science Society. 2011.

[12] Udoh, E. J., and Akpan, S. B. Measuring technical efficiency of water leaf (Talinum triangulare) production in Akwa Ibom State, Nigeria. *American-Eurasian Journal of Agriculture and Environmental Science*, *2*(5), 518-522. 2007.

Marketing of agroforestry products in Bangladesh

Kazi Kamrul Islam[1, 2], Takahiro Fujiwara[2], Masakazu Tani[3], Noriko Sato[2]

[1]Department of agroforestry, Bangladesh Agricultural University, Mymensingh-2202, Bangladesh
[2]Forest Policy Laboratory, Faculty of Agriculture, Kyushu University, Fukuoka, Japan
[3]Faculty of Design, Kyushu University, Fukuoka, Japan

Email address:
kamrulbau@yahoo.com (K. K. Islam), takaf217@gmail.com (T. Fujiwara), mt18916@gmail.com (M. Tani), sato.noriko.842@m.kyushu-u.ac.jp (N. Sato)

Abstract: Bangladesh agriculture, including agroforestry, is dominated by small-scale subsistence farming; however, farmers do not get proper prices for their products due to the involvement of many intermediaries in its value chain. Therefore, the objectives of this study are to analyze the value chain of main agroforestry products and the position of the various intermediaries within it, and to determine the extent of value addition in terms of costs in successive stages of products movements. The study was conducted in Madhupur Sal forests area of Bangladesh and data were collected through questionnaire interviews, focus group discussion, observation and literature review. In analyzing network structures of value chain, the study found that both agroforestry crops and timber marketing have regulated a number of intermediaries which enhanced value addition and created high marketing margins of products. Moreover, farmers had possessed strong negative relationship with intermediaries and that the lack of a farmer organization severely affected farmers' free and fair access to local markets to sell their products. So, there is an immediate need to establish farmers' cooperatives in order to reduce unexpected intermediaries from value chain to get fair prices of their products, and to harmonize and execute marketing legislations.

Keywords: Agroforestry, Value Chain, Intermediaries, Marketing Margin, Sal Forests, Bangladesh

1. Introduction

Marketing systems plays a significant role in enhancing production and consumption, and in accelerating the pace of economic development [1, 2]. At the same time, the marketing of agriculture and small scale forestry products plays a major role in smallholders' economy which are both produced and consumed locally. These products are contributing to a significant role in developing countries because these products make a major contribution to the GDP, and their consumption represents an important part of rural people's expenditure [3, 4].

The analysis of marketing systems, particularly value chain analysis provides the full range of activities which are required to bring a product from conception, through the primary phases of production and delivery to final consumers [5, 6]. As a product goes from the producer to the consumer, a number of transformations and transactions take place along a chain of interconnected activities, and value is added continuously at each level of the chain [7]. Production, standardization, pricing and distribution .of agricultural and small scale forestry products are progressively being organized into value chains, where the movements of products have effected farmers, wholesalers, retailers and other intermediaries. Value chain analysis analyze the organization and behavior of all the actors in the value chain, and also delivers the distribution of value added over various actors. It also express the achievements of all intermediaries/actors in participating the chain [8].

In developing countries including Bangladesh, local product-based marketing channels are characterized by high instability and complexity. Literature also revealed that the structure of value chain has included a number of intermediaries who are playing dominant roles in making the agriculture market unstable and risky by grabbing major portions of the benefit from the farmer and by gaining high profits from the consumer by fixing the high prices of products (9, 10, 11, 12, 13, 14, 15). In developing countries, it is nearly impossible to imagine a market without middlemen [2, 4, 16]. On the contrary, there are many reasons for involving intermediaries in the value chain. Their important functions includes they are specializing at transportation, selling and even processing functions of marketing, the average costs of performing the marketing functions falls as the volume of products handled rises, intermediaries can take the risks, and handling large volume of products and can reduce market search and transactions costs [4]. This does not means that all intermediaries are efficient and necessary, nor that farmers should never implement their own marketing functions. Recent literatures also found that creating farmer cooperatives or institutions would forfeit these economic gains from specialization and scale economies [4, 17, 18, 19, 20].

Bangladesh agriculture is dominated by small scale farming; more than 80% of Bangladeshi people are directly or indirectly involved in agriculture [21, 22]. The agriculture sector contributes 21.11% of the total GDP share, and the non-farm sector, which is also driven primarily by agriculture, has contributed another 33% in Bangladesh [22, 23]. Nowadays, smallholders are more interested in practicing agroforestry (combining agriculture and forestry) in order to produce more integrated, diverse and profitable products. Due to a huge demand of firewood in local markets, agroforestry practices are gaining more preference in Bangladesh (14, 24). Agroforestry programs in Madhupur Sal forest area have been contributing to a substantial economic output to the Bangladesh Forest Department (FD) and local people since 1989 [25, 26, 27] because these programs are considered to be of more environmental and economic importance [14, 25, 28]. Since agroforestry production in Bangladesh is increasing over the recent years, its value chain analysis is very important for farmers, intermediaries and consumers because all are affected due to value addition in marketing systems. There were very few empirical study on value chain analysis of agroforestry products in Bangladesh, so, this study takes the leading steps to analyze value chain of agroforestry products. Therefore, the objectives of this study is to identify different value chains of agroforestry products and the position of the various intermediaries within it, and also to determine the extent of value addition in terms of costs in successive stages of agroforestry products movements.

1.1. Theoretical Frameworks

In the recent decades, there were many extensive theories introducing in the area of value chains, reflected in various concepts and approaches [29, 30]. These approaches vary mainly in their objectives, in the activity that is underlined, and in the pathway in which they have been applied [31]. Value chain analysis in developing countries are characterized by its network structure, its governance form and the way value is added [29, 30, 32] These approaches emphasized value chains as production networks in which actors exploit competitive resources and function within an organized environment. Therefore, the developing countries value chain approaches have three main components: network structure, value added and governance structure.

The structure of value chain network has vertical and horizontal dimensions. The vertical dimension of value chain refers the flow of products from an initial producer up to the last consumer and the horizontal dimension reflects the relationship between stakeholders in the same chain link (e.g. farmer to farmer). Lazzarini et al (2001) developed the concept of netchain (Figure 1) which shows vertical relationships between the several value chain links and horizontal relationships between stakeholders' in the same link. Figure 1 explains that vertical relationship may follow all stages in the value chain or may not, and horizontal relationships between stakeholders have various shapes [29, 30, 33], such as farmer cooperatives or price agreements between wholesalers. Market channels have vertically structured the value chain network and horizontal dimension is shaped by purchase, production and delivery dependencies between stakeholders that are situated in the same value chain [4, 30]. So, the structure of a value chain network is the principle dynamics of all sectors of the economy [30, 34].

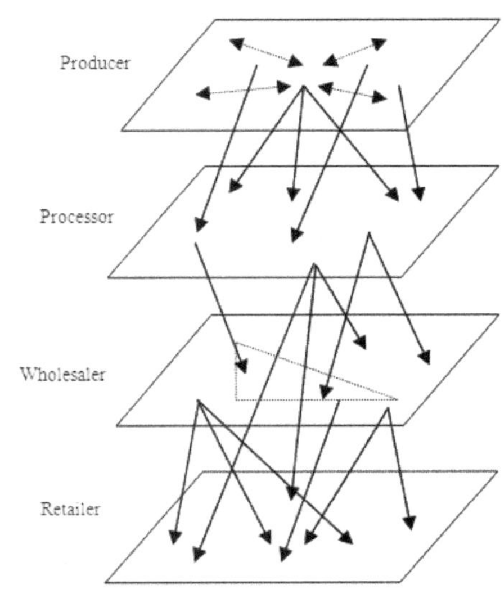

Figure 1. *Netchain (adopted from Lazzarini et al. 2007)*

Value added is the amount by which the value of a product is increased in each stage of its distribution channel, exclusive of initial costs; and is created by different stakeholders throughout the value chain. Value added may relates to quality, costs, delivery frequencies and flexibility, etc. [30]. The amount of value added is decided by the last level consumer's willingness to pay. The value addition in food production mainly depends on the safety and quality of the products, and the quality can be divided in two intrinsic (e.g. color, size, taste. etc.) and extrinsic (organic or inorganic production) characteristics of the product [30]. So, value added is generated in value chains aiming at particular markets and composing a number of stakeholders. Regarding to governance, stakeholders in value chains are linked in a variety of sourcing and contracting relationships, which means the forms of governance [30, 35, 36]. Governance refers as "*authority and power relationships that determine how financial, material and human resources are allocated and flow within a chain*" [34]. There are many dimensions in explaining the governance [37, 38, 39] and a major segment of governance depends on global value chain perspectives which is quite hard to relate within a local value chain analysis. Therefore, the present study mainly focuses on network structure and value addition aspects of agroforestry products value chain analysis.

2. Methodology

2.1. Study Area

Sal forests of Bangladesh cover an area of 0.12 million ha [40] and Madhupur Sal forests are situated in Tangail and Mymensingh Districts possess the majority of the Bangladesh Sal forests [24, 27, 28]. The study was conducted in Madhupur Sal forests (Figure 2) area of Bangladesh where the agroforestry systems were very famous and well established.

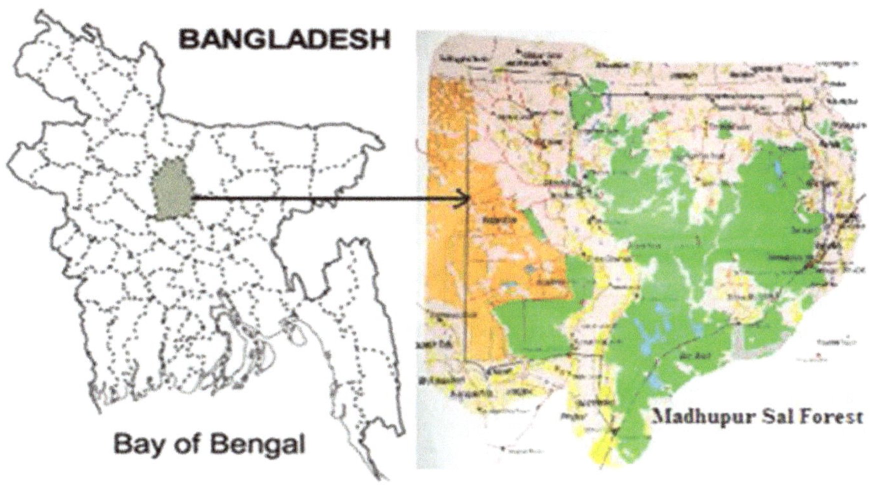

Figure 2. *Study area (Madhupur Sal forests)*

2.2. Description of Agroforestry Programs

Agroforestry programs were very popular and well accepted mainly at the degraded area of Madhupur Sal forests, which were initiated by the Asian Development Bank (ADB) funded project in 1989 [27, 40, 41]. In this program, each participants was allocated 1ha of degraded Sal forest land where they utilized a prescheduled agroforestry model for a 10-year cycle. Each participant can continue up to 3 cycles if they maintain FD prescribed criteria properly. This model provided alternating space for trees and agriculture crops. Approximately 750 trees per ha^{-1} were planted in this process, and a 50% thinning operation was implemented in the 7th year of the cycle (Figure 3b). The remaining 375 trees were finally harvested at the end of the 10-year cycle, and the benefits were equally shared with FD [24]. Mainly farmers (more than 80%) are plating *Acacia auriculiformis* as fast growing fuelwood tree species and pineapple as shade loving crops in tree-crop based agroforestry systems at Madhupur Sal forests area [24, 25]. The most significant aspect of this model was the whole crop benefits belonged to the participants (Figure 3). Agroforestry programs practiced in degraded Sal forests land were mainly controlled by the FD, and the local people were the participants who enjoyed a certain amount of user rights.

a. Common Agroforestry practice b. FD run Agroforestry model

Figure 3. *Agroforestry practices practiced at Madhupur Sal forests area*

Besides FD programs, farmers can practice different tree-crop based agroforestry practices in their homestead, fallow land, cropland, road side and other areas (Figure 3). Other than *Acacia spp.* farmers planted fast growing firewood species such as Bokain (*Melia azedarach*) and Gamar (*Gmelina arboria*) and shade loving crops, Ginger and Turmeric in the study area.

2.3. Sampling and Data Collection

About 3,327 participants were involved in FD run Agroforestry programs in Madhupur Sal forests, and a majority of the participants were poor (Islam and Sato 2012a). So, this study included all types of agroforestry production systems in Madhupur Sal forests area, and in total 110 respondents were randomly selected for interviews. Of them, 80 respondents were from FD run programs, and 30 from other agroforestry programs. The participants were living in and around the Sal forests. Though the participants were already certified as poor by FD [28] but for the purposes of this study, only the small scale farmers having less than two hectare farmland were selected for interviews.

This study covered a local market (Jalsatropur) and another two important urban markets in Mymensingh and Dhaka city, for selecting intermediaries of phoria, bepary, paiker/arathdar, and retailer (see description in Results section). In agroforestry firewood marketing includes some other intermediaries, such as auctioneers, local traders, local saw-mills, brickfields and wholesalers. This study also conducted two focus group discussions (10 farmer in each discussion) and asked farmer to identify 5 intermediaries (next to the farmer link in the value chain). Another 5 intermediaries were sampled randomly in the markets area and along the side of the road where they were buying and selling agroforestry products. This process of sampling was continuing up to the last level of intermediaries. Finally, the study selected 10 intermediaries for each category and 20 consumers for data collection. Data were collected in different months of 2009 to 2014 through practical observations, questionnaire interviews, focus group discussion, FD staff interview and literature review.

3. Results

3.1. Descriptions of the Farmer and Intermediaries

3.1.1. Farmer

The average family size of the farmer was 5.62, which was larger than the national average family size of 4.8 (Table 1) [42]. The literacy rate of the farmers indicated that the majority of the farmers were not literate enough to communicate or get marketing information (Table 1). In the case of farmer ethnicity, the majority of them were non-ethnic (Bengali people) (80.3%) and the remaining were ethnic (29.7%) (Table 1). Historically ethnic people (mainly Garo and a few Koch people) have been living all over the Madhupur Sal forests for many decades [43] were also practiced agroforestry programs. Regarding to farmer household origin; about 96% of the ethnic farmers originated from the same area whereas the majority of non-ethnic farmers (94%) migrated from other areas. (Table 1).

Table 1. *General description of farmer*

Characteristics	Mean, N	Percentage	± SD
Age	42.35	-	9.24
Household size	5.62	-	1.29
Sex of farmer			-
Male	97	88	
Female	13	12	
Ethnicity of farmer			
Ethnic	73	80.3	-
Non-ethnic	27	29.7	
Non-ethnic farmer religion			
Muslim	69	94.5	
...Hindu	3	4.1	
Christian	1	1.4	-
Ethnic farmer religion			
Christian	21	77.8	-
Shangsarek	4	14.8	
Others	2	7.4	
Literacy rate	-	26.36	-
Ethnic household origin			
Migrate	1	3.7	-
From same area	26	96.3	
Non-ethnic household origin			
Migrate	69	94.5	
From same area	4	5.5	-

3.1.2. Intermediaries

Phoria: a phoria purchases products from farmers and deals with small volume of products. They sell products in village markets and sell their whole products to the Bepari (Figure 4). Usually they are small landless farmers/laborers and not involved in full time farming [44]. They possess a small amount of capital, and, in some cases, they will pay farmers after selling products to the Beparies.

Beparies: beparies are professional businessmen who purchase products from the farmers and Phoria. They deal with large amounts of products and often pay phoria for advanced purchasing of local products. Mainly beparies sell their products to paiker/arathdar (Figure 4).

Paiker/Arathdar*:* a paiker acts as a commission agent between beparies and retailers (Figure 4). They are also called wholesalers in some areas. They have the facilities to store the products and always do tricky business on additional value added to the products. That means, in some cases, they have created an artificial product crisis in the market and increase the products price which was also claimed by every farmer.

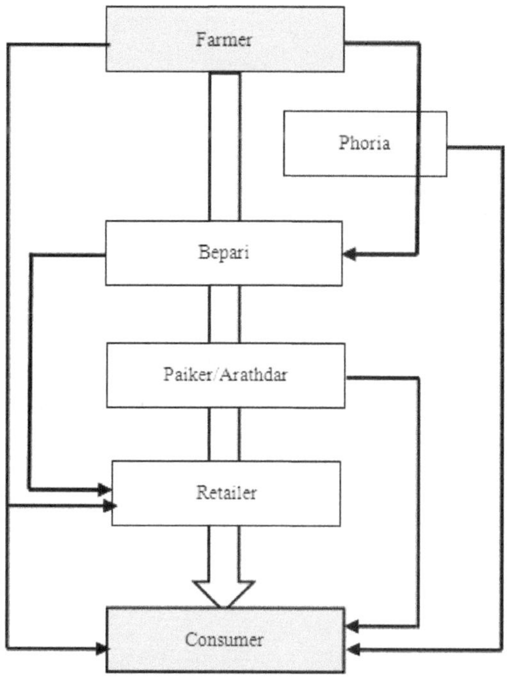

Figure 4. *Agroforestry crops distribution channel in Bangladesh*

Retailer: retailers are the last level intermediaries in the marketing channel of Bangladesh and other developing countries. They usually handle the consumer and buy products from Paiker and Bepary (Figure 4). They have some additional input costs of renting/buying shops and appointing sells-staff and also have the risk of damaged/rotten products.

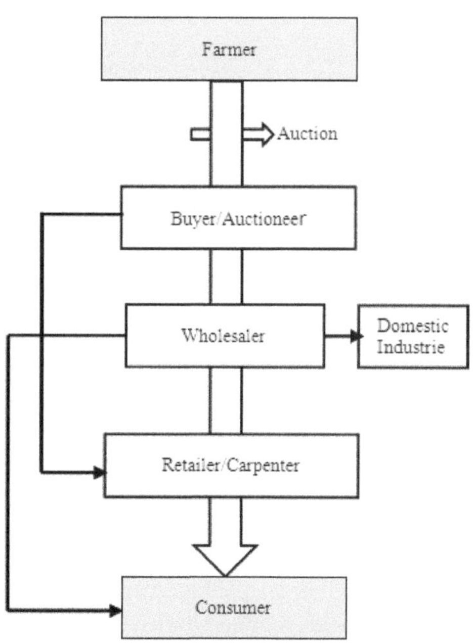

Figure 5. *FD controlled agroforestry timber distribution channel in Bangladesh*

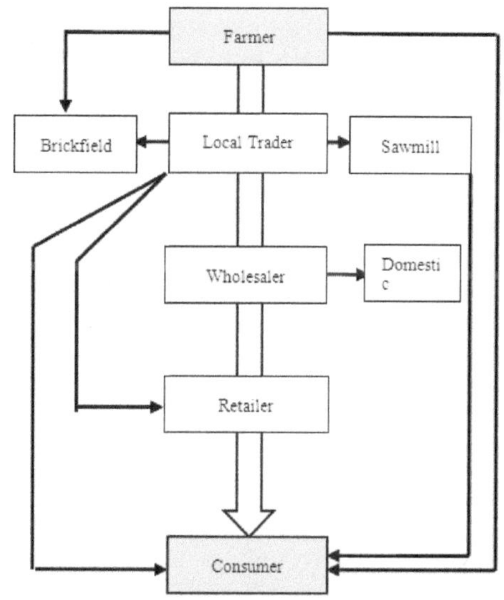

Figure 6. *General agroforestry timber distribution channel in Bangladesh*

Buyer/Auctioneer: These intermediaries only arise in the FD controlled agroforestry timber/firewood marketing channel. Buyers mainly take part in the open timber auction process. Buyers have to register officially through the FD systems and maintain some criteria imposed by FD. Some buyers have storage facilities and can sell timber to the wholesaler or some time to the retailer (Figure 5). They also maintain a good relationship with FD, local political leaders and the elite in order to buy the products from open auction markets.

Local Timber Trader**:** This is a very small group of traders and found in timber marketing channel who deals with a small volume of firewood/timber only. Often they collect firewood from farmer's home and sell it to the wholesaler (Figure 6). They also supply firewood to the local saw-mills and brickfields. Usually they have very temporary and small storage facilities.

3.2. Value Chain Analysis

3.2.1. Network Structure

Network structure of value chain analysis are the possible routes of products flow from farmer to consumers. Vertical structure of value chain may be short or long for a particular products and it depends on product quality, size and the nature of consumers and farmers and the surrounding social and physical environment. Six vertical network structures were identified for agroforestry crops and the main and the most common network structure consists of farmer, phoria, bepary, arathadar, retailer and consumer. Other network structures were: farmer to bepary to arathdar to retailer to consumer; farmer to bepary to retailer to consumer; farmer to bepary to consumer; farmer to retailer to consumer; and farmer to phoria to consumer respectfully. Network structures of agroforestry crops and their mode of action are illustrated in Figure 4. Arrows in Figure 4 show the pathway of agroforestry crop distribution channels and the position of each intermediary in the value chain. In these network structures, the most dominant intermediary was bepary followed by paiker/arathdar; and according to the field data bepary handle the major amount of crop products to regulate marketing functions. For example, pineapple farmers sold 80% of their products to the bepary, 6% to the phoria, 6% to the retailer and 8% to the consumer directly. However, those consumers were situated in local or village areas (Table 2). On the contrary, beparies bought 55% of pineapple from farmers and 15% through phorias; while paikers bought 60% of pineapples from beparies directly. Consumers bought their major percentage of pineapples from retailers (85%) and minor percentages from phorias or directly from farmers (Table 2).

Table 2. Buying and selling percentage of pineapple transacted by different intermediaries

Intermediaries	Buying from (%)						Sold to (%)					
	A	B	C	D	E	F	A	B	C	D	E	F
Farmer(A)								6	80	-	6	8
Phoria(B)	25								5	-	-	1
Bepari(C)	55	15								75	10	-
Paiker(D)	-	-	60								70	5
Retailer(E)	15	2	10	58								86
Consumer(F)	5	8	-	2	85							

Here n=10

Table 3. Buying and selling percentage of agroforestry (FD run) firewood products

Intermediaries	Buying from (%)						Sold to (%)					
	A	Y	W	D	E	H	A	Y	W	D	E	H
Farmer(A)								100	-	-	-	-
Buyer(Y)	100								90	-	10	-
Wholesaler(W)	-	80								5	75	10
D. Industries(I)	-	-	5								-	-
Retailer(E)	-	20	75	-								85
Consumer(H)	-	-	-	-	95							

Here n=10

Table 4. Buying and selling percentage of general agroforestry firewood products

Intermediaries	Buying from (%)								Sold to (%)							
	A	L	K	S	W	I	E	F	A	L	K	S	W	I	E	F
Farmer(A)										75	5	15	-	-	-	5
L.Trader(L)	70										5	8	40	-	5	17
Brickfield(K)	5	5										-	-	-	-	-
Sawmill(S)	15	8	-										-	-	-	23
Wholesaler(W)	-	35	-	-										5	35	-
D. Industry(I)	-	-	-	-	5										-	-
Retailer(E)	2	12		-	30											40
Consumer(H)	8	10	0	5	-	-	44									

Here n=10

Regarding to agroforestry (FD run) timber vertical network structures analysis, only three different channels were identified and the main network structure comprises of farmer, buyer, wholesaler, retailer and consumer. While other network structures were: farmer to buyer to retailer to consumer; and farmer to buyer to wholesaler to consumer. Figure 5 also shows the position of the intermediaries in the network structure and their mode of actions for agroforestry timber marketing. The results revealed that timber buying rights have solely controlled by the buyer or auctioneer in FD run agroforestry programs. They were the dominant intermediaries in this network and Table 3 clearly shows the amount of timber selling and buying percentages of each intermediaries in this network structures. All most every (99%) farmer claimed that buyers had negotiated prior to buy the timber in FD run program.

On the contrary, agroforestry (general) timber products vertical network structure has four different channels and the most common one was farmer to local trader to wholesaler to retailer to consumer. However, other network structures were: farmer to local trader to retailer to consumer; farmer to local trader to consumer; and farmer to local trader to sawmill to consumer. In general agroforestry programs, farmers can directly sold their products to the intermediaries; the position of each intermediary and timer distribution channel are presented in Figure 6. The results also showed that farmers sold their major amount of firewood to the local traders (75%), saw-mills (15%), brickfields (5%) and 5% to the consumers directly (Table 4). However, wholesalers sold 35% of their products to retailers and some minor percentages to domestic industries (e.g. furniture industries, cottage industries) (Table 4).

The study also identify the horizontal network structures of the stakeholders' in agroforestry products value chain analysis in terms of relationship. In a measurement scale of +2, +1, 0, -1, -2 (+2=strongly positive relationship, 0= no relationship or don't know and -2=strongly negative relationship), the results found that farmers did not possess a strong relationship among farmers while farmers had a strong negative relationship with the beparies and other intermediaries (Figure 7). The study also asked for farmer cooperatives or institutions, and the results found that farmer did not have any types of cooperatives among themselves. However, the dominate intermediaries (Bepary) possessed a strong positive relationship among them. Thus, they can easily negotiated among them to fix crop buying prices. Even the other intermediaries (both in crop and timber) had possessed strong positive relationship among themselves (Figure 7). The strong positive relationship among the intermediaries provided a favorable atmosphere towards value addition and fixing price of products in marketing systems. On the contrary, the antagonistic relationship that the farmer had with their next level intermediaries and other intermediaries had a severe impact on free and fair access to sell their products in markets.

Figure 7. *Overall social relationship of the stakeholders*

3.2.2. Value Added

In value chain analysis, value added is created at different levels with the intervention of different intermediaries throughout the whole products distribution channel. To understand the value addition the study first identified the intermediary-wise and total marketing costs of agroforestry products. Marketing costs are incurred when products are dispatched from the farmer to the consumer. Intermediary-wise costs of main agroforestry crop (Pineapple) and timber (Acacia) are presented in Table 5 and Table 6 respectively. Table 5 showed marketing costs of pineapple includes all costs incurred by different intermediaries such as phoria, bepary, paiker, retailer and farmer. Pineapples got added value during their distribution across the value chain network. Results revealed that the share of transportation cost was the highest (50.86%) followed by damaged (15.48%) and labor wages (5.53%) (Table 5). The total value added cost per 100 pineapples was Taka 1815.8 from production to consumption. Among the intermediaries, the retailer total cost was the highest followed by the paiker (Table 5).

Similarly, Table 6 observed that the total cost of 1mon (40kg) acacia timer was Taka 81.5. Like pineapples, the acacia timber transportation costs (40.49%) ranked the top followed by labor wages (14.72%). Regarding the intermediaries, wholesaler's value added costs was the highest followed by Retailer's (Table 6). Furthermore, the value addition scenario of a pineapple is clearly presenting in Figure 8. Famer price of a pineapple was only Taka 10 and when it run through the different marketing channels the total value addition cost was Taka 25; that means the consumer bought the pineapple at Taka 35 (Table 7).

The results showed the marketing margin for pineapple was 71.43%, ginger 44% and Acacia timber 36.84% (Table 7). That means an average pineapple marketer earns a market margin (farmer to retail price spread) of 0.71 Taka for every 1 Taka retail price paid by the ultimate consumer in the marketing channel in Bangladesh. This represents payments for all marketing chargers including intermediaries' commission added to farmer products.

Table 5. *Total marketing costs of different intermediaries involved in Pineapple marketing*

Cost Items (per 100 pineapple)	Intermediaries					Total	% of total costs
	Farmer	Phoria	Bepary	Paiker	Retailer		
Transportation	111.5	-	219	321	272	923.5	50.86
Fruit mulching	-	37	60	-	-	97	5.34
Labor wages	10	15	25.5	20	30	100.5	5.53
Broker commission	20	-	25	30	-	75	4.13
Space/house rent	-	22.5	15	15.5	30	83	4.57
Bazar toll	10	-	10	-	-	20	1.10
Tips and donation	10	10	25.5	30	20	95.5	5.26
Electricity	-	3.3	1	3	5	12.3	0.68
Telephone bill	3	4	5	7.5	7.5	27	1.49
Damage	-	15	40.5	75.5	150	281	15.48
Personal expenses	10	-	12.5	15	17.5	55	3.03
Others	-	3	12.5	15.5	15	46	2.53
Total	174.5	109.8	451.5	533	547	1815.8	100

Here n=10

Table 6. *Total marketing costs of different intermediaries involved in Acacia auriculiformis firewood marketing*

Cost Items 1mon= 40kg	Intermediaries				Total	% of total costs
	Farmer	L. Trader	Wholesaler	Retailer		
Timber harvesting	7.5	-	-	-	7.5	9.20
Transportation	2.5	7.5	12.5	10.5	33	40.49
Labor wages	2	3	4	3	12	14.72
Space/house rent	-	0.5	0.5	1	2	2.45
Bazar toll	0.5	1	1	-	2.5	3.07
Tips and donation	0	1.5	2	3	6.5	7.98
Electricity	-	0.2	0.3	0.5	1	1.23
Telephone bill	0	1	1.5	1.5	4	4.91
Personal expenses	0.5	2	1.5	3	7	8.59
Others	0	1.5	2	2.5	6	7.36
Total	13	18.2	25.3	25	81.5	100.00

Here n=10

Table 7. *Marketing margins of agroforestry products*

Produces	Unit	Producer price (Taka), P	Retail price (Taka), R	Marketing margin =(R-P)/R*100
Pineapple	1no.	10	35	71.43%
Ginger	1kg	70	125	44.00%
Turmeric	1kg	30	45	33.34%
Firewood (Acacia)	1mon (40kg)	120	190	36.84%
Firewood (Gamar/Bokain)	1mon (40kg)	100	175	42.86%

1US$ = 79 Taka (local currency)

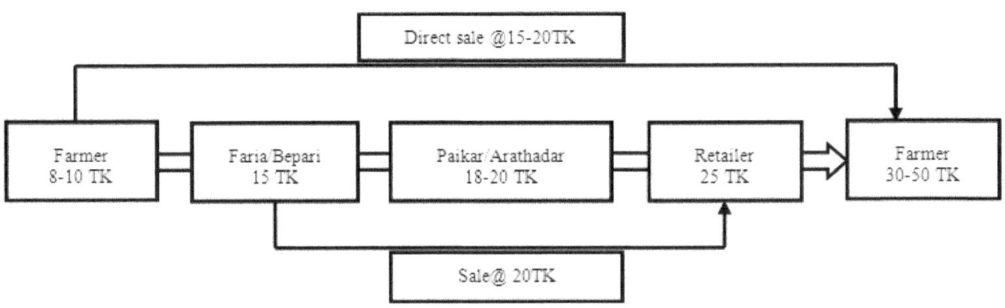

Figure 8. *Value addition of a pineapple in marketing channel*

4. Discussion

Analysis of different types of intermediaries especially network and value added structures have enriched this study's picture of how these channel works and what is needed to improve them further. Based on this study's results it was clear that the presence of many intermediaries made the value chain complex and problematic, especially for the farmer and the consumer. In case of agroforestry crops, the vertical networks identified six different channels and the main and longest channel consists of phoria, bepary, paiker/arathdar and retailer. Farmers sold their 87% pineapples to their next level intermediaries (Table 3) and pineapples passed through the main and longest network structure to reach the ultimate consumer. In a study of Potato, Scott (1988) also observed that farmer were inclined to sell their complete crop to local bepari and in some cases to phoria at home. In a firewood study on Nepal, Dhital (2004) also reported that most of the firewood/timber buying and selling operations are transacted by the wholesalers and farmers often suffered from the low price of their products. The study also identified several long network structures for timber products, and FD run agroforestry timber selling process has solely controlled by the auctioneer/buyer. Almost 99% farmer claimed that buyers have negotiated prior to the auction process in order to set a lowest or minimum price of firewood. Farmer also reported the same for crops selling process where beparies were the main controller. These types of negotiation or coalition was also reported by Islam and Sato (2012b) and Gain (2002) in their studies on Sal forests of Bangladesh. Local government or authority did not pay any attention to this issue or even there was no strong market monitoring cell was formed by local government. Almost every farmer claimed that neither the local government strictly monitored market nor did any action to abolish intermediaries' coalition at local level.

Another important aspect of network structure is its vertical dimension and the existing relationship of the same group/level of stakeholders in the value chain. Results revealed that farmers did not possessed a strong positive relationship among them; thus, they cannot negotiate to fix products selling price at market. Due to a strong positive relationship intermediaries (e.g. Bepary) negotiated among them and controlled the initial buying prices of the agroforestry crops in the local market, which was also reported by Islam and Sato (2012b). Moreover, farmers had possessed the worst relationship to their next level of intermediaries (e.g. Bepary) to whom they sold most of the products (Figure 7). Often farmers faced prefixed prices of their products, such as pineapples or gingers set by the beparies. So, farmers were treated the less powerful actor in every network structures. Similar types of results also reported by Ahmed 2007; Asogwa and Okwoche 2012; Alam 2012; Islam et al 2013; Daital 2004 in their studies. Farmer cooperative organization has been implemented by farmer as one of the best possible solution to exclude unexpected intermediaries from network structures (4, 47; 48, 49. The cooperative should be an independent local association in which the opinion and action of each member have been treated equally. The main tasks of the study area's cooperative would be to collect members' products for sale and also ensure other marketing functions. This will secure the farmer rights and also abolish the unexpected intermediaries' intervention in marketing channel. These cooperatives would bring the farmer one step closer to the final consumer with his products, and both of them will be eventually benefited.

Value added could easily explain the amount of added price for each products in marketing channel. When agroforestry products moves through value chains, every intermediary adds some extra costs together with real costs. So, the higher the number of intermediaries involved in value chain the higher the products price which was clearly visualized in Figure 8 for a pineapple product. Majority of the products value addition incurred to transportation costs which would easily decreased by deducting the intermediaries from value chain. There were also some unexpected costs such as tips and donations, bazar toll, and these would easily minimized by proper execution of marketing regulations. Regarding to marketing margins, pineapple marketing margin (71%) was extremely high compare to other products marketing margin of around 40%. So, the variations of farmer and retail prices, especially for agroforestry crops, clearly indicated that farmers received only a minor share of the final retail prices. Empirical evidence to support this finding is very inadequate in Bangladesh; however, Scott (1988) found similar type of results in potato marketing systems in Bangladesh. In mango marketing, Matin et.al (2008) also found a high marketing margin in several districts level markets of Bangladesh. Nevertheless, other developing countries also showed higher marketing margins for crops marketing systems [15, 45]. So, there is an immediate need to reduce the marketing costs as well as marketing margins in order to stable the agriculture and small scale forestry marketing systems in Bangladesh and other developing countries that possess the similar situations.

5. Conclusion

Agroforestry products value chains examined here are very important in Bangladesh and other developing countries because they offer substantial opportunities for domestic farmers to increase their production and income. Value chain analysis of agroforestry products has been characterized by the involvement of many intermediaries; thus, marketing channels were long and complex in Bangladesh. Often involvement of some intermediaries seems to be redundant; their presence just adds an extra cost to the consumer, creating high marketing margins. Most of the costs incurred in value chain were on transport the agroforestry products from farmer field to retailer places. These transport costs added a significant amount of value addition to products. Moreover, lack of strong positive relationship among the farmers placed them as a less powerful actor in every network structure. It may, thus can be summarized from the study that farmers' free fair access and benefits seemed to be very minimum in the present value chain, while costs of marketing, intermediaries' intervention and value addition were high.

Although the value chain analysis of agroforestry products in Bangladesh is overwhelmed by many complications, there have been a plenty of policy options that are expected to improve the systems. There is an immediate need to establish farmer cooperatives or an independent local associations, especially to help the farmers' products to sell and to help farmers to purchase their needed supplies. These cooperatives can significantly reduce intermediaries from network structure and also integrating farmer horizontally in order to acquire the right and power over powerful intermediaries who dominate and control the price of agroforestry products. The better option to reduce the transportation costs is to adopt the system whereby few experienced individuals (e.g. cooperative members) may transport the products belongings to several stakeholders to markets. The local government should also arrange adequate supports to farmer cooperatives and also play an active role in providing physical facilities. Finally, the government should review and harmonizing existing marketing legislation as well as strictly follow marketing regulation in order to ensure transparency and controlled any types of intermediaries coalition. This will intern ensure free and fair marketing systems of crops and small scale forestry products, and also ensure the pace of economic development of the country.

Acknowledgment

This study was carried out in the context of research funded by the Japan Society for the Promotion of Science (JSPS) fund.

References

[1] K.N. Huda, M.R. Karim, and R.S. Khan. An empirical study on competitive distribution management of tea brands in Bangladesh. International Journal of Marketing Studies, Vol. 4, No. 5, pp. 99-107, 2012.

[2] S. Kapoor, and P. Kansal. Marketing logistics: a supply chain approach. Indian Branch Pearson Education, Singapore, 2003

[3] A. Dorward, J. Kydd, and C. Poulton. Traditional domestic markets and marketing systems for agricultural products. World Development Report Background Paper No. 41359, 2008.

[4] R.L. Kohls, and J.N. Uhl. Marketing for agricultural products. Macmillan publishing, New York, 2002.

[5] R. Kaplinsky, and M. Morris. A Handbook for Value Chain Research. Prepared for International Development Research Centre (IDRC) (Part 1), 2000.

[6] M.E. Porter. The Competitive Advantage of Nations. Simon & Schuster Publication, USA, 1990.

[7] L. Kirimi, N. Sitoko, T.S. Jayne, F. Karin, M. Muyanga, M. Sheahan, J. Flock, and G. Bor G. A Farm Gate-To-Consumer Value Chain Analysis of Kenya's Maize Marketing System. MSU International Development Working Paper No 111, 2011, http://fsg.afre.msu.edu/papers/idwp111.pdf . Accessed 8 April 2014.

[8] E.R. Jacinto. A research framework on value chain analysis in small scale fisheries, 2004. http://ageconsearch.umn.edu/bitstream/139883/2/2006_mul doon.pdf . Accessed 28 March 2014.

[9] A.K. Dey, and B.N. Banerjee. Production and marketing of pulses and oilseeds. New Delhi, India. Mittal Publication, 1991.

[10] R.M. Al-Hassan, D.B. Sarpong, and A. Mensah-Bonsu. Linking smallholders to markets. Ghana Strategy Support Program (GSSP) Background Paper No. 01. Accra, Ghana, 2006.

[11] M.F. Alam, M.S. Palash, M.I.A. Mia and M.M. Dey. Marketing of major fish species in Bangladesh: a value chain analysis. Food and Agriculture Organization Report, Rome, Italy, 2012.

[12] G.J. Scott. Marketing Bangladesh's potato: present patterns and future prospects. CIP- ADAB, Dhaka, Bangladesh, 1988.

[13] K.K. Islam, Kimihiko. Hyakumura, Fujiwara. Takahiro, and Sato. Noriko. Confronting people-oriented forest management realities in Bangladesh: An analysis of actors' perspective. International Journal of Social Forestry, Vol. 4, No. 2, pp.153-179, 2011.

[14] K.K. Islam, and Sato. Noriko. Participatory forestry in Bangladesh: Has it helped to increase livelihoods of Sal forests dependent poor people. Southern Forest, Vol.74, No. 2, pp. 89-101, 2012a.

[15] B.C. Asogwa, V.A. Okwoche. Marketing of agricultural produce among rural farm household in Nigeria: the case of Sorghum marketing in Benue State. International Journal of Business and Social Science, Vol. 3, No. 13, pp. 269-277, 2012.

[16] M. Roy. Agricultural marketing: new challenges. International Journal of Humanities and Applied Sciences, Vol. 1, No. 2, pp.2277-4386, 2012

[17] D. Prakash. Capacity building of agricultural cooperatives to meet the market and human resources development demands. Pamda-Network International, New Delhi, India, 2000.

[18] C. Giagnocavo. The Almeria agricultural cooperative model: creating successful economic and social communities. Report for UN Division for Social Policy and Development, Department of Economic and Social Affairs. pp.1-9, 2012.

[19] T. Term. Smallholders' agricultural cooperatives and rural development in Colombia. Dissertation, Public Policy in Latin America, the University of Oxford, 2011.

[20] OCDC (US Overseas Cooperative Development Council). Cooperatives: Pathway to economic, democratic and social development in the global economy, 2007.http://www.acdivoca.org/site/Lookup/OCDC_Pathway s_Paper/$file/OCDC_Pathways_Paper.pdf . Accessed 2 April 2014.

[21] M.M. Ali, and A.M. Islam. Developing agribusiness strategies for Bangladesh: an analysis. AIUB Bus Econ Working Paper Series No 2011-01. http://orp.aiub.edu/. Accessed 29 November 2013.

[22] World Bank. Agriculture in South Asia, 2011. http://go.worldbank.org/770VR4DIU0. Accessed 22 November 2013.

[23] S. Raihan, and M. Ahmed. Service trade liberation and domestic regulation: perspective from Bangladesh. MPRA (Munich Personal RePEc Archive) Paper No. 37896, 2008.

[24] K.K. Islam, M.O. Ullah, M. Hoogstra, and Sato. Noriko. Economic contribution of participatory Agroforestry program to poverty alleviation: a case from Sal forests, Bangladesh. Journal of Forestry Research, Vol. 23, No. 2, pp. 323-332, 2012.

[25] M.S. Safa. The effect of participatory forest management on the livelihood of the settlers in a rehabilitation program of degraded forest in Bangladesh. Small-scale Forest Economics, Management and Policy, Vol. 3, No. 2, pp. 223-238, 2004.

[26] K.K. Islam, Sato. Noriko, and M. Hoogstra. Poverty alleviation in Bangladesh: The case of the participatory agroforestry program. International Forestry Review, Vol.12, No. 5, pp. 412-415, 2010.

[27] K.K. Islam, G.M. Rahman, Takahiro. Fujiwara, and Sato. Noriko. People's participation in forest conservation and livelihoods improvements: Experience from a forestry project in Bangladesh. International Journal of Biodiversity Science, Ecosystem Services & Management, Vol. 9, No. 1, pp. 30-43, 2013.

[28] N. Muhammed, M. Koike, F. Haque, and M.D. Miah. Quantitative assessment of people oriented forestry in Bangladesh: A case study from Tangail Forest Division. Journal of Environmental Management, Vol. 88, No. 1, pp. 83-92, 2008.

[29] S.L. Lazzarini, F.R. Chaddad, and M.L. Cook. Integrating supply chain and network. Journal on Chain and Network Science, Vol.1, No.1, pp. 7-22, 2001.

[30] J.H. Trienekens. Agricultural value chains in developing countries: a framework for analysis. International Food and Agribusiness Management Review, Vol. 14, No. 2, pp. 51-82, 2001.

[31] C.M. Webber, and P. Labaste. Building competitiveness in Africa's agriculture: a guide to value chain concepts and applications, 2010. https://openknowledge.worldbank.org/bitstream/handle/10986/2401/524610PUB0AFR0101Official0Use0Only1.pdf?sequence=1 . Accessed 29 March 2014.

[32] R.M. Ruben, A. van Boekel, van Tilburg, and J. Trienekens. Governance for quality in tropical food chains. Wageningen Academic Publishers, the Netherlands, 2007.

[33] L.W. Stern, A.I. El-Ansary, and A.T. Coughlan. Marketing channels. Prentice Hall-International, London, 1996.

[34] G. Gereffi, and M. Korzeniewicz. Commodity Chains and Global Capitalism. Praeger Publishers, USA, 1994.

[35] O.E. Williamson. Strategy research: governance and competence perspectives. Strategic Management Journal, Vol. 20, pp. 1087-1108, 1999.

[36] G. Gereffi, and R. Kaplinsky. The value of value chains: spreading the gains from Globalization. IDS Bulletin, Vol. 32, No. 3, 2001.

[37] O.E. Williamson. The economic institutions of capitalism: firms, markets, relational contracting. In: Trienekens JH (ed.) Agricultural value chains in developing countries: a framework for analysis. International Food and Agribusiness Management Review, Vol. 14, No. 2, pp.51-82, 1985.

[38] A. Rindfleisch, and J.B. Heide. Transaction cost analysis: past, present, and future applications. Journal of Marketing, Vol. 61, pp. 30-54, 1977.

[39] P. Gibbon, J. Bair, and S. Ponte. Governing global value chains: an introduction. Economy and Society, Vol. 37, No. 3, pp. 315-338, 2008.

[40] FD (Bangladesh Forest Department). Protected areas of Bangladesh. Ministry of Environment and Forest Government in Bangladesh, 2013. http://www.bforest.gov.bd/ . Accessed 30 November 2013.

[41] K.K. Islam, and Sato. Noriko. Deforestation, land conversion and illegal logging in Bangladesh: The case of the Sal forests. *i*Forest- Biogeoscience and Forestry, Vol. 5, pp. 171-178, 2012.

[42] BBS (Bangladesh Bureau of Statistics). Statistics data, 2006. http://www.bbs.gov.bd. Accessed 2 December 2013.

[43] P. Gain. The last forest of Bangladesh. Society for Environmental and Human Development (SEHD), Dhaka, Bangladesh, 2002.

[44] S. Tasnoova, and I. Iwamoto. Kataribjog rice marketing system in Dinajpur District, Bangladesh. Faculty of Agriculture Kagoshima University, Vol.41, pp.19-50, 2006.

[45] R. Dhital. Rural urban agriculture marketing system: challenges and opportunities: a case study: Eastern Nepal. Dissertation, School of Forestry and Environmental Studies, Yale University, USA, 2004.

[46] N. Ahmed. Value chain analysis of Hilsa marketing in coastal Bangladesh. Agriculture News, Vol. 33, pp.18-28, 2007.

[47] D. Sarker, and B.K. Ghosh. Milk marketing under cooperative and no-cooperative marketing channels: experiences from West Bengal. Economic Analysis, Vol. 55, No. 187, pp. 87-108, 2010.

[48] A.K. Ghosh, and K.I. Maharjan. Impacts of dairy cooperative on rural income generation in Bangladesh. Journal of International Development and Cooperation, Vol. 8, No. 1, pp. 91-105, 2001.

[49] N.A. Kwapong. Restructured agricultural cooperative marketing system in Uganda: study of the tripartite cooperative model. Euricse Working Paper No. 57/13, 2013.

[50] M.A. Matin, M.A. Baset, Q.M. Alam,, M.R. Karim, and M.R. Hasan. Mango marketing system in selected areas of Bangladesh. Bangladesh Journal of Agricultural Research 33(3): 427-438.

Sustainable water use efficiency for rice cultivation in Rajshahi of Bangladesh

Sheikh Helena Bulbul, Md. Redwanur Rahman

Institute of Environmental Science, University of Rajshahi, Rajshahi-6205, Bangladesh

Email address:
hbsplendid07@gmail.com (S. H. Bulbul), redwan_rahman@lycos.com (M. R. Rahman)

Abstract: A field experiment was conducted at the Bangladesh Rice Research Institute, Shampur, Rajshahi to find out possible effects of alternate wetting and drying irrigation (AWDI) on the yield, water use and water use efficiency (WUE) of rice. The experimental plots were laid out with 2 factors RCBD combining two modern varieties of rice *viz.* BRRI dhan28 and BRRI hybrid2, which received four irrigation treatments randomly and was replicated thrice. The treatments ranged from continuous submergence (T_1) of the field to a number of delayed irrigations (T_2, T_3 and T_4) denoting application of 3, 5 and 7 days of disappearance of standing water respectively. The study revealed that treatment T_1 attributed by the highest total water use (49.04 cm) and the lowest WUE (60.62 kg/ha/cm) produced the lowest grain yield (4.59 t/ha). Treatment T_2, on the contrary, gave the highest yield (5.57 t/ha) and consequently the higher WUE (85.55 kg/ha/cm) than T_1 and indicating quite a large water saving (Application of 3 days of disappearance of Standing Water) compared to treatment T_1. The yields in treatments T_3 (5.33 t/ha) and T_4 (5.15 t/ha) were significantly lower at 1% level of significance compared to that of treatment T_1. Significant effect was found either for the treatment or for the varieties on the number of effective and total tillers hill^{-1} 1000 grain weight, plant height, number of effective tillers hill-1, grain yield, straw yield, biological yield and harvest index.

Keywords: Alternate Wetting and Drying (AWD), Sustainable Water Use Efficiency, Rice Cultivation

1. Introduction

The northwest region of Bangladesh broadly covers the whole Rajshahi Division, consisting of 16 districts of the country. The region is classified as medium- to high-land and is normally flood-free. Within the north-western region, the greater districts of Rajshahi, Dinajpur, Pabna, Rangpur and Bogra fall in the Barind Tract. Compared to the rest of Bangladesh, its Northwest has relatively unfavorable climatic conditions for agriculture. Ninety percent of its rainfall of 1200 to 1400mm occurs within the three month period, June to August. Because of the nature of the soils and the monsoonal downpour, much of this rainfall is lost as surface runoff and causes considerable soil erosion. During the dry period of seven to eight months, evapotranspiration exceeds precipitation. In addition, the length of the monsoon varies considerably and extremes of temperature are experienced with many summer days above 400°C and several winter days below 50°C. [1] reported the topsoil is very thin (often sandy) and beneath there is a hard clay pan. Furthermore points out that the organic content of the soil is low (organic matter accounts for only 0.5 to 0.7 % of the soil content) and that the soil is deficient in plant nutrients such as nitrogen, in zinc and sulphur and trace elements. Considering all these aspects, northwest Bangladesh is a marginal area for agriculture, particularly crop growing. The major portion of the cropland of the Barind Tract is used for rice production and most of this rice land is now shown with High Yielding Varieties (HYV) of rice.

But paddy is a high user of water and encouragement of its production in a water-scarce region is risky, especially since it may quickly degrade the poor soil of this region. Government policy has been to encourage intensification of cultivation in this region including the increased of HYV rice. To this end, it established the Barind Integrated Area Development Project (BIADP) within the Ministry of Agriculture. Against advice from foreign experts, deep tube wells are being installed in this region under this project. One problem is that the use of water from underground sources is liable to exceed the rate of recharge of the aquifers. The actual and planned water supplies from these sources are in all probability unsustainable in the long-term. Already water tables have fallen in some areas. For example, mango trees are reported to have died in some areas around Rajshahi due to falling water tables.

Irrigation from underground sources has enabled the area doubled cropped and triple cropped to be substantially increased. This raises the rate of depletion of soil nutrients and accelerates soil erosion. Hence, reports of HYV (rice) yields of 2.6 t/ha decreasing to around 2.0 t/ha or lower (even only 1.2 t/ha) in many areas, are common in recent years [2]. Consequently several reports and papers highlight the lack of sustainability of current agricultural practices in this region and there is significant land degradation with evidence of desertification in some areas [3]. In addition, it seems probable that current and planned supplies of underground water for irrigation particularly in rice cultivation are unsustainable.

Farmers of this region used continuously standing water for rice cultivation. It is true that rice cultivation requires much water than any other crops. It may also be that high yield in this result in a strong time preference for present water use pattern and a belief that the long-term will somewhat take care itself. This raises several questions in political economy. In particular why should the government support agricultural developments which, on the face of it, result in unsustainable increases in income?

This study is significant as it reveals an effective, yet unexplored (in terms of variety) mode of irrigation system (alternate wetting and drying), that can save water and cost and contribute to sustainability. In order to irrigate paddy field, electricity or fuel is needed to bring water to the surface. Water pumping and delivery treatment facilities consume a significant amount of energy. Farmers are currently paying an equivalent of 25% to 30% of their rice harvest for irrigation and these costs are tending to increase [4]. This presents another factor for the economic relevance of water-saving at the farm level.

Experts state that on a national level, the implementation of AWD could save costs for irrigation of up to 56.4 million Euros in electricity or 78.8 million Euros in fuel (30 liter diesel/ha) [5]. In some regions of the world over 15% of total electricity consumption is devoted to water management. However, the demand for energy grossly exceeds the available supply by far [6]. Diesel is not always available in some rural areas and is expensive for poor rice farmers, while farmers connected to a public power grid are subject to frequent power cuts [7]. In order to address farmers' needs to save water, energy and fuel in irrigated rice alternate Wetting and Drying technology, which takes an average of 1,432 litters of evapotranspirated water to produce 1kg of rough rice [8]. The total production costs for the *Boro* season for conventionally irrigated rice ranged in Bangladesh in 2009 from 475 Euros in Rajshahi [10] to 1010 Euros in Camilla [9]. Of these total costs between 12% (Chittagong) and 27% (Rajshahi) were irrigation costs, ranging in 2009 between 54 Euros/ha [10] and 226 Euros/ha [9] through implementing AWD-based irrigation between 15 Euros/ha and 112euros/ha [9] could be theoretically saved, *i.e.* if water and energy savings were transferred to the farmer. It access to contribute to environmental sustainability.

As irrigated rice systems serve as a large sink for atmospheric carbon dioxide and are a significant source of methane [11], it is assumed that reducing the standing water in rice fields will have positive ecological implications.

Water is integrally linked to the health of the environment. Water is vital to the survival of ecosystems and the plants and animals that live in them, and in turn ecosystems help to regulate the quantity and quality of water. Alternate wetting and drying water saving technology can helpful for habitat conservation. Minimizing irrigation water use helps to conserve wetland habitats for indigenous flora and fauna and migrating waterfowl.

Therefore, introducing AWD could have major impacts at the farm level and be important throughout the country by reducing irrigation costs and reducing the excessive use of ground water, leading to a more sustainable use of natural resources.

So, the judicial use of water resources in intensive irrigated area is a crucial need for maintaining both sustainable crop production and accessible water level.

1.1. The Objectives of the Research Were

- To identify the sustainable water use efficiency for rice cultivation by practicing alternate wetting and drying (AWD).
- To determine the water use efficiency and water productivity of AWD for *Boro rice*.
- To analyze impact of wetting and drying on yield and yield contributing characters.

2. Materials and Methods

The Bangladesh Rice Research Institute (BRRI) Farm of Shampur, Rajshahi was selected as the experimental site. Topography of the land being plain was suitable for check basin irrigation. Individual plots were located inside a close growing rice field so that actual growing condition (reception of the direct and diffused fluxes) prevails in the site.

The experimental plots (4×2.5m) were laid out with 2 factors RCBD combining two modern varieties of rice (BRRIdhan 28 and BRRI hybrid2) and four irrigation treatments that were replicated thrice. This resulted in a total of 24 plots in the field with 8 plots in a row. Each of the plots was separated by 1m of transition zone while each of the replications was demarcated by a buffer zone of 1.5m in between. To prevent seepage, polythene sheets were pushed into the edges of the levees along the inner perimeter of all plots. PVC pipes of 4cm in diameter and 40cm in length were installed in the field keeping 7cm above the soil and the remaining 33cm which was perforated underneath to measure the depletion of soil water in the field. Irrigation water was applied when depleting water table inside the pipe reached a certain level.

The first treatment (T1) was continuous submergence (1 to 5cm standing water; maintained throughout the growing season) and the remaining three (T_2, T_3 and T_4) stood for an application of 3, 5 and 7 days of disappearance of standing water respectively. Continuous standing water (5cm) was maintained in all plots up to 28 days after transplantation (DAT) to avoid pre-apprehended weed infestation that could be awesome during crop establishment stage. A bowl of 1.5 liters was used to irrigate the plots from the buffer zones by throwing water in. The seedlings were transplanted maintaining hill to hill distance of 15cm and row to row distance of 25cm. The first and the last hills were kept at 7.5cm away from their nearest levees resulting in 25 hills along the length and 10 hills along the width. Since the grains of BRRI hybrid 2 got ripened earlier than the BRRIdhan 28, the former was harvested (23 April 2011) two weeks earlier than the harvesting date (May 12, 2011) of the latter. Matured plants inside 1m square of land were harvested for subsequent analysis. Moisture content of the grains, however, was adjusted to 14% equivalent moisture content after measuring through digital grain moisture meter for subsequent analysis. Quantitative information related to yield and all the yield contributing characters *viz.* plant height, effective tillers, length of the panicle, no. of spikelets per panicle, no. of filled and unfilled grains per panicle, 1000 grain weight, grain yield, straw yield, harvest index and water use efficiency of the two varieties (BRRIdhan 28 and BRRI hybrid2) were analyzed to obtain the effect for AWDI on rice production.

3. Results and Discussion

Significant consequences of AWDI on the production of Boro rice were observed as given in Table 1. The analysis showed that varietal effect on plant height was statistically significant at 1% probability level. The tallest plant (104.50cm) was found in BRRI hybrid2 (V_2). The shortest plant (101.64cm) was found in BRRI dhan28 (V_1). Variation in plant height might be due to the differences in the genetic make–up of the varieties. The result is in consistent with findings of [12] who also reported a variable plant height existed among the varieties. The highest number of total tillers hill-1(11.75) was found in BRRI hybrid2 and the lowest number of total tiller was found (10.88) in BRRI dhan28. The variation in number of total tillers hill^{-1} might be due to varietal characteristics. Number of effective tillers hill^{-1} was statistically significant at 1 % level of probability. The highest number of effective tillers (9.53) was found in BRRI hybrid2 and the lowest number of effective tiller hill^{-1}(8.79) was found in BRRI dhan28. The highest number of non-effective tillers (2.21) was found in BRRI dhan28 and the lowest number of non-effective tiller hill^{-1}(2.09) was found in BRRI hybrid2. The highest length of panicle (23.65cm) was found in BRRI hybrid2. The lowest length of panicle was (23.35cm) in BRRI dhan28. Varietal differences regarding the panicle length might be due to their differences in genetic make-up. Number of filled grains panicle^{-1} was statistically significant at 1% level of probability. The results showed that the highest grain yield (149.92) was achieved from BRRI hybrid2. The lowest grain yield (118.45) was achieved from BRRI dhan28. These differences occurred due to variations of genetic make-up among the varieties. Number of unfilled grains panicle^{-1} was statistically significant at 1% level of probability. The highest number of unfilled grains panicle^{-1} (30.80) was found in BRRI dhan28. The lowest number of unfilled grains (23.53) panicle^{-1} was found from BRRI hybrid2. The result showed that the highest weight of 1000- grain (25.62g) was obtained from BRRI hybrid2. The lowest weight of 1000-grain (23.93g) was obtained from BRRI dhan28. Grain yield was statistically significant at 1% level of probability. The highest grain yield (5.52t/ha) was achieved from BRRI hybrid2. The lowest grain yield (4.81t/ha) was achieved from BRRI dhan28. These differences occurred due to variations of genetic make-up among the varieties. The result reveals that the highest straw yield (6.58tha^{-1}) was found from BRRI hybrid2. The lowest straw yield (5.71tha^{-1}) was found from BRRI dhan28.The highest yield occurred due to higher plant height, higher total tiller hill^{-1} and lower number non-effective tiller hill^{-1}. These results are consistent with those obtained by [13] who reported differences in straw yield among varieties. The highest biological yield (12.10tha^{-1}) was obtained from BRRI hybrid2 and the lowest one (10.52tha^{-1}) was obtained from BRRI dhan28. Maximum harvest index (45.63%) was obtained from BRRI dhan28 and the minimum harvest index (45.57%) was obtained from BRRI hybrid2.

Table 1. *Varietal (BRRI dhan28 and BRRI hybrid2) Effect on the Yield and Yield Contributing Characters.*

Variety	V_1	V_2	LSD	Level of sig
Plant height (cm)	101.64	104.50	0.28	**
Total tiller hill^{-1}	10.88	11.75	0.08	**
Effective tiller hill^{-1}	8.79	9.53	0.09	NS
Non-effective tiller hill^{-1}	2.09	2.21	0.10	NS
Panicle length (cm)	23.35	23.65	0.18	**
Filled grain panicle^{-1}	126.33	149.92	0.77	**
Unfilled grain panicle^{-1}	30.80	23.53	0.59	**
1000-GW	23.93	25.62	0.18	**
Grain yield (t ha^{-1})	4.81	5.52	0.05	**
Straw yield (t ha^{-1})	5.71	6.58	0.05	**
Biological yield	10.52	12.10	0.07	**
Harvest Index	45.63	45.57	0.33	NS

In a column figures with same letter or without letter do not differ significantly whereas figures with dissimilar letter differ significantly (as per DMRT), V_1 = BRRI dhan28 , V_2 = BRRI hybrid2, NS = Not Significant, * = Significant at 5% level of probability, ** = Significant at 1% level of probability.

The experiment aimed in exploring the possible effects of different irrigation treatments on the production and production related parameters. Different yield contributing characters viz., plant height (cm), number of effective tillers per hill, panicle length (cm), total number of filled grains per panicle, number of unfilled grains per panicle; 1000 seed weight (gm), grain yield (t/ha) and straw yield (t/ha)for each of the varieties were analyzed.

Results obtained from the statistical analysis of the effect of different degrees of delayed irrigation treatments on plant height are shown in table 2. The analysis showed that the irrigation treatments had significant effect on plant height at 1% level of probability. The highest plant height (104.38 cm) was obtained in treatment T_2 (Application of 3 days of disappearance of standing water) and the lowest (101.55 cm) in Treatment T_1 (continuous flooding) as shown in figure 1.

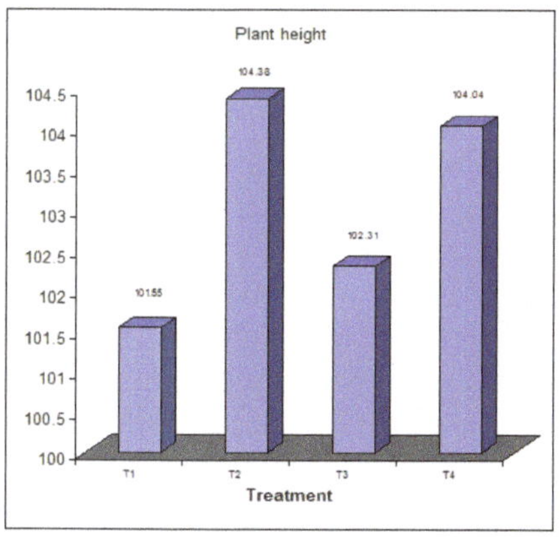

Figure 1. *Variation in Plant Height for Different Degrees of Irrigation Treatments.*

Information obtained after analysis of the experimental findings showed that effect on the number of effective tillers quite significant at 1% level of probability while different degrees of delayed irrigations applied as treatments in the experimental plots. The highest number of effective tillers per hill (11.65) was found in treatment T_2 followed by treatment T_3 (8.69) and treatment T_1 (8.45). The lowest number of effective tiller per hill (7.85) was found in treatment T_4. A little delayed irrigation treatments using alternate wetting and drying irrigating method in rice field showed significant effects on the number of filled grains per panicle compared to one with continuous standing. It was found that the highest number of filled grains (150.55) per panicle was obtained in treatment T_2 (3 days of disappearance of standing water) followed by treatments T_3 (141.65) and T_4 (139.47). The lowest number of filled grains per panicle (120.84) was found for treatment T_1. Thus the result showed that applying irrigation water in rice field when 3 days of disappearance of standing water does not really reduce the total number of filled grains compared to that nursed with 5cm standing water (Figure 1). However, treatment T_1 (continuous standing water) decreased the number of filled grains. Grain yield, the most important character of an agronomic analysis, was found to be significantly influenced by varieties, different degrees of AWDI irrigation treatments. Analysis of the data obtained from the experimental plots resulted in a clear depiction of the scenario. The highest grain yield (5.57 t/ha) was obtained from treatment T_2 (irrigation when water is 3 days of disappearance of standing water) and the lowest yield grain (4.59t/ha) was obtained from treatment T_1 (continuous standing water). The results shows that the grain yield did not decreased when plants suffered little water stress. The second highest yield grain (5.33 t/ha) was found in the treatment T_3 (5 days of disappearance of standing water). [14] and [15] have shown that rice grain yield increased with reduced water use. Straw yield from the experiment was found to be significantly affected by the variety, irrigation treatments. The maximum straw yield (6.45 t/ha) was found from the treatment T_2. The minimum straw yield (6.00t/ha) was found from treatment T_1.

Results obtained from the statistical analysis of the effect of variety and different degrees of delayed irrigation treatments on plant height are shown in table 3. The analysis showed that effect of the interaction between the varieties and the treatments was also found to be statistically significant at 1% level of probability.

Table 2. *Effect of Different Irrigation Treatments on the Yield and Yield Contributing Characters.*

Treatment	T_1	T_2	T_3	T_4	LSD	Level of sig
Plant height (cm)	101.55b	104.38a	102.31b	104.04a	0.57	**
Total tiller hill^{-1}	10.92c	12.94a	11.64b	9.76d	0.16	**
Effective tiller hill^{-1}	8.45b	11.65a	8.69b	7.85c	0.19	**
Non-effective tiller hill^{-1}	2.47a	1.29c	2.95a	1.91b	0.20	**
Panicle length (cm)	23.07b	24.48a	22.91b	23.53b	0.35	**
Filled grain panicle^{-1}	120.84c	150.55a	141.65b	139.47b	1.53	**
Unfilled grain panicle^{-1}	31.72a	23.58b	24.03b	29.34a	1.17	**
1000-GW	23.97b	25.27a	25.27a	24.60ab	0.36	**
Grain yield (t ha^{-1})	4.59b	5.57a	5.33ab	5.15ab	0.09	**
Straw yield (t ha^{-1})	6.00b	6.45a	6.04b	6.10b	0.09	**
Biological yield	10.59c	12.02a	11.37b	11.25b	0.13	**
Harvest Index	43.46b	46.26a	46.90a	45.77a	0.65	**

In a column figures with same letter or without letter do not differ significantly whereas figures with dissimilar letter differ significantly (as per DMRT), T_1 = Continuous Standing water, T_2= 3 days disappearance of standing water, T_3 = 5 days disappearance of standing water, T_4 = 7 days disappearance of standing water, NS = Not Significant, *= Significant at 5% level of probability, ** = Significant at 1% level of probability.

The tallest plant height (106.48cm) was found for the interaction $V_2 \times T_2$. Information obtained after analysis of the experimental findings showed that the interaction effect of the varieties and treatments also came significant at 1% level of probability. The highest number of filled grains (158.67) was, however, marked for the interaction ($V_2 \times T_2$) and the lowest number of filled grains (105.48) was obtained from V_1T_1. Thousand grain weight (1000 grain weight), as it is called the test weight of the desired output, is referred to be considered as one of the most significant agronomic parameters ever trusted that contributes in having a reconnaissance over the possible production of a lot (gain yield). The values of 1000 grain weight were found to be significant in this analysis for the effects of treatments at the 1% level of probability. The highest 1000 grain weight (26.82) was obtained for the interactions ($V_2 \times T_3$) and the lowest 1000 grain weight (23.63) was obtained for the V_2T_4. The study raveled that the varieties V_1 and V_2 and interaction effect between variety × treatments produced statistically insignificant variation in 1000 grain weight among themselves. Thus, it was clear from the interaction effect that AWDI method of irrigation treatments did not reduced the 1000 grain weight as irrigation delayed. Grain yield, the most important character of an agronomic analysis, was found to be significantly influenced by varieties, different degrees of AWDI irrigation treatments and their interactions. Analysis of the data obtained from the experimental plots resulted in a clear depiction of the scenario. It showed that the varietal effect on the grain yield was significant at the 1% level of probability (Table 3).

The interaction between the varieties and treatment also produced significant results for grain yield at the 1% level of probability. The highest grain yield of BRRI hybrid2 (6.16t/ha) was obtained for the interaction ($V_2 \times T_2$) and the lowest grain yield (4.06t/ha) was obtained from the interaction ($V_1 \times T_1$). Straw yield from the experiment was found to be significantly affected by the variety, irrigation treatments and their combination at 1% level of probability. The highest straw yield (6.94t/ha) was obtained for the interaction ($V_2 \times T_2$) and lowest straw yield (5.11t/ha) was obtained for the interaction ($V_1 \times T_1$) (Table 3).

3.1. Irrigation Treatments (AWDI)

The two varieties stated above were subjected to different levels of irrigation developed by delaying the scheduled irrigation in the field. Starting from the very end of the first stage of crop development, four irrigation treatments were started with regular intervals. The time of water application, however, was indicated by the depletion of water level in different perforated pipes measured from the ground surface. Both of the experimental rice varieties, irrespective of their position in the field, receive different levels of irrigation treatments according to the demand of the respective fields in each replication.

Irrigation treatments were applied at different stages of the growing period depending on the depletion of the water level in the perforated pipe. The very first treatment stated at the end of the fourth week after transplantation. During this time 5cm standing water was kept to avoid weed infestation in the plots.

Table 4 shows that the highest number of irrigation (15nos.) was given to the plots with treatment T_1 (continuous flooding) for BRRI dhan28. The other three treatments *viz.*, T_2, T_3 and T_4 received a total of 14, 13 and 13 number of irrigation for BRRI dhan28 while 14, 13, 13 and 12 number of irrigation for BRRI hybrid2, respectively.

Table 3. *Mean effect of the Interaction Between Varieties and Irrigation Treatments on the Yield and Yield Contributing Characters of BRRI dhan28 (V_1) and BRRI hybrid2 (V_2).*

	V_1T_1	V_1T_2	V_1T_3	V_1T_4	V_2T_1	V_2T_2	V_2T_3	V_2T_4	LSD	Level of sig
Plant height (cm)	98.51e	102.28cd	101.29d	104.48ab	104.59ab	106.48a	103.34bcd	103.60bc	1.13	**
Total tiller hill^{-1}	10.45	12.41	11.38	9.30	11.39	13.47	11.90	10.22	0.32	NS
Effective tiller hill^{-1}	8.01	11.43	8.26	7.46	8.90	11.87	9.12	8.23	0.37	NS
Non-effective tiller hill^{-1}	2.44	0.98	3.11	1.84	2.49	1.60	2.78	1.98	0.41	NS
Panicle length (cm)	22.83	24.16	23.02	23.38	23.31	24.80	22.80	23.67	0.70	NS
Filled grain panicle^{-1}	105.48f	142.42c	129.17e	128.26e	136.20d	158.67a	154.13ab	150.67b	3.07	**
Unfilled grain panicle^{-1}	38.31a	25.90c	24.95cd	34.04b	25.13cd	21.27d	23.10cd	24.63cd	2.35	**
1000-GW	23.91	24.46	23.71	23.63	24.03	26.07	26.82	25.57	0.72	NS
Grain yield (t ha^{-1})	4.06c	4.98bc	4.96bc	5.21abc	5.12abc	6.16a	5.70ab	5.09abc	0.19	**
Straw yield (t ha^{-1})	5.14e	5.96c	5.63d	6.11c	6.85a	6.94a	6.45b	6.09c	0.19	**
Biological yield	9.20e	10.94cd	10.59d	11.33c	11.97b	13.09a	12.15b	11.17c	0.26	**
Harvest Index	44.13	45.49	46.86	46.02	42.79	47.02	46.93	45.52	1.30	NS

In a column figures with same letter or without letter do not differ significantly whereas figures with dissimilar letter differ significantly (as per DMRT), * = Significant at 5% level of probability, ** = Significant at 1% level of probability, NS = Not significant.

Table 4. *Total number of irrigation required for different irrigation treatments.*

Treatment	*No. of Irrigation		Rainfall+ water for crop established (cm)	Total water required(cm)	
	BRRIdhan28	BRRI hybrid2	42.1	BRRI dhan28	BRRI hybrid2
T_1	15	14	42.1	53.27	44.89
T_2	14	13	42.1	52.67	43.79
T_3	13	13	42.1	51.73	42.69
T_4	13	12	42.1	51.42	42.69

*One irrigation means application of 5cm irrigation water

Water required for crop establishment and water received from the rainfall was estimated to be 42.1cm during the growing period for each of the treatments. For BRRI hybrid2, maximum amount of water (44.89cm) was required for T_1, while, second maximum (43.79cm) for T_2 was followed by other two treatments, T_3 (42.69cm) and T_4 (42.69cm). For BRRI dhan28 the treatments T_1, T_2, T_3 and T_4 required 53.27, 52.67, 51.73 and 51.42cm of water, respectively. The graphical representation of water usage by different treatments after transplantation is shown in figure 2.

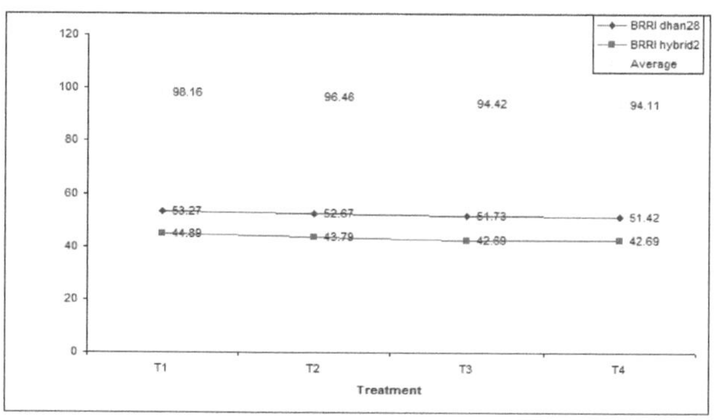

Figure 2. *Water Usage of Different Treatments for the Production of BRRI dhan28 and BRRI hybrid2.*

Water use efficiencies for the individual effect of different treatments were derived along with the values of WUE for the 8 interactions between treatments and varieties (Table 5). The highest water use efficiency, WUE was found to be 81.67kg/ha/cm ($V_1 \times T_4$). All the highest water use efficiencies were found in the combinations having variety V_1 (BRRI dhan28).

The lowest WUE 56.60 kg/ha/cm was obtained in the treatment T_1 for V_2 (Fig. 4.19). In case of BRRI dhan28 (V_1) the highest WU was found to be 81.67 kg/ha/cm in treatment T_4 and the lowest WUE 64.64 kg/ha/cm in treatment T_1. The second highest WUE (87.38 kg/ha/cm) was found in the treatment T_2 though it gave poor yield (4.98 t/ha). Treatment T_4 gave high yield with high water use efficiency (81.67kg/ha/cm) among the others (Figure 2).

Water productivity was found to be the highest (0.088 t/ha/cm) in treatment T_4 (irrigation 5 days disappearance of standing) followed by treatment T_4 and T_2 (0.082t/ha/cm) (irrigation 7 days and 3 days disappearance of standing water) and a minimum of 0.049 t/ha/cm treatment T_1 (continuous flooding). From these results, it can be seen that the water productivity decreased with the increase of irrigation water. Above result reveals on-farm strategies for reducing water input in irrigated rice [16].

This study shows that the alternate wetting and drying irrigation system has significant economic and environmental benefit for a farmer, the environment and the society. The system has better performance in generating profit even under worst case scenario. A little amount of input requirement and high yield capacity are the major reasons, enhancing the system's performance in a sustainable way. The practices of high-input agriculture are causing concerns about the sustainability of crop production.

Table 5. *Water use Efficiency for Different Treatments and Interactions.*

Interactions	V_1T_1	V_1T_2	V_1T_3	V_1T_4	V_2T_1	V_2T_2	V_2T_3	V_2T_4
Total water required (cm)	53.2	52.66	51.70	51.46	44.89	43.79	42.69	42.69
Water applied (cm)	82.3	64.6	64.1	63.3	79.3	74.6	64.1	63.1
Grain Yield (ton/ha)	4.06	4.98	4.96	5.21	5.12	6.16	5.70	5.09
Water use efficiency (kg/ha/cm)	64.64	81.42	80.65	81.67	56.60	58.69	66.59	66.59
Treatments	T_1		T_2		T_3		T_4	
Average total water required (cm)	49.04		48.22		47.19		47.07	
Average grain yield (ton/ha)	4.59		5.57		5.33		5.15	
Water use efficiency kg/ha/cm)	60.62		70.05		73.62		74.13	
Water productivity (t/ha/cm)	0.049	0.077		0.082	0.064	0.082	0.088	0.080

There are many negative effects on the environment, including pollution by pesticides, emission of greenhouse gases, soil degradation, air pollution by dust, and loss of landraces and other biodiversity. People need to develop new techniques that will keep agriculture both profitable for the farmer and make it sustainable for the future.

There is a concern that the AWDI method of water management promotes greater weed populations, thus requiring more labor for weed management. [17] emphasized that even on coarse textured soils continuous submergence is only required for 3 week period after transplanting to ensure seedling establishment. The field may be drained for 1 to 5 days, depending on the weather, before the next irrigation. There will be no adverse effect on yield and 25 to 50% of water can be saved. [18] studied the effect of applying improved water management technology during the wet season in canal-irrigated rice fields in Bihar, India and found that irrigation to a depth of 7cm, 3 days after ponded water disappeared gave 34% more grain yield than the usual farmer's practice (more than 10 cm irrigation water continuously). A field experiment at the Bangladesh Agricultural University farm was conducted by [19] to compare the traditional water management practices in the rice field with the modern water saving techniques. The experiment was figured using split plot design and consisted of 5 irrigation treatments *viz.*, Treatment T_1, T_2, T_3, T_4 and T_5 meaning continuous submergence, application of irrigation water 3 days, 5 days, 7 days and 10 days after the disappearance of standing water from the plots, respectively. They also reported an increase in plant height with the increase in water requirement though its effect on the production unit remained insignificant. He recommended that 5cm irrigation water should be applied 3 days after the disappearance of standing water from the soil surface to obtain the maximum water use efficiency. An interval of 5 to 7 days between consecutive irrigations recommended [20]. It was also found that AFDI technique reduces continuous decline of water table.

However, a range of options are available for increasing the productivity and efficiency of water in surface irrigated rice ecologies. More radical options departing from traditional systems are also available and may be required.

The most appropriate strategies to adopt will vary over time and space and will have to be designed carefully with the involvement of the farmers, but will need to be resolutely forward-looking and perhaps revolutionary. Identifying the policies, management practices and technologies needed at farm, system and basin level will require a multi-disciplinary approach

4. Conclusion

The results of this research support the thesis that alternate wetting and drying treatment. The practice of keeping rice fields dry for some period of time during the growing season resulted in a significant reduction in water use. Maintaining the rice field dry during mid season not only reduced water use and significantly increased grain yield as compared to continuous flooding treatment if alternate wetting and drying practice is adopted.

Countrywide, particularly in the water major rice growing areas where groundwater is major source of irrigation and the farmers grow 2-3 rice crops per year, water use would significantly decrease. Using water more efficiently over several agro-ecological zones of the country would help alleviate water shortages allowing water to be used for other needs.

References

[1] Zuberi, M.I. "Natural Resource Management and Utilization in Bangladesh: on Irrigation and Drainage" Tokyo, Japan, 15-25 October 1989. I-C, pp. 480-Geological Survey Water-Resources Investigation Report 84-4012, 1992, 47p.

[2] Zuberi, M.I. "A Description of North Western Bangladesh' (mimeo.) Centre for Environmental Research" 1993, University of Rajshahi, Rajshahi.

[3] Ministry of Forestry and IUCN. "Towards Sustainable Development: The National Conservation Strategy of Bangladesh" Ministry of Environment and Forestry, Dhaka, Bangladesh and IUCN, 1991, Glauds, Switzerland.

[4] M.A. Sattar, M. Maniruzzaman, and M.A. Kashem, (ed.) "National Workshop Proceedings on AWD Technology for Rice Production in Bangladesh" 2009, Bangladesh Rice Research Institute. Gazipur.

[5] M. Hamid; T.P. Tuong, B. Bouman and H. Elizabeth, "Dissemination, Spread and Adoption of Awd in Bangladesh" 2009, International Rice Research Institute, Dhaka, Bangladesh.

[6] Economist Intelligence Unit, "Bangladesh Country Profile" 2008. Economist. London.

[7] CPD (Centre for Policy Dialogue) "State of the Bangladesh Economy in FY 2009/10 From Stability to Accelerated Growth" 2010, Dhaka, Bangladesh.

[8] IRRI (International Rice Research Institute). "Site-Specific Nutrient Management" 2010, www.irri.org/ssnm.

[9] L. Kabir, "Benefit-cost Analysis of Demonstration on AWD Irrigation Method" 2009

[10] M. Hossain "Pumping up production: shallow tube wells and rice in Bangladesh" In: Spielman DJ, Pandaya-Lorch R(eds) Millions fed: proven successes in agricultural development. International Food Policy Research Institute, 2009, Washington, DC

[11] B.A. Bouman, M.L. Feng, T.P. Tuong, G. Lu, H. Wang and Y. Feng, "Exploring options to grow rice under water-short conditions in northern China using a modeling approach" II: Quantifying yield, water balance components, and water productivity. Agricultural Water Management, 2007, vol. 88, pp. 23-33.

[12] A.M. Shamsuddin, M.A. Islam, and A. Hossain, "Comparative study on the yield and agronomic characters of nine cultivars of aus rice" Bangladesh J. Agril. Sci. 1988, vol. 15, 1, pp. 121-124.

[13] M.J.U. Chowdhury, A.U. Sarker, M.A.R. Sarker and M.A. Kashem, "Effect of variety and number of seedlings hill-1 on the yield and its components on late transplanted *aman* rice" Bangladesh J. Agril. Sci. 1993, vol. 20, 2. pp. 311-316.

[14] Q. Shi, X. Zeng, M.Li, X. Tan, and F. Xu. "Effect of different water management practices on rice growth" *In:* " Water-wise rice production" Bouman BAM, Hengisdijk H, Hardy B, Bindraban PS, Tuong TP, Ladha JK, (editors). 2002.

[15] I.P. Wardana, P.S. Bindraban, A. Gani, A.K. Makarim, and L. Isrel, "Biophysical and economic implications of integrated crop and resource management for rice in Indonesia". *In*: "Water-wise rice production" Bouman BAM, H Hengisdijk, Water-wise rice production. Proceedings of the international workshop on water wise rice production, 8-11 April 2002, Los Banos, Philippines. Los Banos (Philippines): International Rice1Research Institute. 2002, pp 3-13.

[16] D.F. Tabbal, B.A.M. Bouman, S.I. Bhuiyan, E.B. Sibayan and M. A. Sattar, "On-farm strategies for reducing water input in irrigated rice; case studies in the Philippine" Agric. Water Mange. 2002, vol. 56, pp. 93-112.

[17] K.L. Khera, B.S. Sandhu, and T.S. Aujla "Judicious irrigation saves water in rice cultivation" Indian Farming 1991, vol. 41, 2 pp. 27-28.

[18] K. Prasad, A.R.P. Bally, E.F. Schmid and Levy, J.C. Benoit, Barthou and P. Benalloul, "Ce-doped TiO_2 insulators in thin film electroluminescent devices" Japanese J. Applied Phys., 1997 vol. 36, pp. 5696-5702.

[19] S.K. Jha, M.K. Hossain and M.S.U. Talukder, "Response of transplanted Boro rice under different irrigation practices" Bangladesh J. Agril. Unv. Bangladesh. 2007, vol 5, 2, pp. 393-398.

[20] Bangladesh Institute of Nuclear Agriculture (BINA). "Achievements of Non Commodity Technologies: Water saving in rice cultivation" Online publication ("http://hyperlink%20%22http//www.bina.gov.bd/%22www.bina.gov.bd") of BINA, 2007, Mymensingh, Bangladesh.

Are small scale eggplant producers efficient in resource use?

Nsikak-Abasi A. Etim

Department of Agricultural Economics and Extension, University of Uyo, P.M.B. 1017, Uyo, Akwa Ibom State, Nigeria

Email address:
etimbobo@gmail.com

Abstract: An empirical study to investigate the efficiency of resource use by small scale eggplant producers was conducted. A stochastic production frontier function which incorporates a model for the technical efficiency effect was applied. Through a multi-stage sampling method, 60 egg plant farmers were selected with the aid of questionnaire. Using the maximum likelihood estimation method, asymptotic parameters were evaluated to describe efficiency determinants. Results revealed that the mean resource use efficiency is 0.71 leaving inefficiency gap of 0.29 implying that 29 percent higher eggplant output could be achieved using the same resource combination. Land, hired labour and seeds were identified as the most critical and important efficiency determinants. Access to credit and market, and age were estimated as inefficiency determinants. Results underscore the need to formulate policies aimed at making land more accessible to farmers, provision of credit facilities and good road networks.

Keywords: EggPlant, Resources, Efficiency, Farmers

1. Introduction

According to the World Health Organization, adequate fruit and vegetable intake entails a consumption of at least 400g of fruits and vegetables per day per capita (an equivalent of 146kg per year per capita) [28]; [7]. The under consumption of fruits and vegetable is largely responsible for deaths globally. Fruits and vegetables play significant roles in human nutrition, especially as sources of vitamins (C, A, B_6, thiamine, niacin; E), mineral and dietary fiber [26];[27] and [36]. These food items are rich sources of vitamins and minerals for the normal functioning of the human body [28]. Most deaths from chronic disease in low and middle income countries arise from micronutrient deficiencies causing vitamin A, iron and iodine disorder. But the production and consumption of certain fruits and vegetables can improve mineral and vitamin intake.

One of such fruits capable of solving these problems is eggplant (*Solanum spp.* L). Eggplant also known as garden egg belong to the family solanaceae. According to [17], the crop develops a strong taproot with a branded root system that does not spread widely but responds positively to well drained soils of medium texture that are moderately deep and is well adapted to the tropic. The crop is bushy and grows to a height of 120cm. Eggplant may be boiled, fried or stiffed in curries or eaten fresh. It may be prepared in form of sauce or stew and consumed with yam, plantain or rice. Nutritionally, garden egg contains water (89%), protein (14%), fibre (1.5%), carbohydrate (8.0%) and good level of vitamin B, Calcium, Phosphorus and Iron. The low consumption of fruits (eggplant inclusive) can result from low productivity due to inefficiency of resource use in production. Earlier and empirical studies by [14]; [13]; [10] and [11] suggest that efficient utilization of resources in agriculture can increase production. This study was conducted to empirically investigate efficiency of resource use by small scale eggplant producers.

The term efficiency of a firm can be defined as its ability to provide the largest possible amount of output from a given set of inputs. The modern theory of efficiency dates back to the pioneering work of [15] who proposed that the efficiency of a firm consist of technical and allocative components and the combination of these two components provide a measure of total economic efficiency (overall efficiency). As noted by [15] technical efficiency, which is the main focus of this study, is the ability to produce a given level of output with a minimum quantity of inputs and can be measured either as input conserving oriented technical efficiency or output-expanding oriented technical efficiency. Output-expanding oriented technical efficiency is the ratio of observed to maximum feasible output, conditional on technical and observed input usage [18] and [2]. This study aims at using output-expanding orientation to measure technical efficiency effects.

The term frontier involves the concept of maximality in which the function sets a limit to the range of possible observations [16]. It is therefore, possible to observe points below the production frontier for firms producing below the maximum possible output, but there cannot be any point above the production frontier, given the available technology. Deviations from the frontier are attributed to inefficiency. The need to measure inefficiency effects is the major motivation for the study of frontiers. Frontier studies are classified according to the method of estimation. [19] grouped these methods into two broad categories-parametric and non-parametric methods. For the parametric methods, it can be deterministic, programming and stochastic depending on how the frontier model is specified. Many researchers, including [29] have argued that efficiency measures from deterministic models are affected by statistical noise. This however, led to the alternative methodology involving the use of the stochastic production frontier models. [1] and [21] independently proposed the idea of stochastic measurement. The major feature of the stochastic production frontier is that the disturbance term is a composite error consisting of two components, one symmetric and the other one-sided. The symmetric component, V_i, captures the random effects due to measurement error, statistical noise and other influences outside the control of the firm and it is assumed to be normally distributed. The one-sided component U_i, captures randomness under the control of the firm. It gives the derivation from the frontier attributed or exponential. The major weakness of the stochastic frontier model is its failure to provide an explicit distribution assumption for the inefficiency term [30].

By definition, stochastic frontier production function is

$$Yi = F(X_i; \beta) \exp(V_i - U_i)\ I = 1, 2, \ldots, N \quad (1)$$

Where Y_i is the output of the ith farm; X_i is the corresponding (MX2) vector of conventional physical inputs; β is a vector of unknown parameter to be estimated; F(.) denotes an appropriate functional form, V_i is the symmetric error component that accounts for random effects and exogenous shock; while, $U_i<0$ is a one sided error component that measures technical inefficiency.

In recent times, econometric modeling of stochastic frontier methodology associated with efficiency estimation has been important aspect of economics research. Both time varying and cross-sectional data have been used in studies based mostly on Cobb-Douglas function and transcendental logarithmic functions that are specified either as production function or cost function to estimate individual firm efficiency [5]; [6]; [2]; [3]; [37]; [31]; [32]; [35]; [14] and [33]. However, this study uses a Cobb-Douglas production function to estimate technical efficiency effect at farm levels by assuming a stochastic nature of production.

2. Methodology

2.1. The Study Area, Sampling and Data Collection Procedure

The study was conducted in Uyo Local Government Area, the capital city of Akwa Ibom State, Nigeria. Uyo is situated 55 kilometers inland from the coastal plain of South-East Nigeria. The area lies within the humid tropical rainforest zone with two distinct seasons – the rainy and short dry season. The annual precipitation ranges from 2000 – 3000mm per annum. According to [9], this rainfall regime received in most parts of the State encourages farming throughout the year. The area is located between latitude 5°17' and 5°27'N and longitude 7°27' and 7°58' and covers an area of approximately 35 square kilometers. The occupation of the inhabitants reflects the economic activity of the residents. The settlement pattern in Uyo is nucleated and being an administrative headquarters, majority of civil and public servants and political office holders reside there. [12] noted that these people engage in part-time farming activities and other commercial ventures within and around their homes as a way of augmenting and supplementing family income and food supplies.

Data used for this study are mainly primary and were obtained from the waterleaf farmers using questionnaire during 2013 farming season. Simple random sampling technique was employed to select a total of 60 egg plant farmers for study. Baseline information on socio-economic characteristics, input use and output levels were collected and analyzed.

2.2. The Empirical Model

The study utilized stochastic production frontier, which builds hypothesized efficiency determinants into the inefficiency error components [8]. Assuming we specified a Cobb-Douglas functional form as:

$$\text{Ln (Qty)} = \beta_o + \beta_1\text{Ln (Land)} + \beta_2\text{Ln (Hired Labour)} + \beta_3\text{Ln (Inorganic fertilizer)} + \beta_4\text{Ln (Seeds)} + \beta_5\text{Ln (Capital)} + Vi - Ui \quad (2)$$

Where Qty is the quantity of output measured in kg; Land is the farm size measured in square meters; Hired labour is the labour employed in farm operations measured in mandays; Inorganic fertilizer is fertilizer applied on the soil measured in kg;Organic fertilizer is farm yard manure applied on the soil measured in kg; Seeds is planting materials measured in naira; Capital is the depreciation value of the implement used measured in naira.

With $Vi \sim N(0, v^2)$; and

$$e^{-ui} = \alpha_o + \alpha_1\text{(Tech)} + \alpha_2\text{(Age)} + \alpha_3\text{(FamS)} + \alpha_4\text{(Gender)} +$$

α_5 (Cred) + α_6 (Mkt) + Zi (3)

Where Tech is access to extension contact (dummy), Age is the age of the farmer (years); FamS is the number of persons in a household who share the same dwelling and meals; Gender is the sex of the farmer (dummy); Credit is access to credit facilities (dummy); and Mkt is access to market (dummy); Zi is an error term assumed to be randomly and normally distributed. The value of the unknown coefficients in equations (1) and (2) are jointly estimated by maximizing the likelihood function [37] and [32].

3. Results and Discussion

3.1. Maximum Likelihood and Inefficiency Estimate Results

Results of the maximum likelihood and inefficiency estimate is presented in Table 1. Farm size is aimed at capturing the effect of scale production on the technical efficiency of the farm. Earlier and empirical study by [20] established a varied relationship between farm size and technical inefficiency in developing countries using the frontier production function. The sign of the land variable in this study was negatively significant (P<.01). This is explained by the fact that increased farm size diminishes the timeliness of resource use thus leading to decline in technical efficiency. [22];[4] and [11] obtained similar findings that showed inverse relationships. Results suggest the need to design policies aimed at encouraging small scale egg plant farmers to continue in production as they are the backbone of agricultural production in developing countries.

In this study, hired labour provided for various farming operations seems to be the most critical input with a coefficient of 2.1802 and is positively significant (P< .01). The relative large elasticity for hired labour is an indication of the labour intensive nature of egg plant production especially during weeding. Similar finding was reported in a recent study by [11].

Planting material is the seed used for planting. This variable is positively significant as expected. Result underscores the importance of seed as its ready availability for production tends to increase efficiency of use. Finding however suggests the need to encourage proper storage and preservation of seeds for use by local farmers. According to [11] in a recent study, such decisions will not only ensure timely availability of planting materials to farmers but will reduce the additional cost and inefficiency which would have been incurred in purchasing the seeds.

The effect of age of technical efficiency may be either positive or negative. Older farmers seem to be more experienced and would be more technically efficient than younger farmers. But regarding innovations and farming techniques, older farmers are less likely to adopt innovations and would thus, be less technically efficient than younger farmers. This study reveals that age has a positive sign and significantly impacts on technical efficiency in the model, hence, age indexes experience and serves as a proxy for human capital implying that egg plant farmers with more years of experience in farming will be more technically skilled in management and policy decisions, and thus a higher efficiency than younger farmers. This result is synonymous with recent empirical study by [11] and [10] who posited that increased experience in cultivation may also enhance critical evaluation of the relevance of better production decisions, including utilization of productive resources.

Credit is positively signed as expected. The variable has an elasticity 0.1653 and is significant (P< .05). Result implies that accessibility and availability of credit to egg plant farmers seems to eliminate the various production constraints thus making it easier for timely and ready purchase of inputs thereby increasing productivity through efficiency. Result agree with earlier empirical studies by [23]; [4]; [13] and [11]. These studies have positively linked access to credit to agricultural productivity. [25] however posited that one of the key problem associated with small holder farmers inaccessibility to agricultural credit.

Market captures egg plant farmers' access to market and serves as a proxy for development. The variable was positive and significant (P< .10). Farms located farer from the market are believed to be less technically efficient than farms nearer the market. This is because farms located farer from the market will not only add to production and marketing cost but also impacts on various farming operations. Result conforms with earlier findings by [10] and [11].

Table 1. *Maximum likelihood Estimates and inefficiency function*

Variable	Coefficient	Asymptotic t-value
Production Function		
Constant terms (β_o)	2.0011	1.7354*
Land (β_1)	-1.8131	-2.0158**
Hired Labour (β_2)	2.1802	3.1427***
Inorganic Fertilizer (β_3)	1.0455	1.4105
Seeds (β_4)	0.3878	2.2571**
Capital (β_5)	1.9344	1.0202
Explainers of Inefficiency		
Intercept (α_o)	1.2501	1.7343*
Technical Assistance (α_1)	0.0952	1.5308
Age (α_2)	0.1828	1.9351*
Family Size (α_3)	0.5667	1.6177
Gender (α_4)	0.6891	1.3540
Credit (α_5)	0.1653	2.0973**
Market (α_6)	0.2151	1.9938**
Diagnostic statistics		
Sigma-square 5^2	0.0824	2.3562**
Gamma (λ)	0.7481	1.9924
Ln (likelihood)	16.5202	
LR Test	7.3589	
Quasi Function	1.5102	
Number of observations	60	

3.2. Resource-Use Efficiency

One important feature of the stochastic production frontier is its ability to estimate individual farm specific technical efficiency. The farm-specific resource use efficiency indices is shown in figure 1. The efficiency indices across eggplant farms vary considerably as the technical efficiencies of all the sampled eggplant farms are less than one. This means that no egg plant farm reached the frontier threshold. From the figure, the most efficient egg plant producer in terms of resource use has an efficiency index 0.92 and the least efficient an index of 0.04. The fact that none of the producers reached the frontier threshold suggest that producers may have encountered some technical, production and/or environmental problems, they were unable to completely overcome [34] and [14]. In small scale farming, resources are mostly allocated to various uses on the basis of their shadow values, which is the amount by which the contribution could be raised if an additional unit of the input is used, thereby preventing the producers from reaching the maximum production efficiency, [24]; [2] and more recently [11]. The mean resource-use efficiency is 0.71 leaving an inefficiency gap of 0.29 meaning that about 29 percent increase in output could be achieved using the same input combination.

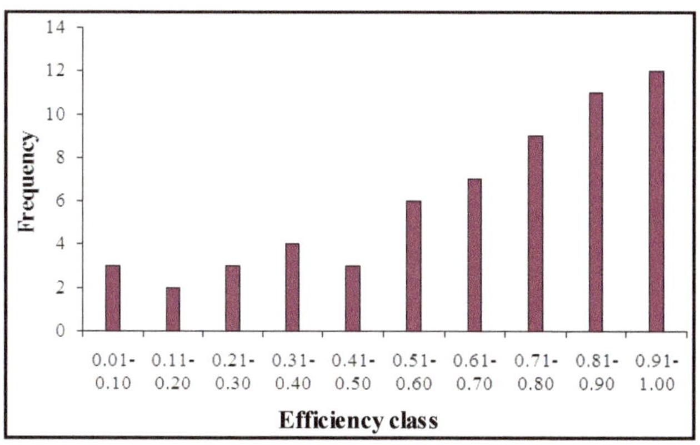

Figure 1. *Farm specific technical efficiency*

4. Conclusion

This study identified the factors affecting the use of resources by eggplant farmers. Specifically, it estimated the determinants of technical efficiency among small scale eggplant producers. The mean efficiency index was 0.71 suggesting that egg plant output could be increased by producers using the same input mix. Findings also reveal that the size of farmland; hired labour and planting materials were the major efficiency determinants. Results show that all the sampled eggplant farms were operating below the frontier threshold. Result suggests the need to formulate policies aimed at providing credit facilities to egg plant farmers. Roads leading to farms should also be more accessible to allow for ready and easy disposal of farm products.

References

[1] Aigner, D. J., Lovell, C. A. K. & Schmidt, P. (1977). Formulation and estimation of stochastic frontier production function study. J. Agric. Econ., 41:61-74.

[2] Ali, M. (1996). Quantifying the socio-economic determinants of sustainable crop production: An application of wheat cultivation in the Tarui of Nepal. Agric Econs., 14:45-60.

[3] Apezteguia, B. I. & Garate, M. P. (1997). Technical efficiency in the Spanish agro and food industry. Agric. Econ., 16:185-92.

[4] Aye, G. C. & Mungatana, E. D. (2010). Technical efficiency of traditional and hybrid maize farmers in Nigeria: Comparison of alternative approaches. African Journal of Agricultural Research, 5(21):2909-2917.

[5] Bagi, F. S. &Huang, C. I. (1983). Estimating production technical efficiency for individual farms in Tenesse. Can. J. Agric. Econ., 31:249-56.

[6] Bagi, F. S. (1984). Stochastic frontier production function and farm-level technical efficiency in West Tenessee. J. Agric. Econ., 6:48-55.

[7] Banwat, M. E., Lar, L. A., Daboer, J., Audu, S. and Lassa, S. (2012). Knowledge and intake of fruit and vegetables consumption among adults in an urban community in North Central Nigeria. The Nigerian Health Journal, 12(1):12-15.

[8] Coelli, J. T. & Battese, G. (1996). Identification of factors that influence the technical inefficiency of Indian farmers. Australian Journal of Agricultural Economics, 40(2):103-128.

[9] Etim, N. A. & Ofem, B. (2005). Urban farming and household waste management in Uyo urban: Implications for environmental harmony. Nigerian Journal of Agriculture, Food and Environment, 2(2):74-58.

[10] Etim, N. A. & Okon, S. (2013). Sources of technical efficiency among subsistence maize farmers in Uyo, Nigeria. Journal of Agricultural and Food Sciences, 1(4):48-53.

[11] Etim, N. A. & Udoh, E. J. (2014). Identifying Sources of Efficiency among Resource Poor Indigenous Vegetable Farmers in Uyo, Nigeria. International Journal of Food and Agricultural Economics, 2(1): 33-39.

[12] Etim, N. A., Azeez, A. A. & Asa, U. A. (2006). Determinants of urban and peri-urban farming in Akwa Ibom State. Global Journal of Agricultural Sciences, 5(1):13-16.

[13] Etim, N. A., Thompson, D. & Onyenweaku, C. E. (2013). Measuring efficiency of yam (*Dioscorea spp*) production among resource poor farmers in rural Nigeria. Journal of Agricultural and Food Sciences, 1(3):42-47.

[14] Etim, N. A., Udoh, E. J. & Awoyemi, T. T. (2005). Measuring technical efficiency of urban farms in Uyo metropolis. Global Journal of Agricultural Sciences, 4(1):91-95.

[15] Farrel, M. (1957). The measurement of productive efficiency. J. Royal Stat. Soc. ACXX Part, 3:253-90.

[16] Forsund, F. R., Lovell, C. A. K. & Schmidt, P. (1980). A survey of frontier production functions and their relationships to efficiency measurement. J. Econ., 134:54-25.

[17] IFA (International Fertilizer Industry Association). (1992). International World Fertilizer Use Manual Limburgerhof, Germany.

[18] Jondrow, J. C. A., Lovell, L. S. & Schmidt, P. (1982). On the estimation of technical inefficiency in the stochastic frontier production function model. J. Econ., 19:233-238.

[19] Kalaitzandonakes, N. G., Xiang, W. S. & Jianahun, M. (1992). Relationship between technical efficiency and farm size revisited. Can. J. Agric. Econ., 40:427-442.

[20] Lundvall, K. & Battese, G. E. (2000). Farm size, age and efficiency: Evidence from Kenyan manufacturing firms. Journal of Development Studies, 35(3):146-163.

[21] Meeusen, W. & Van den Broek, J. (1977). Efficiency estimation from Cobb-Douglas production function with composed error. Int. Eco. Rev., 18:435-44.

[22] Msuya, E. E., Hisano, S. & Nariu, T. (2008). Explaining productivity variation among small holder maize farmers in Tanzania.

[23] Muhammed, I. J. (2009). Efficiency analysis of cotton-wheat and rice: Wheat systems in Punjab, Pakistan. Unpublished Ph.D. Thesis, University of Agriculture, Faisalabad.

[24] Parikh, A., Ali, F. & Shah, M. K. (1995). Measurement of economic efficiency in agriculture. American Journal of Agricultural Economics, 77:675-85.

[25] Philip, D., Nkonya, E., Pender, J. & Oni, O. A. (2009). Constraints to increasing agricultural productivity in Nigeria: A review. Nigeria Strategy support Program. Paper No. NSSP 006. International Food Policy Research Institute, Washington DC.

[26] Quebedeaux, B. & Bliss, F. A. (1988). Horticulture and Human health. Contributions of fruits and vegetables. Proc. 1st Intl. Symp. Hort and Human Health. Prentice Hall, Englewood, NJ.

[27] Quebedeaux, B. & Eisa, H. M. (1990). Horticulture and Human health. Contributions of fruits and vegetables. Proc. 2nd Intl. Symp. Hort. And Human Health. Hortscience 25:1473-1532.

[28] Ruel, M. T., Nicholas M. & Lisa, S. (2011). Patterns of fruit and vegetable consumption in sub-Saharan Africa. FAO/WHO Workshop on Fruits and Vegetables for Health, 1st – 3rd September, 2004, Japan (Online). Available at www.who.mt../en/. Accessed 21/5/2011.

[29] Schmidt, P. (1976). On the statistical estimation of parametric frontier production functions. Rev. Econ. Stat., 58:238-239.

[30] Sharma, K. R., Leung, P. S. & Zaleski, H. M. (1999). Technicalities allocative and economic efficiencies in swine production in Hawaii. A comparison of parametric and non parametric approaches. Agric. Econ., 20:23-35.

[31] Udoh, E. J. & Akintola, J. O. (2001a). Measurement of the technical efficiency of crop farms in the South Eastern region of Nigeria. Nigerian Journal of Economics and Social Studies, 43(1):93-104.

[32] Udoh, E. J. & Akintola, J. O. (2001b). land management and resource-use efficiency among farmers in South-Eastern Nigeria. Elshaddai Global Limited, Ibadan, Nigeria.

[33] Udoh, E. J. & Etim, N. A. (2006). Cocoyam Farms in Akwa Ibom State, Nigeria. A Stochastic Production Frontier Approach. Journal of Sustainable Development in Agriculture and Environment, 2:41-48.

[34] Udoh, E. J. (2000). Land Management and Resource – use efficiency among farmers in South-Eastern Nigeria. Unpublished Ph.D. Dissertation. Department of Agricultural Economics, University of Ibadan. Nigeria.

[35] Udoh, E. J. (2005). Technical inefficiency in Vegetable Farms of Humid Region: An Analysis of Dry Season Farming by Urban Women in South-south Zone, Nigeria. Journal of Agriculture & Social Science, 1(2): 80-85.

[36] Wargovich, M. J. (2000). Anticancer properties of fruits and vegetables. Hort Science, 35:573-575.

[37] Yao, S. & Liu, Z. (1998). Determinant of grain production and technical efficiency in China. Journal of Agricultural Economics, 49:171-84.

Effect of T&V innovation on income and farmers performance in Edo State Nigeria

Abiola Matthew Oladipupo[1], Omoregbee Friday Egbenayabuwa[2], Caroline Sede[2]

[1]Department of Agricultural Economics and Extension, Faculty of Agriculture, Federal University of Oye Ekiti State, Nigeria
[2]Department of Agricultural Economics and Extension Services, University of Benin, Edo State, Nigeria

Email address:
maidenjournal@yahoo.com(A. M. Oladipupo) , fridayomoregbee@yahoo.com (O. F. Egbenayabuwa)

Abstract: The study examined the effect of training and visit innovation on the income and famer performance in two out of three agro ecological zones of Edo State, Nigeria. This study focused on contact crop farmers drawn from two out of three agro-ecological zones of Edo State, namely Edo central and Edo south. A multi-stage random sampling method was used both for economic reasons and especially because the sampling units occur in strata such as zones, blocks and cells. Data were collected from randomly sampled respondents from two senatorial districts in Edo State. Edo state is divided into three agro-ecological zones- Edo North, Edo Central and Edo South, out of which Edo Central and Edo south were randomly selected. Two blocks each were selected randomly from a total of five and seven blocks in Edo central and Edo south zones respectively. These blocks were Igueben and Esan North East in Edo central and Oredo and Egor in Edo south senatorial zone. Each block consists of eight cells. A random selection of four cells was made from each block and total of sixteen cells were chosen and eight from each zone and ten farmers were randomly selected from each of the 16 cells, giving a total of one hundred and sixty farmers/respondents for the study but only a hundred and fifty six copies of a questionnaire were found useful for the study. Data collected were subjected to descriptive statistics and budgetary tools. The outcome of the study indicated that training and visit innovation had increasing effect on farm size and income of farmers in the study areas.

Keywords: Income, Training, Visits, Innovation, Edo

1. Introduction

1.1. The T&V Extension System and Agricultural Food Production

Daniel Benor's Training and Visit (T&V) extension system was vigorously introduced to Nigeria in 1986 by the World Bank and was subsequently adopted in all the 36 states and the federal capital territory. It is regarded as a potent weapon for heralding a true agrarian revolution. (Ilevbaoje, 2004).Agricultural extension system is the primary delivery system for information to farmers (Munyua *et al.* 2002). Improving agricultural productivity may not be achieved without relevant and reliable agricultural information. The objective of any extension system is to be effective in communicating information that helps people in decision-making. The T&V extension system modified to suit each country's agricultural goals is a dominant extension approach used in reaching farmers with information on agricultural production.

Agricultural extension has the potential to stimulate agricultural development and is often used as a tool for implementing government policy. It is recognized as the main link between the farmers and research and crucial in communicating improved practices needed in agricultural development (Rivera and Carey, 1997 and Van den Ban and Hawkins, 1996). As indicated by the World Bank (1997), supply of appropriate technology is essential if extension investments are to be worthwhile and especially for public sector extension.

According to Sappho (1993) research and extension should be closely linked in order to contribute appreciable and sustainable increase in food production. The goal of agricultural extension is to facilitate farmers' acceptance of innovative practices from research which should lead to increased output, productivity and incomes. This agrees with FAO (1997) which states that improved agricultural extension management system is recognized as a central mechanism to achieving increased food productivity through technology transfer. The T & V is one of the foreign innovations introduced to improve famer's performance in Nigeria. What is the effect of this innovation on famer's income and performance? .The specific objectives of the study include:

A To examine the socio-economic characteristics of famers in Edo State

B Determine the effect of T&V innovation on farmer's income.

C Identified the perceived benefits derived by famers from T&V innovation in Edo State, Nigeria

Nigeria is one of the 60 countries now using the T&V extension system. Like many developing countries, Nigeria adopted the T&V extension system in 1986 because of its promise of improving extension management which is the key to increased agricultural production and national development. It was formally introduced through pilot testing in Oyo, Kaduna, Imo and Plateau states. Guided by experience from the pilot ADPs, implementation assistance terms (IATs) were mounted by Federal Agricultural Coordination Unit (now Project Coordination Unit (PCU)) to rapidly and systematically replicate the T&V extension model in all the state ADPs (Ilevbaoje, 2004). There is a general consensus that farm level utilization of agricultural research institutes' results (developed technologies) has not made the expected impact on agricultural production in Nigeria. Research Institutes' results differ markedly from actual farm yields indicative of development-adoptive (use) gap. For instance, in the Nigerian Institute for Oil palm Research (NIFOR), the yield per hectare on experimental plots range from 15 to 18 tonnes per annum, while yield on farmers' field is 8-10 tonnes per annum (Omokhudu, 1998). The difference between experimental station yield, the potential farm yields and the actual farm yields has been attributed to biological and socio-economic constraints as well as the communication channels through which these technologies get to the (end users) farmers (Akinbode, 1976).

Many valuable results have been obtained from most of the agricultural research institutes, but these agricultural practices do not get to the farmers.

This situation has created a wide gap between research and the utilization of the research results. To enable effective service, the ideal ratio of extension agent to farm family should be 1:250 (Benor and Baxter, 1984).

This clearly suggests that one of the strategies of developing the agricultural sector in Nigeria is to ensure that these developed technologies get to the farmers through the Training and Visit system. But unfortunately, because the farmers are scattered and isolated, extension agents are very few to be able to carry out this effectively by individual contact, a communication channel that has be regarded to be the most effective in disseminating technologies to the target audience (Laogun 1985).

Table 2.1 below shows the extension agent/ farm family ratio as at the year 2000 in 27 States in Nigeria. Majority (96%) of the states have more than 1000 farm families to one extension agent. The state with the highest ratio is Lagos with 1:5800, while the state with the least ratio is Gombe with 1:826. None of the 27 states achieved the recommended extension agent\farm family ratio of 1:250.

Table 1. *Extension agent/farm family ratio in Agricultural Development Programmes in 27 states in Nigeria*

S/N	State ADP	Extension Agent- Farm Family Ratio
1	Abia	1:1000
2	Adamawa	1:1452
3	Akwa-IBOM	1:2320
4	Anambra	1:3084
5	Bauchi	1:1238
6	Bayelsa	1:2893
7	Benue	1:1460
8	Borno	1:1268
9	Ebonyi	1:5320
10	Edo	1:1500
11	Ekiti	1:2800
12	Enugu	1:4041
13	Gombe	1:826
14	Imo	1:2861
15	Jigawa	1:1075
16	Kaduna	1:3500
17	Kano	1:2267
18	Katsina	1:1750
19	Kebbi	1:1000
20	Kwara	1:1000
21	Lagos	1:5800
22	Nasarawa	1:1381
23	Niger	1:5000
24	Ogun	1:1857
25	Ondo	1:1900
26	Taraba	1:1065
27	Yobe	1:1120

Source: Project Coordinating Unit, Abuja (2000)

The nation-wide adoption of T&V extension approach is perhaps the most outstanding development in agricultural extension in Nigeria over the past two decades or so. According to Ilevbaoje (2004), some of the problems which the T&V extension system attempt to solve are:

- to improve the organization of extension by introducing a singular direct line of technical support and administrative control,
- to change the multi-purpose nature of many extension workers to a clearly defined, single-purpose role involving only education and communication activities,
- to improve coverage by limiting the number of farm families or households one extension worker is expected to visit
- to improve mobility by providing appropriate transport so that each worker can regularly visit his/her contact farmers

- to improve each extension worker's technical skill and knowledge about improved agricultural technology by providing in-service training sessions
- to improve extension's ties with agricultural research through an increased number of subject matter specialists, who are expected to maintain regular contact with their research counterparts and to secure a continuous flow of information that transmits technology to farmers and farmers' problems back to research personnel
- to improve the status of extension personnel by giving them a relatively clear-cut extension job with reasonable expectations that they can successfully carry out; this will increase their level of respect in the community and begin to build their self-confidence, and
- to reduce the duplication of services that occurs when extension is fragmented among different ministries.

The T&V System of Extension management has been adopted in several developing countries to revitalize their agricultural extension services. Studies of agricultural extension services in these countries reveal various experiences. Ada-Okungbowa and Ojukoko (2003) found significant differences in farm size and income of farmers at both 95% and 99% confidence limit of adopters of improved cassava varieties in Delta states as against that of non-adopters.

2. Methodology

2.1. Area of Study

The study was carried out in Edo State. Edo state has an area of about 19,794 Kilometers and a provisional population of 2,159,848. It lies roughly between longitude 050 04'E and 060 43'E and latitudes 050 44'N and 070 43'N. It is bounded on the North by Kogi State, on the West by Ondo State, on the South by Delta State and on the East by Kogi and Anambra States (Iloeje, 2002).

Edo State has a tropical climate characterized by two distinct seasons, the wet and dry seasons. The wet season occurs between April and October with a break in August and an average rainfall ranging from 150cm in the extreme north of the state to 250cm in the south. The dry season lasts from November to April while a cold harmmattan is between December and January. The temperature averages about 25^{0}C in the rainy reason and 28°C in the dry season (Iloeje, 2002).

The climate is humid tropical in Edo South agro-ecological zone which covers the Benin City area. It is mainly characterized by heavy rains almost all year round and thick rain forest. Edo North agro-ecological zone covers the Auchi area of the State which has a sub-humid climate and is characterized by light rainfall and semi-savannah vegetation. Edo Central agro-ecological zone covers the Ekpoma area of Edo State which lies between the south and the north.

More than 70% of the population of Edo State make their living from agriculture. They cultivate cash crops such as rubber (*Hevea brasiliensis*), oil palm (*Elaeis guineensis*) and cocoa (*Theobroma cacao*), food crops such as yam (*Dioscorea specie*), cashew (*Anacardium occi*dentale),

Edo State Map Showing the three Senatorial Districts

Figure 1. *Map of Edo State showing the three agro-ecological zones.*

Cassava (*Manihot esculenta*), rice (*Oryza species*), plantain (*Plantago media*), maize (*Zea mays*), sugar cane (*Saccharum officinarum*), groundnuts (*Arachis hypogea*), soya-beans (*Glycine max*) and coco yams (*Xanthosoma sagittifolium*). Fruits like pine –apple (*Ananas Specie*), coconut (*Cocos nucifera*), Oranges (*Citrus species*), paw-paw (*Carica papaya*) and mango (*Mangifera indica*).

2.2. Study Population and Sample Size

This study focused on contact crop farmers drawn from two out of three agro-ecological zones of Edo State, namely Edo central and Edo south. The population of farmers in this area is about 1,007,929.

A multi-stage random sampling method was used both for economic reasons and especially because the sampling units occur in strata such as zones, blocks and cells. The sampling process proceeded as follows:

Zonal Level- Edo state is divided into three agro-ecological zones- Edo North, Edo Central and Edo South, out of which Edo Central and Edo south were randomly selected.

Block Level- Two blocks each were selected randomly from a total of 5 and 7 blocks in Edo central and Edo south zones respectively. These blocks were Igueben and Esan North East in Edo central and Oredo and Egor in Edo south senatorial zone.

Cell Level- Each block consists of 8 cells. A random selection of 4 cells was made from each block. A total of 16 cells were thus chosen, with 8 from each zone.

Farmers' level- 10 farmers were randomly selected from each of the 16 cells, giving a total of 160 farmers/respondents for the study. However, only 156 copies of respondents' interview schedule were useful for analysis.

Zones	Blocks	Cells
Edo Central	Igueben	Ekekhen/Idumugo
		Eguare/Oyomon
		Afuda/Idumuka
		Uhe/Idumogbo
	Esan North East	Uzeha
		Emendokhian
		Ivue/Obedu
		Isua/Arue
Edo South	Oredo	Ogbe-ebuya
		Iyeko-Ogba
		Uzebu
		Esigie
	Egor	Uselu 1
		Uselu 2
		Ohoro
		Ugbowo

This study made use of both primary and secondary sources to obtain data. An interview schedule was employed in getting primary data. The secondary data were sourced from textbooks, journals, previous articles, State Agricultural Development Programme, the internet and related literature on the topic.

A structured interview schedule was employed in gathering primary information from the respondents. The instrument utilized a combination of close-ended and open-ended questions but with emphasis on the former to facilitate the analysis of the results. The structured interview guide was divided into four sections. The first section was on the socio-economic characteristics of farmers while the second was on the improved practices disseminated and adopted by farmers. The third section attempted to determine the impact of adopted practices on farmers' food productivity. The judgments and opinions of agricultural extension workers who were regarded as experts were used to validate the instrument. The experts selected for this task included

Lecturers in Department of Agricultural Economics and Extension Services.

Experienced extension agents from ADP.

The experts independently judged the instrument based on the objectives of the study and made correction. The reliability of the instrument was obtained from pre-test results collected from 20 farmers outside the sample group, specifically from Jattu and Agenebode in Edo North Agro-ecological Zone. Using the split-half method, the researcher obtained a reliability coefficient (r) of 0.84. This result certified the instrument good to be used for the research.

Data collected was analyzed using both descriptive statistics and budgetary tools.

3. Results

3.1. Socio Economic Characteristics of the Respondents

3.1.1. Sex

Farmers' personal and socio-economic characteristics are presented in Tables 2 and 3. More than half (57.7%) of the respondents were female and the remaining 42% were male.

3.1.2. Age

The age distribution as shown in Table 2 indicates that majority (60.2%) of the respondents were above 41 years of age.

3.1.3. Marital Status

Table 2 shows that a high proportion (62.2%) of the respondents was married while 9.6% were single. The divorce rate seemed low at 7.1% but the proportion of respondents that was widowed was significantly high (21.2%).

3.1.4. Educational Qualification

The level of education of the respondents is shown in Table 2. Majority of the respondents, (56.4%) had only primary education and about 23.7% had no formal education. About 13.5% of the respondents had secondary education while only about 6.4% had tertiary education. This result shows that a majority of the respondents have a low level of education

3.1.5. Number of Dependents

As shown in Table 2, a significantly high proportion of respondents (69%) had more than 6 dependents; while only 7.7% of the respondents had 2 dependents.

3.1.6. Type of farmer

As shown in table 3, majority of the respondents (73.7%) were subsistence farmers while only 26.6% were commercial farmers.

3.1.7. Experience in Farming (in Years)

Only 20% of the respondents had less than 5 years' experience in farming. The rest had more than 5 years' experience, with 32.4% having more than 15 years' experience.

3.1.8. Farm Size (Ha)

About half of the population (49.4%) owned farms less than one hectare large, 25% owned farms 2-3 hectares large, while those that owned farms 4-5 hectares large were 19.2%. Only 6.4% of the respondents owned farms more than 5 hectares large.

3.1.9. Type of Crops Planted

Majority of the farmers (72.4%) planted food crops while only 27.6% planted export crops.

3.1.10. Holding Type

Sixty-five percent of the farmers claimed to own the land on which they are farming. However, 21.7% and 13.3% rented and leased the land they were using for farming.

3.1.11. Type of Labor

The farmers and their families did most of the work in the farms. Only 21.7% of the labor used by the farmers was hired.

4. Discussion

Farmers' personal and socio-economic characteristics are presented in Tables 2 and 3. More than half (57.7%) of the respondents were female. Sixty percent of the respondents were above 41 years old, while only 6.4% were less than 21 years and 33.3% were between 21-40 years. This result shows the aged nature of most respondents and the need to encourage most young people to take up agriculture as a profession.

Sixty-five percent of the respondents claim to own their farmlands, while 21.7% and 13.3% rented and leased land for farming. Majority (72.4%) of the respondents planted food crops while only 27.6% dealt with export crops. About half (49.4%) of the respondents, owned farms less than one hectare large. The real big timers with farms larger than five hectares constitute only 6.4% of the farmers. This result shows the subsistence nature of agriculture in Edo State. This is further corroborated by the number of dependents, which shows that most of the farmers (69.2%) have more than six dependents. Most of the work in the farms is done by the farmers themselves and their families. Hired labour constitutes only 21.7%. About half (56.4%) of the respondents had primary education, while 13.5% had secondary education. However, 23.7% had no formal education and only 6.4% had tertiary education. This result shows that the educational level of the farmers is quite low. This finding is in agreement with that of Fabiyi and Doma (2001) who reported a low level of education among rural women in Dass Local Government Area. Tologbonse and Osanyintade (2001) also reported that the low level of education among the farmers in the Federal Capital Territory, Abuja affected their level of adoption of cowpea crop protection recommendations.

Education places farmers in a position to be receptive to innovative ideas, as it is known to create a positive mental attitude towards adoption of modern farming techniques (1976). Egbugara (1993) noted that the level of a farmers' education is related to the level of innovations he or she adopts. This is also in line with Adams (1982) who proffered that the main way to improve farm efficiency and to increase agricultural production is to educate farmers. Ogunbameru and Pandey (1992) working in Adamawa and Taraba States saw education as a tool for realizing fullest potentials for contribution to agricultural development.

Table 2. *Social characteristics of Respondents (n=156)*

Variable	Frequency	Percentage
Sex		
Male	66	42.3
Female	90	57.7
Age		
Less than 21 years	10	6.4
31-40 years	23	14.7
41-50 years	37	23.7
51-60 years	58	37.2
Above 60 years	28	17.9
Marital status		
Married	97	62.2
Single	15	9.6
Divorced	11	7.1
Widowed	33	21.1
Educational Qualification		
No formal education	37	23.7
Primary Education	88	56.4
Secondary Education	21	13.5
Tertiary Education	10	6.4
Number of dependents		
2	12	7.7
3-5	36	23.1
6-8	64	41.0
Above 8	44	28.2

Table 3. Economic characteristics of respondents (n=156)

Variable	Frequency	Percentage
Type of farmer		
Subsistence farmer	115	73.7
Commercial farmer	41	26.3
Experience in farming (in years)		
Less than 5 years	31	20.0
6-10	27	17.3
11-15	47	30.3
More than 15 years	51	32.4
Farm size (Ha)		
Less than hectare	77	49.4
2-3	39	25.0
4-5	30	19.2
Above 5 hectares	10	6.4
Type of crops planted		
Food crops	113	72.4
Export crops	43	27.6
Holding type		
Rented	34	21.7
Leased	21	13.3
Owned	101	65.0
Type of labor		
Self-only	60	38.3
Family labor	62	40.0
Hired labor	34	21.7

Source: Field survey data (2%)

4.1. Effect of T&V on Farmers Farm Size

As shown in table 4, before the commencement of T&V, majority (63.5%) of the respondents owned farms less than one hectare in size. Only 2.6% of the respondents own owned farms larger than five hectares in size. The rest of the respondents owned farms with sizes between one and five hectares.

Table 4 shows that after the commencement of T&V the percentage of respondents owning farms less than one hectare dropped from 63.5% to 49.4%. Consequently, the percentage of respondents owning farms 2-3 hectares large increased from18.6% - 25%, those owning farms 4-5 hectares large from 15.4% - 19.2% and those owning farms greater than five hectares increased from 2.6% -6.4%.

Table 4. Farm sizes of respondents before and after the commencement of T&V

Farm size (ha)	Before		After	
	F	%	F	%
<1	99	63.5	77	49.4
2-3	29	18.8	39	25.0
4-5	24	15.4	30	19.2
>5	4	2.6	10	6.4

4.2. Effect of T&V on the Incomes of Farmers in The study Areas

As shown in table 5, majority of the respondents (70.5%) earn less than 5000 Naira monthly from their farms, while only 29.5% earn more than 5000 Naira monthly from their farms.

After the commencement of T&V, as shown in table 5, the percentage of respondents who earn less than 5000 Naira monthly dropped from 70.5%-54.5%. On the other hand, the percentage of farmers earning more than 5000 Naira monthly from their farms increased from 29.5%-45.5%. The results show that there was a general increase in income of the farmers after the commencement of T&V. Logically, usage of improved varieties coupled with best practices in the farms led to increased yield which brought more money to the farmers.

Table 5. Monthly Income of farmers before and after the commencement of T&V (Naira)

Monthly Income	Before		After	
	F	%	F	%
>1000:00	42	26.9	15	9.6
1001-5000	68	43.6	70	44.9
5001-9000	39	25.0	47	30.1
>9000	7	4.5	24	15.4

4.3. Respondents' Perceived Benefits from Training and Visit Extension System

The result in Table 6 shows the respondents' perceived benefits from Training and Visit. The respondent's perceived increased yield at harvest as the first benefit and this was followed by increased income. Other benefits include improved household food security and nutrition, proper use of improved varieties, purchase of articles of convenience, increased access to research information, purchase of more farming equipment, improved housing conditions, improved educational status, increased farm sizes, reduced farm losses, enhanced social status in community, access to credit facilities and more labor employed

The distribution of respondents by their perception of the impact of adopted farming practices on farm productivity before and after adoption of improved practices is shown in table 6. Majority of the farmers had positive perception that the adopted farming practices had increased their income (84.6%), increased their yield at harvest (84.0%), improved their feeding habits/household food security (83.3%), and they were able to make use of improved varieties (75.2%). Many of the respondents had positive perception of increased ability to purchase articles of convenience such as cars and cooking gas (70.1%) due to their increased income. They also felt positively that their social status in community had been enhanced (73.1%). Most of the farmers (82.6%) had positive perception that they experienced reduced farm losses as a result their adoption of improved farming practices. This finding agrees with Ogunwale et al. (2006) who reported that contact with extension agents and the use of various recommendations had positive impact on the standard of living of farmers as a result of increased income. Majority of the respondents (78.4%); however, perceived that access to credit facilities had not been improved, which was evident in the negative perception indicated by farmers. This development is worrisome and calls for concern by the authorities concerned.

Table 6. *Respondents' perceived benefits from T&V Extension System*

T&V strategy	Mean	Standard deviation	Rank
Increased yield at harvest	2.17	1.33	1
Increased income	2.55	1.15	2
Improved household food Security and nutrition	2.30	0.95	3
Purchase of farming equipment	2.20	0.97	4
Increased access to research Information	2.19	0.93	5
Purchase of articles of Convenience	2.71	0.89	6
Use of improved varieties	2.16	0.85	7
Increased farm sizes	2.15	0.98	8
Improved housing condition	2.12	0.99	9
Improved educational status	2.01	1.00	10
Reduced farm lose	2.01	1.00	11
Enhanced social status in Community	1.81	0.97	12
More labour employed	1.68	0.91	13
Access to credit facilitates	1.51	0.82	14

5. Conclusion

The study established the following findings from the study:

1. Women dominated the farming population in the study area and majority of the farmers were above the age of 41 years.
2. The level of education of the respondents was low with majority having only primary education while some had no formal education.
3. Most of the respondents were married with a household size of more than six.
4. Majority of the respondents were subsistence farmers.
5. The findings also showed that respondents generally increased their farm sizes after the commencement of T&V.
6. There was a general increase in the income of respondents after the commencement of T&V compared to before. Many of the farmers now earn monthly salaries above the 7500 Naira minimum wage set by the Government for all federal civil servants in Nigeria.

References

[1] Adams, M. E. (1982), Agricultural Extension in DevelopingCountries, Longman Group Ltd. Singapore, pp 108.

[2] Ada-Okungbowa, C.I. and Ojukoko, O. (2003) "Comparison of Adopters and Non-Adopters of Improved Cassava Varieties in Delta State", *J. of Agriculture, Forestry and Fisheries,* (3&4): 15-18.

[3] Adebayo, O.O. and Adeola, R.G. (2005), Socio-economic factors affecting Poultry farmers in Ejigbo Local Government Area of Osun State.*Journal of Human Ecology, 18:39-41.*

[4] Adekoya, A.E. and Tologbonse E.B. (2005), adoption and diffusion of Innovations. Agricultural Extension in Nigeria. Adedoyin S.F. (ed), pp 28-37.

[5] Agbamu, J. U. (2000). Agricultural Research- Extension Linkage Systems: An International Perspective. Agricultural Research & Extension (AgREN) Network Paper. No. 106.

[6] Ajayi, M.T. and Okafor, C. (2006): "Extension Agents' perception of Participatory Agricultural Extension Approaches adopted by Agricultural Development Programme (ADP) in Ondo State, Nigeria." *International Journal of Agricultural and Biological Sciences, 4(1): 20-25.*

[7] Akinbode, I. A. (1976) The Organisation and effectiveness of Extension Services in Nigeria. Agricultural Administration, Applied Science Publishers Ltd. England.

[8] Ali, M. M. and Halim, A. (1985). The effects of T&V system of Extension at Farmers level in Borga. In GTI publication, Bangladesh Agric University Press, No. 52.

[9] Apantaku, S. O., (2004). Analysis of Senior Secondary Agricultural Science Students' Attitude towards Agricultue as a Career. *Journal of Extension Systems Vol. 20(1), June.*

[10] Atala, T.K., T. Arokoyo and P.A. Omata, 1992. The impact of the Training and Visit (T&V) System of extension on adoption of farm innovations And farm output in Kaduna State, Nigeria. *The Nig. J. Agric. Ext., 7: 59-70.*

[11] Benor, .D and Baxter .M (1984) Training and Visit Extension. The World Bank, Washington D. C.

[12] Benor, D. and Harrison, J. Q (1977) Agricultural Extension: The Training and Visit System", the World Bank, Washington D. C.

[13] Blankenburg, P. V. and Hoeper B. (1985) Organisational changes and Persistence: Observations in the introduction of the T&V system of Extension in two Indian States. *Quarterly J. of International Agric. Vol. 24, No. 4, pp 306-321.*

[14] Brech, E. F. L. (1983), the Principles and Practices of Management, 3rd edition, London, Longman.

[15] Cernea, M.M. and Tapping, B. (1977) A system for Monitoring and Evaluating Agricultural Extension Projects. World Bank staff Working Paper No. 277.World Bank, Washington D. C.

[16] Dejene, A. (1985) The T&V Extension for small Farmers; Lessons in the Design of Extension Programmes for high cereal production Regions in Ethiopia, Development Discussion Papers No. 199, N. Y. University Press, pp 39.

[17] Edo State Agricultural Development Project (2001) Annual Report of Activities, pp 5-10.

[18] Edo State Agricultural Development Project/On Farm Adaptive Research (2002), Annual Report of Activities.

[19] Egbugara, C.A. (1993) Economic Status and Rural women's involvement in Agriculture in Imo State. Journal of Rural Extension and Development. Vol. 1, pp 87-93.

[20] Egharevba, P. N. (1996) Food- The sustainers of life. Inaugural Lecture Series 44, University of Benin, Nigeria. Pp 7.

[21] Ekumankama O.O. and Onuekwusi G.C. (2008). Comparative Analysis of the Constraints to Effective Extension Services in Abia and Akwa Ibom States. Pakistan J. of Soc. Sc. 5(1): 70-75.

[22] Fabiyi, E. F. and Doma U.D. (2001), Status of Rural women's education in Dass Local Government Area and its implication in Agricultural Research, T. A. Olowo (ed), Proceedings of the 7th Annual National Conference of AESON in ARTMI, Ilorin.

[23] Food and Agriculture Organization (FAO 1997)1. Improving Soil fertility in Africa. FAO soils Bulletin 14, Rome.

[24] Feder, G., Slade, R. H. and Sundaram, A. K. (1986). The T&V Extension System; an analysis of operation and effects. *Journal of Agric Admin, Vol. 12, No. 1, pp 33-59.*

[25] Flippo, E. W. (1965), Principles of Personnel Management. Prentice Hall Inc. Eaglewood Cliffs, New Jersey, U.S.A.

[26] Garforth, C. (1997) Supporting Sustainable Agriculture through Extension in Asia; New resource perspectives, O. D. L. London, No. 21, June, pp 1-5.

[27] Gautam M. (1999). Agricultural Extension, the Kenya Experience. World Bank Agricultural Extension in Kenya. Report No. 19523.

[28] Ilevbaoje, I. E. (2004), Training and Visit Extension System Flourishes in Nigeria. BeraterInnen News 1/2004.

[29] Iloeje, N.P (2002). A New Geography of Nigeria. New Revised Edition, Longman Nigeria Ltd. Pp 155.

[30] Isiaka B.T. (1998). Livestock rearing practices and problems in Lagos. In Oduguwa, O.O., Fanimo, A.O. and Osinwo, O.A., (Eds). Proceedings of the Silver Anniversary Conference, Nigerian Society for Animal Production. Gateway Hotel Abeokuta, pp 223-225.

[31] Jaja, J. S., Chukuigwe, E. C. and Ekine, D. I. (1998) Stimulating Sustainable Agricultural Development through youth mobilisation Schemes: The case of the School-to-land Programme in Rivers State, Nigeria. In Nwosu, A. C. and Mbanassor, J. A. (eds.); Sustainable Agricultural Investment in Nigeria. Proceedings of the 13th Annual Conference of Farm Management Association of Nigeria. Alphabet Nigeria Publishers, Oweri, pp 294-301.

[32] Jibowo, A.A. (2005): History of Agricultural Extension in Nigeria, in Adedoyin, S.F. (ed.) Agricultural extension in Nigeria, Ilorin Agricultural Society of Nigeria. Pp. 1-12.

[33] La-Anyane, S. A. (1976) Agricultural Research and Extension in National Development in West Africa. *Journal of the Association for Advancement of Agricultural Science in Africa. Vol. 3, No. 2,pp 7-12.*

[34] Laogun, E. A. (1985) Perception of Farmers' Training Needs: A Case of Isoya Rural Development Project in Oyo State. *The Nigerian Journal of Agricultural Extension. Vol. 3, Nos. 1&2, pp 12-17.*

[35] Lawal B.O., Torimiro D.O. and Makanjuola B.A. (2009), Impact of Agricultural Extension Practices on the Nigerian poultry farmers' Standard of living: A percept ional Analysis. Tropical and Subtropical Agro ecosystems, 10:465-473.

[36] Matata, P. Z., Ajayi O. C., Oduol P.A. and Aggrey Agumya (2008), Socio-Economic factors influencing adoption of improved fallow practices Among smallholder farmers in Western Tanzania. *International NGO Journal Vol. 3 (4), pp. 068-073.*

[37] Mehta, N. N. (1983), Developing Technical Recommendations and Research – Extension Linkages: Experiences in India. In Michael M. et al (eds.); Agricultural Extension by Training and Visit – The Asian Experience, U. N. D. P. and World Bank, Washington D. C.

[38] Mosher, A. T. (1976) Thinking about Rural Development, ADC Inc. Washington DC.

[39] Munyua, C. N., Adams P. F. And Thomson J. S. (2002) Designing Effective Linkages for Sustainable Agricultural Extension Information SystemAmong Developing Countries in Sub-Saharan Africa. Proceedings of The 18th Annual Conference, Durban, South Africa.

[40] Odebode S.O and Chia H.M. (2001) Poverty alleviation Strategies of rural Household in Mezam Division, Northwest Province, Cameroon.*Moore Journal of Agricultural Research. Vol. 2 No 2 pp 179-185.*

[41] Odeyinka, S. M., Torimiro, D.O., Oyedele, J.O. and Asaolu, V.O. (2007) Farmers' Awareness and Knowledge of Moringa oleifera In South-western Nigeria. A Percept ional Analysis. *Asian Journal of Plant Science. 6:320-325.*

[42] Ogunbameru, B. O. and Pandey, I. M. (1992), Nigeria Rural Women's Participation in Agriculture and decision-making: Focus on Adamawa and Taraba States, *Nigerian Journal Of Agricultural Extension. Vol. 7 pp 71 - 76.*

[43] Ogunfiditimi, T.O. and Ewuola, S.O. (1995): The Synthesis of Corporative Agricultural Extension Systems. Emmi Press Ltd. Ibadan, Nigeria.

[44] Ogunsumi, L. O., (2008).Analysis of extension activities on farmers' Productivity in Southwest, Nigeria. African Journal of Agricultural Research Vol. 3 (6), pp. 469-476.

[45] Ogunwale, A.B., Ayode, A.R. and Ayansina, S.O. (2006). Impact of Extension Service on farmers' production activities in Ogbomoso Agricultural Zone of Oyo State, Nigeria. *Journal of Agricultural Extension, 9:150-158.*

[46] Omogbemi, O. (1998): Private Sector Involvement in Extension: The Shell Petroleum Development Company of Nigeria Limited example, Proc. of the fourth Annual Conference of the Agricultural Extension Society of Nigeria, 17-19 June, 1998. Pp. 94-98.

[47] Omokhudu, C.A. (1998), Agencies disseminating agricultural Information To farmers in the Northern Agricultural Zone of Edo State, Nigeria, the Nigerian Rural Sociologist, Vol. 2 pp. 92-98.

[48] Omonona, B.T., Yusuf, S.A., Olatunji, O.B. 2004. Determinants of Adoption of White egg shelled layers by poultry farmers in Ogun State,Nigeria. http://www.cababstractsplus.org/abstracts/Abstracts.aspx

[49] Omoregbee, F. E., (2006). Evaluation of the effect of Extension activities Of the Edo State Agricultural Development Programme (ADP) on small scale farmers, Nigeria. (PhD Thesis), University of Benin, Benin City.

[50] Onazi, O. C (1973) Analysis of the Training Needs of Agricultural Extension Workers in the Northern States of Nigeria. *Journal of Rural Economics and Development, Vol. 19, No. 2, pp 133-140.*

[51] Opeke, L. K (1976) Agricultural Research and Food Production for Africa. *Journal of the Association for Advancement of Agricultural Science in Africa. Vol. 3, No. 2, pp17-22.*

[52] Osuntogun, A. and Fabiyi, Y. L. (1980) Training for Agricultural and Rural Development in the ECOWAS. *J. of Agric Admin Vol. 8, No. 1, pp 3-13.*

[53] Patel, A. U. (1982) an Appraisal of Agricultural Extension as a Strategy forAgricultural Development in Nigeria. In T. A. Olowo (ed.); Workshop on effective transfer of technology to Nigerian Farmers, UNIFE/IITA, Ibadan

[54] Phiri D., Franzel S. Mafongoya P., Jere I., Katanga R. Phiri S. (2003). Who is? Using the new technology? The association of wealth status and gender with the planting of improved tree fallows in Eastern province, Zambia. Agroforestry System.

[55] Pickering, D. C (1983) Agricultural Extension: A tool for Rural Development. In Michael et al (eds.); Agricultural Extension by Training and Visit – The Asian Experience. U. N. D.P. and World Bank, Washington D. C.

[56] Project Coordinating Unit (2000) Minutes of meeting of Agricultural Development Projects, Umuahia, Abia State.

[57] Ray, A. K., Atteri, B. R., Sen, A. C. and Mathur, P. W. (1979) Quantitative and Qualitative Impact of T&V system in different groups of Farmers. A case study of Hooghly District, West Benghal. *Indian J. of Agric Econs. Vol. 36, No. 4, pp 11-12.*

[58] Rivera, W.M. & Cary, J. W (1997). Privatizing Agricultural Extension. In Swanson, B. E., Bentz, R. P., & Sofranko, A. J. (Eds). Improving Agricultural Extension: A Reference Manual. FAO, Rome, 1997. Pp 203-211.

[59] Rucks, C. (1981) what kind of Extension Worker do we need? In Swanson E. et al (eds.); Training for Agriculture and Rural Development. No. 21, Rome, FAO, pp 81-85.

[60] Sappho, H. (1993) Profile of Rural Poverty in Africa: The State of World Rural Poverty. N. Y. Harvard University Press, pp 39.

[61] Sonii D. (1992). Adoption dynamics incentives and constraints the case of Agroforestry. Paper presented at ICRAF DSO training Course.

[62] Tologbonse, E. B. and Osanyintade, M. (2001) Adoption of Cowpea Crop Protection Recommendations by farmers in the Federal Capital Territory, Abuja. In T. A. Olowo (ed.); Proceedings of the 7th \ Annual National Conference of AESON, in ARMTI, Ilorin.

[63] Torimiro, D.O., Alao, J.A. and Fapojuwo, O.E. (1999) Relationship between Socio-economic characteristics of farmers and Adoption of improved agricultural technologies in Ogun State, Nigeria. The Nigerian Rural Sociologist 3:44-51.

[64] Van den Ban, A. W. & Hawkins, H. S. (1996). Agricultural Extension. Second Edition. Cambridge, USA: Blackwell Science Ltd.

[65] Vanlauwe B, Giller K. E. (2006). Popular myths around soil fertility Management in sub- Saharan Africa. Agricultural Ecosystem and Environment, 116:34-46.

[66] Williams, S. K. T. (1978) Rural Development in Nigeria. Obafemi Awolowo University Press Ltd., Ile-Ife.

[67] World Bank (1997). Agricultural Extension and Research: Achievements and Problems in National Systems. World Development Report, 1997. Oxford University press. Oxford.

24

Analysis of causes of top-dying in Sundarbans natural forest resources

Awal, Mohd. Abdul

Environmental Scientist, Founder & Chief Advisor of Health & Pollution Research Farm, Address: Long Island City, New York, USA

Email address:
abdul_awal2004@yahoo.com

Abstract: The Sundarbans is the largest, biologically richest, and most extensive mangrove forest in the world, which is also a world heritage saite. *Heritierafomes* (Sundri) tree is the most important ecologically dominant and economically valuable tree species in the Sundarbans. However, a serious disease (top dying) of *H. fomes* in Sundarbans is affecting millions of the trees. An inventory by Chaffey *et al.,* (1985) revealed that there were about 45.2 million top dying trees in the Sundarbans of which 20 million had more than 50% of their crown affected by top dying. The loss of *H. fomes* will have a major impact on the Sundarbans mangrove ecosystem, as well as lead to economic losses.Despite various hypotheses as to the causes of this top dying, the underlying causes are still not well understood. The present work has explored some of the possible factors involved, focusing particularly on the relationship between the amount of top-dying in different places and the concentrations of a number of chemical elements present in the soil and water, in order to test the hypothesis that chemical pollution might be responsible. Other factors such as the pH, salinity and nutrient status were also assessed. Nine plots were selected for sampling of soil, water, and vegetation in order to categorize different areas in terms of their intensity of top-dying.63 soil samples and 9 water samples were tested, mainly by ICPMS, to investigate certain parameters of the soil and water, such as Sn, Exchangeable K, Soil pH, Pb, Zn, Ni, soil pH, CEC, soil nutrients, soil moisture content, and elemental concentrations of 32 other elements. Most of the elements studied had no significant correlation with the top dying of *Heritierafomes*.However, Sn, Exchangeable K, and soil pH were significantly related, and three elements, namely Pb, Zn, Sr, were also close to significance.Of the hypotheses previously put forward to explain top-dying, the present results do not support enhanced salinity as the cause. It is likely that several of the above-mentioned environmental factors interact with each other to induce top dying in Sundri.But, the present results have showed thatSn, Exchangeable K, Pb, Srand Zn could be directlylinked with top-dying of *Heritierafomes* (Sundri) in Sundarbans, probably particularly by weakening the vigour of the trees and allowing other factors such as pathological agents to attack the plants.

Keywords: Causal Factors, Heavy Metal Concentrations, *HeritieraFomes,* Mangrove, Sundarbans, Top-Dying

1.Introduction

Coastal lands cover 6% of the world's land surface (Tiner, 1984).Coastal and wetlands everywhere are under threat from agricultural intensification, pollution (Awal, 2007. 2009, 2014), major engineering schemes and urban development, (UN-ESCAP 1987; 1988). Mangroves in Asia including Bangladesh, India, and East Africa previously contained a much fuller range of species (Seidensticker, and Hai, 1983; Khan, 1997).

In the Southeast Asian region, species diversity of mangroves was previously much higher, where approximately two-thirds of all species and 70% of the major vegetation types with 15% of terrestrial species in the Bangladesh-India-Malayan realm have already been destroyed (Ellison, 1998, 2000). The Indo-Pacific region is known for its luxuriant mangroves. Some earlier estimates of the distribution of mangroves include Bangladesh, Australia, Burma, Brunei, Fiji, India, Indonesia, Kampuchea, Malaysia, Pakistan, Papua New Guinea, Peninsular Malaysia, Philippines, Sabah, Sarawak, Sri Lanka, Thailand, Vietnam (Government of India, 1990). The distribution of mangroves in the Indo-West Pacific bio-geographical region is outlined in Macnae (1968).But half of these countries mangroves had already been destroyed through various pollution and population pressure.

Sundarbans mangrove forests usually create thick, organically rich sediments as their substrata. Most of the substrata in the tropics except under deltaic environments consist of mangrove peat that mainly derives from mangrove roots (Fujimoto *et al.,* 1996; and Siddiqi*et al.,* 1993).

Sundarban mangrove forests have great belowground productivity and play a significant role in carbon concentration not only above ground but also below ground. Mangrove ecosystems therefore provide a unique and valuable range of resources and services. Thus a great number of birds, mammals, fishes and invertebrates utilize mangroves like Sundarbans during at least one part of their life cycle (Siddiqi*et al.,* 1993). Sundarban also provides a critical habitat for a diverse marine and terrestrial flora and fauna, and acts as a natural wind break, protecting low lands of the Khulna division and other parts of the country along with 3.5 million surrounding people, with their resources, from hurricanes, tidal surges and natural catastrophes. Also, several hundred thousand people of Bangladesh and India depend directly for their livelihood upon the Sundarbans (Field, 1995).

Sundarbans is the breeding ground for juvenile fish, crabs, shrimps, and mollusks, and is also a prime nesting site for hundreds of birds, animals, reptiles, insects, and species of micro organism. In a healthy coastal ecosystem the forest detritus, consisting of leaves and branches from the mangroves, provides nutrients for the marine environment, and is the basis of intricate food webs and food chains associated directly through detritus to the planktonic and epiphytic algal food chains. Sundarban attracts and shelters fishes under the shade of mangrove, protects all water resources and terrestrial resources from tidal storms, cyclones, strong waves and winds, and also provides food as leaf-litter to fishes.

Many of the countries in Asia and In the U.S.A, and Australia also, mangroves are being adversely affected by both nearby developments and pollution problems (Peters *et al.,* 1985).Coastal biological resources have been depleted by commercial fishing, including poison and blast fishing (Jones, 1992).In addition, pollution from shipping, in particular oil, and in some areas the discharge of toxic wastes, had adversely affected the marine environment (UNESCAP 1988). According the present result, the combined pollution of soil, in Sundarbans as a long term consequences, top-dying of *Heritierafomes* might have created in Sundarbans.

1.1. Location

TheSundarban mangrove forest is located mainly at the southern portion of the Gangetic delta bordering on the Bay of Bengal of Bangladesh (Blasco, 1977; Naskar 1999; and Bakshi, 1954; Thom, 1982, *Figure 1.0*). It occupies a flat deltaic swamp rarely exceeding 0.9 to 2.1 m above the sea level and most of the area is under water during the high spring tides of the monsoon and the major portion of land is a low plain with maximum 10 m height above mean sea level (FAO, 1994).

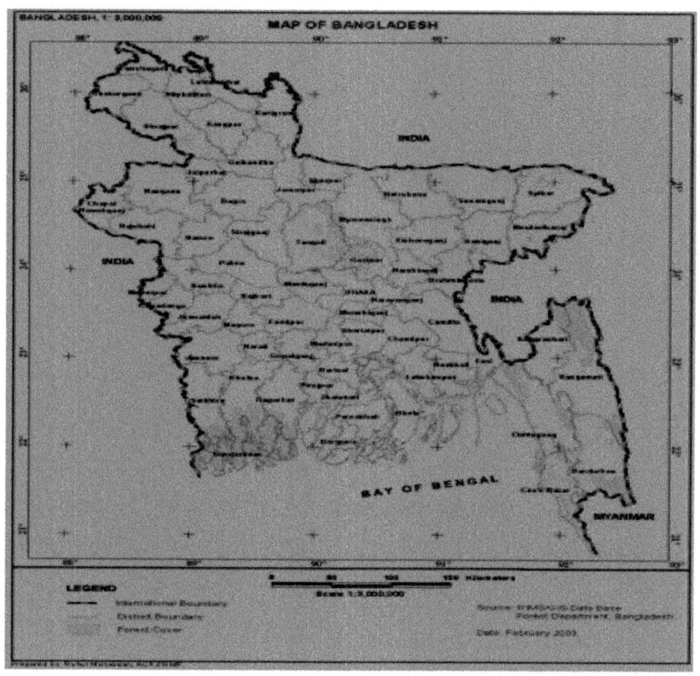

Figure 1.0. *Map showing the administrative districts of Bangladesh, including the location of the Sundarbans (the shaded area in the south-west of the country).*

Sundarbans is subdivided into 8 blocks (Rahman, 2003) and further subdivided into 55 compartments (Figure 2.0), varying in size from 4000-16,000 ha (Tamang, 1993), depending on the height classes & stocking conditions of the forest trees.

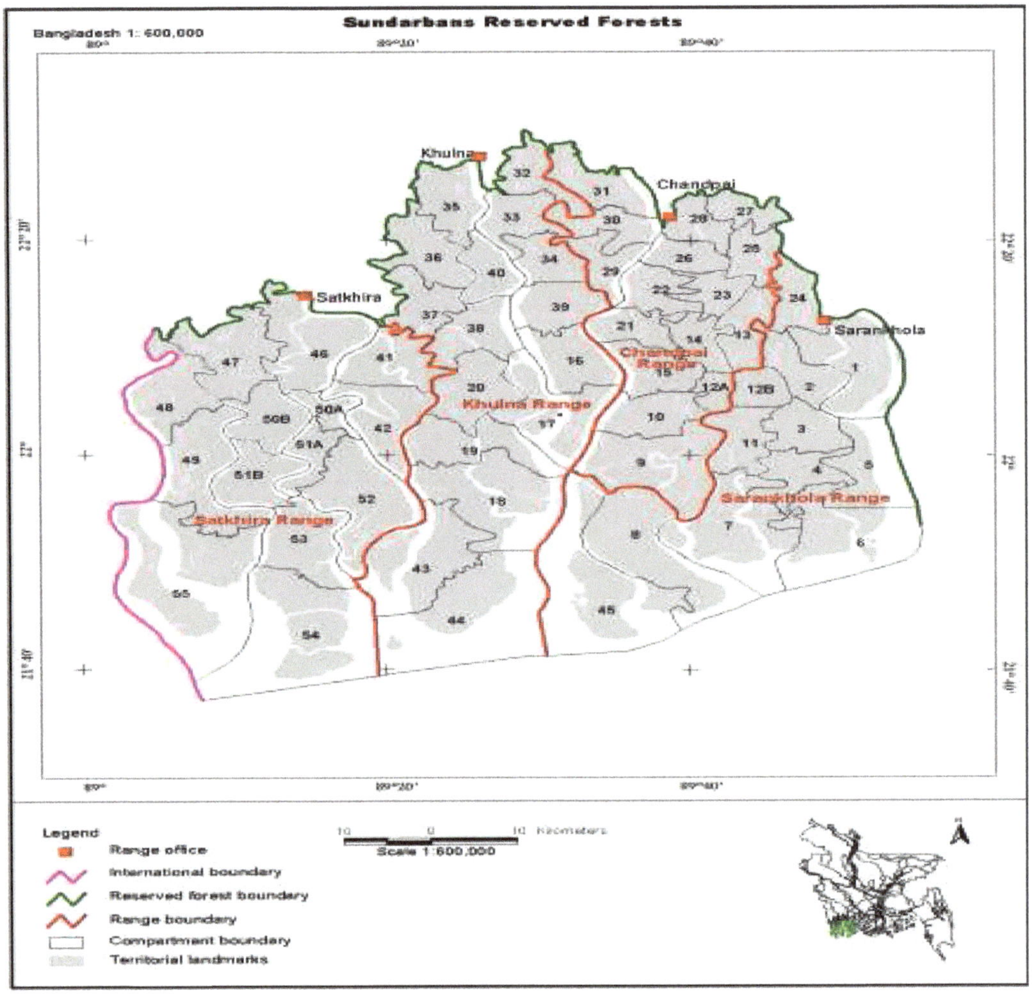

Figure 2.0. Shows compartment numbers and Ranges of Sundarbans Reserved Forests.

The forest lies just south of the topic between 21°38′-22°30′ North latitudes and 89°0′-90°0′East longitudes. Within the total areas of the Sundarbans, *Heritierafomes*constituted 63.8% among the total tree vegetation (Chaffey *et al.,* 1985).

1.2. Top Dying Disease and Vegetation Changes

Top-dying or dieback leads to the death of Sundri (*Heritierafomes)* trees and consequently to the decline of Sundri in the Sundarbans. This condition starts with a decline or death of foliage and twigs in the upper part of the tree, and gradually the symptoms extend downward. This gives a stag-headed condition to the affected trees (Gibson, 1975). This has led to increasing concern of the management of the forest. Sporadic or patchy occurrence of top dying of Sundri was mentioned early on by Troup (1921). Sundritop-dying was also reported by Curtis (1933) in the Sundarbans working plan, and this plan drew attention to the occurrence and severity of the problem (Chowdhury, 1983). The top dying of Sundri has been of particular concern since the 1970s and it has aggravated in recent years (Chowdhury, 1984; Gibson, 1975). Quantitative data regarding the extent of top dying was presented by Chaffey *et al.* (1985). The decline of Sundri at present due to top dying is in the range of 13-25%, but 13-17 % unsound (top dying) trees was reported in the Sundarbansin the 1920s (Ahmed, 1957). It is, therefore, suggested that the level of top dying (13-17%) of Sundri over 5 cm d.b.h. during 1926-28 was only slightly lower than the 13-25% found during the enumeration of 1983-84. Top dying and the accumulation of dead Sundri trees reached such a proportion that it was clearly visible even to casual visitors (Sattar, 1977; Chowdhury, 1984).

In April 1973, a seven member team from different Departments of the Government inspected eleven compartments of Sundarbans to study the problem of top dying of Sundri, and prepared a report in May of that year. The principal observations of the team were the following: Top dying was scattered and limited to individual trees or small groups of trees; (ii) The incidence of top-dying was more in the northern compartments and reducedconsiderably in the southern ones; (iii) Top dying was more on the banks of rivers than inside the islands; (vi) The old large–size trees showed dead tops while the younger trees had good and healthy crowns. *Loranthus* (a fungal) attack was very common to the trees with dead tops;

this was noticed towards the bottom of the dead trees. Canker was found to be very common irrespective of whether the trees were dead, dying or apparently healthy (Chowdhury, 1984). In the north central part of the Sundarbans, influenced by the Sibsa and Passur rivers, top dying of Sundri was severe in 1984. In that area seedling mortality is also very high. Representatives of young age classes are largely absent. The tree species Kakra, Passur, Bean and Keora also have shown signs of mortality in addition to Sundri in the same area (Chowdhury, 1984). These problems are still evident today.

2. Methodology

In this research work the various field and laboratory methods used in this study will be discussed.

2.1. Field Sampling Methods

The Sundarbans Reserved Forest is located at the southwest corner of the Ganges River Delta close to the Bay of Bengal, mainly on the sea-shore line, riverbanks, channels, and small creeks. The location of the Sundarbans within Bangladesh has been shown in Figure 1.0.

2.2. Site Selection and Location of the Study Area

Visiting all the possible regional areas before categorizing and selecting plots for sampling made general reconnaissance of possible sites.

It was decided to sample from the Chandpaiarea, which is the mostly human accessible and ecologically polluted area. Three compartments from this regional area (range), namely numbers 26, 28 and 31, were selected because they were believed to represent a range of severity of top dying disease, based on relevant maps, documents, literature, consultations with forest professionals, and surrounding peoples. The location of these compartments within the Chandpai area, and the location of this area in the wider Sundarbans is shown in *Figure 2.0.*

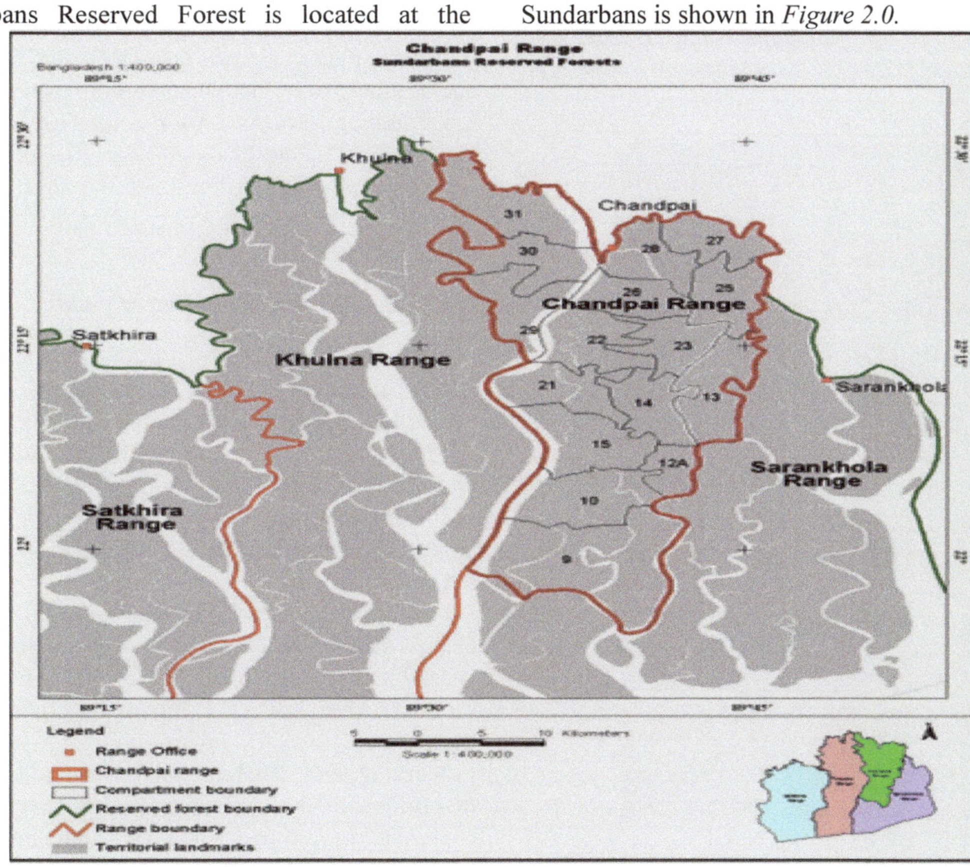

Figure 3.0. Map showing the location of the numbered compartments within the Chandpai area, and the position of this area relative to other parts of the Sundarbans in Bangladesh (darker shaded area).

Among the three compartments, compartment number 26 was selected as an area highly affected by top-dying, where most of the trees were affected severely (Figure 3.0). Compartment 26 had pronounced human activities, and also in places is undergoing rapid housing development involving extensive construction activities due to the presence nearby of the Range HQ office in Chandpai. Compartment number 28 was selected as a moderately affected area.This compartment has various human activities including boat making grounds, football-playing grounds, and cattle-grazing fields, all types of major soil erosion, a moderate amount of construction activities and the presence of communities of fishermen. Compartment number 31 was chosen as being relatively little affected by top-dying disease. Of the three chosen compartments, the nearest compartment to Mongla port is compartment 31 (Figure 3.0), with comparatively modest human activities, but which nonetheless involve clear-cutting of natural vegetation, replanting with other species rather than mangrove or other native species, all types of soil erosion, and construction activities present. Within each of the three compartments, detailed observations of the regeneration and sampling of soil and water took place within three20 m x 20 mplots, chosen to reflect a range of top dying intensities (High, medium and Low for that area). The sampling was conducted in a randomized block design, in that a plot was sited within a particular top-dying intensity block, but the precise location of that plot was randomized so as not to bias the detailed data collection. Thus in total nine plots were sampled, representing a range of top-dying intensities. Intensive field data collection was made among these nine selected plots. Observations were performed from observation towers during low and high tides, also traversing the forest floor and vegetation on foot, as well as using a speed boat, trawlers, country-boats, and a launch as required to gain access. All sampling was accompanied by Deputy Rangers, Foresters, and Forest guards armed with a rifle from the local office, Chandpai rang, Bangladesh Forest Department, to prevent a fatal attack by a Bengal tiger (*Pantheratigris*).

Figure 4.0. *Photograph showing the Sundarbans forest trees and understory vegetation. The adult trees behind the author are of Heritierafomes. The forest floor shows dead leaves from trees affected by top-dying.*

Fieldwork was performed in October, 2003 to March, 2004. Locations of sampling points were determined using a Global Positioning System with a precision of 5-10 m. For one typical plot, in compartment 31, the altitude was recorded as 4.4 m above sea level.

2.3. VegetationRecording Methods in the Field

Within each of the nine 20mx20m plots, each adult tree was assessed for three parameters. The diameter at 1 m height was recorded (in cm) by using a measuring tape or slide calipers depending on girth. The tree height to the top of the crown was determined mainly by ocular estimation but some heights were checked by using Clinometers at a set distance of 20 m to test the accuracy of such ocular estimations. Finally, the status of the tree in respect of the amount of top-dying was assessed by using a four point qualitative scale of intensity, namely; not affected, little affected, moderately affected or highly affected by top-dying. This was later expressed as a semi-quantitative or rank scale of 0 to 3 respectively, so that a median rank value could be calculated and used as an index of top-dying intensity in that plot.

After that, the total number of seedlings (individuals of the tree species <1 m tall), and saplings (young trees >1 m tall with a diameter of trunk of < 10 cm), were counted within the plots. Care was taken to ensure that trees, saplings and seedlings were not counted more than once or missed in the counting process. After recording, adult trees were marked with white chalk to segregate those marked trees from other trees, seedlings and saplings; red paints were applied to all seedlings and saplings as they were recorded (Figure 4.0).

2.4. Soil and Water Sampling Methods

As stated above, from the three selected compartments, a total of nine plots of 20m x 20m were selected. From each of these plots, seven soil samples were collected; one from the centre of the plot, four (one each) from all the corners, and two from the middle sides of the plot. Therefore a total of 63 soil samples were taken. Also nine water samples were collected from nearby rivers, creeks or channels, one from the area of each of the sampled plots.

Soil samples were collected from 0-30 cm soil depth by using a stainless steel spatula and steel cylinder (d=5.25 cm), and all soil samples were kept in sealed plastic bags. Water samples were collected directly in pre-cleaned plastic-containers.

Marking and labeling was performed with a detailed description of the selected sampling site on both the soil-containing plastic bags and water containers, and preserved in portable coolers until arrival at the laboratory at Dhaka University for initial chemical analysis. This field sampling method followed the W.H.O, U.K, and E.P.A systems of standard laboratory and field sampling principles, rules and regulations. Rainfall for the area during sample collection was not notably different from the respective monthly averages for the Sundarbans of recent years (Awal, 2007); there was no heavy intensity of rainfall within one month before sampling.

Figure 5.0. Photograph showing the Sundarbans forest, near compartment number 31, taken from one of the water channels. Note the evidence of top-dying in some of the trees.

Top dying is therefore seen to occur in the entire area of the Sundarbans (Figure 5.0), but concentrated in the central and northern part. Gibson (1975), on the basis of a visit to the Sundarbans during April 1975 noted that top dying of Sundri had become increasingly serious, particularly in the north-eastern region. He noticed top dying generally in older trees, but it was found at all ages beyond the sapling stage. He reported the disease as most severe in the trees growing on the banks of channels. The incidence of epicormic branching and formation of adventitious crowns in the older mature stands has been reported as high at Chandpai while it was hardly noticeable in the younger or immature stands at Kalabogi (Balmforth, 1985).

The incidence of top dying varies considerably within a block or a compartment (Figure 5.0). Severely affected areas may be in compartment centers as in the case of the compartments 23, 26, and 28 or may be quite inland e.g. inland to the north of Kotka in compartment 6 or along the bank of rivers, e.g. on the eastern side of the Passure river in compartment 45 of Kalabogi, along the Sibsa river, and east of Kalabogi along the Hatdura canal in compartment 33 (Chaffey *et al.,* 1985). Chaffey *et al.,* (1985) reported the results of their survey in randomly located sample units of the Sundarbans during 1983-84, finding that 18% Sundri of 5 cm d.b.h class and above were defective.

The status of natural regeneration of Sundri in the Sundarbans for the period 1972 to 1986 has been found to be reasonably consistent and satisfactory. An inventory by Chaffey *et al.,* (1985) revealed that there were about 45.2 million top dying trees in the Sundarbans of which 20 million had more than 50% of their crown affected by top dying. If it is decided to fell trees having more than 50% of their crown affected by top dying, which is currently being proposed by the Bangladesh Government, then about 20 million Sundri trees might have to be removed from Sundarbans, which is not sensible. Before carrying out such an operation on the Sundarbans ecosystem, all possible adverse effects must be taken under critical consideration to save all the species. If all affected Sundri trees were removed from Sundarbans, then the whole forest ecosystem might be lost (Personal observation, 1993-1998).

According to Curtis (1993), having a canker or stag-headed condition on Sundri of above 132 years is quite acceptable, because catabolism naturally exceeds anabolism in old age. This is true for most of the tree species.Moreover, the extensive enumeration data provided by Chaffey *et al.,* (1985) did not show such a pattern either in Sundarbans in general or in the Sundri dominated areas; there are only a few mature Sundri of the above age class, whichwere found to be affected by top dying. The highest percentage of top dying Sundri laying in the 5cm.d.b.h class, and this percentage gradually reduces and becomes quite insignificant in trees of 20 cm.d.b.h. It is, therefore, suggested that top dying is not associated with over-maturity and old age. Most of the Sundri trees occupy the upper canopy, where scattered tops and branches of Sundri may be broken by abiotic factors such as atmospheric drying, oxidation and biotic infestation, pathogens and bacterial attacks. Only a limited number of tops and branches of Sundri are likely to be killed in these ways. But in the Sundarbans, trees with quite a few meters of dead tops and branches are seen; only rarely is the whole tree found dead. Moreover, the large proportion of Sundri having d.b.h 5 cm and above is unlikely to be affected by cyclones, except the exposed once on the banks of the rivers. Hence, cyclone damage as a main cause of top dying of Sundri is also ruled out.

Direct evidence on the causes of death of trees due to the top dying and its mechanism, particularly in the Sundarbans, is of course desirable butto the present there are hardly any such studies which directly explain the cause of top dying as well as the decline of Sundri from the Sundarbans (Rahman, *et al.,* 1995). Since Sundri is one of the main trees in the Sundarbans, and its loss is therefore of significant detriment to the area, the possible causes of this top dying and the evidence supporting such hypotheses are presented in more detail in sections 1.3; 2.5.

2.5. Possible Causes of Top Dying of Trees

Various authors have reported the top-dying of Sundri trees in the Sundarbans at different times, notably Toup (1921), Curtis (1933), Forestal (1960), Gibson (1975).On the basis of their observations, these and other authors have at various times suggested several possible hypotheses to explain the cause of the observed top-dying. The several suggested mechanisms, and the evidence available to support each of them, will now be presented.

2.6. Changed Water Flow and Increased Salinity

A wide range of arguments had been put forward concerning edaphic changes in Sundarbans. These included reduction in flow of fresh water throughout the Sundarbans because of withdrawal of water at Farakka Barrage and resultant increase in soil and water salinity (Shafi, 1982), reduction in siltation (Chowdhury, 1984), increase in siltation (Sattar, 1977), etc. These edaphic changes have been suggested to be responsible for top dying. It may be noted that most of the suggestions were not based on experimental data, for example Banik (2000).Chowdhury (1984) stated that the reduction of fresh water discharges through the Sundarbans was due to (a) the diversion of the upstream Ganges water and (b) more use of upstream water for irrigation and industrial use. Previously the rivers of the area were very active and used to lay freshly deposited silts on top soil every year where Sundri would survive well. But with the reduction in discharge of fresh water, silt carrying capacity of the river has also been reduced. Previously these areas were well drained at ebb tides but now due to change in river regime due to the Farakka barrage, the drainage patterns including flow characteristics have been changed. Consequently the trees are becoming thinner and shorter.Sundri (*Heritierafomes*) lacks the ability to exract fresh water from salt water at even a relatively low salt concentration (Hoffman *et al.,* 1986). Balmforth (1985), based on the report of Chaffey *et al.,* (1985), stated that the primary cause of Sundri top dying was the increase in soil salinity. Withdrawal of Ganges water at Farrakka because of the dam construction in 1976 was, in his opinion, one of the possible causes for the increase in salinity in the Sundarbans. Balmforth's statement regarding the causes of Sundri top dying due to an increase in salinity was, however, not supported by his data. Moreover, the data presented by Chaffey *et al.,* (1985), from examined soil samples collected from 20 sites including top dying and healthy Sundri areas in the Sundarbans, are not in agreement with Balmforth's statement, nor with the conclusion of Chaffey *et al.,* (1985) themselves. These authors had suggested that soil salinity was higher on sites affected by top dying than on unaffected sites, and that therefore an increase in soil salinity is an important factor in the onset of top dying. However, severely affected top dying areas are, in fact, distributed in slightly saline to non-saline zones as presented by Chaffey *et al.* (1985). Thus a total of 18% of Sundri trees with d.b.h above 5 cm had been affected by top dying of Sundri in Sundarbans (Chaffey *et al.,* 1985). Only a very small number of top-dying affected Sundri were present in the salt water zone (Chaffey *et al.,* 1985); 13-15% in moderately salt water zone, 21-24% in salt water zones and 25-47% in the fresh water zone respectively. This contrast with results of Ismail (1987), who stated that soil salinity was higher on sites affected by top-dying than on unaffected sites, suggesting that soil salinity is an important factor in the onset of top dying.Sattar (1977) inferred that mortality of Sundri is possibly due to edaphic changes such as heavy siltation, increase in salinity, less flooding by fresh water, and alteration in the frequency and duration of inundation. However, the decline of more salt tolerant species suggests that the cause of decline of the Sundri is not primarily due to insufficient fresh water flow during the dry season.Christensen (1984) stated that salinity may not be a direct cause of top dying but is an indicator of many geochemical processes that have an impact on living organisms (e.g. by affecting nutrient availability).Hassan (1988) provided data on soil and water salinity of a number of sites in the Sundarbans. He also studied the effect of salinity of Sundri seedlings in a simulated experiment.The data on salinity was compared with those of Chaffey *et al.,* (1985). It was observed that the top dying did not show any consistent relationship with the salinity. Hassan (1988) reported that soil salinity in the Sundarbans varies between less than 2000 micro mhos to more than 4000 micro mhos, which rises up to about 6000 micro mohs during the lean period in April. Hassan (1988) also observed that Sundri top dying did not show any consistent relationships with salinity. Moreover, regeneration and survival of Sundri seedlings did not show any change due to water salinity variation between the range 1,000 to 49,000 micro mhos under experimental conditions. Therefore, Hassan *et al.,* (1984) provided a detailed report on soil and salinity of the Sundarbans in relation to top- dying, regeneration and survival of Sundri. They did not observe any consistent relationships between top dying and salinity. Regeneration and survival of Sundri seedlings also did not show any relationships to the salinity variation between 10,000 to 49000 micro mohs under simulated conditions (Hassan, *et al.,* 1984).

Accordingly Shell and Ross(1987)studied the Na+ and K+ levels in Sundri tissues and found that: (a) the Na+ / K+ ratios in the tissues of healthy trees appear to be related to the salinity of their local environment; (b) the average percentage ofNa+ and K+ was not significantly different in healthy and dying Sundritrees;(c)though somelink was indicated between top dying and ionic balance in plant tissue, it was not possible to resolve the causal factor and its effect. According to Lugo and Snedaker (1974), gross primary productivity of mangroves increases with the availability of fresh water. The fresh water input to the estuary is associated with physical, chemical, geological and biological melange, fauna and flora of the estuary / coastal zone. This input tends to enrich the sites (Snedaker*et al.,* 1977).Imam (1982) noted that a system analysis and simulated studies on the ecology of mangrove forests in Florida, USA, had demonstrated that terrestrial water plays a vital role in the development of mangrove biomass, which in turn reflects the gross photosynthesis. It has been construed that both mangrove zonation and vigour are functions of nutrient availability rather than salinity alone. On the basis of the above findings Sheil and Ross (1987), concluded that these observations are compatible with a pathogenic cause of top dying as forwarded by Rahman*et al.,* (1983; see next section). This means that in their opinion salinity is not the causal factor for the top dying (Awa, 2007).

More recent research on the association between salinity and top dying of *Hetierafomes*has concluded that it is not the causal factor previously thought. Hassan *et al.* (1990) suggested that there is no relationship between salinity and top dying of Sundri, and data of Rahaman*et al.* (2003), where soil salinity was estimated in terms of the level of sodium (Na) and electrical conductivity, revealed that there is no significant variation in the mean value of Na and EC in soil of the Sundarbans.

According to Gupta *et al.* (1986), reporting on the salt tolerance of some tree species at the seedling stage, the salinity ranges which prevail in the Sundarbans are tolerable even to some mesophytes such as *Eucalyptus camaldulensis* and *Acacia nilotica*. This observation contradicts with the inference that a species such as Sundri should be seriously affected by the salinity observed. Moreover, without a governing cause top dying could not merely increase with time, as suggested by Chaffey *et al.* (1985), indicating that salinity alone can not be the cause of top dying of Sundri.

2.7. Altered Pneumatophores

Chaffey (1985) reported that reduction in the number of lenticels on the pneumatophores might be associated with top dying of Sundri. Accordingly, a study was carried out in which the numbers of lenticels were counted under a magnifying lens on 90 sq. cm area of the pneumatophores of nine top dying and nine healthy Sundri trees, selected randomly from relatively fresh water zone at Burigolini. Actually, the results indicated that there was no relation between the number of lenticels on the pneumatophores and the top dying.

2.8. Soil Factors

According to Siobhan (1973), who compared the chemical and physical properties of triplicate soil samples collected from severely affected top-dying and healthy Sundri areas, there was little difference in chemical properties of the two regions; however, the sand component in the healthy Sundri area was much higher than in top dying area. No consistent difference in chemical properties was observed but the sand component in samples from areas of healthy Sundri was much higher than that of top dying affected areas. According to Ismail (1988, 1987) when a pronounced dry season is followed by heavy showers, some critical nutrients are washed away. He argues that a poorer-quality forest in the Sundarbans is due to the loss of critical nutrients by sporadic heavy showers of recent years and not due to high salinity. The effect of individual ions present in saline soil, particularly sodium, has been studied by a few workers. The most important effect of sodium may be dispersing the soil and preventing root ramification (Plice, 1949) but very little is known about the effect of chloride ions (Francy*et al.*, 1959). Blanquet (1932) stated that sodium chloride is the determining factor in the distribution of plant species and communities on low level coasts. In general, salinity is a controlling factor in the floristic ecology of the Sundarbans (Snedakar, 1984). The soil analysis by Chaffey *et al.,* (1985) showed that soils of a number of sites contain lime which is a normal component of the Himalayan alluvium. Islam (1992, 1982) mentioned that the presence of lime would be expected to alleviate to some extent the adverse effect of high levels of soluble sodium. Soil analysis did not show that lime is a factor in the presence or absence of Sundri top dying. No evidence was found by Chaffey *et al.,* (1985) of acid sulphate which adversely affected the plant growth. No evidence of acid sulphate was found by Ismail (1987) which adversely affected the plant growth. The results of soil analysis strongly indicated a relationship between the presence of top dying and the concentration of pH in the soil profile however (Ismail, 1987).Boron deficiency is known to cause large scale dieback in Forestry (Bangash and Gardiner, 1985) but the saturation extract analyses by Chaffey *et al.,* (1985) show very low boron levels (below 0.5 ppm) in some sites. These low levels of boron are not correlated with the incidence of top dying (Awal, 2007).

Heavy metal concentrations in the Sundarbans soil have not yet been measured systematically, despite the potential contamination from effluent discharged, as discussed on section 1.3. The information of heavy metals in the soil with possible impacts upon the Sundarbans Reserved Forests is scanty (Awal, 2007). All kinds of wastes or pollutants dumped into the rivers can produce a change of physical characteristics of the soils and ecosystems (Awal, 2007). Mangrove plants are generally considered as filters for many pollutants, but such materials can be passed along a food-chain, ultimately with a potential of attaining unacceptable high levels in species consumed by humans.

2.9. Entomological Causes

As the heart rot fungi decay the wood and beetle larval tunnels increase, the commercial value of the wood is largely diminished (Chowdhury and Baksha, 1983). A study of the state of health of the cable, anchor, and nutrition roots of healthy and top dying Sundri trees was conducted at Sutarkhali. The results showed no significant difference in the root of healthy and top dying trees (Singh, 1986, Chaudhury, 1968; and Gibson, 1975).Chowdhury (1973) observed plenty of *Chrysocroa*species in the twig crotches of dead and dying Sundri trees in the compartment Nos.26, 28, 29, 30 and 31 at Marapassur in adjacent areas. The role of *Chrysocroa*species in the initiation of top dying of Sundri was subsequently studied and reported by Chowdhury and Baksha (1983). For the study, three diseased trees at various stages of attack and one healthy tree were selected at random in each of compartments 31, 32 and 46 representing relatively fresh water, moderately saline water and highly saline water zones respectively. The trees were felled and split thoroughly from top to bottom. The position of the dead branches, larval tunnels and the portions damaged by heart rot were recorded. There were in all 19 dead branches of which the relative positions of 15 branches were from the sites of the larval tunnels and heart rot. Heart rot was observed in 11 out of 12 trees. According to them, it became obvious that larval tunnel and /or heart rot did not have any contiguity with the dead branches. Even the tissues at the base of these branches were found to be healthy and the entrance holes of the larvae were healed up. They concluded that *Chrysocroa*species.did not have any role in the initiation of top dying of Sundri (Chowdhury and Baksha, 1983).

2.10. Pathological Causes

Pathological studies on top dying of Sundri were conducted by Rahman*et al.* (1983, 1986). Based on these and other studies conducted so far (reviewed in Rahman*et al.*, 1986) there does not appear to an association between pathological factors and top-dying. According to Rahman et *al.* (2001), a precise description of symptoms of any plant is important for further study leading to disease prevention. The same is true in case of top dying of Sundri as well.Therefore, trees were labelled in each of the damage categories (Rahman *et al.*, 2001). After that the labelled trees were observed to identify the symptoms of top-dying on the various branches at the time of recording within the selected plots. Subsequently, observations at monthly intervals were recorded of the symptoms of top dying on the branches and progressive changes were recorded for 12 months. In this way Rahman*et al.* (2001) tried to identify the first symptom expression and its progressive changes over time. But there is a big gap between primary and secondary causes of infection and symptoms for each particular Sundri tree. There is no scientific proof that pathogens are causing this top dying of Sundri. Many of the scientists had made technical comments about pathological studies, but they did not find any scientific base for the correlation between pathogens and top dying. Previous research had tried to find out the causal agents of top dying of Sundri, and tried to determine the infection-biology of the causal factors (pathogens) for top dying of Sundri (Hartung*et al.* 1998; Chowdhury and Baksha, 1983, 1986; Rahman*et al.* 1983). Rahman*et al.* (2001) also tried to find out the biotic and abiotic factors responsible for aggravating the top dying disease of Sundri.However, they failed to find out the real cause of top dying of Sundri. Hartung*et al.* (1998) reported their findings about the putative biotic disease agents of top dying of Sundri in the Sundarbans. For the study, they collected 20 samples of wood, bark, and leaves from compartment 19, 22, 26, 32, and 39 in 1997. This study was conducted for the isolation of bacteria, and eight types of bacterial colonies were recorded. The colour of bacteria colonies werewhite, yellow, red, white-yellow, yellow-white, yellow-brown, grey-white, shiny- white, etc. There was no yield of bacterial colonies from wood tissues taken from un-affected Sundri trees (3 samples). Samples of affected-dead-standing Sundri (2 samples), 4 unaffected wood samples of other tree species (2 from Gewa, 1 from Passur, and 1 from Goran) were taken. Out of the 6 to 8 types of bacterial colonies, two isolates have been identified as *Arthrobacternicotianae* and *Pseudomonas* species due to their sequence homologies in a PCR assay. Both the species of bacteria were isolated from samples of severely top dying affected living Sundri trees. The bacterial species showed phytopathogenic ability in a bioassay. The authors concluded a close association of phytopathgenic bacteria with top dying disease of Sundri (*Hartung et al., 1998*). But a through search of literature does not reveal any association of *A. nicotianane* with any other tree disease problem anywhere in the world. Moreover the data presented do not provide enough evidence of consistent association of *Pseudomonas* species with top dying of Sundri wood.

2.11. Gall Cankers

Occurrence of cankers on top dying Sundri trees has been recorded by Curtis (1933), Chowdhury (1973). It may be noted that the cankers on apparently healthy trees could be an early stage of deterioration and death of twigs or branches. Rahman*et al.* (1983) reported that a fungus identified as *Botryosphaeriaribis* was consistently isolated from Sundri canker.An early stage of gall canker develops on Sundri saplings. To investigate the proper cause of canker, the fungi were isolated from the bark and woody tissues of Sundri Canker samples following standard phytopathological method (Booth, 1971). One fungus was isolated as no. BFRI 230 and identified as *Botryosphaeriaribis* (Grosenb and Duggar), and occurred consistently. A number of other fungi were obtained only sporadically and hence disregarded.

The distribution and severity of perennial canker development on healthy and dying twigs, branches and stems of Sundri were studied at SutarkhaliSarbotkhali, Kalabogi and Mankiduania (Rahman*et al.*, 1983, 1986). It was found that there were top dying trees which had virtually no canker. The results suggested that there existed an association between top dying of Sundri and twigs and branches with gall canker in nearly 50% of the top dying Sundri trees surveyed, but in some trees the top dying condition was independent of gall canker development. There is therefore no conclusive link between the canker and top dying.

3. Results Section

Table 1.0. *Comparison of tested heavy-metals element concentrations (ppb) in Sundarbans. (Key to other sources: PL= Peach leaf reference material used; M in I = Ministry of India, 1987, Nayak, 1993; Nig = Nigerian information (AJEAM- RAGEE, 2004); GGC = Generation for Good chemistry {MallincKradt Chemicals; HNo(6623)}; England and Wales= (E.W); ES= Estuarine sediments reference material; SRM=Standard Reference Material). Compareable data could not be found for all elements.*

Element	Sundarbans (Present data)		Values reported elsewhere	
Identified heavy-metals in Sundarbans.	**Minimummean values(in ppb)**	**Maximummean values(in ppb)**	**Minimummean values(in ppb)**	**Maximummean values(in ppb)**
Al (Aluminium)	15.62	37,570.00	(Min I): 3.13	(Min I) 46.10
As(Arsenic)	(-) 0.57	10.06	0.00006(Pl) 0.00005(Nig)	(ES): 0.0116 0.00111 (Nig)
B(Boron)	4.81	77.21	0.029 (SRM)	(GGC):10.00

Element	Sundarbans (Present data)		Values reported elsewhere	
Identified heavy-metals in Sundarbans.	**Minimummean values(in ppb)**	**Maximummean values(in ppb)**	**Minimummean values(in ppb)**	**Maximummean values(in ppb)**
Ba(Barium)	0.41	141.80	0.124(Pl) 0.011(E.W)	(GGC):1.00 2.973 (E.W)
Bi(Bismuth)	(-) 0.01	0.74	—	(GGC):10.00
Ca(Calcium)	44.77	17,600.00	0.00005 (E.W)	(GGC):200.0
Cd(Cadmium)	0.14	1.62	0.002(Thai) (Pl)0.00003 (Min I):0.0003	0.016 (Thai) (ES): 0.0004 (Min I): 0.014
Co(Cobalt)	5.93	143.6	- (Min I): 0.003	(ES): 0.0105 (Min I): 0.026
Cr(Chromium)	3.14	114.90	0.0000016 (Nig)	(ES): 0.076 0.000017(Nig)
Cu(Copper)	1.85	43.76	0.131(Thai) (Pl): 0.004 (Min I): 0.006	0.260(Thai) (ES): 0.018 (Min I): 0.086
Fe(Iron)	25.82	248200.0	16.00 (Thai) (PL): 0.218 (Min I): 5.30	48.00 (Thai) (Min I): 52.00
Hg (Mercury)	(-)1.43	83.30	0.029 (Min I) (PL): 0.00003 0.000023 (Nig)	0.089 (Min I) (ES):0.00007 0.03(Nig)
In(Indium)	0.15	8.84	0.00005(E.W)	—
K(Potassium)	186.50	7495.00	—	—
Mg(Magnesium)	5.14	18280.00	—	(GGC):10.0
Mo(Molybdenum)	0.20	26.15	(PL): 0.00006 0.000008 (Nig)	0.00005(Nig)
Mn(Manganese)	0.70	698.00	(PL): 0.098 (Min I): 0.005	(ES):0.375 (Min I): 0.44
Na(Sodium)	75.83	6065.00	(PL): 0.024 0.3 (UK)	(GGC):500.0 -
Ni(Nickel)	7.58	1127.00	(PL): 0.0007 (Min I): 0.008	(ES):0.032 (Min I): 0.04 (GGC):5.00
P(Phosphorus)	(-) 62.83	1094.00	—	(GGC):20.0
Pb(Lead)	0.32	34.19	0.032 (Thai) (PL): 0.0009 (M in I): 0.0015	0.079(Thai) (ES):0.0282 (Min I): 0.024
Rb(Rubidium)	0.15	76.19	(PL): 0.0197 0.02 (E.W)	0.0209 (SRM)
Sb(Antimony)	(-)0.23	2.92	0.00003(Nig) 0.0005 (E.W)	0.000107(Nig) 0.00002(SRM)
Sc(Scandinium)	(-)1.15	8.98	0.0003 (E.W)	0.00004(SRM)
Se(Selenium)	(-)2.98	5.71	(PL): 0.00012 0.00003 (Nig)	0.00013(SRM) 0.0014(Nig)
Si (Silicon)	(-)2033.o	3978.00	—	(GGC):50.0
Sn (Tin)	(-)0.41	9.68	<0.0002 (SRM)	(GGC):10.0
Sr (Strontium)	0.18	44.17	(PL):0.053	(GGC):1.00 0.057(SRM)
Ti (Titanium)	1.37	1350.00	1.0 (E.W)	(GGC):0.30
V(Vanadium)	0.09	42.41	(PL):0.0004 0.004 (Nig)	(ES):0.094 0.298(Nig)
Y (Yttrium)	0.03	16.69	—	27000.00(E.W)
Zn(Zinc)	2.30	112.60	(PL):0.0179 (Min I): 9.64 0.029 (Thai0	(ES):0.138 (Min I):56.2 0.088(Thai)

4. Discussion

Each individual tree, when it was being recorded, had an estimate made of the severity of top-dying affecting it. In order to be able to calculate an average value per plot, these qualitative assessment categories of severity were assigned a numerical score of 0--3 (table 1.0).

Table 6.0. *Average index scores of the severity of top-dying per tree for each of the plots sampled.*

Compartment / Plot	26	28	31
1	1.88	2.51	2.00
2	1.85	1.57	1.78
3	1.58	1.45	1.07

This score was then summed to give an aggregate score which, when divided by the number of trees included, gave an average index score per plot. These values are included in Table 6.0. It should be remembered that the plots had been selected from within each of the compartments to reflect an apparent increase in severity of top-dying, and the compartments had also been selected to reflect differences in this aspect, so it is not surprising that there appears to be differences in this parameter in the table. However, when this was tested by a 2-factor Analysis of Variance, with Compartment as one factor and Plot as the other, the results were not significant for either compartment and plot ($F_{compartment} = 0.49$, df= 2,p= >0.10 ; F_{plot} =5.36, df = 2, p= <0.10), although the result for plots came close to significance. The fact that these results for top-dying did not prove to be significant, despite selection of the compartments and plots to reflect a gradient in the prevalence of top-dying may be because of the limited range of values that the top-dying index could take. Also, further replicates of the number of plots would have probably helped make it more likely that a significant result was achieved. At least the difference between plots in their top-dying severity is believed to be valid despite it not quite being proven statistically, and differences between compartments are also thought to be real, if not detectable in this analysis.Treating the average index scores as one variable, the top- dying index can be correlated with the various variables in order to test whether there is any relationship between the incidence of top-dying and parameters such as tree height or bole diameter. Since the average top-dying index is a variable that may not be meeting all the assumptions for a parametric test it is better to use the Spearman's Rank Correlation Coefficient for these analyses. A summary of the outcome for these analyses is given in Table 2.0 below.

Table 2.0. *Outcomes from the analyses using Spearman's Rank Correlation Coefficient of the relationship between severity of top-dying and various parameters of tree structure and growth (Key : NS= p>0.05; * = p<0.05).*

Parameters comparing withTop-dying.	Correlation(r-value, with7df).	Probability(p-value)	Sig. orN.S
Tree height (m)	- 0.113	0.773	N.S
Tallest tree height (m)	0.072	0.854	N.S
Bole diameter (cm)	0.067	0.865	N.S
Tree density (m^2)	0.021	0.956	N.S
Soil Moisture Content Percentage (M.C. %).	-0.736	0.024	*

None of the growth parameters were significantly related to amount of top-dying, but the percentage of soil moisture content was; the lower the amount of moisture, the greater the top-dying(Table 2.0).

In order to describe the vegetation structure more fully at each of the location it is desirable not only to record the adult trees but also the potential regeneration of new trees by recording the abundance of seedlings (newly germinated plants) and saplings (young trees less than 10 cm diameter). This information is presented in Tables 3.0 and 4.0.

Table 3.0. *Frequency of seedlings of all tree species combined in each of the nine plots sampled.*

Compartment / Plot	26	28	31	Totals
1	1261	1440	625	3326
2	1326	1501	1430	4257
3	1455	1575	1830	4860
Totals	4042	4516	3885	12443

Table 4.0. *Frequency of saplings (trees in m tall) of all tree species combined in each of the nine plots sampled.*

Compartment / Plot	26	28	31	Totals
1	282	228	167	677
2	266	172	396	834
3	566	215	343	1124
Totals	1114	615	906	2635

In order to test whether there is any association between the frequency of seedlings or saplings and the compartment or plot in which they were recorded, a chi-square analysis was performed for the data in each of Table 4.0 and 5.0. The result for seedlings was that $\Sigma\aleph^2$ gave a value of 344.7 which was very highly significant (p<0.001, df = 4), while the value of $\Sigma\aleph^2$ for saplings was 145.9 and was also highly significant (p<0.001, df = 4). Clearly, both sets of data for regenerating trees were highly influenced by location, with clear evidence of a reduction in abundance in sites which had been selected as showing high incidences of top-dying, with least regeneration in Compartment 31 Plot 1 and, particularly with the sapling data, with the greatest numbers of regenerating saplings in Compartment 26 Plot 3.

5. Final Conclusions

The overall conclusions from the results presented in this section are that the selection of sites has not produced clear statistical differences in the amount of top-dying evident; probably because of the way the data were collected. However, it is believed that there is notable variation between plots and compartments, and certainly this seems to be reflected in the ability of the trees to regenerate. However, the link between top-dying and the size of the trees is not clear, with tree height and diameter not being directly related consistently to amount oftop-dying, although moisture content of soil was inversely related. Since the great majority of trees present in all plots is the species *Heritierafomes*, this means that the comments above are essentially referring to the response of this species rather than that of any others.

Comparing figures in the table 1.0, it would suggest that about two thirds of the elements have concentrations which are elevated compare to other reference sources in the

Sundarbans. This would be consistent with the evidence that heavy metals were having an influence on top-dying intensity (Awal, 2007, 2009, 2014). The elements Pb, Sn, and Zn were highlighted earlier in this discussion, and although not all of them quite reached statistical significance (Awal, 2007), the positive trend linking two of them to top-dying suggests a likely mechanism of influence, namely that greater concentration of the heavy-metal weaken the resistance of the tree to attack by the pathogenic fungi (Awal, 2007, 2009, 20014). This might well be a process that other elements contribute to as well(Awal, 2007, 2009, 2014), but has not been picked out by the analysis as showing a link because of the variability between samples inherent in the data (Awal, 2007). In this respect, the anomaly of the negative relationship indicated for Sn is harder to explain (Awal, 2007), but a possible process might be an antagonistic response of Sn and another element(Awal, 2007), so that when Sn is less abundant the other element can have a stronger (deleterious) effect on the trees (Awal, 2007), thus allowing more top-dying to occur (Awal, 2007). A further point is that variations in soil pH from site to site (shown to be significant) will also have a marked effect on the bio-availability of some of these heavy metals (Awal, 2007), and thus perhaps influence top-dying (Awal, 2007).

Table 5.0. *Data on locations of plots, as indicated by Geographical Position System (G.P.S).*

Compart ment.Num ber.	Plot Numbe r.	Geographical Position Systems(GPS) reading.	
		North latitudes	East longitudes
26	1	N 22° 32347′	E 089° 64470′
26	2	N22° 32977′	E089° 64276′
26	3	N 22° 33820′	E 089° 63504′
28	4	N 22° 36728′	E 089° 66538′
28	5	N 28° 36378′	E 089° 66918′
28	6	N 22° 36878′	E 089° 66291′
31	7	N 22° 42821′	E 089° 57611′
31	8	N 22° 42715	E 089°580440′
31	9	N 22° 44201′	E 089° 58219′

The present data also contribute evidence regarding other suggested hypotheses for the causes of top-dying (outlined in the sections 1& 2). The non-significant relationship between soluble or exchangeable Na (Awal, 2007) and the amount of top-dying (Awal, 2007), or with the amount of tree growth(Awal, 2007), strongly suggests that the hypothesis of salinity changes being the cause of top-dying is not correct(Awal, 2007,2009,2014). However, the data on moisture content percentage do indicate that the Sundarbans has lower values than many other mangrove forests(Awal, 2007), and the values obtained are rather low even compared to others obtained from the Sundarbans (Faizuddin, 2003). This suggests that the Sundarbans is experiencing problems with its water supply (Awal, 2007, 2014), which may be linked(Awal, 2007), at least in part, to the construction of the Farraka barrage ((Awal, 2007; FAO report, 1984). If the drainage system of the water basin is poor, the tendency towards more acidic soil and water may increase ((Awal, 2007;Christensen, 1984).

The evidence presented in this thesis therefore points not to one specific factor as being entirely responsible for the increasing amount of top-dying (Personal observation staying in Sundarbans as Head of East Wildlife Sanctuary from 1993-1998, Awal, 2007, 2009, 2014), but rather to a combination of factors(Personal observation staying in Sundarbans as Head of East Wildlife Sanctuary from 1993-1998, Awal, 2007, 2009, 2014); changing water supply and heavy metal contamination in particular(Personal observation staying in Sundarbans as Head of East Wildlife Sanctuary from 1993-1998, Awal, 2007, 2009, 2014), coupled with possible pH changes, lead to a weakening of the trees(Personal observation staying in Sundarbans as Head of East Wildlife Sanctuary from 1993-1998, Awal, 2007, 2009, 2014),, which are then more susceptible to greater attack by pathogens(Personal observation staying in Sundarbans as Head of East Wildlife Sanctuary from 1993-1998, Awal, 2007, 2009, 2014),and lowered resistance to them by the trees (Personal observation staying in Sundarbans as Head of East Wildlife Sanctuary and also as ACF from 1993-1998, Awal, 2007, 2009, 2014),. This combination of factors is also supported by the comments of the people interviewed, (Personal observation staying in Sundarbans as Head of East Wildlife Sanctuary from 1993-1998, Awal, 2007, 2009, 2014), who suggested many different factors as the cause as well as implying that the change was gradual by not being able to pick out a specific starting date to the problem (Awal, 2007). This would be precisely the pattern that might be expected where no one factor was solely responsible, (Personal observation staying in Sundarbans as Head of East Wildlife Sanctuary from 1993-1998, Awal, 2007, 2009, 2014), and the combinations of factors might be somewhat different in different areas(Personal observation staying in Sundarbans as Head of East Wildlife Sanctuary from 1993-1998, Awal, 2007, 2009, 2014). Of course, having a number of interlinked factors as believed likely causes of the problem makes a solution to the problem more difficult (Personal observation staying in Sundarbans as Head of East Wildlife Sanctuary from 1993-1998, Awal, 2007, 2009, 2014). Some suggestions for ways of alleviating the problems, as well as further research that could be done(Personal observation staying in Sundarbans as Head of East Wildlife Sanctuary from 1993-1998, Awal, 2007, 2009, 2014), are included in the section below.

Although there were only a few individual chemical elements which showed significant correlations with the amount of top-dying (Personal observation staying in Sundarbans as Head of East Wildlife Sanctuary from 1993-1998, Awal, 2007, 2009, 2014), there was evidence of elevated heavy-metal concentrations of many elements in the Sundarbans (Personal observation staying in Sundarbans as Head of East Wildlife Sanctuary from 1993-1998, Awal, 2007, 2009, 2014), which are likely to contribute to the increase in top-dying observed(Personal observation staying in Sundarbans as Head of East Wildlife Sanctuary from

1993-1998, Awal, 2007, 2009, 2014). There were also significant differences between different areas in the Sundarbans (Awal, 2007), and evidence of water supply problems to the system(Personal observation staying in Sundarbans as Head of East Wildlife Sanctuary from 1993-1998, Awal, 2007, 2009, 2014). Therefore, it is concluded that several factors are responsible for this increase in top-dying, rather than one specific factor only (Personal observation staying in Sundarbans as Head of East Wildlife Sanctuary from 1993-1998, Awal, 2007, 2009, 2014).

Acknowledgements

The Peoples' Republic of Bangladesh and the Asian Development Bank (ADB), whom I thank, supported my research work financially. Particular thanks are due to the ADB head office, Manila for their support and help. I thank sincerely Dr. W.G.H. Hale (Principal Supervisor), University of Bradford, UK, Professor Mike Ashmore (Technical Supervisor), University of York, UK and Dr. P.J. Hogarth for their advice and comments on the work; Dr. Ben Stern and the staff at the Analytical Centre, Bradford, for their help with the ICPMS analyses; Professor SirajulHoque, Mustafa (lab Technician) and staff at Dhaka University for providing facilities; and staff of the Forestry Service, Bangladesh Government, for field assistance. Moreover I indebted to my beloved parents (Munshi Aowlad Hossain, Mrs. Ashrafunness), my beloved grand father (Munshi Bellal Hossain Biswas and Athaher Hossain Muktar), and mother (Rohima Khatun, and Alimoonnessa), my maternal uncle (Aftabul Azim, Azizul Azim, Amerul Azim, Anowarul Azim, Aminul Azim, Monu, (my beloved wife (Dr. Shahanaj Khatun), my beloved son (Munshi Tasneem Redowan), my beloved daughter (Marwa Ashra), my beloved brothers (Munshi Abul Kalam Azad), and (Munshi Abdus Salam) and my beloved 6 sisters (Layla Anjumand Banu (Chandu), Akter Rashida Banu (Turi), Saleha Pervin (Lili), Azmeri Ferdowsi (Dolly), Jannatul Ferdowsi (Polly) and my beloved youngest sister BedowraFerdowsi (Jolly) as well as all family members for their inspiration and help.

References

[1] ADB, 1992. *Non-Wood forests Products*. Project 372001/9; forest master Plan, Bangladesh. Asian development Bank, Manila, Philippines: 46-50.

[2] Asian Development Bank, 1993-95. Main Plan-1993/2012. Vol.1. *Forestry Master Plan*. Asian Development Bank, Manila, Philippines.

[3] Ahmed, K.J. 1957. Tidal forests of East Pakistan, their growth and regeneration. *The Pakistan Journal of Forestry*, Volume, Deradun, Pakistan.

[4] Bakshi, B. K. 1954. Wilt disease of *Dalbergiasissoo*Roxb due to *Fusariumsolani*Sensu Snyder and Hensen. *Natural Nature* 174: 278-291.

[5] Balmforth, E. B. 1985. *Observation on Sundri top dying-descriptive Sampling. In the Sundarbans reserved forest.* Draft Working Paper, UNDP/FAO Project BGD/77/017, and Dhaka.32 pp.

[6] Banik, H. 2000. *Bangladesh Sundarbans*. Hitutumi Book House and Publishers.TalpukurPar, Comilla, (in Bangla), Bangladesh.

[7] Blasco, F. (1977). Outlines of Ecology, Botany and Forestry of the Mangals of the Indian sub-continent. In: *Wet coastal ecosystem* (ed. V.J. Chapman), *Elsevier, Amsterdam* 13(1), 19(2).

[8] Chaffey, D. R; Miller, F.R; Sandom, J. H. 1985. *A forest inventory of the Sundarbans,* Bangladesh, *Main report*, Project Report No.140, 196 pp; Overseas Development Administration, London, U.K:195-196.

[9] Chaudhury, A.M. 1968. *Working Plan of Sundarbans Forest Division for the period from 1960-61 to 1979-80,* 82: East Pakistan Government Press, Dacca.

[10] Chowdhury, J. H. 1973. *Preliminary report of the investigation of the top dying of Sundri in the Sundarbans.* BFRI, Chittagong, Bangladesh, 2 p.

[11] Chowdhury, J. H. and Baksha, M.W. 1983. Studies on the role of Chrysocroa sp. in the initiation of top dying of Sundry. *Bano-BiggyanPatrika,* 12 (1&2): 30-34.

[12] Chowdhury, A.M. 1984. *Integrated Development of the Sundarbans, Bangladesh: Silvicultural Aspects of the Sundarbans*. FAO Report No / TCP/ BGD/ 2309 (Mf), W / R003.

[13] Chowdhury, M. I. 1984. *Morphological, hydrological and ecological aspects of the Sundarbans*. FAO report N0. FO: TCP/BGD/2309(Mf) W /R0027, 32 P.

[14] Curtis, S.J. 1933. *Working Plan for the forests of the Sundarbans Division, for the period from 1st April 1931 to 31st March, 1951*, Vol.1, 175 p. Bengal government press, Calcutta, India.

[15] Ellison, J.C. 1998. Impacts of sediment burial on mangroves. *Marine Pollution Bulletin*37: 420-426.

[16] Ellison, J.C. 2000. *Mangrove restoration: Do we know enough? Restoration Ecology.*8: 219-229.

[17] Faizuddin, M. 2003. *Research on the Top Dying of Sundri in Bangladesh*: 43, Mangrove Silviculture Division, Bangladesh Forest Research Institute, Khulna, Bangladesh.

[18] Faizuddin, M. and Islam, S.A. 2003. *Generated Technology and Usable Information of the Mangrove Silviculture*. Mangrove Silviculture Division, Bangladesh Forest Research Institute, Khulna, p. 17.

[19] FAO, 1994. Review of the state of world marine fisheries resources. *FAO Fisheries resources. FAO Fisheries Technical Approach Paper* 335:143.

[20] Forestal, 1960. *Forest inventory 1958-59, Sundarban Forests*. Forestal International Incorporated, Oregon, Canada.

[21] Fujimoto, K; Miyagi, T; Kikuchi, T; and kawana, T. (1996). *Mangrove habitat formation and response to Holocene sea-level changes on kosrae Island*, Micronesia. Mangrove and salt marshes, 1:47-57.

[22] Gibson, I.A.S. 1975. *Reports on a visit to the People's Republic of Bangladesh,* 28 February to 1April 1975.Unpublished Report, ODA, London, 28pp.

[23] Government of India, 1990. *Conservation of Mangroves in India,* Ministry of Environment and Forests, New Delhi, August, 1990.

[24] Gupta, D.P. (1982). *The control of Resources*. Cambridge, M.A: Harvard University Press.

[25] Hassan, M. M. 1984. Soil formation in the recent deltaic region of Bangladesh. *Bangladesh Journal of Soil Science* 20:37-45.

[26] Hassan, M. M. 1988. Soil and Salinity of the Sundarbans in relation to top dying, regeneration and survival of Sundri. *UPR of Bangladesh Forest Research Institute*, Chittagong: 9.

[27] Hassan, M. M.; Mazumder, A. H.; Islam, A. T. M. N. and Hossain, A.T. M. E. 1990. Soil, hydrology and salinity of Sundarbans in relation to top dying, regeneration and survival of Sundri (*Heritierafomes*) trees. In: Rahaman, M. A.; Khandakar, K.; Ahmed, F.U. and Ali, M.O. (Eds.). *Proceedings of the Seminar on Top Dying of Sundri (Heritierafomes) Trees*. Bangladesh Agricultural Research Council, Dhaka: 1- 11.

[28] Hoffman J. A. Katz, J. AndBertness, M. D 1986. Mangrove ecosystem. *Journal of Ecology*, 72 (2): 621-625.

[29] Howlader, A. S. 1973. *Report on the Preliminary investigation of Probable causes of Top-Dying, Regeneration and Survival of Sundri. Unpublished Report,* BFRI, Chittagong, 9 pp.

[30] Islam, A. K. M. N. 1982. Physico-chemical properties of Sundarbans mangrove forest. *Proceedings of the Second Bangladesh national Conference on Forestry* 50-54 p. Dhaka, Bangladesh.

[31] Islam, M. A. 1992. *Some relevant information about the Sundarbans Forest Division*, Khulna, Bangladesh.

[32] Jimenez, C. Y. 1993. Soil compaction as a constraint to tree growth in tropical and subtropical urban habitats. *Environmental Conservation.* 20: 35-49.

[33] Jones, G; Battershill, C; and Cole, R.(1992). Marine reserves: do they work? In: C. Battershill (ed.) *Proceedings of the International Temperate Reef Symposium*, Auckland, New Zealand, p.29-45.

[34] Khan, J. A. 1997. Estimation of unregulated densities by line transects method in Gir forest, India. *Tropical Ecology*. 38(1): 65-72.

[35] Macnae, W. 1968. A general account of the fauna and flora of mangrove swamps and forests in the Indo-West Pacific region. *Advance Marine Biology,* 6: 73-270.

[36] Naskar, K. and Mandal, R. 1999. *Ecology and Biodiversity of Indian Mangroves*, Publishing House, New Delhi, India, Volume, I & II, p.21.

[37] Paul, S.P; and Hossain, A. T. M. E. 1997. *Final report on the SDC-VFFP and SSD-BFRI collaborative studies on soil fertility status under cropland agroforestry systems.* SDC-VFFP, Dhaka, Bangladesh.

[38] Peters, R.L; and Darling, J. D. (1985). The greenhouse effect and nature reserves. *Biological Science*, 35 (11): 707- 717.

[39] Phillips, O. L; Gentry, A.H: 1994. Increasing turnover through timein tropical forests. *Science*, 263: 954-958.

[40] Plice, M.J. 1949. Some effects of salt water on soil fertility. *Proceedings of the Soil Science Society of America*, 24: 275-279.

[41] Rahman, M.A.; Khisa, S.K. and Basak, A.C. 1983. Top dying of Sundri in Sundarbans, *Bono BiggyanPatrica* 12 (1& 2): 69-71.

[42] Rahman, M. A. 1986. Present understanding of top dying of Sundri in the Sundarbans. *Paper presented at the 5th Biennial Botanical Conference held at the Department of Botany*, Dhaka University: 3.

[43] Rahman, M. A. 1995. *Mangrove plant pathology of the Sundarbans reserved forest in Bangladesh*, field document No.3 of FAO/UNDP Project BGD/84/056-Integrated Resources Development of the Sundarbans Reserved Forest, Khulna, Bangladesh, p.83.

[44] Rahman, M.A. *2001.* Diseases and disorders of tree species in the Sundarbans and their management. In: Siddiqi, N.A. and Baksa, M.W.(eds.). *Mangrove research and Development*, Bangladesh Forest Research institute, Chittagong: 1986-1997, 3(1 and 2):86-93.

[45] Rahman, M.A. 2003. *Genetic Approach to mitigate the top Dying Problem of Heritierafomes in the Mangrove Forests*, Khulna University, Bangladesh, 87pp.

[46] Rahman, M.A. 2003. *Mid-term Report on Top Dying of Sundri (Heritierafomes) and Its Management in the Sundarbans* Biodiversity Conservation Project, Khulna.109pp.

[47] Ross and Shell, 1987. *Na and K level in Sundri tissue*, Cambridge University press, London.

[48] Seidensticker, J. Hai; A. 1983. *The Sundarbans wildlife managemenplan*: conservation in the Bangladesh (cited in Chaffey et al., 1985), Bangladesh.

[49] Seidensticker, J. 1986. Large Carnivores and the consequences of habitat insularization, *Ecology and conservation of tigers in Indonesia and Bangladesh.* In: Cats of the World: Biology, Conservation & management. (Eds. S. D. Miller & D.D. Eveutt). Wetidal Wildlife Federation, Washington DC, p. 1-41.

[50] Seidensticker, J. & Hai, M.A. 1983. *The Sundarbans Wildlife Management.* Bangladesh coastal zone. International Union for Conservation of Nature. Gland, Switzerland.

[51] Seidensticker, J.; Christie, S.& Jackson, P.1999. *Introducing the tiger*. Cambridge University press, London, p.1-3.

[52] Shahidullah, M.1980. Salinity penetration on the south-west of Bangladesh: *Impact of reduced low flow of the major rivers of Bangladesh* (Seminar).Mimeographed report.10 pp.

[53] Siddiqi, N. A; Islam, M. R; Khan, M.A.S. and Shahidullah, M. (1993). Mangrove nurseries in Bangladesh. *Mangrove ecosystems occasio-nal papers. International Society for Mangrove Ecosystems*, Japan: 1-14.

[54] Singh, V.P; Garge, A; Mail, L.P; and Pathak, S.M. (1986). Some ecological aspects of mangrove forest of Andaman Islands. *Journal of Bombay Natural History Society*. 83:525-537.

[55] Smith et al., *2000.Heavy Metals in Nigerian crude Oils.* AJEAM-RAGEE, Volume 9 October 2004: 78-87.

[56] Snedakar, S .C. 1984. *General Ecology*: Resource Utilization Potential for the Bangladesh, Dhaka, Bangladesh: 31.

[57] Snedaker, S.C. 1977. *General Ecology and Resource Utilization Potential of the Sundarbans,* FAO Report No.Fo:TCP/BGD/2309 (MF) W/R, 28 prepared for the Government of Bangladesh, Dhaka: 31.

[58] Sobhan, A.1973. *Report on the preliminary investigation of potable causes of top dying of Sundri trees in the Sundarbans.* Unpublished report, Soil Science Division, Bangladesh Forest Research Institute, Chittagong.7 pp.

[59] Tamang, K. M. 1993. *Wildlife Management Plan for the Sundarbans Reserved Forest.* Integrated Resource Development of the Sundarbans Reserve Forest. UNDP /FAO, BGD/84/056, Khulna, Bangladesh.

[60] Thom, B.G.1982. *Mangrove ecology- a geo-morphological perspective*. 1982, p.3-17.

[61] Tiner, R. W; J R. (1984). *Wetlands of the United States*: Current Status and Recent Trends. Newton Corner, Massachusetts: U.S. Fish and Wildlife Service, Habitat Resources.

[62] Troup, R.S. 1921. *The Silviculture of Indian Trees*—Volume 1. Clarendon Press, Oxford: 153-544.

[63] UN-ESCAP, 1987. Final Report: Volume 2. *Coastal environment management plan for Bangladesh.*Bangkok, Thailand.

[64] UN-ESCAP, 1988. *Coastal environment management plan for Bangladesh.*Bangkok, Thailand: 7-34.

[65] UNESCO, 1997. *Convention Concerning the Protection of the World Cultural and Natural Heritage*, France, Paris.

[66] WHO, 1981. *Resistance of Disease vectors to pesticides.* World Health

Extreme events (flood & drought) and food security measures in the aspects of local perception

M. A. Rakib[1, *], M. S. Akter[2], M. B. Hossain[3]

[1]Department of Disaster Management, Begum Rokeya University, Rangpur, Bangladesh
[2]Department of Statistics, Jahangirnagar University, Savar, Dhaka, Bangladesh
[3]Department of Zoology, Rangpur Government College, Rangpur, Bangladesh

Email address:
rakibmamun_ju@yahoo.com (M. A. Rakib), abdur.rakib84@yahoo.com (M. A. Rakib)

Abstract: Natural hazards and its negative consequences on agricultural activities revealed that, crop production rate is gradually dwindled. Climatic hazards were identified as a critical factor, whereas it may emerge a barrier to get better crop yielding options. Local perception on natural hazards directly coincided with the observational factors. A number of people were observed food insecure during flooding and drought hazards in case of improper cultivation and marketing approach. Now-a-days food security is a great challenge in rural and disaster prone area, which is indicated to local impacts, occupational stress and livelihood diversification. In addition, a number of climatic factors were found to be significantly negative correlation with the crop production rate. However, compartmental research would be very helpful to build up coping capacity, adaptation and recovery on time measures along with the proper action.

Keywords: Extreme Events, Food Security, Agriculture and Local Perception

1. Introduction

Natural hazard is an emerging issue whereas it may pose potential threat to survive in an ordinary livelihood pattern. Most of the peoples are directly or indirectly being victimized due to uncertain natural hazards. Natural hazards (flood and drought) frequency is directly influenced by climate change [1]. It is a critical point to find out the real cause of hazards and make out the controlling outline as a mitigation approach. In any situational case, the social development depends on sustainable managerial capacity and knowledge empowerment at the regional and global scale. But the overall performance helps to reduce livelihood crisis and food insecurity. Food security, where does it ensure all access of safe food and nutrition lead to active life [2]. Food insecurity indicates to scarcity and/or unavailability of nutritionally acceptable adequate safe food leads to healthy life [3]. In the developing countries, most of the peoples are sufferings from serious health problem due to food insecurity [4].

The developed and developing country is encountered with several types of natural and human induced hazards. But most of the developing country is faced with worst condition due to unavailability of resource and lack of mitigation technique. Nevertheless, Bangladesh is a hazard prone country which is located at the geographically hazards settings area. A number of hazards like flood, tornado, drought, cyclone and etc hits every year in a cyclic order. Gradually increasing drought, flood, salinity and sea level rise, its frequency and intensity has been negatively impacted on natural resource along with human system and ecology [5]. The river bank erosion, flood, drought, cyclone, salinity intrusion and tornados are a common natural event which is frequently occurred in Bangladesh [6]. But its frequency and magnitude directly emphasize on local people and increase vulnerability. Consecutively, it is negatively attributed to worsen socio-economic condition and social crisis. In addition, socioeconomic crisis has been coincided with the uncertain natural and anthropogenic stress in a specific area and cause. However, food security and insecurity directly make a linkage between livelihood pattern and its availability along with household activities and resource availability. But it varies from place to place which is depends on development treads and its continuity.

Sometimes social crisis indicate to the socioeconomic

imbalance and food scarcity. Food security deals with food access (including income and ability to buying food in relation to marketing parameter), availability (depends on crop productivity along with food waste) and its proper utilization (containing nutritional quality and safety) [7, 8]. It varies from rural to urban area because of financial activities and source. However, agriculture based economy and development goals one of the better way to secure food in the rural area. But in some cases it is impaired and/or broken-down due to the seasonal climatic uncertainty and lack of consciousness. Consecutively, natural hazard and its negative consequences act as a key factor to split the economical trends and enhancing poverty. The aim of the study is to investigate the real setting of natural hazards occurring area and the local perception with special reference to fulfillment of daily dietary intake. It also deals with regional feature along with traditional expertise of the local people in relation to adaptation at the time of crisis.

1.1. Food Security and Climate Change

Bangladesh as an agro-economy depended country, its food comes from agricultural sector but agriculture activities depends on climatic variables such as temperature, precipitation, pressure, wind, humidity and etc. Consecutively, Agricultural production rate gradually decreases due to imbalance condition of climate and food becomes insecure. Food security depends on regional natural activities and its variation, such as – natural hazards, climate changing pattern, distribution and quality of agricultural land, education, employment opportunities and food habits. Climatic variables and its changing pattern are expected negatively attribute to crop growth and take into further challenge to fulfill local and household food security [9-13]. Extreme events such as droughts, floods, cyclones etc. causes loss of crop yield that affects specially in food security and agricultural livelihood. As a result, agriculture dependent people do not get enough facilities of fundamental needs income, food, safe water, habitation, proper education and etc. and they are diverted themselves into other occupations for the better living standard. Climate change is directly focused to food security and social livelihood pattern. In summer season, temperature remains so high, it is one of the main causes of drought occurring. However, the soil becomes dry and does not able to make availability of necessary nutrients and reduces soil fertility. As a result, agricultural production profit is less than farmer expenditure. A number of factors such as low soil fertility, trade relationship, feeble infrastructure and institution, population growth and future climatic variation act as a key factor to remove poverty and achieve food security [14]. In addition, yearly drought damages enormous amount of crops, crop lands and fail to reach the target production. Mainly in char area like “chilmari” in Kurigram district of Bangladesh, agriculture dependent people lost their crop lands because of river bank erosion due to unexpected and/or uncertain flooding hazards. Nevertheless, Cyclone in coastal area increases salinity problem of agricultural field by flooding which causes agricultural productivity damage. Different categorized of farmers such as tenant, owner cum-tenant and owner tends to change their livelihood pattern. Many of them become unemployed, so that tenant people are diverted into other occupation such as rickshaw puller, hawker, garments worker, poultry worker, van driver, day labor etc. Owner cum-tenant shift to involve in various business. A number of peoples sometimes do not get any works then they involve to different crimes. At the same time, the price of essential commodities is rising with decreasing food availability. Food crisis occurs in the month of Katrik (October-November), Aswin (September to October), Choitra (March to April) and Boishakh (April to May) in the Northern part of Bangladesh. When the food crisis occurs, the villagers also use to change their food habit, villagers eats various types of vegetables and fruits such as-Lalshak (Red amaranth), Dhekishak (fern vegetables), Khalmiashak (Water Spinach), Palongshak (Spinach), Alu (Potato), Alushak (Potato leaf), Misty alu (Sweet potato), Patshak (Jute leaf), Bangi (Muskmelon), Tarmuj (Watermelon) etc (local name).

2. Research Methods

This qualitative research was carried out at Chilmari Upazila (Char Land Area) in the Kurigram district of Bangladesh. Most of peoples were depended on agro-economy. It was located at the northern part of Bangladesh. However, natural hazard is a common issue in the aspects of geographical location.

This research work was carried out by the applying four types of methods which were adopted using different types of scholarly cited papers. It was carefully followed in the study area and kept more precision and accuracy for the Questionnaire Survey (Qs), Focus Group Discussion (FGD), Informal Dialogue (ID), Sitting Attitudinal Observation (SAOT) and Statistical Technique (ST).

The study was performed through structural questionnaire survey in the study area. It was consist of four parts (part one contained demographic information, part two contained perceptional information on climatic stress and agricultural activities, part three contained social crisis and part four livelihood pattern changing information). The questionnaire was administrated randomly [15] to serve our study purpose. The respondents were willingly agreed with interviewer persuasion and gave information politely. The information was recorded using feedback form and also written down in an extra note book which was preserved for the future coding and analysis.

Focus group discussion is a valuable technique to accumulate different opinion on a specific field of interest and to get clear concept on socioeconomic status and vulnerability [16]. It has an option to compare with another and get majority opinion in one thing. The FGD were performed within two groups, one was consisted of 21 members and another was contained 14 people. The participants were local people who are directly involved with agricultural activities and also fighting with natural hazards regularly. They were congenial to conversation with the interviewer and also agreed to give thematic information on climatic issues and its negative aspects of study area. The FGD was very much effective to enrich clear concept on climatic stress and livelihood changing pattern.

The informal dialogue (ID) is one of the suitable techniques to get lucid information from the local people. It is very much important which is done by behind of the people concern. It was contained instantly realistic view of the study purpose and objectives. Nevertheless, it was performed informally in various location and/or places like as on the street, local shop, tea stall and people gathering site in the community. This is confirmed to better output and inside view of the respondents.

Respondent's attitude test is the best approach to find out the real perception, core views, thinking and future plan. It is a new technique which was applied in this study. After the FGD, all participant were gossiping with each other then it was observed with the deep concentration by the sitting aside of them. They discussed, what was the feedback of FGD, actual problematic area and their future expectation?

The questionnaire was designed with using a number of likert-scale questions for getting clear concept on local environmental stress. It was confirmed real views on environmental hazards in the aspects of daily life style and pragmatic experiences in the study area. In categorical aspects, knowledge variability has been drawn from the participants. It was cleared to regional environmental aspects and its presents trends compare to previous condition. A number of aspects were included like as "Rainfall" "Average Temperature" "Flood Frequency" "Drought" "Wind Storm" "Average Humidity" "Groundwater Level" "River Discharge" "Food Security" and "Food Insecurity". It was stated with the scoring system in the reverse (positive and negative) order like as "agree"-5 " "Strongly agree"-4 "Either agree or disagree"-3 " disagree"-2 and "No comments"-1.

Correlation coefficient matrix was performed to know the relationship among the measured variables. This statistical analysis was carried out using statistical (MS Excel 2007 version and SPSS 11.5 version) windows package software. It is important to data coding, gathering and arrangement system for the statistical analysis [17]. This analysis has been revealed that, the better understanding feature and its coexisting controlling parameter.

3. Results and Discussion

3.1. Socio-Economic Status and Existing Feature

Socioeconomic status of a society indicates to the life style and professional achievement. It acts as primary indicator for one's to assess the life style and social acceptance. However, the studied population age was varied range from 12-67 years old. Most of the respondents (56 percent) were capable to write their name and address as a signatory level of education. It is divided into four groups depends on income level like as (US$ 38.69) 3000-(US$ 64.48) 5000tk for one, (US$ 64.48) 5000-(US$ 128.95) 10000tk for two, (US$128.95) 10000-(US$193.43) 15000tk for three and (US$193.43) 15000-(US$257.90) 20000tk for four. Most of the respondents are observed who (both male and female) were directly involved with agricultural activities. Agricultural activities are a main source of income in this community. A number of peoples were engaged in teaching profession at primary level education.

3.2. Climatic Stress and Local Perception

Climatic variables act as a controlling parameter to make balance in the atmospheric environment. Now, the discontinuation of climatic variables certainty is a concerning issues for the living organism in the aspects of survive. It is emerged from natural and anthropogenic induced hazards.

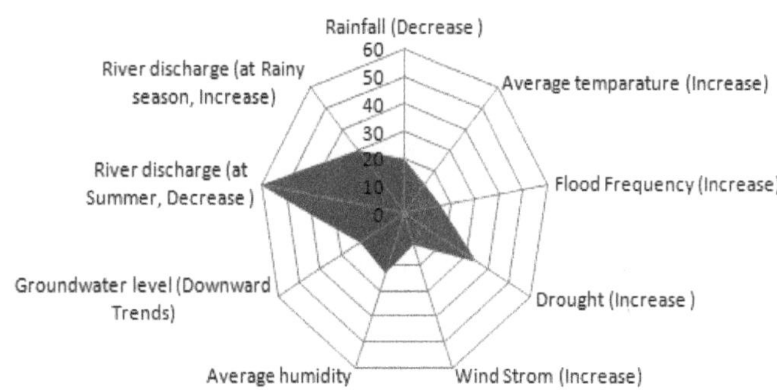

Figure 1. *Local Perception and climatic variables (It is drawn primary concept on climatic variables and its negative aspects on agricultural sector. It was studied to find out the realistic point of environmental crisis and feasibility. Most of the respondents were known on this aspects and its probable theme on impacts at the local scale.)*

A number of respondents reported that, overall climatic condition has been changed in the long term measure and existing evidence. According to previous experience 12 percent, 35 percent, 16 percent, 14 percent and 30 percent respondents were reported that, the average wind storm, drought, flood frequency, average temperature, river discharge (at rainy season) has increased and ground water level (35 percent) has decreased (Figure 1), but average rainfall (20 percent) and river discharge (at summer season, 60 percent) decreased (Figure 1) due to unexpected events. The average changing pattern of climatic variables will influence to the crop production rate and hydrological cycle [13] along with land degradation [18]. People perception and its long term impacts have been overlapped with the observational feature of the study area. Perceptional views of farmers on drought and secondary data of climatic variables indicate to resemblance in the aspects of extreme drought event [19]. Most of the people's livelihood pattern is being centralized around the agricultural activities. But climatic variables certainty and co-existing with the agricultural needs is very important to get high yield. Its discontinuation is negatively attributed to the average crop production rate and demand based needs fulfillment in the community people. Regarding with farmer perception, climatic variables (e.g. rainfall and temperature variability) are gradually changing and its negative impacts appear in their farming activities while a few number of resilient verities of crops are used to make balance but not more. Schlenker and Lobell, [20] reported that farmers have brought change in their farming activities against with more frequently occurring extreme events.

3.3. Food Crisis and Natural Hazards

Local people of the study area are continuously fighting with natural calamity for their present existence. It is negatively impacted on their socioeconomic status due to agricultural damage and production failure. Food security is directly indicated to nutritional status of human health which may relate to socioeconomic status [21]. Nevertheless, it was identified that, in particularly two types of natural hazards is directly made a linkage with the vulnerability and social crisis. Drought and Flooding hazards are frequently occurred in two times in a year. Soil moisture deficiency, duration and the total area of the affected region are considered to be important criteria for the drought severity measurement [19]. At the time of events, most of inhabitants are fallen in the worst condition which is associated with food shortage, shelter problem and domestic animal management. Regarding with respondents perception *".....Actually we are victimized due to the flooding hazards in our char land because, at the time [Ashar (June to July), Srabon (July to August), Vadro (August to September)] of flooding hazards we did not get enough food and support to cope with this problem. Just we are instantly developed our adaptive capacity through bearing patients and fasting one or two times in a day. On the other hand, we are scared in night because any types of disaster may arise in this time like as death".* A number of respondents has been reported that *"Drought is one of our curse, because many of our agricultural crops spoil at that time due to overheating, burning, land exposure, fertility decrease, irrigation problem and ground water depletion, it is the main cause of food chain breakdown and poverty in our society. In addition, our nearest river has been lost its own current and water body owing to make a dam at the upstream point".*

Regarding with farmer perception, February, March and April is critical time to adapt and ensure food availability at the household level (figure 2). Most of the family members remain under threat and risky position, which makes more vulnerable condition due to drought hazards. However, food shortage and local demand directly implies to high level of production rate at the time of hazards but on time it is not possible due to water scarcity (ground water level depletion, lack of rainfall, river dry out and scorching hot) and climatic variables instability. After a certain period of time, June, July and August are another time to occur flooding hazards regarding with farmer perception. Sometimes, it is occurred as catastrophic measure whereas it is not possible to predict of the flood frequency and magnitude using pragmatic experience due to climate change. At the time agricultural damage directly make a linkage with the adverse environmental condition and crop production failures which may negatively attributed to local food crisis and enhance poverty. Farmer perception only one factor, which is leading to alter of farming activities but extreme events like as flood and drought act as a key factor of food security. In addition, it is expected to formal credit and labor [14].

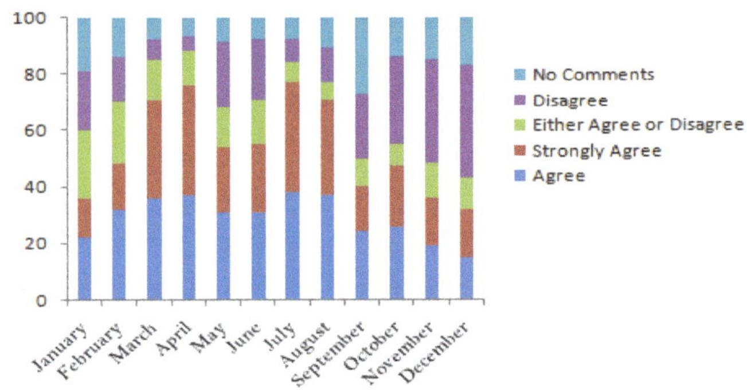

Figure 2. *Shows the respondents opinion (in percentage) on natural hazards and food security (according to the respondents information, February, March and April is a very critical time for drought hazards and June, July and August is for flooding hazards. At this time, they do not able to lead their life style in a proper way).*

3.4. Land Use Pattern and Agricultural Activities

Land use pattern of the study area was recorded whereas it was exposed as a char land at the Bramaputra river site. A major portion was used to agricultural activities after and before flooding hazards. A number of authors has emphasized on agricultural adaptations in the aspects of farming and land use management using and/or response of many climatic drivers [22-27]. Most of the land is used to different types of crop cultivation regarding with respondents views. But the production rate is varied from one place to another and crop to crop in a same site of the river bank. The cultivation item of crop was recorded in the char land like as rice, wheat, mustard, onion, garlic, pea, pumpkin, maize, sugarcane, millet, ladies finger, jute, tobacco, ground nut, potato and etc. However, a number of crops were not being so well production rate like as rice, mustard, sugarcane, garlic, banana and etc. In addition, a few respondents has been reported that a number of crops like as jute, maize, ground nut and wheat production rate is comparatively high due to the settlement of alluvial sediments after the flooding hazards. Sometimes, farmers apply of their won strategy (e.g. crop and water management, agriculture input, income diversification, migration and expenditure adjustment) to mitigate drought hazards and agro-based activities [19]. But long term crop diversification is a risk spreading matter by farming at the household level and in their field [28]. In some cases, uncertainly of the flooding hazards and heavy rainfall exposes as an unexpected catastrophic events. They reported that *"we did not see again in our life because flood will be occurring but its discontinuation has been appeared for the last thirty years. That's why, we are scared about our life and its dynamism to future"*. It may be emerged from the long term climatic stress and its variable change.

3.5. Why Food Insecure? Crop Production and Marketing System

Food security and its abundance is a reflection of social security which is embodied on fundamental rights. In some cases, it may break chain of supply due to the interruption of proper management and resource unavailability. According to the respondent's perception, *"we (producer) cultivate different type's crop. But it is very important to look after farming activities until harvesting. On the other hand, now-a-days all sorts of agricultural accessories have to buy at high level of cost compare to selling our agricultural commodity to primary, secondary buyer and retailer"*. Consecutively, regarding with root level consumer *"regularly we buy our necessary commodities and vegetables from the nearest market but I did not observe any remarkable price difference for a specific goods. So, it may not be depends on high production rate and availability. In my long term experience, we have to control at the price hiking step from primary buyer to retailer. It is so much than farmers selling at low cost"*. Have potential impacts on non-market and marketing system to secure food at the poor household level [29]. At the time of food crisis, it can have negative impacts on farmers and agricultural market price [29, 30].

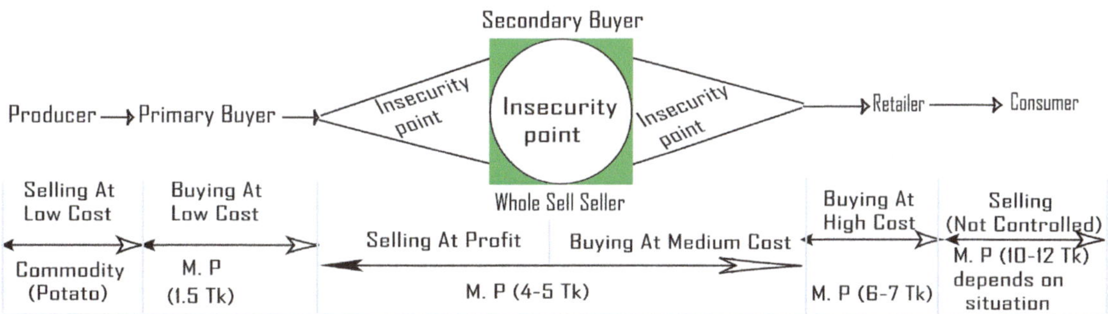

Figure 3. *Shows commodity supply chain and market price [M.P, Tk (1USD=77.55 Tk, date: 24-06-2014)] trends (e.g. sometimes it acts as a great barrier to price control among the social people, food producers, vendors and retailers. Insecure point denoted that, market price instability and food crisis. In few cases, it may follow upward trend without any demarcation due to lack of proper monitoring and assessment).*

Agricultural land reducing along with frequent natural disaster occurring at the study area is made food insecure in Bangladesh. Now-a-days food security is a great challenge in rural and disaster prone area which is indicated to local impacts, occupational stress and diversification. In the consequences of natural hazards, crop yielding rate is hampered and local respondents cited that *"we are now facing climatic effects and food shortage problem"*. Further, frequently occurred extreme events (e.g. drought or heat wave) impede to proper crop production and its maintenance, it reduces ability to cope with food security, capital and productive assets and so on [31, 32]. On the other hand, a number of respondents reported that *"After harvesting, we do not get actual price of our yielding crop"*. It may create supply chain impairment due to imbalance condition from producer to consumer (Figure 3). Ringler et al. [33] predicted that, 1.3% food availability would be decreased across Sub-Saharan Africa by middle of the century due to negative impacts of climate change and global food price hiking on key crops. Local consumer bought agricultural commodity at three to five times higher than first hand of buyer. In the aspects of expenditure, a number of smallholder farmers who are net buyers as they frequently purchase crop grain in hunger period and are compelled to sell at the harvesting time when prices are follow downward trends [34]. The government and support organization should take some steps in response of climatic effects which can have major impacts on regional agricultural markets and farmers. Climate change act as a key factor to control food price and it make food insecure who spend their most of the incomes to buy on food [35].

3.6. Livelihood Change and Crisis

Livelihood pattern is an indicator of socioeconomic structure and regional development in a specific area. It actually depends on social structure and resource availability which is act as a controlling parameter. But in some case, in presence of external factor (e.g. natural and human induced hazards and/or crisis) it may be changed from one point to another. According to respondents, most (90 percent) of people were depends on farming activities. However, at the time of hazards they do not able to crop cultivation due to uprising flooding and drought hazards. It is a critical point to find out the real feature of the study area crisis. Nevertheless, they do not get enough facility and/or support to lead their life at the crisis point. That's why, major part of respondents were reported that *"At the time of flooding and drought hazards, we do not get job at the local place. So we migrate from our village to town to find out a job like as Rickshaw polar, day labor and in the Garments"*. A number of people were identified who were interested to business at the crisis moment at the small scale like as poultry farm, cloth, dry food, peanut selling and etc. Overall livelihood pattern has been changed and it may be expedited due to long term climate change. Climatic variables uncertainty has become a potential threat to char land population.

3.7. Perceptional and Attitudinal Observation

Perceptional views sometimes reveal real point of crisis which may explain individual experience and its probable scenario. The respondent's attitude has been demonstrated that, regional feature and the problematic site in a long term way of their life history. Local views and their practical evaluation indicate to the positive and negative aspects in between natural hazards and external initiatives to support (government and non-government) at the upcoming and recovery stage. A group of people were cited that *"All sorts of initiatives are right but if we get instantly at the time of hazards occurring then it would be better to manage the situational condition. For this late support, a number of people do not get any way to cope with against with crisis"*. It is one of the negative aspects to ensure sustainable management approach. Regarding with respondents perception *"Firstly we are human being then poor, we know everything, how to face and adapt at the time of hazard but financial crisis is our main problem, if get support then it will be great to survive against with all together"*. Pragmatic experience acts as a key parameter to learn something which is embodied on traditional life style in society.

3.8. Food Security and Local Impact

From the ancient period, subconsciously peoples are trying to secure their food. Most of the part in Bangladesh, Peoples are mainly depended on agro-economy. When agriculture production is gradually decreased, it's very tough to secure daily food habit in the community. Peoples are obliged to change their food habits due to crop yielding uncertainty. In the study area, most of the peoples were generally changed their traditional dietary and taking potato, maize instead of rice. People also take various vegetables like as chichinga (Snake gourd), dhekishak (Fern vegetable), lalshak (Red amaranth), alushak (Potato leaf) and etc as a usual food due to production failure and unavailability. However, for securing food, people also tends to change their livelihood patterns. Maximum people of the study area were farmer. Consequently, mainly three types of farmers were identified like as tenant, owner-cum-tenant and owner [36]. Tenants are main victim of food crisis. They were frequently used agricultural land to cultivate through leasing system. In addition, flooding hazards, drought and/or heat wave act as key factor to loss of life, tends to loss of livelihood while it will affect to human health and infrastructure [37, 38]. However, the daily wages of tenant farmers were gone to follow downward trends. It was not sufficient to lead normal life style. Consequently, they have to change their present activities to better one and also migrated to nearest town, involving other business like as rickshaw puller, day labor, poultry farming and so on. On the other hand, Maximum respondents of them tend to go nearest town and/or Dhaka for more income as well as secure their foods. Unfortunately some of them involve various crimes like as hijacking, picketing, gambling, murder and so on. As a result, they lose their social status in the society. In addition, Owner cum tenant farmers reported that, they were involving in other business such as stock business of various grains, different shops, poultry business, hatchery business etc. But it is very difficult to adapt in a new income generating sector. They have been faced adverse situational and financial case. In the same way, owners were facing threat to overcome regional crisis which is evolved due to natural hazards. Sometimes their lands remain unproductive, and they willingly want to sell their land to make out the new way to survive. These diversifications are mainly occurs to maintain socio-economic status and sustain in a minimum living standard.

3.9. Correlation Coefficient Matrix

The correlation coefficient revealed the better understanding among the variables dependency and independency characteristics. This correlation study was performed regarding with the respondent perception on climatic variables and its negative effects in the study area. It was found to be a significant positive correlation among the groundwater (r=0.21; P=.01) and crop production rate (r=0.19; P=*0.01*) considering with rainfall intensity. On the other hand, a significant negative correlation was observed both of flooding hazards (r=-0.031; P=0.01) and drought (r=-0.15; P=0.01) intensity with respect to crop yielding status. But, after flooding hazards it was showed a significant positive correlation with the production rate regarding with respondents information based on pragmatic experiences. In addition, a number of negative significant correlations were observed among the variables like temperature (r=-0.19; P=0.01) and river discharge level (r=-0.31; P=0.01) in relation to local impacts on agricultural activities.

3.10. Recommended Thematic Outline

The problematic site and its negative consequences has been drawn some outline in the study area. Most of peoples are affected due to unwanted natural hazards but it is very much important to control and/or manage the situation against with adverse environmental condition. In addition, compartmental researches step by step, need to find out some crucial points for coping mechanism build up, need to ensure fundamental rights at the time of natural hazards and further potential research (component wise) would be helpful to communicate within a short time among the victim (figure 4). It may be the outstanding way to get sustainability in a society against with natural and human induced hazards. Now-a-days food security is a great challenge in rural and disaster prone area which is indicated to local impacts, occupational stress and diversification. Now climate change is a critical point for the food security but it is possible to ensure food availability for the growing population using united efforts of policy and science. It should be needed to change dietary pattern on time resource scarcity [39, 40]. At the local and global policy linked research along with institutional knowledge shearing attitude helps to accelerate make out a new way for the mitigation and adaptation approach [41].

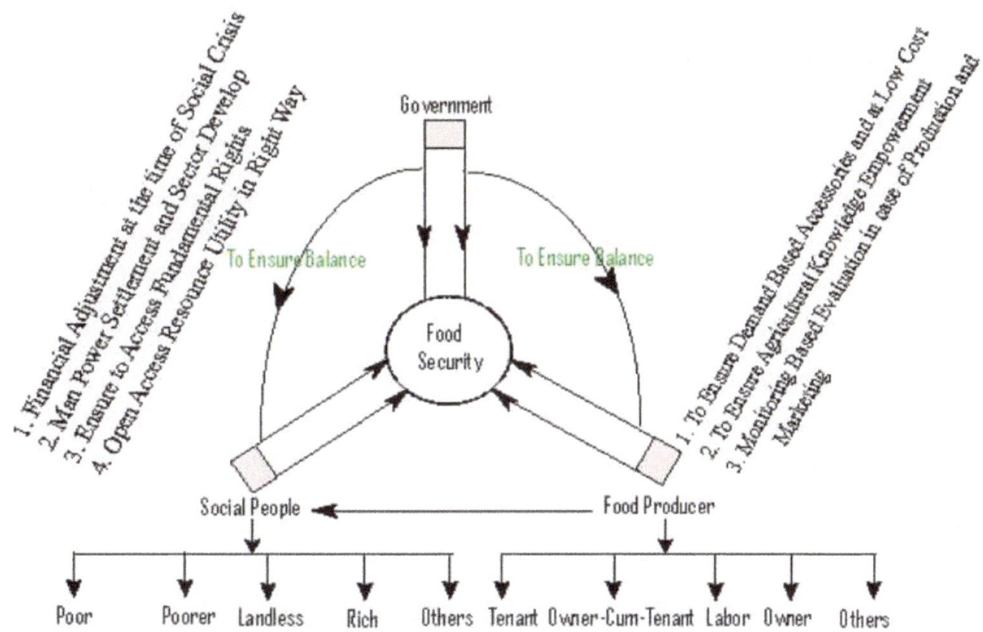

Figure 4. *shows the real feature of agriculture depended food crisis in presence of natural hazards (flood and drought) and its probable outline to recover. It consists of perceptional views of local people and their expectation from government where does it may play a vital role to ensure community sustainability. It is noted that, food security goes on full fill the criteria, meet up government crisis to social people and food producer in the aspects of natural hazards.*

4. Conclusion

People perception was observed owing to find out the real scenario along with adverse environmental circumstances in the study area. Local attitude was revealed towards worsen socioeconomic condition due to unexpected events (e.g. flood and drought). Regarding with respondents, climatic parameter frequently appeared with the changing pattern (showing discontinuation) in different times at the present age. However, People perception and its long term local impacts have been overlapped with the observational feature of the study area. It was found to be climatic parameter significantly negatively correlate with the crop production rate. Consequently, environmental condition and commodity marketing approach negatively attributed the local food crisis and/or insecurity. Also it was found to be majority of studied people tends to change their livelihood pattern from present farming activities to better one. Actually, climatic variables uncertainty identified as a potential threat for the char land people. In addition, compartmental researches step by step, need to find out some crucial points for coping mechanism build up, need to ensure fundamental rights at the time of natural hazards and further potential research (component wise) would be helpful to communicate within a short time among the victims.

Acknowledgement

The author acknowledged to the respondents who were willing to agree to give information regarding with research question. Author also acknowledged to the local people named Royal for his intensive support and cooperation.

Conflict of Interest

Author declares, it has no conflict of interest.

References

[1] M. Mirza, "Climate change, adaptation and adaptive governance in water sector in South Asia", Phys. Sci. Basis, pp. 1–19, 2007.

[2] Jarvis, C. Lau, S. Cook, E. Wollenberg, J. Hansen, O. Bonilla, and A. Challinor, "An integrated adaptation and mitigation framework for developing agricultural research: synergies and trade-offs", Exp. Agric., vol. 47, pp. 185-220, 2011. doi: 10.1017/ S0014479711000123.

[3] P. Ericksen, P.Thornton, A. Notenbaert, L. Cramer, P. Jones, and M. Herrero, "Mapping hotspots of climate change and food insecurity in the global tropics", CCAFS Report No. 5. CGIAR Research Program on Climate Change, Agriculture and Food Security (CCAFS). Copenhagen, Denmark, 2011.

[4] CCAFS. Proposal for CGIAR Research Program 7: Climate Change, Agriculture and Food Security (CCAFS). Coordinating Unit of the CGIAR Research Program on Climate Change, Agriculture and Food Security, Copenhagen, Denmark; 2011. http://ccafs.cgiar.org/sites/default/files/ ccafs_final_proposal_jan2011.pdf.

[5] IPCC Fourth Assessment Summary Report 2007, UNEP/UNDP.

[6] Asian Disaster Preparedness Centre, (ADPC) and Bangladesh Centre for Advanced Studies (BCAS), Draft Disaster Management Information Link Report, Comprehensive Disaster Management Programme (CDMP)", Component 5b, Ministry of Food and Disaster Management (MoFDM), Government of Peoples Republic of Bangladesh, Dhaka, Bangladesh 2008.

[7] Food and Agricultural Organization, "The State of Food Insecurity in the World 2001", FAO, Rome, 2002.

[8] FAO, IFAD and WFP, "The State of Food Insecurity in the World 2013. The Multiple Dimensions of Food Security", FAO, Rome, 2013.

[9] T. Wheeler, and J. Von Braun, "Climate change impacts on global food security", Science, vol. 341, pp. 508–513, 2013.

[10] O. Mertz, C. Mbow, J. Ostergaard Nielsen, A. Maiga, D. Diallo, A. Reenberg, A. Diouf, B. Barbier, I. Bouzou Moussa, M. Zorom, I. Ouattara, and D. Dabi, "Climate factors play a limited role for past adaptation strategies in West Africa", Ecol. Soc., vol.15, pp. 25–39, 2010.

[11] H. Meinke, S.M. Howden, P.C. Struik, R. Nelson, D. Rodriguez, and S.C. Chapman, "Adaptation science for agriculture and natural resource management—urgency and theoretical basis", Curr. Opin. Environ. Sustain., vol. 1, pp. 69–76, 2009.

[12] I.M. Goklany, "Integrated strategies to reduce vulnerability and advance adaptation, mitigation, and sustainable development", Mitig. Adapt. Strat. Glob. Change, vol. 12, pp. 755–778, 2007.

[13] W.E. Easterling, P.K. Aggarwal, P. Batima, K.M. Brander, L. Erda, S.M. Howden, A. Kirilenko, J. Morton, J.F. Soussana, J. Schmidhuber, and F.N. Tubiello, "Food, fibre and forest products", In: Parry, M.L., O.F. Canziani, J.P. Palutikof, P.J. vander Linden and C.E. Hanson, (Eds.), Climate Change 2007: Impacts, Adaptation and Vulnerability. Contribution of Working Group II to the Fourth Assessment Report of the Intergovernmental Panel on Climate Change. Cambridge University Press, Cambridge, UK, pp. 273–313, 2007.

[14] H. Webber, T. Gaiser, and F. Ewert, "What role can crop models play in supporting climate change adaptation decisions to enhance food security in Sub-Saharan Africa?", Agricultural Systems, xxx xxx–xxx 2014 (in press).

[15] M. A. Rakib, and M. A. H. Bhuiyan, "Arsenic Contamination: Food Toxicity and Local Perception", Int. J. Sci. Res. Env. Sci., vol. 2(1), pp. 1-7, 2014.

[16] R. Chambers, "The origins and practice of participatory rural appraisal", World Dev., vol. 22, pp. 953–69, 1994.

[17] J. Corbin, and A. Strauss, "Grounded theory research: procedures, cannons, and evaluative criteria", Qualitative Sociology, vol. 13(1), pp. 3–21, 1990.

[18] M.V.K. Sivakumar, and N. Ndiang'ui, "Climate and Land Degradation", Environmental Science and Engineering, Springer, Berlin, 2007.

[19] M. Ashraf, and J.K. Routray, "Perception and Understanding of Drought and Coping Strategies of Farming households in north-west Balochistan", International Journal of Disaster Risk Reduction, vol. 5, pp. 49-60, 2013.

[20] W. Schlenker, and D.B. Lobell, "Robust negative impacts of climate change on African agriculture", Environ. Res. Lett., vol. 5, pp. 1–8, 2010.

[21] Bishwajit, R. Barmon, and S. Ghosh, "Reviewing the status of agricultural production in bangladesh from a food security perspective", Russian Journal of Agricultural and Socio-Economic Sciences, vol. 1(25), pp. 19, 2014.

[22] E. Bryan, T.T. Deressa, G.A. Gbetibouo, and C. Ringler, "Adaptation to climate change in Ethiopia and South Africa: options and constraints", Environ. Sci. Policy, vol. 12, pp. 413–426, 2009.

[23] E. Bryan, C. Ringler, B. Okoba, C. Roncoli, S. Silvestri, and M. Herrero, "Adapting agriculture to climate change in Kenya: household strategies and determinants", J. Environ. Manage., vol. 114, pp. 26–35, 2013.

[24] P. Tschakert, "Views from the vulnerable: understanding climatic and other stressors in the Sahel", Glob. Environ. Change, vol. 17, pp. 381–396, 2007.

[25] G.A. Gbetibouo, R.M. Hassan, and C. Ringler, "Modeling farmers' adaptation strategies for climate change and variability: the case of the Limpopo Basin, South Africa", Agric. Econ. Res. Policy Pract. South. Afr., vol. 49, pp. 217–345, 2010.

[26] O. Mertz, C. Mbow, A. Reenberg, and A. Diouf, "Farmers' perceptions of climate change and agricultural adaptation strategies in rural Sahel", Environ. Manage., vol. 43, pp. 804–816, 2009.

[27] O. Mertz, C. Mbow, A. Reenberg, L. Genesio, E.F. Lambin, S. D'haen, M. Zorom, K. Rasmussen, D. Diallo, B. Barbier, I.B. Moussa, A. Diouf, J.O. Nielsen, and I. Sandholt, "Adaptation strategies and climate vulnerability in the Sudano-Sahelian region of West Africa", Atmos. Sci. Lett., vol. 12, pp. 104–108, 2011.

[28] W.N. Adger, S. Huq, K, Brown, D. Conway, and M. Hulme, "Adaptation to climate change in the developing world", Progress in Development studies, vol. 3, pp. 179-95, 2003.

[29] C.B. Barrett, "Food security and food assistance programs", In: B.L. Gardner, G.C. Rausser, (Eds.), Handbook of Agricultural Economics, Elsevier Science, Amsterdam, 2002.

[30] A.Abdulai, C.B. Barrett, and P. Hazell, "Food Aid for Market Development in Sub-Saharan Africa", DSGD Discussion Paper #5. International Food Policy Research Institute, Washington, DC, 2004.

[31] P.K. Thornton, P.G. Jones, P.J. Ericksen, and A.J. Challinor, "Agriculture and food systems in sub-Saharan Africa in a 4 C+ world", Philos. Trans. R. Soc. A., vol. 369, pp. 117– 136, 2011.

[32] Devereux, "Livelihood insecurity and social protection: a re-emerging issue in rural development", Develop. Policy Rev., vol.19, pp. 507–519, 2001.

[33] Ringler, T. Zhu, X. Cai, J. Koo, and D. Wang, "How can African Agriculture Adapt to Climate Change: Climate Change Impacts on Food Security in Sub-Saharan Africa: Insights from Comprehensive Climate Change Modeling", International Food Policy Research Institute (IFPRI), 2011.

[34] M.E. Brown, "Markets, climate change, and food security in West Africa", Environ. Sci. Technol., vol. 43, pp. 8016–8020, 2009.

[35] S. Vermeulen, R. Zougmore, E. Wollenberg, P. Thornton, G. Nelson, P. Kristjanson, J. Kinyangi, A. Jarvis, J. Hansen, A. Challinor, B. Campbell, and P. Aggarwal, "Climate change, agriculture and food security: a global partnership to link research and action for low-income agricultural producers and consumers", Current Opinion in Environmental Sustainability, vol. 4, pp. 128–133, 2012a.

[36] U. Habiba, R. Shaw, Y. Takeuchi, "Farmer's perception and adaptation practices to cope with drought: Perspectives from Northwestern Bangladesh", Inte. J. of Disaster Risk Reduction,vol. 1,pp 72–84,2012

[37] S. Dercon, "Growth and shocks: evidence from rural Ethiopia", Journal of Development Economics, vol. 74 (2), pp. 309–329, 2004.

[38] M.R. Carter, P.D. Little, T. Mogues, and W. Negatu, "Poverty traps and natural disasters in Ethiopia and Honduras", World Development, vol. 35(5), pp. 835–856, 2007.

[39] H.C.J. Godfray, J. Pretty, S.M. Thomas, E.J. Warham, and J.R. Beddington "Linking policy on climate and food", Science, vol. 331, pp. 1013-1014, 2011.

[40] J.A. Foley, N. Ramankutty, K.A. Brauman, E.S. Cassidy, J.S. Gerber, M. Johnston, N.D. Mueller, C. O'Connell, D.K. Ray, and P.C. West, et al. "Solutions for a cultivated planet", Nature, vol. 478, pp. 337-342, 2011, doi: 10.1038/nature10452.

[41] S.J. Vermeulen, P.K. Aggarwal, A. Ainslie, C. Angelone, B.M. Campbell, A.J. Challinor, J.W. Hansen, J.S.I. Ingram, A. Jarvis, P. Kristjanson, C. Lau, G.C. Nelson, P.K. Thornton, and E. Wollenberg "Options for support to agriculture and food security under climate change", Environmental Science and Policy, vol. 15, pp. 136-144, 2012b.

Production economics of sustainable soil management based cauliflower (*Brassica oleracea* L. var. *botrytis*) in Dhading district of Nepal

Binod Ghimire[1,*], **Shiva Chandra Dhakal**[2]

[1]Agriculture Extension Officer, District Agriculture Development Office, Mustang, Ministry of Agricultural Development, Government of Nepal, Nepal
[2]Department of Agricultural Economics, Institute of Agriculture and Animal Science, Tribhuvan University, Rampur, Chitwan, Nepal

Email address:
binodghim@gmail.com (B. Ghimire)

Abstract: Study investigated the production economics of sustainable soil management based cauliflower conducted at Nalang VDC of Dhading District in Nepal. It was based on primary data for the year 2012 obtained from randomly selected 60 sustainable soil management based cauliflower producers using pre-tested semi-structured interview schedule. Findings of the study revealed that, per ropani total cost of sustainable soil management based cauliflower production was lower whereas, gross income and gross margin were significantly higher by NRs. 1749.58 and 2173.36 respectively compared to before sustainable soil management practices in the study area. Also, comparatively higher benefit-cost ratio (4.26) and higher return to scale (0.802) was observed from sustainable soil management based cauliflower production. Similarly, production function analysis revealed that one percent increase in the use of hired human labor, expenditure on seed and expenditure on chemical inputs would increase the yield of cauliflower by 0.027, 0.362 and 0.177 percent respectively. Resource use efficiency analysis resulted that the resources were not utilized to their optimum economic level. For optimum allocation of resources expenditure on seed, chemical inputs and organic manure is required to increase by 96.1 percent, 87.5 percent and 72.0 percent respectively whereas, expenditure on human labor is required to decrease by 116.0 percent.

Keywords: Benefit-Cost Ratio, Cauliflower, Production Economics, Production Function, Resource Use Efficiency, Sustainable Soil Management, Nepal

1. Introduction

Cauliflower (*Brassica oleracea* L. var. botrytis) is an annual highly nutritive vegetable belong to family Brassicacae and reproduce by seeds. It originated from northeast Mediterranean and presently cultivated all over the world. It is an important commercial vegetable grown in Nepal and is also one of the highly preferred vegetables in Nepalese kitchen. Also, in Dhading district cauliflower is prioritized in production and marketing. The total production of vegetables in Dhading district in year 2011/12 was 74797 metric tons under total area of 6051 hectare with the yield of 12361 kg/ha. and cauliflower contributes 7054 mt. from 495 hectares (MoAD, 2012) of land. Contribution of the district in total vegetable supply to *Kalimati* vegetable market, Kathmandu was 17.68 percent in 2012/13 with total supply of 41,555 Mt. including 7,560 tons of tomato, 8,533 tons of cauliflower, 2,588 tons of beans and 2,752 tons of bitter gourd (KFVMDB, 2012).

The total area, production and productivity of vegetable in Nepal are 245037 hectare, 3298816 mt. and 13.46 ton/ha. respectively and the contribution of vegetables in AGDP is 20.74% (MoAD, 2012). Tomato, cauliflower, cabbage, cucumber, onion and chili are the major off-season vegetables of Nepal which generates income for household and among them, the cultivation of tomato, cauliflower, and cabbage are the most popular and the most profitable (NARC, 2006). Awasthi (2003) also reported cultivation of more than 247 vegetable crops in Nepal and out of which 50 species and their varieties are grown at commercial scale.

Sustainable Soil Management Program (SSMP) targets improvements in soil fertility and productivity in rain fed upland-dominated farming systems in the mid hills of Nepal with the aim of increasing food production, food security and farm incomes (SSMP, 2009). SSM practices includes improved farm yard manure, improved cattle sheds and urine use, legume integration, use of bio-pesticides, botanical pesticides, integrated plant nutrient systems, fodder promotion for livestock, SSM-based vegetable production etc. Moreover, the introduction of SSM practices has a direct role in improving the nutrition of many mid-hill households, in terms of both quantity and quality. These practices tends to reduce the use of fertilizer, pesticides and maximal tillage but rely more on crop rotation, crop residues, animal manure, green manure, leguminous crops, appropriate mechanical control or minimal tillage to optimize soil conservation activity (Noorivandi *et al.*, 2009). But the farmers have not adequately applied these measures to enhance overall system productivity because they are deprived of the access to resources, knowledge and skills on sustainable soil fertility management (FORWARD, 2006). Thus in-depth economic analysis of SSM based production system is urgent.

In Nepal, very little research has been carried out on the use, effect and economics of such sustainable soil management based crop production. This study will assess the economic significance of SSM-based cauliflower production compared with before.

The general objective of this research was to assess the comparative economics of sustainable soil management based cauliflower production. However, the specific objectives were;

i. To illustrate socio-demographic characteristics of selected SSM based cauliflower producers.
ii. To analyze benefit cost of SSM based cauliflower production compared with conventional system along with determinants factors of production.
iii. To analyze resource use efficiency in SSM based cauliflower production.

2. Materials and Methods

2.1. Study Area

The research was conducted at Nalang Village Development Committee (VDC) of Dhading district in Nepal where sustainable soil management program was conducted successfully for about five years and is also one of the major vegetable producing districts of the country. Dhading district spreads from 27'40"E to 28'17"E and 80'17"N to 84'35"N.

2.2. Sampling Method and Data Collection Procedure

The cauliflower producers adopting SSM practices were taken as the sample population for this study. A sample size of 60 is generally regarded as the minimum requirement for larger population that yields a sufficient level of certainty for decision-making (Poate and Daplyn, 1993). Such that, 60 SSM based cauliflower producers from Nalang VDC of Dhading district were selected for the study. Primary data was collected with the aid of a semi-structured interview schedule using simple random sampling technique in June, 2012. Also, focus group discussion and key informant survey were carried out. The final analysis was done by using computer software Microsoft Excel and STATA 12.

2.3. Techniques of Data Analysis

After the collection of necessary information it was coded and entered to computer for analysis. Descriptive statistics using percentage and frequency tables were used in the analysis for the socio-economic characteristics of the sampled farmers.

2.3.1. Gross Margin and Benefit Cost Analysis

Important variable cost items were taken under study and to make comparisons more scientific prevailing market price was used for each input used and also for price of output.

$$\text{Cost of cauliflower production} = C_{labor} + C_{seed} + C_{fert} + C_{pesticide} + C_{manure}$$

Where,

C_{labor} = Cost of labor used

C_{seed} = Expenditure on seed

C_{fert} = Expenditure on chemical fertilizer

$C_{pesticide}$ = Expenditure on chemical pesticides

C_{manure} = Cost of organic manure

Gross margin analysis was done as given by Olukosi *et al.*, (2006) by using formula;

Gross Margin (NRs.) = Gross farm income (NRs.) - Total variable cost (NRs.)

Where, Gross income (NRs.) = Price of cauliflower (NRs.) × total quantity sold (Kg.)

Total variable cost (NRs.) = Summation of cost on all variable inputs.

Benefit cost ratio is the quick and easiest method to determine the economic performance of farm business. Similarly, benefit cost analysis was done using the total variable cost and gross return from the cauliflower. Cost of production was calculated by summing all the variable cost items in the production process. For calculating gross return, income from the sale was accounted. Thus, the benefit cost analysis was carried out by using formula;

$$B/C\ ratio = \frac{Gross\ return\ (NRs.)}{Total\ variable\ cost\ (NRs.)}$$

For comparison, before-after approach was used and paired t-test was applied to test the mean difference of cost and return items in SSM based and conventional system of cauliflower production in the study site.

2.3.2. Production Function and Return to Scale Analysis

Cobb-Douglas type of production function was used to determine the contribution of different factors on production and to estimate the efficiency of the variable factors of producing cauliflower before and after the adoption of SSM practices.

$$Y= aX_1^{b1}X_2^{b2}X_3^{b3}X_4^{b4}e^u$$

Where,
Y= Total cauliflower production (Kg.)
X_1= Labor (Man-days)
X_2= Expenditure on seed (NRs.)
X_3= Expenditure on Chemical inputs (NRs.)
X_4= Organic manure (*Doko*)
u = Random disturbance term and $b_1, b_2, \ldots b_4$ are the coefficient to be estimated.

The Cobb-Douglas production function in the form expressed above was linearised in to a logarithmic function with a view of getting a form amenable to practical purposes as expresses below;

$$\ln Y= \ln a+b_1\ln X_1+b_2\ln X_2+b_3\ln X_3+b_4\ln X_4+u$$

Where,
ln= Natural logarithm, a= constant and u= Error term

For the calculation of return to scale from sustainable soil management based cauliflower production and before project, Cobb-Douglas production function was used and calculated using formula;

Return to scale (RTS)= $\sum b_i$
Where, b_i = Coefficient of ith variables.

2.3.3. Resource Use Efficiency

The efficiency of a resource used was determined by the ratio of MVP of inputs (based on the estimated regression coefficients) and the MFC. Following Goni *et al.* (2007) the efficiency of resource was calculated as;

$$r= MVP/MFC$$

Where,
r= Efficiency ratio
MVP= Marginal value product of a variable input.
MFC= Marginal factor cost (Price per unit input)

The value of MVP was estimated using the regression coefficient of each input and the price of the output.

$$MVP= MPPxi \times Py \text{ (Unit price of output)}$$

But, MPPxi= dy/dxi = $bi\bar{y}/\bar{Xi}$
Where; bi= Estimated regression coefficient of input Xi
$\bar{y}$= Arithmetic mean value of output.
$\bar{xi}$ = Arithmetic mean value of input being considered.

The prevailing market price of input was used as the Marginal Factor Cost (MFC).

MFC= Pxi Where, Pxi= Unit price of input xi.

The decision rule for the efficiency analysis was as;
r=1; Efficient use of a resource
r>1; Underutilization of a resource
r<1; Overutilization of a resource

Again, the relative percentage change in MVP of each resource required so as to obtain optimal resource allocation i.e r=1 or MVP= MFC was estimated using the following equation below;

$$D= (1-MFC/MVP) \times 100$$

Or, $D= (1-1/r)\times 100$
Where, D= absolute value of percentage change in MVP of each resource and r = efficiency ratio.

3. Results and Discussion

3.1. Socio-Demographic Characteristics of Respondent

Total sampled population in the study area was 357 of which male population was 53.22 percent. Average family size was 5.95 and was higher than the national average 4.5 (CBS, 2011). About 50 percent of the study population was economically active as classified by the Government of Nepal. About 83 percent of household were male headed and 72 percent of the household were found with nuclear family. About 62 percent were found literate and higher illiteracy rate may be due to poor economic status and geographical situation of the study area. Total area owned in an average was 11.90 ropani of which 10.43 ropani was cultivated and only 3.40 ropani was under irrigation. Farmers in the study area perceived irrigation as the major problem hindering the vegetable production. Among the total respondents about 67 percent had received trainings related to sustainable soil management based vegetable farming in the study area.

3.2. Benefit-Cost and Gross Margin Analysis

Table 1 depicted that the area of cauliflower under cultivation was significantly increased by 0.52 ropani after adopting SSM practices. Cost of labor and organic manure was significantly increased by NRs. 386.03 and NRs. 651.52 per ropani compared to before. Per ropani cost on chemical fertilizers and pesticides was significantly decreased by NRs.735.31 and NRs. 728.89 respectively. Although, total cost of cultivation and average cost were insignificant and were lower by NRs. 423.72 and NRS 0.46 respectively. The result exactly agree with the findings of Satpute *et al.* (2009) who studied on costs, returns and resource use efficiency in organic and inorganic cotton farming in Parbhani district, India. Finding of Kawasaki and Fujimoto (2009) also support this result which concluded lower total cost of production of organic Yard long bean compared to conventional system.

Table 1. *Comparative cost of cauliflower cultivation before and after SSM practices in the study area*

Particulars	SSM based	Before SSMP	Mean Difference	t-value	Level of Significance
Area (Ropani)	2.15	1.63	0.52***	4.025	0.0001
Labor Cost(NRs./Ropani)	1812.92	1426.88	386.03***	3.289	0.0008
Cost of Seed(NRs./Ropani)	331.5	320.05	11.44	0.573	0.2841
Total Manure Cost(NRs./Ropani)	2092.75	1441.22	651.52***	7.253	0.0000
Cost of Chemical Fertilizer(NRs./Ropani)	471.15	1206.47	-735.31***	-9.710	0.0000
Cost of Pesticides(NRs./Ropani)	110.21	839.11	-728.89***	-10.603	0.0000
Total cost of cultivation(NRs./Ropani)	5605.606	6029.33	-423.72	-1.357	0.0900
Average cost (NRs./Kg)	8.76	9.22	-0.46	-0.763	0.2243

*** Significant at 1% level *Source: Field survey, 2012*
** Significant at 5% level
SSMP= Sustainable soil management program
Buying rate of 1 US$= 95.91. NRs. (Nepalese Rupees) as of July 29, 2014.
19.66 Ropani =1 Hectare

Similarly, table 2 revealed that gross income, gross margin and benefit cost ratio were significant different before and after adopting SSM Practices. Result showed that the productivity of cauliflower decreased by 0.31 quintal on SSM-based production system. This finding is in line with the finding of Kshirsagar (2006) who reported that the organic farmers realized 7.17 percent lower yield than inorganic farmers in India. Gross income and gross margin were increased by NRs. 1749.58 and 2173.36 respectively whereas benefit cost ratio was higher by 0.41 compared to before SSM practice and were significant at 5 percent level (Table 4). Result is in line with Adhikari (2009) who reported higher cost and higher revenue in inorganic production system but higher benefit cost ratio in organic production system of carrot in Nepal. Also the findings are in conformity with the study conducted by Bharadwaj *et al.* (2000) and Patil (2008).

Table 2. *Comparative return from Cauliflower cultivation before and after SSM practices in the study area*

Particulars	SSM based	Before SSMP	Mean Difference	t-value	Level of Significance
Productivity(qtl/Ropani)	7.00	7.31	-0.31	-1.181	0.1211
Gross Income(NRs./Ropani)	22565.56	20815.97	1749.58**	1.934	0.0289
Gross margin (NRs./Ropani)	16959.95	14786.64	2173.36***	2.668	0.0049
B:C Ratio	4.26	3.85	0.41**	2.311	0.0121

*** Significant at 1% level *Source: Field survey, 2012*
** Significant at 5% level

3.3. Production Function Analysis

Result of production function analysis revealed that the values of coefficient of multiple determinations were 0.60 and 0.56 indicating that 60% and 56% of the variations in production of cauliflower was explained by the explanatory variables included in the model for SSM-based and before SSMP respectively. The f-values were 21.23 and 18.03 for after and before SSMP respectively which were highly significant at 1% confidence level implying that the variation of yield mainly depends on the explanatory variables included in the model. The estimated value of coefficients and related statistics of Cobb-Douglas production function were presented in table 3.

Model resulted that the coefficient of hired human labor for both before and SSM-based production were 0.027 and 0.029 respectively, and were positive and significant at 5% level of confidence indicating that 1 percent increase in the use of hired human labor, keeping other factors constant would increase the yield of cauliflower by 0.027 and 0.029 percent. Result is similar with that of Chithra *et al.* (2008) and Bakhsh *et al.* (2007).

The coefficients of expenditure on seed were positive and significant for both before and for SSM-based production. Keeping all other factor constant, 1% increase in expenditure on seed would increase the yield of cauliflower by 0.376 and 0.362 percent respectively (Table 3). Result agrees to that of Hassan *et al.* (2005).

Coefficient of expenditure on chemical inputs was significant for SSM-based production system and 1% increase in expenditure on chemical inputs, keeping all other factors constant there would be increase in production by 0.177 percent. But for before SSMP the value of coefficient was 0.196 and was negative indicating that yield would decrease by 0.196 if further increased in chemical inputs (Table 3).

The regression coefficient of organic manure was positive and significant at 1% level of confidence for before SSM practices. It indicates that keeping all other factors constant, a unit increase in organic manure would increase yield of cauliflower by 0.390 percent. Whereas, in case of SSM based cauliflower production use of organic manure was positive and insignificant (Table 3). Similar to this, Akter *et al.*, (2011) using revenue type of Cobb-Douglas production function resulted manure cost as significant factor in tomato production.

Table 3. Estimated coefficients and their related statistics of production function for cauliflower in the study area

Explanatory variables	SSM based		Before SSMP	
	Coefficient	t-value	Coefficient	t-value
Constant	2.716(0.680)	3.99	4.953(1.181)	4.19
Human labor (X_1)	0.027(0.012)**	2.18	0.029(0.014)**	2.12
Expenditure on seed (X_2)	0.362(0.135)**	2.68	0.376(0.134)***	2.80
Chemical inputs (X_3)	0.177(0.089)**	1.98	-0.196(0.135)	-1.45
Organic manure (X_4)	0.236(0.155)	1.52	0.390(0.145)**	2.69
R^2	0.60		0.56	
F-value	21.23		18.03	
Return to Scale (∑bi	0.802		0.599	

Figure in parentheses indicate standard error. *Source: Field survey, 2012*
SSMP= Sustainable soil management program
*** Significant at 1% level
** Significant at 5% level

3.4. Return to Scale Analysis

Regression coefficients of the estimated production function of cauliflower for SSM-based and before SSM practices were 0.802 and 0.599 respectively (Table 3). This result implies that the production function exhibits decreasing return to scale. In this case, if all the variables specified in the production function were increased by one percent, production of cauliflower would increase by 0.802 percent and 0.599 percent for SSM-based and before SSM practices respectively. Similar to this Wosor and Nimoh, (2012) reported decreasing return to scale in chilli pepper production with value 0.304. Also, Akter *et al.*, (2011) reported decreasing return to scale in tomato production with value 0.571.

3.5. Analysis of Resource Use Efficiency

The value of r was greater than unity for all the variable inputs used in cauliflower production except in expenditure on human labor as presented in table 4. This implies that expenditure on human labor in SSM based cauliflower production was over utilized whereas, expenditure on chemical inputs, expenditure on seed and organic manure were underutilized in SSM-based production system. Study result showed that all the inputs were not utilized to optimum economic advantage. Result agrees with the findings of Goni *et al.* (2013).

Result revealed that chemical inputs, seed and organic manure were underutilized whereas human labor was over utilized by the producers. To ensure optimum use of resources in SSM-based cauliflower production, expenditure on chemical inputs, seed and organic manure is required to increase by 87.5 percent, 96.1 percent and 72.0 percent respectively whereas, expenditure on human labor is required to decrease by 116.0 percent (Table 4). Producers can decrease the cost of production by reducing human labor and can increase the production by increasing the use of other inputs. Result is similar with the findings of Kehinde *et al.,* (2012). Similarly, Chapke *et* al., (2011) also reported that for optimum allocation of resources more than 88 percent increase in fertilizer, 76 percent in labor and 30 percent in agrochemicals was needed whereas, 127 percent and 114 percent reduction was required for seed and irrigation for sorghum production in India.

Table 4. Estimated resource use efficiency and required adjustment in Marginal Value Product (MVP) in percentage for SSM based cauliflower production in the study area

Variable inputs	Coefficient value	MVP	MFC	r	D
Human labor (Man-days)	0.027	128.04	276.6	0.46	116.0
Expenditure on seed (NRs.)	0.362	25.85	1	25.85	96.1
Expenditure on chemical inputs (NRs.)	0.177	8.02	1	8.02	87.5
Organic Manure (*Doko*)	0.236	164.58	46	3.58	72.0

MVP= Marginal Value Product *Source: Field survey, 2012*
MFC= Marginal Factor Cost
r= Efficiency ratio (MVP/MFC)
D= Absolute value of percentage change in MVP of each resource

4. Conclusion and Recommendations

The study has concluded that sustainable soil management based cauliflower production is a highly profitable enterprise in the study area as it showed decreased cost of production, higher benefit cost ratio and higher return to scale compared with before. Human labor, expenditure on seed, expenditure on chemical inputs and organic manure were the most determining factors in cauliflower production. Findings of this study revealed that

SSM based cauliflower producers in the study area were technically inefficient in the use of farm resources. Most of the inputs in SSM based cauliflower production were underutilized and it shows great opportunity of increasing productivity and income. On the basis of the findings, the study recommends that the farmers should reduce the level of those inputs that are being over used. Thus, to obtain economic advantage, farmers are recommended to reduce labor force and increase expenditure in seed and organic manure. Although chemical inputs are also underutilized, judicious use of such inputs should be done to ensure sustainability of soil and cauliflower production. Since, Nepal is facing the problem of scarcity, inaccessibility, toxicity and timely unavailability of chemical fertilizers these practices will be a best alternatives. Based on the discoveries of this study, it is recommended that, Government of Nepal should implement the intensive program related to sustainable soil management for producing diverse crops in different districts of the country. Provision of providing subsidized inputs, trainings and other demonstration program from Government level may also help farmer for better adoption of SSM practices and on the other hand also decreases the pressure on demand for chemical inputs.

Acknowledgements

We are thankful to IAAS (Institute of Agriculture and Animal Science), Department of Agricultural Economics, (IAAS) and Directorate of Research, (IAAS) Rampur, Chitwan, Nepal. We are cordially indebted to the financial support by Sustainable Soil Management Program (SSMP)-HELVETAS, Nepal for this study. We would also like to express our deepest gratitude to our colleagues for their direct and indirect help for the completion of research and heartwarming appreciation goes to respondents of the study site for their cooperation, time and valuable information.

References

[1] Adhikari R.K. (2009). Economics of organic vs inorganic carrot production in Nepal Journal of Agriculture and Environment.10:23-28.

[2] Akter S., M.S. Islam and M.S. Rahman. (2011). An economic analysis of winter vegetable production in some selected areas of Narsingdi district. Journal of Bangladesh Agriculture University. 9(2): 241-246.

[3] Awasthi B.D. (2003). Vegetable production and marketing in Kathmandu valley. Community Rural Development Society, Devinagar, Kathmandu, Nepal.

[4] Bakhsh K, B. Ahmad, S. Hassan and Z.A. Gill. (2007). An analysis of technical efficiency of growing bitter gourd in Pakistani Punjab. Pakistan Journal of Agriculture Science. 44(2):350-355.

[5] Bhardwaj M.L, R. Harender and B.L. Koul. (2000). Yield response and economics of organic sources of nutrients as substitute to inorganic sources in tomato, okra, cabbage and cauliflower. Journal of Agriculture Science. 70(10):653-656.

[6] CBS. (2011). Population census report. Central Bureau of Statistics, Singha Durbar, Kathmandu Nepal. pp. 120-268.

[7] Chapke R.R., B. Mondal and J.S. Mishra. (2011). Resource use efficiency of sorghum production in Rice- fallows in Andhra Pradesh. Indian Journal of Human Ecology. 34(2): 87-90.

[8] Chithra M.S., K.J. Thomas, J.C. Laly and E.K. Thomas. (2008). Technical efficiency in bitter gourd production. The ICFAI Journal of Agriculture Economics. 5(2):41-47.

[9] FORWARD. (2006). Annual Progress Report on "Promotion of integrated plant nutrition system (IPNS) in Maize-based Systems in the Hilly Areas of Makawanpur District". Bharatpur: FORWARD, Nepal. 17p.

[10] Goni M., M. Mohammed, B.A. Sandbaba. (2007). Analysis of resource use efficiency in rice production in the Lake Chad Area of Bomo State, Nigeria. Journal of Sustainable Development and Agriculture Environment. 3: 31-37.

[11] Goni M., M. Umar, A.S. S and U. Sman. (2013). Analysis of resource-use efficiency in dry season vegetable production in Jere Borno State, Nigeria. Journal of Biology and Agriculture Health. 3(19): 18-23.

[12] Hassan I., K. Bakhsh, M.H. Salik, M. Khalil and N. Ahmad. (2005). Profitability of winter vegetables in Faisalabad (Pakistan). International Journal of Agriculture Biology. 2:321-322.

[13] Kawasaki J. and F. Akimi. (2009). Economic and technical assessment of organic vegetable farming in comparison with other production systems in Chiang Mai, Thiland. Journal of ISSAAS 15(1):144-169.

[14] Kehinde F.T., J.O. Olukosi, A..L Ala, M.A. Maikasuwa and A.A. Odunsi. (2012). Determinants of the level of resource use efficiency in quality protein maize production in Kaduna State, Nigeria. International Journal of Applied Agriculture and Apiculture Research. 8(2): 24-30.

[15] KFVMDB. (2012). Annual report. Kalimati Fruit and Vegetable Market Development Board. Ministry of Agricultural Development, Kathmandu, Nepal.

[16] Kshirsagar K.G. (2006). Organic sugarcane farming for development of sustainable agriculture in Maharastra. Journal of Agriculture Ecology. 19: 145-153.

[17] MoAD. (2012). Statistical information on Nepalese agriculture. Ministry of Agriculture and Development. Agri-Business Promotion and Statistics Division (ABPSD), Kathmandu, Nepal.

[18] NARC. (2006). A Quarterly Newsletter. Nepal Agricultural Research Council. April-June 2006. 13(2).

[19] Noorivandi A.N., A. Ajili, M. Chizari and M. Bijani. (2009). The socio-economic characteristics of wheat farmers regarding adoption of sustainable soil management. Journal of Human Ecology. 27(3): 201-205.

[20] Olukosi J.O., S.U. Isitor and M.O. Ode. (2006). Introduction to agricultural marketing and prices: principle and application. 115 p. Living Book Series, GU publications Abuja.

[21] Patil M. (2008). A study on production and marketing management behavior of organic vegetable growers in Belgaum district. *Master Thesis*. University of Agricultural Science, Dharwad (India).

[22] Poate, C.D. and P.F. Daplyn. (1993). Data for Agrarian Development. Cambridge University Press.

[23] Satpute T.G., S.S. More and D.J. Sanap. (2009). Costs, returns and resource use efficiency in organic and inorganic cotton farming in Parbhani district. Journal Agriculture Update. 4(1):138-142.

[24] SSMP. (2009). Farmer Profiles from the Mid-hills of Nepal. Sustainable Soil Management Program, Kathmandu, Nepal. 112p.

Molecular and biological studies of *Papaya ringspot virus* isolates from Brazil and Cuba

Douglas Rodríguez Martínez[1, *], Priscilla de Sousa Geraldino Duarte[1], Justo González Olmedo[2], Antonia dos Reis Figueira[1]

[1]Department of Plant Pathology, Federal University of Lavras, Lavras, Brazil
[2]Bioplants Center, University of Ciego de Avila, Ciego de Avila, Cuba

Email address:
rodriguezdouglas1975@yahoo.com.br (D. R. Martínez), psgeraldino@yahoo.com.br (P. d. S. Geraldino), justo@bioplantas.cu (J. G. Olmedo), antonia@dfp.ufla.br (A. d. R. Figueira)

Abstract: The coat protein genes of 21 *Papaya ringspot virus* (PRSV) isolates from Brazil and 7 isolates from Cuba were sequenced and analyzed, and the symptoms induced by the virus isolates were evaluated in papaya (*Carica papaya* L. cv. Solo) and squash (*Cucurbita pepo* L. cv. Caserta). The virus isolates were grouped in phylogenetic trees by geographic region, presenting the highest nucleotide and amino acid identities with those isolates collected in the same regions. The American and Indian isolates, from GenBank were grouped together in the tree, and the Brazilian isolates were grouped by state; the Cuban isolates from the eastern region were closer to the American isolates than to those from the central-west region. Typical mosaic symptoms, grouped according to intensity, were observed in the inoculated papaya and squash seedlings. The variability of the coat protein genes confirmed that virus control using cross-protection and transgenic plants requires the selection of region-specific virus isolates in each country.

Keywords: PRSV-P, Coat protein, Phylogeny, Papaya, Squash, Symptom intensity.

1. Introduction

Ringspot disease caused by *Papaya ringspot virus* (PRSV) is considered the major disease of papaya (*Carica papaya* L.) worldwide [1, 2]. This virus is transmitted naturally to plants of the *Cucurbitaceae* and *Caricaceae* families by many species of aphids [3, 4]. This fact, in addition to the absence of resistant varieties of papaya, makes the control of this virus and its management extremely difficult in all regions where papaya is grown.

PRSV belongs to the family *Potyviridae*, genus *Potyvirus*, and has elongated, flexuous particles measuring 760-800 x 12 nm, with positive-sense ssRNA [1]. The viral genome contains approximately 10,326 nucleotides, with an ORF encoding a polyprotein of 350 kDa, starting at positions 86/88 and ending at positions 10,118/10,120; a poly-A tail is present at the 3' terminus [5, 6]. After synthesis, the polyprotein is proteolytically cleaved into nine final products: 5'-P1 (63 K), HC-Pro (52 K), P3 (46 K), CI (72 K), K 6, VPG (21 K), NIaPro (27 K), NIb (59 K), and CP (35 K) [4, 6, 7]. PRSV isolates have been divided into two biotypes, PRSV-P and PRSV-W [1]; both are able to naturally infect plants belonging to the *Cucurbitaceae* family, though only PRSV-P can infect the *Caricaceae* family ([1, 8].

Many strategies have been adopted as control measures of PRSV; however, the results have not been satisfactory, or have been limited to local viral isolates [8]. Currently, research is focused on the demand for transgenic plants that express a PRSV gene, typically the gene encoding the coat protein (CP), and also on the selection of weak or attenuated strains that could be used in cross-protection [9]. Studies on the transformation of papaya have demonstrated that the genetic resistance to PRSV obtained through the expression of the CP gene in papaya plants depends on several factors, such as the degree of homology between the protecting virus gene and the challenging virus, a property that is considered dependent on the virus isolate [4, 10, 11]. Moreover, cross-protection with mild strains has shown limited success in protecting against severe strains, thus precluding a durable, economic, and safe control, mainly due to problems of the protection loss from instability and a lack of specificity of mild isolates [4, 12, 13, 14, 15, 16].

There are numerous areas in Cuba with favorable conditions for papaya cultivation, and papaya crops represent 7% of the cultivated area and account for 15% to 20% of the income from fruit production in the country [17]. Currently, papaya cultivation is an alternative to the diversification of fruit crops as part of a national program for the development of the Cuban fruit agro industry, which aims to support the increasing fruit demand of its population and to expand the possibilities of the domestic market for export. In Brazil, the papaya crop is also greatly important, as evidenced by statistics that show this country as the largest producer and third largest exporter of papaya in the world. In 2011, Brazil produced more than 1.8 million tons from 35.531 hectares [18].

In both Brazil and Cuba, as also observed in the largest papaya-producing areas worldwide, PRSV-P is the main virus infecting this plant and has been a limiting factor to the establishment and development of the crop. It is very common to find papaya plants infected with this virus in commercial plantations, smallholder areas, urban household gardens, and along roadsides. These observations indicate that there are abundant sources of PRSV-P throughout the national territories, highlighting the need to develop and implement strategies for its management and control, which would reduce the harmful effects caused by this pathogen, thus increasing the income and quality of the fruits produced in the cultivation areas.

Within this context, it is necessary to know the degree of genetic variability of the isolates that occur in each specific geographic region as a basis for the development of transgenic plants and/or cross-protection programs. To assess this variability, Lima *et al.* [19] performed phylogenetic studies on the CP gene of twelve isolates of PRSV-P from different regions of Brazil, identifying a medium degree of homology of 97.3% between them. However, few studies have been performed in Cuba, with only partial CP sequences of PRSV-P available in GenBank [20, 21].

In this work, we present and discuss, for the first time, the results of sequence analyses and the comparison of the complete CP gene of seven PRSV-P isolates collected in several provinces in Cuba, the analysis of 21 PRSV-P isolates from several regions of Brazil, and the symptoms induced by the 28 isolates in papaya cv. Solo and squash cv. Caserta plants.

2. Material and Methods

2.1. Selecting a Collection and Maintenance of Isolates

Between 2010 and 2011, 28 samples were collected from papaya plants exhibiting the typical symptoms of PRSV-P in commercial plantations of ten Brazilian states and five Cuban provinces. Details of the origins of the strains and the sequences from GenBank used for comparison are shown in Table 1. The isolates were collected and mechanically inoculated onto papaya seedlings and maintained under greenhouse conditions, and the obtained infected leaves were dried and stored at -20 and -80 °C.

Table 1. *Origin and denomination of Papaya ringspot virus (PRSV-P) isolates used in the study*

Studied Isolates (collected from Brazil and Cuba)					**Isolates from GenBank**		
Country	**Isolate**	**State/Province**	**Municipality**	**GB Access**	**Country**	**Isolate**	**GB Access**
Brazil	BrMG-1	Minas Gerais	Lavras	KC662372	Brazil	Br-PRSV-W-1	DQ374153
Brazil	BrMG-2	Minas Gerais	Lavras	KC748204	Brazil	Br-PRSV-W-C	DQ374152
Brazil	BrMG-3	Minas Gerais	Lavras	KC748205	Brazil	Brazil-CE	AF344647
Brazil	BrMG-4	Minas Gerais	Lavras	KC748206	Brazil	Brazil-PE	AF344646
Brazil	BrBA-1	Bahia	Cruz das Almas	KC748212	Brazil	Brazil-DF	AF344650
Brazil	BrBA-2	Bahia	Cruz das Almas	KC748213	Mexico	Mx-VrPro	AY231130
Brazil	BrSP-1	São Paulo	Itapira	KC748207	USA	USA-Haw-PG	EU126128
Brazil	BrSP-2	São Paulo	Itapira	KC748208	USA	USA-B9	JN132432
Brazil	BrSP-3	São Paulo	Itapira	KC748209	USA	USA-T14b	JN132470
Brazil	BrSP-4	São Paulo	Itapira	KC748210	Venezuela	Ve-Merida-6	EF189736
Brazil	BrSP-5	São Paulo	Mococa	KC748211	Venezuela	Ve-Merida-8	EF189733
Brazil	BrRJ-1	Rio de Janeiro	Rio de Janeiro	KC748214	Venezuela	Ve-Trujillo-1	EF189734
Brazil	BrPR-1	Paraná	Maringá	KC748222	Venezuela	Ve-Trujillo-5	EF189735
Brazil	BrPR-2	Paraná	Maringá	KC748223	Venezuela	Ve-Zulia-7	EF189732
Brazil	BrES-1	Espírito Santo	Linhares	KC748216	Taiwan	Ta-PRSV-SLM	DQ340771
Brazil	BrES-3	Espírito Santo	Linhares	KC748217	Taiwan	Ta-PRSV-1	X67672
Brazil	BrAM-1	Amazonas	Manaus	KC748220	Taiwan	Ta-PRSV-2	X97251
Brazil	BrMA-7	Maranhão	São Luís	KC748218	Taiwan	Ta-PRSV-SLD	DQ340769
Brazil	BrMA-8	Maranhão	São Luís	KC748219	Taiwan	Ta-PRSV-P-5-19	EU882728
Brazil	BrPA-1	Paraíba	Pombal	KC748221	Taiwan	Ta-CI	AY027810
Brazil	BrDF-1	D. Federal	DF	KC748215	China	Ch-HN-1	HQ424465
Cuba	CbMY-1	Mayabeque	Nueva Paz	KC748229	China	Ch-PRSV-P	EF183499
Cuba	CbMT-1	Matanzas	Colón	KC748227	India	In-PRSV	EU475877
Cuba	CbMT-2	Matanzas	Jagüey Grande	KC748228	India	PRSV-DEL	EF017707
Cuba	CbVC-1	Villa Clara	Santa Clara	KC748230	South Korea	SK-PRSV-W	AB369277
Cuba	CbGR-1	Granma	Jiguaní	KC748224	Thailand	Th-PRSV-P	AY162218
Cuba	CbGR-2	Granma	Jiguaní	KC748225	Thailand	Th-PRSV-W	AY010722
Cuba	CbHG-1	Holguín	Holguín	KC748226	Malaysia	Ma-PRSV	AB044342

2.2. RNA Extraction, RT-PCR, and Sequence Analysis

Total RNA was extracted from the young leaves of PRSV-infected papaya plants using the Trizol method [22]. The primers used for the amplification of a 1,238 pb fragment containing the capsid gene were designed based on the sequences available in GenBank for PRSV (http://www.ncbi.nlm.nih.gov/genbank/): PRSV9016-F (5'-CTTGARCARGCTCCATTC-3') and PRSV10253-R (5'-CTAAAAGCACGGAGG-3').

Complementary DNA (cDNA) was synthesized from the total RNA using the primer PRSV10253-R and M-MLV reverse transcriptase (Promega Corp. Madison, WI, USA). The PCR was performed with GoTaq® Flexi DNA Polymerase (Promega, Corp. Madison, WI, USA) and both primers using the following cycling conditions: 95°C for 2 minutes, followed by 35 cycles of 95°C for 45 seconds, 42°C for 1 minute, and 72°C for 1 minute, with a final extension at 72°C for 5 minutes. The product was analyzed by electrophoresis through 0.7% agarose, purified using a purification kit (QIAGEN Sample & Assay Technologies – Venlo, The Netherlands), and sequenced by Genewiz (South Plainfield, NJ, USA).

2.3. Sequence Analysis

Based on the sequences obtained, contigs were constructed using the program DNA Baser Sequence Assembler v2 (www.dnabaser.com) and analyzed with the Basic Local Alignment Search Tool (BLAST), which is available online at the NCBI (National Center for Biotechnology Information) website (http://www.ncbi.nlm.nih.gov/).

The final sequences were analyzed with the BLAST program, and the multiple alignment was performed with the CLUSTALW program included in Molecular Evolutionary Genetics Analysis (MEGA5) [23], using, for comparison, the PRSV isolates available in GenBank (Table 1). The phylogenetic trees were constructed with the MEGA5 program using bootstrapping with 2,000 repetitions.

2.4. Survey of Symptoms

Two tests were performed to evaluate the symptoms induced by the virus isolates in papaya (*C. papaya* L. cv. Solo) and squash (*Cucurbita pepo* L. cv. Caserta) seedlings under different temperature conditions: winter (July 2010), with a minimum temperature of 4.8°C and a maximum of 25.2°C (average 15°C), and summer (February 2012), with values of minimum and maximum temperatures of 19.8°C and 34°C, respectively (mean of 26.9°C).

The studied isolates were mechanically inoculated using friction with 600 mesh silicon carbide of papaya plants (15 to 20 cm high) under greenhouse conditions. Three seedlings were used in each experiment, with three replications. Healthy plants inoculated only with phosphate buffer were used as the control. The evaluation of symptoms was performed up to 45 days after inoculation.

All the inoculated plants were maintained in a greenhouse until the end of symptom assessment, which was performed in accordance with the descriptive scale used by Rodriguez *et al.* [24] to evaluate papaya seedlings inoculated with PRSV-P.

3. Results and Discussion

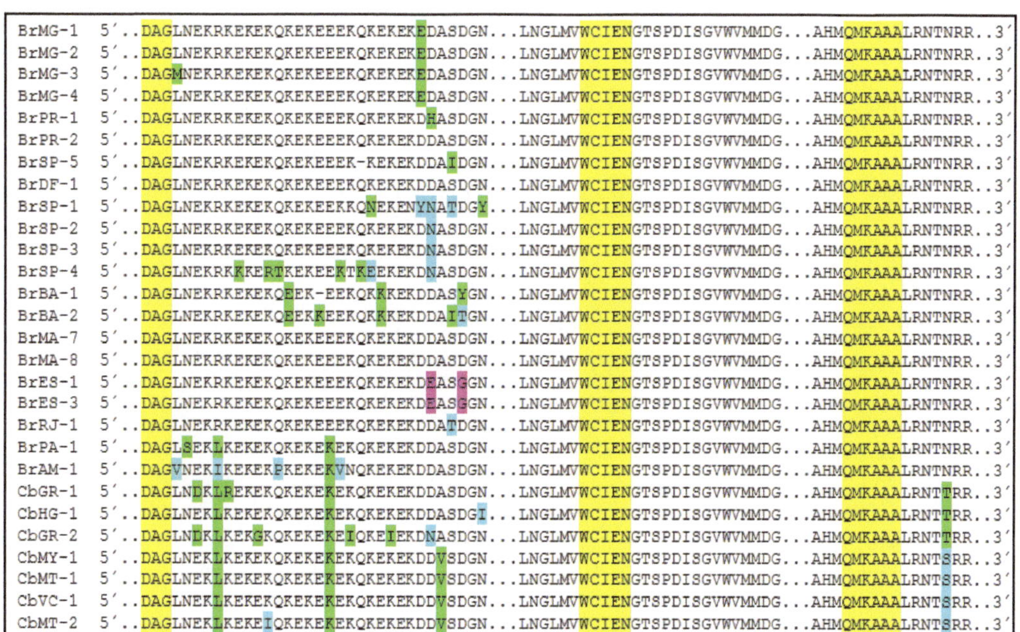

Figure 1. *Amino acids sequences of different regions of CP protein of PRSV-P isolates from Brazil and Cuba showing the deletions in the BrSP-5 and BrBA-1 isolates (-), conserved DAG, WCIEN & QMKAA domains and more conserve sequences on the 3′ region than the 5′region (*). Different colors in the same column indicate different amino acids in that position. Multiple alignments were performed with the CLUSTALW program included in MEGA5.*

A total of 26 of the 28 isolates for which the PRSV CP gene was sequenced in this study produced a band with 924 nucleotides. Brazilian isolates BrBA-1 and BrSP-5 generated a 921 nucleotide band; the deletion of three nucleotides corresponds to amino acid 22 in the BrSP-5 isolate and to amino acid 18 in the BrBA-1 isolate, after the conserved DAG domain (Fig. 1). Such differences with regard to the size of the CP gene were previously reported in Brazil by Lima *et al.* [19]. These authors found 10 Brazilians isolates with a CP gene containing 924 nucleotides and 2 isolates from Bahia and Paraná with only 921 nucleotides.

Inoue-Nagata *et al.* [25] compared the genetic sequences from two strains of PRSV-W (a mild strain and a severe strain) and observed that a mild strain genome presented six additional nucleotides, located in the CP region, generating two additional amino acids (Asn and Asp). Variations in the nucleotide and amino acid lengths of PRSV CP have also been described by other authors [26, 27]. These variations are consistently found in the first 50 amino acids of the N terminus, particularly in the EK repetitive region, suggesting that this region is variable and, thus, likely to exhibit nucleotide insertions and deletions [28]. Therefore, it has been suggested that the 3' terminus of the CP gene should be used to obtain transgenic plants because this region is more conserved than the 5' region [29]. We also found a greater conservation in the 3' region (Fig. 1), confirming that this region is the most suitable for plant transformation.

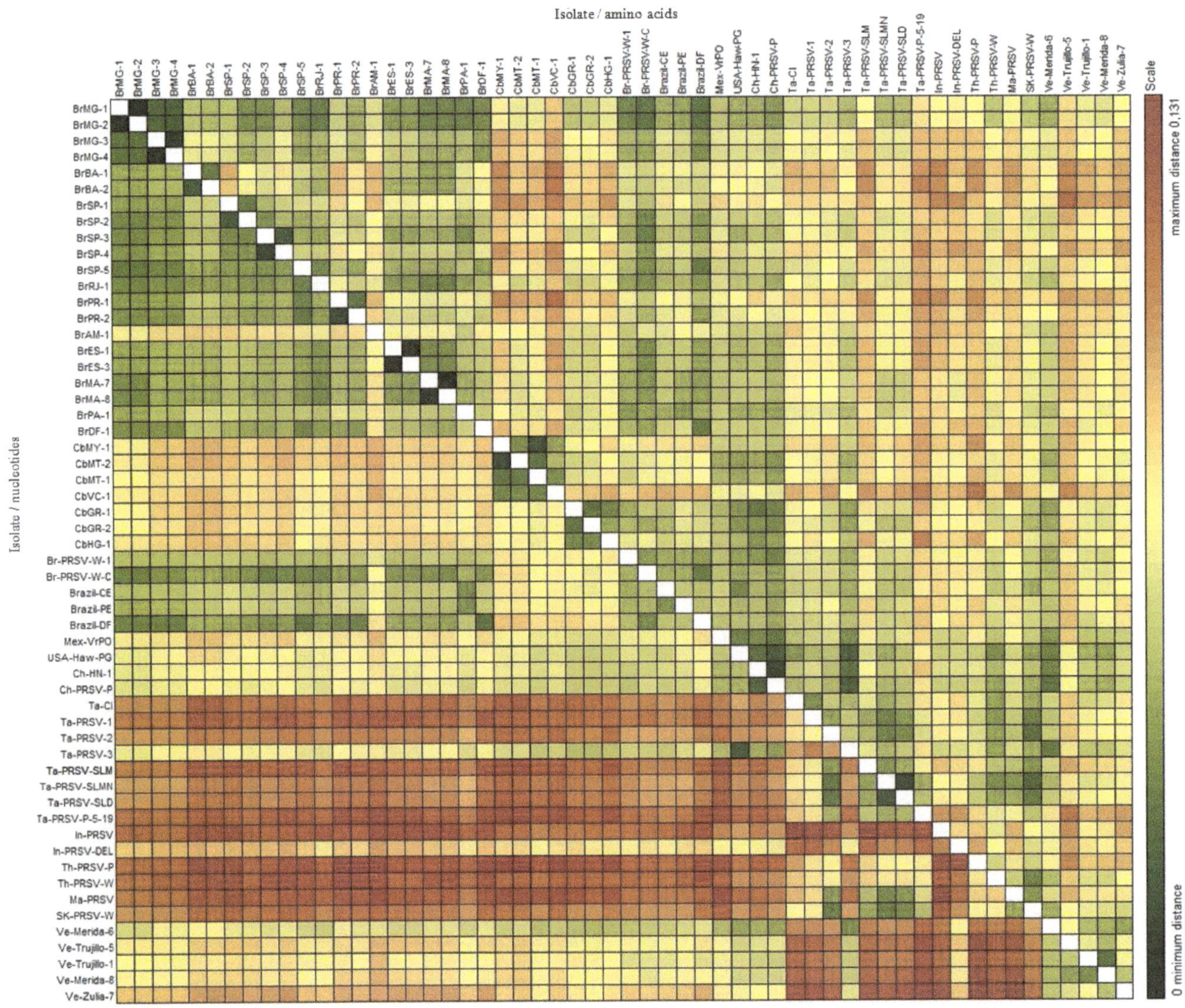

Figure 2. *Estimated genetic distance (pairwise analysis by deletion of gaps) among nucleotide sequences and among amino acid sequences of the coat protein gen of PRSV-P isolates from Brazil, Cuba and remaining isolates available in GenBank. Multiple alignments of sequences and evolutionary analyses were performed by the MEGA 5 program.*

As shown in Figure 2, the comparison using the pairwise tool of the genetic distance among the nucleotide sequences of the 28 isolates analyzed in this study and those from GenBank (n=56) indicated a range from 0 to 0.131. However, when we compared the isolates from Brazil and Cuba sequenced in this study (n=28), the genetic distance was lower, between 0 and 0.092, which is most likely because the isolates were derived from only two countries.

In this case, a zero value was observed among the isolates from the same states and municipalities in Brazil, (BrMG-1 and BrMG-2, BrES-1 and BrES-3), which shows the genetic similarity among the isolates from the same region. Moreover, the isolates that were more distant (0.092) are from the two different countries (BrAM-1 with CbMT-2 and CbVC-1). When only the Brazilian isolates were analyzed (n=21), there was a smaller distance, of up to 0.082, and the highest values (0.66 to 0.82) were found among the isolated BrMA-1, from the state of Amazonas, with the other isolates. This result is most likely due to the geographical distance between Amazonas and the other states, where the isolates were collected, because the distances varied between 0 and 0.063 among the other Brazilian isolates.

The analysis of the Cuban isolates (n=7) showed a smaller genetic distance, with values of up to 0.073, which suggests that the isolated conditions of Cuba and also its smaller territory (when compared to Brazil) allow the PRSV-P isolates to retain a higher level of conservation. In fact, the longest distance was observed between two Cuban isolates from geographic regions more distant within the archipelago (CbMT-2 and CbHG-1), whereas the lowest distance was observed in virus isolates from neighboring provinces (CbMY-1 and CBMT-1 and 2), reinforcing the hypothesis that genetic conservation is greater among isolates from nearby regions.

The identities among the gene sequences (data not shown) corroborated the results of the pairwise analyses, showing the lowest values among the isolates from America and Asia. Lower values (below 86%) were observed when the Brazilian isolates BrBA-2, BrSP-1, BrSP-4, BrPR-1, and BrMA-8 were compared with the Asian Ch-HN-1, Th-PRSV-P, and Th-PRSV-W isolates. The identities of nucleotides among the Brazilian isolates varied from 91% (among BrAM-1 and the isolates BrMA-8 and BrDF-1) to 100% (between BrMG-1 and BrMG-2 and between BrES-1 and BrES-3). Among the Cuban isolates, the identity ranged from 99% (among isolates CbMY-1 and CbMT-2) to 92% among isolates from the western region (Mayabeque, Matanzas, and Villa Clara) and the eastern region (Granma and Holguín).

Unlike the results obtained in this study, the study carried out by Lima *et al.* [19] showed higher identities between the CP sequences of PRSV isolates from different regions of Brazil, which is interesting because of the large geographical distance among the isolates evaluated by these authors. The greatest differences observed in the present study are likely due to the inclusion of an isolate from Amazonas, which was more diverse and more distant from the rest of the Brazilian isolates in both the pairwise and phylogenetic tree analyses.

When we analyzed the amino acid sequences of the studied isolates with the sequences obtained from GenBank, the conserved DAG, WCIEN, and QMKAAA domains were present (Fig. 1); these domains were also identified by other authors, who postulated that they are related to virus transmission by aphids [29, 30, 31]. Following the third amino acid after the triple block DAG domain, glutamic acid and lysine repeats known as the "EK region" [30] were observed and were more variable in the tested virus isolates.

The pairwise analysis of the amino acid sequences showed genetic distances similar to those observed for the nucleotide sequences (Fig. 2), with the lowest distance values detected among the isolates from the same states in Brazil (BrMG-1 and BrMG-2, BrES-1 and BrES-3) and the longest distances observed among the isolates from the two countries, and the highest value was between CbVC-1 and BrPR-1. In this case, the distance among all the sequences (n=56) ranged from 0 to 0.114 and among the studied isolates (n=28) were between 0 and 0.111.

Confirming the previous results, the identities between amino acids were higher than among nucleotides (data not shown). The lowest identity, 88%, occurred when all the sequences were compared (n=56) among the isolates from America and Asia. It is important to highlight the low identity (88%) that was registered between BrSP-1 and Ve-Trujillo-5, both from South America, which also showed higher distances by the pairwise method. Among the isolates sequenced in this study (n=28), the identities ranged from 89% (between CbVC-1 and BrBA-1, BrBA-2, BrSP-1, BrSP-4 and BrPR-1) to 94% (between CbGR-1 and BrMG-1, BrMG-2, BrRJ-1, BrES-1, BrES-3, BrPA-1; CbMT-1 and BrSP-2, BrMA-7, BrMA-8, BrPA-1; CbMT-2 and BrPA-1). Among the isolates from Brazil (n=21), identities varied from 90% (among the isolate from Amazonas BrAM-1, BrBA-1, BrBA-2 and BrSP-1) to 100% (among BrMG-1, and BrMG-2; BrMG-3, and BrMG-4; and also between the two isolates from Espirito Santo state and the two isolates from Maranhão state); the identities among the Cuban isolates ranged from 91% (between isolates from different regions; CbVC-1, CbGR-2, and CbHG-1) to 99% (CbMY-1 and CbMT-2).

The genetic distances among the CP amino acid sequences of the 28 isolates were slightly lower than those obtained for the nucleotide sequences, suggesting that the amino acid sequences of CP are affected by changes in the nucleotide sequences, in agreement with the results obtained by Silva-Rosales *et al.* [29]. Comparing the CP sequences of 7 Asian isolates with 3 Mexican isolates, these authors also found a higher degree of identity among nucleotides than among amino acid sequences. However, Jain *et al.* [32] detected higher degrees of identity among amino acid sequences than nucleotide sequences when comparing CP from two Indian isolates of PRSV-P, though this may have occurred because only two isolates from the same country were included.

In the phylogenetic tree (Fig. 3A), the different isolates were grouped into two clusters based on their CP nucleotide sequence: one large cluster (I) containing the isolates from America, one from Taiwan, and two from India and one minor of the Brazilian isolates (Ia). The Ia group showed a greater similarity among the isolates from the same region, except for the Amazonas isolate, which was far from all of the isolates, confirming the pairwise analyses. In the first cluster (I), we can observe the formation of another even smaller cluster (Ib) formed by isolates from the central western region of Cuba (CbMT-2, CbVC-1, CbMY-1, and CbMT-1) (Ib1), isolates from Venezuela (Ib2), those from the eastern region of Cuba (CbGr-1, CbGR-2, and CbHG-1) (Ib3), and the isolate from Taiwan (Ta-PRSV-1) with the U.S. and Mexican (Mex-VrPo) (Ib4) isolates. Finally, the isolates from India in the first cluster were significantly more distant (In-PRSV-DEL and In-PRSV) (Ic). The remaining isolates from Asia, including China, Thailand, Taiwan, South Korea, and Malaysia, were grouped into the second largest cluster (II).

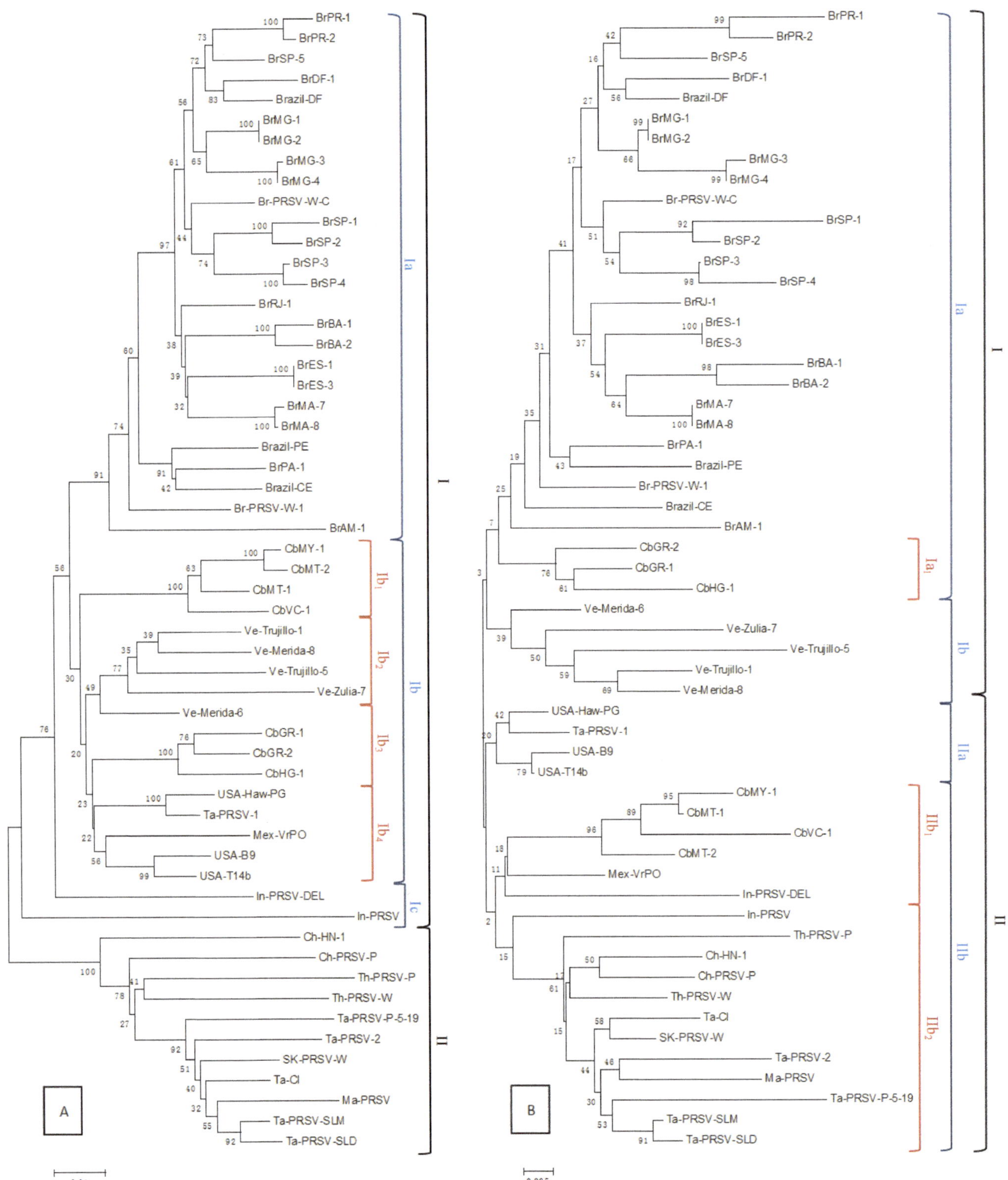

Figure 3. Phylogenetic trees based on the nucleotide (A) and amino acids (B) sequences of the coat protein gene of PRSV isolates from Brazil, Cuba and the remaining isolates available in GenBank. Bootstrap values were obtained by MEGA5 program, using Neighbor Joining with 2.000 replicates. Equal numbers, letters and/or sub-indexes indicate common groupings.

The phylogenetic tree based on the amino acid sequences of CP (Fig. 3B) revealed a grouping different from that observed in the tree constructed based on the nucleotide sequences. In this case, we observed two major clusters. The first cluster (I) contained the isolates from Brazil, the eastern region of Cuba (Ia), and Venezuela (Ib), which were subgrouped by country, and the Brazilian isolate from Amazon (BrAM-1) was more distant from the other isolates from Brazil, as observed in the nucleotide-based tree. The second cluster (II), containing the rest of the isolates, showed the American isolates close to each other, forming a smaller cluster (IIa), and the isolates from the western region of Cuba, Mexico (Mex-VrPo), and India (In-PRSV-DEL) (IIb1) separate from the other Asian isolates (IIb2).

Lima *et al.* [19] and Inoue-Nagata *et al.* [25] also observed that Brazilian isolates grouped into the same cluster, with few differences among them, which could be due to the use of isolates from other states that were not studied here. The results based on the CP nucleotide sequences also showed that the isolates were distributed into different groups according to the region of origin. Chin *et al.* [33] also detected a high genetic variability among the PRSV isolates from Venezuela compared to isolates from Jamaica. Such differences could be related, among other factors, to geographic isolation [34].

The phylogenetic groups obtained when the sequences studied were compared to the virus isolates available in GenBank were similar to those obtained in studies performed by Fernandez-Rodríguez *et al.* [35]. These authors observed one group containing PRSV-P isolates from Mexico, The United States, Cuba, Puerto Rico, Brazil, and Australia, as has been previously reported by several other authors, with groups of America-Australia isolates [2, 19, 25, 34, 36].

Phylogenetic relationships between the isolates from India and the isolates from other continents were also observed by some authors [28, 29, 37, 38]. These authors also noted a phylogenetic relationship among the isolates from India and America but always with a certain genetic distance among them, thus strengthening the hypothesis that PRSV may have been dispersed from India to America, via Brazil, Venezuela, or Mexico [34, 38].

The results obtained with the isolates from eastern and central-western Cuba were interesting, as they remained in separate clusters, in the trees based on both the nucleotide and on amino acid sequences. This finding is in agreement with the data obtained when Arocha *et al.* [20] analyzed the partial sequences of the CP gene of one virus isolated from the central region of Cuba, which grouped separately from the isolate collected in the eastern region. This fact might suggest different origins for these groups of isolates in Cuba.

Another fact to be noted in this study is that there were no differences in the groups containing strains PRSV-P and PRSV-W. These results were also observed by other authors when analyzing the CP sequence of isolates of these two strains from different regions of the world [25, 28, 37]. In addition to indicating possible recombination events between the two strains, the results suggested that either PRSV-P arose from PRSV-W or vice versa [34, 38].

The incubation period of PRSV-P in the papaya plants was variable among the different isolates and duration of the experiment (Table 2). In the winter, more time was required for the plants to display the first symptoms (incubation period from 10 to 38 days) when compared to the symptoms in summer (5-26 days), which suggests that summer temperatures (19.8°C and 34°C) favor viral multiplication and disease development. When different isolates were compared, it was observed that the BrRJ-1 and BrMA-8 isolates required a longer time period to induce symptoms under both experimental conditions, suggesting that these isolates are less aggressive than the other isolates.

Table 2. *Incubation period and intensity of symptoms induced by PRSV-P isolates on Carica papaya* L. cv. Solo *and Cucurbita pepo* L. cv. Caserta *45 and 21 days after inoculation, respectively.*

Isolate	*Carica papaya*					*Cucurbita pepo*		
	I.P. (days)		Foliar symptoms					
	Exp. 1	Exp. 2	M	B	D	S.S	I.P.	S.I
BrMG-1	15-37	9-17	++	++	+	++	9	+
BrMG-2	13-17	9-18	+++	++	-	++	9	+
BrMG-3	14-22	7-14	+++	++	++	++	9	+
BrMG-4	23-35	8-15	++	++	+	+	8	++
BrBA-1	14-17	7-10	+++	++	+++	+++	12	+
BrBA-2	11-16	6-11	+++	+++	+++	++	9	+
BrSP-1	15-27	8-23	+++	++	+	++	7	++
BrSP-2	12-35	9-15	++	+	++	++	7	++
BrSP-3	11-15	7-10	++	+	++	++	7	+++
BrSP-4	11-15	7-10	+++	+	++	++	13	++
BrSP-5	11-38	11-24	++	+	+	+	9	++
BrRJ-1	21-32	13-24	++	+	++	+	10	++
BrPR-1	12-14	7-13	++	++	++	-	11	+
BrPR-2	10-15	5-17	+++	++	+++	++	7	++
BrAM-1	13-32	7-12	++	+	+	+++	9	+
BrES-1	11-32	19-26	++	++	+	++	8	++
BrES-3	15-19	9-13	+++	++	+++	++	8	++
BrMA-7	15-30	10-21	++	+	+	+	7	+++
BrMA-8	18-20	13-23	++	+	+	+	7	++
BrPA-1	10-16	7-10	++	++	++	+	12	+
BrDF-1	11-16	8-16	+++	++	++	++	9	++
CbMY-1	11-19	12-22	++	++	+	+	12	++
CbMT-1	15-21	11-17	+	+	++	+	12	+
CbMT-2	16-31	9-15	++	++	++	++	13	+
CbVC-1	16-35	10-20	+++	++	++	+	12	+++
CbGR-1	14-20	13-21	++	+	+	+	8	++
CbGR-2	15-17	8-21	+++	++	+++	+	10	++
CbHG-1	11-22	8-14	++	++	++	+	10	++

I.P. Incubation period, Exp. experiment, M. mosaic, B. bubbles, D. distortion, S.S. stem spots, S.I. symptoms intensity

The plants inoculated with isolate BrMG-4 showed symptoms later in the winter, but the plants inoculated with BrES-1 and CbGR-1 showed symptoms later in the summer. This behavior suggests that the differential aggressiveness of these isolates could be influenced by the ambient conditions, particularly the temperature, as suggested Silva-Rosales *et al.* [29]. The same results were reported by Mangrahutia *et al.* [15], who found a maximum accumulation and expression of viral symptoms at temperatures between 26°C and 31°C.

In most of the isolates, the first symptoms observed were vein clearing and mild mosaic patterning (Fig. 4A). However, the BrMG-2, BrMG-3, BrSP-1, BrSP-3, BrSP-4, BrPR-1, and BrPR-2 isolates and also induced symptoms of wrinkled leaves with vein thickening (Fig. 4B), and BrMG-3, BrMG-4, and BrRJ-1 induced oily streaks in the stem (Fig. 4C). The symptoms intensified over time. In addition to the above-described symptoms, at 45 days after inoculation, moderate or severe (Fig. 4D, E) bubbles (Fig. 4F) and leaf distortion, with strong thickening of the ribs (Fig. 4G), were also observed.

Figure 4. *Initial symptoms of vein clearing and mild mosaic (A), curling (B) and small oily streaks in the stem (C) and severe symptoms of oily streaks on the stem (D), severe mosaic (E), blistering (F) and leaf distortion (G), in plants of Carica papaya L. cv. Solo, 45 days after mechanical inoculation with PRSV-P isolates.*

The type and severity of symptoms at 45 days after inoculation varied among the isolates (Table 2): BrBA-1, BrBA-2, BrPR-2, BrES-3, and CbGR-2 induced severe symptoms; BrSP-5, BrMA-7, BrMA-8, CbMT-1, and CbGR-1 induced mild symptoms; and the remaining isolates induced moderate symptoms. In general, the less aggressive isolates induced symptoms that ranged from mild to moderate, though none induced severe symptoms, suggesting the possible existence of a relationship between symptom severity and the aggressiveness of isolates. However, it is recommended that the severity of symptoms produced by these isolates be evaluated in other cultivars of commercial interest to both countries. Rodríguez *et al.* [24] also observed variability in the intensity of symptoms in different genotypes of papaya inoculated with the same isolate of PRSV-P.

The BrMG-1 e BrMG-2, BrES-1 e BrES-3 squash plants developed chlorotic spots all over the leaf surface (Fig. 5A) and vein clearing, with larger chlorotic spots (Fig. 5D). However, not all the leaves showed these symptoms, suggesting that, although the infection is systemic, the plants demonstrated some resistance to the movement of the virus. *C. quinoa* and *C. amaranticolor* plants did not show any symptoms and were not infected by these isolates of PRSV-P. The absent of symptoms in these hosts could be caused because sometimes the latex of papaya plants inhibits the viral infection.

Figure 5. *Symptoms induced by PRSV-P in leaves of Cucurbita pepo L. cv. Caserta, 21 days after mechanical inoculation. Healthy plant (A), chlorotic spots (B and C), ribs yellowing (C), curling deformation (D) and plant growth reduction (F).*

The initial symptoms in the squash plants were observed at approximately seven days after inoculation. Most isolates induced symptoms of chlorotic spots, as described above, but the BrES-1, BrES-3, BrDF-1, and BrMG-4 isolates and Cuban CbVC-1 caused leaf distortion and severe curling (Fig. 5E), stunting in plant growth (Fig. 5F) and yellow vein coloration (Fig. 5D) that was also induced by BrSP-5 in the squash plants.

Silva-Rosales *et al.* [29] reported similar results, observing symptoms only in papaya and *Cucumis metuliferus* plants, among the 13 species inoculated with PRSV-P isolates, including *C. quinoa* and *C. amaranticolor*. These results confirm that the PRSV-P isolates have a very narrow host range and a high specificity for *C. papaya* and some species of cucurbits [1, 39, 40].

The results based on the CP sequences of isolates from different regions of Brazil and Cuba and the symptoms induced in papaya and squash seedlings help to understand the genetic diversity of PRSV-P and offer valuable information for papaya ringspot management strategies, employing cross-protection and transgenic plants, in both countries. Therefore, it is recommended that biological studies should be conducted on papaya cultivars with commercial importance for both countries, before choosing the best control strategy to be employed.

Acknowledgements

This work was supported by Conselho Nacional de Desenvolvimento Científico e Tecnológico (CNPq), Coordenação de Aperfeiçoamento de Pessoal de Nível Superior (CAPES) from Federal Government of Brazil and Fundação de Amparo à Pesquisa do Estado de Minas Gerais (FAPEMIG). The authors thank Aurivan Soares (UFLA), Cristiane de Jesús Barbosa (Embrapa), Emí Lorenzeti (UFLA), Hermes Peixoto (Embrapa), João Eduardo de Melo (UFLA), Justo González (UNICA), Luis Gasparotto (Embrapa), Flávio H. R. Moraes (CEUMA), Enilton Santana (INCAPER), Daniel Schurt (Embrapa), Keize Pereira Junqueira (Embrapa) e Maita Ávila (UNICA), by collecting isolates of PRSV-P from several Brazilian and Cuban regions.

References

[1] D. E. Purcifull, J. R. Edwardson, E. Hiebert, and D. Gonsalves, "Papaya ringspot virus," CMI/AAB. Description of plant viruses, 292. 8p, 1984.

[2] P. F. Tennant, G. Fermin, and M. Roye, "Viruses infecting papaya (Carica papaya L.): etiology, pathogenesis and molecular biology," Plant Viruses. Wellesbourne, vol. 1, pp. 178-188, Abril 2007.

[3] C. M. Kalleshwaraswamy, and N. K. Kumar, "Transmission efficiency of Papaya ringspot virus by three aphid species," Phytopathology. Saint Paul, vol. 98, pp. 541-546, May 2008.

[4] S. Tripathi, J. Y. Suzuki, S. A. Ferreira, and D. Gonsalves, "Papaya ringspot virus-P: characteristics, pathogenicity, sequence variability and control. Mol. Plant Pathol. London, vol. 9, pp. 269-280, May 2008.

[5] M. J. Adams, J. F. Antoniw, and F. Beaudoin, "Review: overview and analysis of the polyprotein cleavage sites in the family Potyviridae," Mol. Plant Pathol. London, vol. 6, pp. 471–487, July 2005.

[6] S. D. Yeh, F. J. Jan, C. H. Chiang, T. J. Doong, M. C. Chen, P. H. Chung, and H. J. Bau, "Complete nucleotide sequence and genetic organization of Papaya ringspot virus RNA," J. Gen. Virol. London, vol. 73, pp. 2531-2541, October 1992.

[7] C. H. Wang, and S. D. Yeh, "Divergence and conservation of the genomic RNAs of Taiwan and Hawaii strains of papaya ringspot potyvirus," Arch. Virol. Vienna, vol. 142, pp. 271–285, May 1997.

[8] D. Gonsalves, "Control of Papaya rings-pot virus in papaya: A case study," in Annu. Rev. Phytopathol., vol. XXXVI, R.K. Webster, G. Shaner, N.K. Van Alfen, Eds. Palo Alto: Annual Reviews, 1998, pp. 415-437.

[9] G. Fermin, L. T. Castro, and P. F. Tennant, "CP-transgenic and non-transgenic approaches for the control of papaya ringspot: current situation and challenges," Transgenic Plant J. Kagawa, vol. 4, pp. 1-15, August 2010.

[10] M. T. Souza, "Analysis of the resistance in genetically engineered papaya against Papaya ringspot potyvirus, partial characterization of the PRSV. Brazil. Bahia isolate, and development of transgenic papaya for Brazil," PhD. These, Cornell University, 1999.

[11] P. Tennant, G. Fermin, M. Fitch, R. Manshardt, J. Slightom, and D. Gonsalves, "Papaya ringspot virus resistance of transgenic Rainbow and SunUp is affected by gene dosage, plant development, and coat protein homology," Eur. J. Plant Pathol. Dordrecht, vol. 107, pp. 645-653, July 2001.

[12] H. L. Wang, S. D. Yeh, R. J. Chiu, and D. Gonsalves, "Effectiveness of cross-protection by mild mutants of Papaya ringspot virus for control of ringspot disease of papaya in Taiwan. Plant Disease. Quebec, vol. 71, pp. 491-497, June 1987.

[13] J. A. M. Rezende, and G. W. Muller, "Mecanismos de proteção entre vírus e controle de viroses de vegetais por premunização," Revisão Anual de Patologia de Plantas. Passo Fundo, vol. 3, pp.185-226, 1995.

[14] R. C. A. Lima, J. A. A. Lima, Jr. M. T. Souza, G. Pio-Ribeiro, and G. P. Andrade, "Etiologia e estratégias de controle de viroses do mamoeiro no Brasil," Fitopatol. Bras. Brasilia, vol. 26, pp. 689-702, December 2001.

[15] S. K. Mangrauthia, P. Singh, and S. Praveen, "Genomics of helper component proteinase reveals effective strategy for Papaya ringspot virus resistance," Mol. Biotechnol. New York, vol. 44, pp. 22-29, January 2010.

[16] S. D. Yeh, and D. Gonsalves, "Practices and perspective of control of Papaya ringspot virus by cross protection," Adv. Dis.Vector Res. New York, vol. 10, pp. 237-257, 1994.

[17] IPS-Inter Press Service in Cuba, "Frutales en ascenso," 2009. http://www.ipscuba.net/. Accessed 26 June 2009.

[18] FAO Faostat (classic), "Production/crops primary," 2012, http://faostat.fao.org/site/2012. Accessed 16 January 2013.

[19] R. C. A. Lima, Jr. M. T. Souza, G. Pio-Ribeiro, J. A. A. Lima, "Sequences of the coat protein gene from Brazilian isolates of Papaya ringspot virus," Fitopatol. Bras. Brasilia, vol. 27, pp. 174-180, March/April 2002.

[20] Y. Arocha, B. Piñol, K. Acosta, R. Almeida, J. Devonshire, A. Van de Meene, E. Boa, and J. Lucas, "Detection of phytoplasma and potyvirus pathogens in papaya (Carica papaya L.) affected with 'Bunchy Top Symptom' (BTS) in eastern Cuba," Crop Protection. Guildford, vol. 28, pp. 640-646.

[21] O. Portal, D. Cabrera, A. Sánchez, A. L. Darías, J. E. González, and R. Gómez, "Molecular characterization of two Cuban isolates of Papaya ringspot virus by means of coat protein analysis," Commun Agric. Appl. Biol. Sci. Berlin, vol. 71, pp. 1203-1205, September 2006.

[22] AFGC: Arabidopsis Functional Genomics Consortium, "Total RNA isolation," 2002. http://www.arabidopsis.org/portals/masc/AFGC/RevisedAFGC/site2RnaL.htm>. Accessed 09 September 2010.

[23] K. Tamura, D. Peterson, N. Peterson, G. Stecher, M. Nei, and S. Kumar, "MEGA5: Molecular Evolutionary Genetics Analysis using Maximum likelihood, Evolutionary Distance, and Maximum Parsimony Methods," Mol. Biol. Evol. Oxford, vol. 28, pp. 2731-272739, October 2011.

[24] D. Rodríguez, Y. Tornet, M. Alonso, L. Valero, I. Peña, A. R. Figueira, And R. Ramos, "Severidade da mancha anelar do mamoeiro em diferentes genótipos do grupo Solo introduzidos em Cuba," J. of Biotec. Biodivers. Tocantins, vol. 2, pp. 28-36, Novembre 2011.

[25] A. K. Inoue-Nagata, C. M. Franco, D. P. Martin, J. A. M. Rezende, G. B. Ferreira, L. S. Dutra, and T. Nagata, "Genome analysis of a severe and a mild isolate of Papaya ringspot virus-type W found in Brazil," Virus Genes. Amsterdam, vol. 35, pp.119-127, August 2007.

[26] M. F. Bateson, R. E. Lines, P. Revill, W. Chaleeprom, C. V. Ha, A. J. Gibbs, and J. L. Dale, "On the evolution and molecular epidemiology of the potyvirus Papaya ringspot virus," J Gen. Virol. London, vol. 83, pp. 2575-2585, October 2002.

[27] R. K. Jain, J. Sharma, A. S. Sivakumar, P. K. Sharma, A. S. Byadgi, A. K. Verma, and A. Varma, "Variability in the coat protein gene of Papaya ringspot virus isolates from multiple locations in India," Arch. Virol. Vienna, vol. 149, pp. 2435-2442, December 2004.

[28] Y. W. Lu, W. T. Shen, P. Zhou, Q. J. Tang, Y. M. Niu, M. Peng, and Z. Xiong, "Complete genomic sequence of a Papaya ringspot virus isolate from Hainan Island, China," Arch. Virol. Vienna, vol. 153, pp. 991-993, May 2008.

[29] L. Silva-Rosales, N. Becerra-Leor, S. Ruiz-Castro, D. Teliz-Ortiz, and J. C. Noa-Carrazana, "Coat protein sequence comparisons of three Mexican isolates of Papaya ringspot virus with other geographical isolates reveal a close relationship to American and Australian isolates," Arch. Virol. Vienna, vol. 145, pp. 835-843, April 2000.

[30] D. D. Shukla, and C. W. Ward, "Structure of potyvirus coat proteins and its application in the taxonomy of the potyvirus group," Adv. Virus. Res. New York, vol. 36, pp. 273-314, June 1989.

[31] C. D. Atreya, B. Raccah, and T. P. Pirone, "A point mutation in the coat protein abolishes aphid transmisibility of a potyvirus," Virology. New York, vol. 178, pp. 161-165, September 1990.

[32] R. K. Jain, H. R. Pappu, S. S. Pappu, A. Varma, and R. D. Ram, "Molecular characterization of Papaya ringspot potyvirus isolates from India," Annals of Applied Biology. Worwick, vol. 132, pp. 413-425, June 1998.

[33] M. Chin, Y. Rojas, J. Moret, G. Fermin, P. Tennant, and D. Gonsalves, "Varying genetic diversity of Papaya ringspot virus isolates from two time-separated outbreaks in Jamaica and Venezuela," Arch. Virol. Vienna, vol. 152, pp. 2101-2106, August 2007.

[34] X. A. Olarte Castillo, G. Fermin, J. Tabima, Y. Rojas, P. F. Tennant, M. Fuchs, R. Sierra, A. J. Bernal, and S. Restrepo, "Phylogeography and molecular epidemiology of Papaya ringspot virus," Virus Res. Amsterdam, vol. 159, pp. 132-140, August 2011.

[35] T. Fernández-Rodríguez, L. Rubio, O. Carballo, and E. Marys, "Genetic variation of Papaya ringspot virus in Venezuela," Arch. Virol. New York, vol. 153, pp. 343-349, November 2008.

[36] O. A. Abdalla, and A. Ali, "Genetic diversity in the 3'-terminal region of Papaya ringspot virus (PRSV-W) isolates from watermelon in Oklahoma," Arch. Virol. New York, vol. 157, pp. 405-12, March 2012.

[37] S. K. Mangrauthia, B. Parameswari, R. K. Jain, and S. Praveen, "Role of genetic recombination in the molecular architecture of Papaya ringspot virus," Biochem. Genet. New York, vol. 46, pp. 835-846, December 2008.

[38] S. K. Mangrauthia, VP. Singh Shakya, RK. Jain, and S. Praveen, "Ambient temperature perception in papaya for Papaya ringspot virus interaction," Virus Genes. Amsterdam, vol. 38, pp. 429-434, June 2009.

[39] M. F. Bateson, J. Henderson, W. Chaleeprom, A. J. Gibbs, and J. L. Dale, "Papaya ringspot potyvirus: Isolate variability and origin of PRSV type P (Australia)", J. Gen. Virol. London, vol. 75, pp. 3547-3553, 1994.

[40] S. D. Yeh, and D. Gonsalves, "Evaluation of induced mutants of Papaya ringspot virus for control by cross protection," Phytopathology. Saint Paul, vol. 74, pp. 1086-109, September 1984.

Optimization of fermentation medium compositions from dewatered wastewater sludge of beer manufactory for *Bacilus thuringiensis* delta endotoxin production

Nguyen Thi Hoa[1], Tang Thi Chinh[1, *], Dang Thi Mai Anh[1], Ngo Dinh Binh[2], Le Thi Minh Thanh[2]

[1]Dept of Environmental Microbiology, Institute of Environmental Technology, Vietnamese Academy of Science and Technology, Hanoi, Vietnam

[2]Institute of Biotechnology, Vietnam Academy of Science and Technology, Hanoi, Vietnam

Email address:

hoabio78@gmail.com (N. T. Hoa), tangthichinh@gmail.com (T. T. Chinh), maianh058@gmail.com (D. T. M. Anh), binh.gen@gmail.com (N. D. Binh), phucphat@gmail.com (L. T. M. Thanh)

Abstract: Optimization of medium compositions from dewatered wastewater sludge of beer manufactory for delta endotoxin production by *Bacillus thuringiensis* was investigated in flask fermentation. Its composition consisted of wastewater sludge hydrolysis broth with several agricultural byproducts (rice bran, soybean meal, corn flour) and mineral salts ($MgSO_4.7H_2O$; K_2HPO_4; KH_2PO_4; $MnSO_4$; $CaCl_2$; NaCl). Rice bran and $MgSO_4.7H_2O$, $CaCl_2$ were all found to have a significant influence on delta endotoxin production. The optimal concentration of these three factors were then sequentially investigated using the response surface methodology with a central composite design. The resulting optimal medium components for delta endotoxin production were determined as follows: WWS (20 g/l, dry weight), $CaCl_2$ (0.3 g/l), $MgSO_4.7H_2O$ (0.32 g/l), and rice bran (5.8 g/l). Using this optimized fermentation media, the yield of delta endotoxin was increased by 34% to 565 mg/l compared with unoptimal medium. Viable cell and spore counts obtained in optimum fermented broth were $1.25x10^9$ CFU/ml, $1.07x10^9$ CFU/ml, respectively. The LC50 value of *Bacillus thuringiensis* serovar *isralensis* against *Culex quinquefasciatus* was 0.056 mg/l and there was no significant difference between LC50 value of bacterium grown in the medium.

Keywords: *Bacillus thuringiensis*, Delta endotoxin, Dewatered Wastewater Sludge, Bio-Pesticide

1. Introduction

Last century, *Bacillus thuringiensis* (Bt) was the most studied in biological agents for control of pest insects. Moreover in last four decades *Bacillus thuringiensis* was used to forming bio-pesticide to control pest insects in agriculture, forestry and mosquito that is vector causes many diseases such as: malaria, yellow fever, Japanese encephalitis, and dengue fever. *Bacillus thuringiesis* – based bio-pesticides have utmost importance and occupy more than 90% of the world bio-pesticide market and not harmful for people, animals and environment [4]. Today, bio-pesticides are world wide applied generously to protect plant in the world. However, using bio-pesticides still limited as chemical pesticides so high cost. Stanbury et al. (1995) estimated cost of materials for producing bio-pesticides about 35 – 59%. So finding cheap and popular materials to reduce product cost is target of productions [18].

Today, in the world there are a lot of experiments to reuse industry, agriculture waste to produce bio-pesticide. In Canada, there are a lot of studies about treating, reusing slaughterhouse, starch industry wastewater sludge as material for production of bio-pesticides to provide for agriculture. Reusing wastewater sludge can reduce about 30% cost as TSB [1, 11]. Besides that, there are also studies reusing other organic waste as material for *Bacillus thuringiensis* fermentation: experiment of Subbiab Poopathi and Archana shown that use chicken feather from poultry farm to culture *B. thuringiensis* subsp. *israelensis* maybe reduce 49 times cost of product as TSB [19]. Recently, Songqing Wu et al (2014) studied to reuse spent mushroom substrate to culture *Bacillus thuringiensis* by the solid – state fermentation method. The toxicity of this product against fourth instars larvae of *Culex quinquefasciatus* was 1.487IU/mg and cost about US$ 0.075/kg [17]. In this paper we show about optimization of medium composition for delta – endotoxin production by *Bacillus thuringiensis* used dewatered wastewater sludge from beer manufactory.

2. Material and Methods

2.1. Bacterial Strain and Media

- *Bacillus thuringiensis* serovar *israelensis,* H14 (Bti) used in this study provide by Department of Microbial Genetics, Institute of Biotechnology, Vietnam Academy Science and Technology, Hanoi Vietnam.
- *Reference medium* — Tryptic Soya broth (TSB) was used as a control media (g/l): soybean meal 5.0; glucose 5.0; starch 5.0; K_2HPO_4 1.0; KH_2PO_4 1.0; $MgSO_4.7H_2O$, 0.3; $FeSO_4.7H_2O$ 0.02; $ZnSO_4.7H_2O$ 0.02; $CaCO_3$ 1.0
- *Inoculum preparation*: The inoculum was prepared follow reported by Lachhab et al [13]. The Bti was subcultured on tryptic soya agar (TSA) tube, incubated for 48 h at 30 ± 0.1°C and then preserved at 4 ± 0.1°C for future use.
- Media used for inoculum preparation were adjusted to pH 7.0 ± 0.1 before autoclaving. A loopful of Bti grown on TSA tube was used to inoculate a 200 ml Erlenmeyer flask containing 50 ml of sterilized tryptic soya broth (TSB) medium. The flask was incubated in a rotary shaker at 200 revolutions per min (rpm) and at 30 ± 0.1°C for 8 – 12 h. A 2% (v/v) inoculum from this flask was then used to inoculate 500-ml Erlenmeyer flasks containing 100 ml of sterilized medium [12].

2.2. Sludge

The dewatered wastewater sludge (DWS) of Saigon Hanoi brewing factory (Hanoi, Vietnam) was used as a raw material for Bti growth. DWS was sampled and stored at 4 ± 0.1°C to minimize microbial degradation.

2.3. Sludge Pre-Treatments

The dewatered wastewater sludge was pre-treatment by thermal alkaline hydrolysis [21]. The total solid of DWS was approximately 20%, was diluted to 2% and raised to pH 10±1 and treated at 121°C for 30 minutes. The chemical composition of DWS hydrolysis broth was showed at Table 1. In which, the chemical composition was determined by standard methods [2]. The DWS hydrolysis broth was adjusted to pH7±0.1 before using for studies.

2.4. Fermentation

Erlenmeyer flasks (500 ml) containing 100 ml of DWS hydrolysis broth were inoculated with 2% (v/v) inoculum and incubated in a shaker at 200 rpm and 30 ± 0.1°C for 48 hrs.

2.5. Estimation of Total Cell and Spore Count

The total cell count and spore count were performed by counting colonies grown on TSA plates [11]. The samples were serially diluted with sterile saline solution (0.85% NaCl) to determine total cell count and spore count, 0.1 ml appropriately diluted samples were plated on TSA plates. All samples were incubated at 30°C for 24 hrs to form fully developed colonies. To determine spore count, appropriately dilute samples were heated in a silicon bath at 80°C for 10 min. and then chilled on ice for 5 min. [20]. Counting colonies grown on nutrient medium to estimate total cell and spore count. Each dilution was repeated three times and took the average of three replicate.

Table 1. *Characteristics of DWS hydrolysis broth*

Characteristics	Concentration (mg/l)
Total organic carbon (TOC)	306
Total nitrogen (TN)	16.7
Total phosphorous (TP)	1.94
Al	21.8
Ca	38.9
Cd	0.0029
Cr	0.0171
Cu	0.242
K	27.8
Mn	0.155
Ni	1.070
Pb	0.021
Fe	20.80
Mg	8.760
Zn	6.3
Na	172

2.6. Estimation of Delta-Endotoxin

Delta-endotoxin concentration was estimated according to Khanh Dang vu et al 2008 based on the solubility of insecticidal crystal proteins in alkaline condition[12, 22-24]: 1 ml of sample was centrifuged at 10,000g for 10 min. at 4°C. The pellet containing a mixture of insecticidal crystal proteins, spores, cell debris and residual suspended solids was used to determine the concentration of alkali soluble insecticidal crystal proteins (delta-endotoxin). These pellets were washed with 1 ml of 0.14 M NaCl - 0.0 1% Triton X-100 solutions three times to eliminate the soluble proteins and proteases that might be added on to the pellet and could affect the integrity of the crystal protein. The delta endotoxin in the pellet was dissolved with 0.05 N Na0H (pH 12.5) at 30°C for 3 h with stirring. The suspension was centrifuged at 10,000g for 10 min at 4°C and discard the pellet which containing spores, cell debris and residual suspended solids. The supernatant was used to determine the delta-endotoxin concentration by Bradford method using bovine serum albumin as standard protein [6].

2.7. Medium Optimization Using One-Factor-at-a-Time

The production of delta endotoxin was optimized base on

adding nutrients to the hydrolysis sludge broth. Three organic compounds: rice bran, corn bran, soybean meal were added at a concentration of 5 g/l each; Mineral salt: $CaCl_2$, $MgSO_47H_2O$, NaCl, K_2HPO_4, KH_2PO_4 was added at a concentration of 0.1 g/l and 0.5 g/l.

2.8. Central Composite Design and Response Surface Methodology

A central composite design was adopted to optimize the effect of the major variables (rice brand, $CaCl_2$, and $MgSO_4.7H_2O$) on delta endotoxin production. The effect of each variable was studied at five different levels, namely -α, 1, 0, +1, +α, and the set of 20 experiments was performed in triplicate (table 3). The central coded value for all the variables was set at zero. The data obtained from the RSM on delta endotoxin production were subjected to analysis of variance (ANOVA). The results of the RSM were used to fit a second-order polynomial, Eq. (1), to represent the behavior of the system:

$$Y = b_0 + b_1X_1 + b_2X_2 + b_3X_3 + b_1b_1X_1^2 + b_2b_2X_2^2 + b_3b_3X_3^2 + b_1b_2X_1X_2 + b_1b_3X_1X_3 + b_2b_3X_2X_3 \quad (1)$$

Where Y is the response variable representing delta endotoxin concentration, b_0 is the intercept, b_1, b_2, b_3, are the linear coefficients, b_1b_1, b_2b_2, b_3b_3 are the squared coefficients, b_1b_2, b_2b_3, b_1b_3 are the interaction coefficients, and X_1, X_2, X_3, X_1^2, X_2^2, X_3^2, X_1X_2, X_2X_3, X_1X_3 are the levels of the independent variables. The data analysis and generation of the response surface graphs were conducted using the statistical software Design Expert (Design-Expert 9.0.2).

2.9. Validation of the Model

The validation experiment was carried out in triplicate tests under the optimized condition to verify the predicted results.

2.10. Insect Bioassays

The larvae Culex quinquefasciatus (Diptera: Culicidae) received from National institute of Malariology Parasitology and Entomology, Hanoi, Vietnam.

The bioassays were carried out as described previously [3]. Preparing a stock solution of delta endotoxin (100 mg/l) and serial dilution were made (0.003 – 0.1 mg/l). Bioassays were carried out in disposable cups, containing 100 ml of water with appropriate dosage. 10 third instar larvae were taken into the paper cups separately. Bioassays were carried out at room temperature (28 – 30°C) and larvae mortality was monitored in 24 hrs. The LC50 value was estimated by probit analysis calculations [3]. The experiments were carried out simultaneously with control samples and also observed in 24 hrs.

3. Results

3.1. Effects of Mineral and Organic Matter Added to the DWSB Thermal Alkaline Hydrolysis Broth on Delta Endotoxin Production

The effects of mineral salts and organic matter were investigated by adding $MgSO_4.7H_2O$, $CaCl_2$, NaCl, KH_2PO_4, K_2HPO_4, $MnCl_2$, rice brand, soya meal to the DWS hydrolysis broth on delta endotoxin production was shown on Table 2.

Table 2. *Cell, spore counts and delta endotoxin concentration of fermented broth when adding minerals and organic matter*

Culture medium	Cell count ($x10^8$CFU/ml)	Spore count ($x10^8$CFU/ml)	Con. of delta endotoxin (mg/l)
DWSHB	1.5	1.3	420
Mineral salts			
DWSHB + $MgSO_4.7H_2O$ (0.5g/l)	3.1	2.7	510
DWSHB + $CaCl_2$(0.5g/l)	3.2	2.5	505
DWSHB + NaCl(0.5g/l)	2.5	2.0	448
DWSHB + KH_2PO_4 (0.5g/l)	2.2	2.1	442
DWSHB BTSH + K_2HPO_4 (0.5g/l)	2.5	2.4	438
DWSHB + $MnCl_2$ (0.01g/l)	0.02	0.001	-
Agro. byproduct			
DWSHB + rice bran (5g/l)	5.1	4.8	509
DWSHB + corn flour (5g/l)	3.8	3.3	452
DWSHB + soya meal (5g/l)	5.5	3.9	504

As shown in Table 2, Out of six inorganic salts examined, $MgSO_4.7H_2O$ and $CaCl_2$ were found to have a significant influence on delta endotoxin production. Viable cell counts and spore counts increased by two times, the concentration of delta endotoxin also increased by 20% compared to other media. Meanwhile, NaCl, KH_2PO_4, K_2HPO_4 had no obvious effect and $MnCl_2$ even had a negative effect on growth potential or delta endotoxin production of Bt.

All of organic matters were significant influence on delta endotoxin. There was no significant difference between delta endotoxin concentration obtained from DSW + soya meal and DSWHB + rice bran fermented broth. A maximum delta endotoxin was observed in the medium DSWHB + rice bran (509 mg/l) increased about 25%, viable cell counts and spore counts also increased by 3 times compared to DSWHB media (Table 2). However, rice bran is an agro-industrial, cheap, popularity so rice bran was chosen for further studies. Interestingly, it was observed that the medium containing all of $MgSO_47H_2O$, $CaCl_2$, rice bran and DWS yielded the highest delta endotoxin production. Thus, $MgSO_4$, $CaCl_2$, rice bran were chosen to optimize the medium for delta endotoxin.

3.2. Optimization Using Central Composite Design

Table 3. Experimental design of the model

No.	X_1: $CaCl_2$ (g/l)	X_2: $MgSO_4.7H_2O$ (g/l)	X_3: rice brand (g/l)
1	1	0.6	0
2	0.5	0.3	13.40896
3	0.5	0.3	0
4	1	0	10
5	0.5	0.804538	5
6	0	0.3	5
7	1	0	0
8	0.5	0.3	5
9	1	0.6	10
10	1.340896	0.3	5
11	0	0	0
12	0.5	0.3	5
13	0.5	0	5
14	0	0.6	10
15	0.5	0.3	5
16	0.5	0.3	5
17	0	0.6	0
18	0	0	10
19	0.5	0.3	5
20	0.5	0.3	5

The effect of each variable on delta endotoxin was characterized at five different levels, namely -1.68, -1, 0, +1, +1.68. Thus, a set of three factors with five levels and a total of 20 treatments were performed (Table 3). The experiments were repeated three times. The significance of each term in the model is presented in Table 4. The ANOVA for the selected quadratic model showed that the model was significant with a Model F = 32.92 and P > F-value < 0.0001. Values of "Prob > F" less than 0.0500 indicate model terms are significant. In this case, X_3, X_1X_3, X_1^2, X_2^2, X_3^2 are significant model terms.

The F-value and P-value of 'lack of fit' were 2.6604 and 0.1367, respectively, which indicated that 'lack of fit' was non significant. A non-significant value of lack of fit indicates that model was significant [5, 14]. It was also suggested that the quadratic and linear terms of rice bran, $MgSO_4.7H_2O$ and $CaCl_2$ of the model primarily determined endotoxin production by Bti. Thus, the response of delta endotoxin production (Y) by *Bacillus thurringiensis* serovar *israelensis,* H14 could be expressed in terms of the following regression equation 2:

Table 4. Analysis of variance (ANOVA) for the model regression representing delta endotoxin production

Source	Sum of Squares	df	Mean Square	F Value	p-value Prob > F
Model	188417.1	9	20935.24	304.255	5.39E-11
X_1-Canxi clorua	19391.3	1	19391.3	281.8167	1.18E-08
X_2-Magie sulfate	113.995	1	113.995	1.656707	0.227042
X_3-Rice bran	180.1918	1	180.1918	2.618755	0.136677
X_1X_2	162	1	162	2.354371	0.155941
X_1X_3	24.5	1	24.5	0.356062	0.563965
X_2X_3	1300.5	1	1300.5	18.90037	0.001449
X_1^2	81250.14	1	81250.14	1180.821	1.03E-11
X_2^2	13011.62	1	13011.62	189.0999	8.04E-08
X_3^2	13125.63	1	13125.63	190.7568	7.71E-08
Residual	688.0819	10	68.80819		
Lack of Fit	512.7486	5	102.5497	2.924422	0.131945
Pure Error	175.3333	5	35.06667		
Cor Total	189105.2	19			

$$\text{Delta endotoxin (Y)} = 409.00502 + 84.7006*X_1 + 4407.36222*X_2 + 36.38446*X_3 + 142.79838*X_1*X_2 + 17.68827*X_1*X_3 - 17.20373*X_2*X_3 - 245.53815*X_1^2 - 740.09354*X_2^2 - 4.49948*X_3^2 \quad (2)$$

Where, X1 is concentration of $CaCl_2$, X2 is concentration of $MgSO_4$, and X3 is concentration of rice bran.

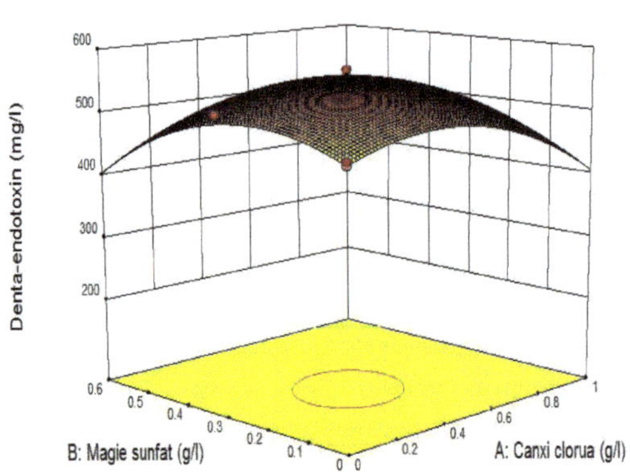

(a)

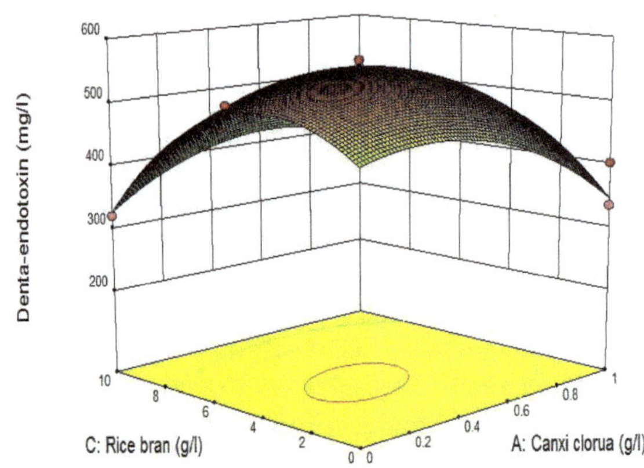

(b)

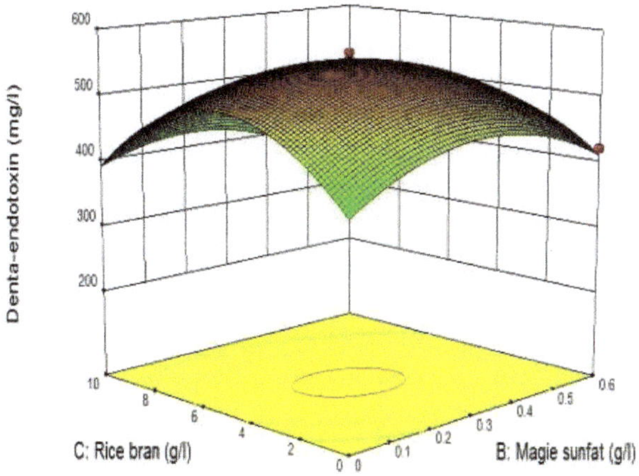

(c)

Figure 1. *Three-dimensional response surface plots of delta endotoxin production by Bti under conditions optimized using RSM*

(a) Effect of interaction of Magie sulfat and canxi clorua; (b) effect of interaction of rice bran and canxi clorua; (c) effect of interaction of rice bran and magie sulfat

A 3D response surface was drawn based on the model equation to investigate the interaction among the variables and determine the optimum concentration of each factor for maximum delta endotoxin production by Bti (Fig.1).

The model predicted that the maximum delta endotoxin production was up to 559 mg/l U/ml when the medium was DWSHB + $CaCl_2$ 0.3 g/l, $MgSO_47H_2O$ 0.32 g/l and rice bran 5.8 g/l.

3.3. Validation of the Model

On the basis of medium optimization, the quadratic model predicted that the maximum production of delta endotoxin by *Bacillus thuringiensis* serovar *israelensis*, H14 was 559 mg/l, when the DWS hydrolysis broth was added $CaCl_2$ 0.3 g/l, $MgSO_4$ 0.32 g/l and rice bran 5.8 g/l (Table 5). To confirm the accuracy of the model, validation experiment was performed in triplicate tests. The results presented in table 5 shown that the difference between concentration of delta endotoxin predicted and experimental was insignificant, therefore, the model was validation. Thus, the optimal medium was DWS hydrolysis broth + $CaCl_2$ 0.3 g/l, $MgSO_4$ 0.32 g/l and rice bran 5.8 g/l. When using this medium, delta endotoxin concentration of fermented broth increased by 34% compared to DWSHB media at the end of fermentation (Table 5).

Table 5. *Experimental verification of combined effect of optimized medium on delta endotoxin production*

Medium	Delta endotoxin (mg/l)		Cell count (CFU/ml)	Spore count (CFU/ml)
	Predicted	observed		
DWSHB+ rice bran 5.8 g/l + $MgSO_4.7H_2O$ 0.32 g/l + $CaCl_2$ 0.3 g/l	559	565	$1,25x10^9$	$1,07x10^9$
DWSHB		422	$1.8x10^8$	$1.3x10^8$

3.4. Toxicity of Bt Product on Culex Quinquefasciatus

The toxicity of Bti produced from two culture media (TSB, DWSHB + rice bran + $CaCl_2$ + $MgSO_4.7H_2O$) was tested with larvae Cx *quinquefasciatus* larvae. The results showed that, there was no significant difference between LC50 value of Bti grown in the TSB media or optimum DWS media. The LC50 value of Bti against Cx *quinquefasciatus* larvae was 0.056mg/l

4. Discussion

Microorganisms require nutrients and minerals for their growth and metabolic product formation. Nutrient has a strong effect on delta endotoxin synthesis. According Zouari N (1999) when changing the nutrient composition of the fermentation medium will alter the ability to synthesize delta endotoxin of Bt. In this study, concentration of delta endotoxin increase by 20 – 30% when adding some minerals and organic matters to DWS hydrolysis broth. Içgen et al. (2002) and Braun (2000) showed that deficiency of Mg or Mn reduced growth and sporulation, but the most noteworthy effect was a decrease in toxin biosynthesis [7, 9]. Magnesium acts as an enzymatic regulator, by activating some enzymes involved in spore formation [1, 9]. Ozkan et al. (2003) indicated that Mg increased toxin biosynthesis when provided at $8x10^{-3}M$ concentrations [16]. In addition, Içgen et al. (2002) shown that growth, sporulation and delta endotoxin synthesis were all very effective when Mg was provided at $8x10^{-5}$ M to $4x10^{-3}M$. In our study, delta endotoxin concentration of fermented broth achieved highest when Mg was provided at $1.3x10^{-3}M$, the lower or higher concentrations both reduced delta endotoxin production. Thus, $MgSO_4.7H_2O$ $1.66x10^{-3}M$ (original Mg concentration of DWS hydrolysis broth $0.33^{-3}M$, table 1) was best effect on delta endotoxin production.

In Ozkan study, canxi also plays an important role in the composition of the Bt fermentation medium at concentrations of 5.5 $x10^{-4}M$. However, our study results shown that Mg of $3,67x10^{-3}M$ had a best effect on delta endotoxin production. Subbial Poopathi et al. (2012) shown production of Bti from chicken feather medium was increased significantly by adding $MnCl_2$ at concentration of $3.9x10^{-2}M$. Mn^{2+} concentration of DWS hydrolysis broth was $2.8x10^{-6}M$ (Table 1) when adding $MnCl_2$ at concentration of $7.9x10^{-5}M$ (0.01 g/l) had a negative effect on Bti production [19]. In agreement with this result, Ozkan et al. (2003) also found that Mn^{+2} were the most critical element for the biosynthesis of toxins at 10^{-6} M concentration [16].

Traditional methods to optimize fermentation medium composition change one independent variable, while the variables are keeping fixed at a certain level. However, the single-dimensional search not only is laborious and time-consuming but also incapable of reaching a true optimum owing to the interactions among the variables [8]. The response surface method was first described by Box and Wilson in 1951 [5]. This is an effective strategy for seeking the optimum conditions for a multivariable system. Jaekoo Lee et al. (2013) shown that galactosidase activity of *Bacillus* sp. LX-1 increased by 6.3 times compared to that value previously obtained under basal conditions [10]. Gao et al. (2013) found that using response surface method to optimize the culture medium for laccase production by *Trichoderma harziamum* ZF-2, the yield of laccase was increased 59.68 times compare with the laccase production with an unopitimized medium [8]. Linna Du et al. (2012) also shown that the biomass, adenosine, polysaccharide and cordyceps acid yields were enhanced by 8.200, 3.580, 23.170 and 31.510%, respectively, when optimizing culture medium by response surface method [15]. In this study, optimization of culture medium for delta endotoxin production by Bti using response surface method, the yield of delta endotoxin production was increased by 34% compare with the delta endotoxin productional obtained from unoptimum medium.

Acknowledgements

This research was supported by the Vietnam Academy of Science and Technology (VAST). Grant number VAST.TD.02/13-14.

References

[1] A., Yezza, R.D.Tyagi, J.R.Valero, and R.Y.Surampalli (2005) *Bioconversion of industrial wastewater and wastewater sludge in Bacillus Thuringiensis based biopesticides in pilot fermentor,* Bioresource Technology. 97, pp. 1850-1857.

[2] Apha, Awwa, and Wpcf (1998) *Standard methods for examination of water and wastewaters (sections 2541B-E, 5310B, 4500¬NH3 H, 4500-N B and 4500-P H), 20th edn. In: Clesceri LS, Greenberg AE, Eaton AD (eds)DC* American Public Health Association, Washington.

[3] Binh, Ngo Dinh, Tuan Nguyen Dinh, Ha Trinh Thu, Duong Pham Thuy, Huong Pham Minh, and Luy Nguyen Thi (2009) *Production and Evaluation of Mosquitocidal Efficacy of Bacillus thuringensis subsp. israelensis Based Formulations in Viet Nam,* Proceedings of the 2nd International Conference of Integrated Pest Management in Asia and Africa. Science and Technics Publishing House. 2, pp. 279-283.

[4] Binh, Ngo Dinh, Thanh Le Thi Minh, Ha Trinh Thi Thu, Thuy Pham Kieu, Nhung Le Thi Hong, and Tien Dang Van (2011) *35 years – Research and Development of Bacillus thuringiensis Biopesticides in Vietnam,* Proceedings of the 4th International Conference of Integrated Pest Management in Asia and Africa. Bangladesh Agricultural University, pp. 87-91.

[5] Box, Gep and Wilson Kb. (1951) *On the experimental attainment of optimum conditions,* J. Roy. Stat. Soc., pp. 13-145.

[6] Bradford, Mm (1976) *A rapid and sensitive method for the quantitation of microgram quantitites of protein utilizing the principle of protein-dye binding,* Analytical Biochem. 72, pp. 248-254.

[7] Braun, S. (2000) *Bioassays of Bacillus thuringiensis 1D. Production of Bacillus thuringiensis insecticides for experimental uses,* In: Navon, A., Ascher, K.R.S (Eds.), Bioassays of Entomopathogenic Microbes and Nematodes. CAB International, pp. 49-72.

[8] Gao, Huiju , Xiang Chu, Yanwen Wang, Fei Zhou, Kai Zhao, Zhimei Mu, and Qingxin Liu (2013) *Media Optimization for Laccase Production by Trichoderma harzianum ZF-2 Using Response Surface Methodology,* J. Microbiol. Biotechnol. 23(12), pp. 1757-1764.

[9] Içgen, Y., B. Içgen, and Ozcengiz (2002) *Regulation of crystal protein biosynthesis byBacillus thuringiensis: I. EVects of mineral elements and pH,* Research in Microbiology 153(9), pp. 599-603.

[10] Jaekoo, Lee, Inkyung Park, Sangbuem Cho, Sooki Kim, Yunjaie Choi, and Jaiesoon Cho (2013) *Optimization of medium composition for α-galactosidase production by Antarctic bacterial isolate, Bacillus sp. LX-1 using response surface methodology,* African Journal of Microbiology Research. 7(27), pp. 3494-3500.

[11] Khanh, Dang Vu, R.D. Tyagi, S.K. Brar, J.R. Valero, and R.Y. Surampali (2009) *Starch industry wastewater for production of biopesticides – ramifications of solids concentrations,* Environmental Technology. 3(4), pp. 393-405.

[12] Khanh, Vu Dang, Rajeshwar Dayal Tyagi, José R. Valéro, and Rao Y. Surampalli (2009) *Bath and fed-bath fermentation of Bacillus thuringiensis using starch industry wastewater as fermentation substrate,* Bioprocess and Biosystems Engineering. 33, pp. 691-700.

[13] Lachhab, K., Tyagi R.D., and Valero J.R (2001) *Production of Bacillus thuringiensis biopesticides using wastewater sludge as a raw material: eVect of inoculum and sludge solids concentration,* Process Biochemistry. 37(2), pp. 197-208.

[14] Li, Y., Liu Z., Zhao H., Xu Y., and Cui F. (2007) *Statistical optimization of xylanase production from new isolated Penicillium oxalicum ZH-30 in submerged fermentation,* Biochem. Eng. J. 34, pp. 82-86.

[15] Linna, Du, Jia Song, Hongbin Wang, Peng Li, Zhongzhou Yang, Lingjun Meng, Fanqing Meng, Jiahui Lu, and Lirong Teng (2012) *Optimization of the fermentation medium for Paecilomyces tenuipes N45 using statistical approach,* African Journal of Microbiology Research. 6(32), pp. 6130-6141.

[16] Ozkan, M., F.B. Dilek, U. Yetis, and G. Ozcenzig (2003) *Nutritional and cultural parameters in Xuencing antidipteran delta-endotoxin production,* Research in Microbiology. 154, pp. 49-53.

[17] Songqing, Wu, Yanjiao Lan, Dongmei Huang, Yan Peng, Zhipeng Huang, Lei Xu, Ivan Gelbic, Rebeca Carballar Lejarazu, Xiong Guan, Lingling Zhang, and Shuangquan Zou (2014) *Use of Spent Mushroom Substrate for Production of Bacillus thuringiensis by Solid-State Fermentation,* Journal of Economic Entomology. 107(1), pp. 137-143.

[18] Stanbury, P.F., A. Whitaker, and S.J. Hall (1995) *Principles of Fermentation Technology,* Second ed. Elsevier Science Ltd., New York.

[19] Subbiah, Poopathi and B. Archana (2012) *Optimization of medium composition of mosquitocidal toxins from Bacillus thuringiensis Subsp. Isrealensis,* Indian Journal of Experimental Biology. 50, pp. 65-71.

[20] Vidyarthi, A.S., Tyagi R.D., Valéro J.R., and Surampalli R.Y. (2002) *Studies on the production of B. thuringiensis based biopesticides using wastewater sludge as raw material,* Water Research. 36(19), pp. 4850-4860.

[21] Vlyssides, A.G. and P.K. Karlis (2004) *Thermal-alkaline solubilization of waste activated sludge as a pre-treatment stage for anaerobic digestion,* Bioresource Technology. 91, pp. 201-206.

[22] Zouari, N. and Jaoua S. (1999) *The effect of complex carbon and nitrogen, salt, Tween-80 and acetate on delta-endotoxin production by a Bacillus thuringiensis subsp. kurstaki,* J Ind Microbiol Biotechnol. 23, pp. 497-502.

[23] Zouari, N. and Jaoua S. (1999) *Production and characterization of metalloproteases synthesized concomitantly with delta-endotoxin by Bacillus thuringiensis subsp. kurstaki strain grown on gruelbased media,* Enzyme Microb Technol. 25, pp. 364-371.

[24] Zouari, Z., Ben Sik Ali S., and Jaoua S. (2002) *Production of delta-endotoxin by several Bacillus thuringiensis strains exhibiting various entomocidal activities towards lepidoptera and diptera in gruel and fish-meal media,* Enzyme Microb Technol. 31, pp. 411-418.

Pollen analysis of *Apis mellifera* honey collected from Nigeria

Joshua Kayode, Sunday Dele Oyeyemi*

Department of Plant Science, Ekiti State University, Ado-Ekiti, Nigeria

Email address:
delejohnson6817@yahoo.com(S. D. Oyeyemi)

Abstract: Pollen analyses of twelve honey samples collected in two successive years (2011-2012) from different towns and villages in Ekiti State, South western Nigeria were carried out in order to ascertain their pollen compositions. The results from the analysis revealed eighty-five taxa belong to thirty-three botanical families. Of these, thirty-two were identified to generic level and forty-five plants were identified to the specific level. The number of pollen grains in the honey samples varied significantly (between 4,818 - 85,087 pollen grains), indicating their richness in pollen grains. Also in this study, some important honey plants such as *Spondias mombin, Alchornea cordifolia, Lannea sp*, Asteraceae, *Alchornea sp*, Rubiaceae, *Elaeis guineensis, Pavetta sp, Oldenlandia corymbosa, Triplochiton scleroxylon, Mimosops warneckii, Blighia sapinda, Piptadeniastrum africanum, Entada gigas, Tithonia diversifolia*, and Combretaceae/Melastomataceae have been identified to be characteristics of vegetation typical of Ekiti State and reflection of common pollen load and nectar sources.

Keywords: Honey Sample, Pollen Composition, Nectar Source, Ekiti State

1. Introduction

Honey is defined as the natural sweet substance produced by honey bees from the nectar of plants or from secretion of living parts of plants or excretions of plant sucking insects on the living parts of plants [1,2]. Honeybees collect this material, transform and combine it with specific substance of their own, store and leave to ripen and mature inside the honey combs. Honey contains pollen grains and other microscopic particles such as fungi and spores originating from the plants from which the nectar has been collected by the bee. Therefore the pollen composition of a honey sample reflects the vegetation type where the honey has been produced and is useful for the determination of the geographical as well as botanical origin of honey [3, 4]. Honey composition, flavor, and color varies considerably depending on the floral sources [5]. Other external factors such as seasonal and environmental factors and processing methods play an important role in honey composition. Honey is unique in its compositions and uses. Because of this unique and complex nature, honey is proved to be useful in the treatments of burns, wounds, skin ulcers, skin rashes, as an antioxidant as well as in treatment of external eye diseases [6, 7]. Increasing interest in the therapeutic uses of certain honey varieties may contribute to the demand of a reliable determination of their botanical origin. The various variety of honey may be grouped into unifloral or multifloral depending basically on whether a dominant pollen grain originated from one particular plant or no dominant pollen type in the honey sample [8, 9]. Pollen analysis is an indispensable method to authenticate honey origin and honey characteristics. Assessment of honey botanical source is of great importance in food analysis, since authenticity guarantee the quality of honey [10]. From economic point of view, the assessment of floral origin and other parameters usually add to quality and commercial value of honey. Palynological examination of honey provides some important information about honey extraction methods, filtration, fermentation [11] and some kind of adulteration [12].

In Southwestern Nigeria, the geographical and prevailing climatic conditions provide a suitable environment for beekeeping and honey production. In Ekiti State, beekeeping activities have undergone a noticeable development with the formation of Bee Farmers Association, training and financial assistances from government to the beekeepers. All these activities have boosted honey production in the state. Recently, [13, 14] reported that there is dearth of information on palynological investigations of honey produced in the area. Hence, this study is to establish the botanical origin of honey in the area and to identify the pollen and nectar sources important for the bees.

2. Materials and Methods

2.1. Area and Material of the Study

Ekiti state is situated entirely within the tropics. It lies between longitudes $7^0$25 North and $5^0$00 and $6^0$00 East in the rain forest belt of Nigeria. It lies South of Kwara and Kogi State, East of Osun State and bounded by Ondo State in the East and in the South. The state enjoys tropical climate with two distinct seasons; rainy season (April-October) and dry season (November-March). Temperature ranged between 21 ^{0}C and 28 ^{0}C with high humidity. Tropical forest exists in the South while Guinea savanna occupies the Northern peripheries of the State. The land enjoys luxuriant vegetation as a result of the favourable climatic condition thus, has abundant resources of different species of timber such as *Daniellia oliverii* (Rolfe), *Vitex doniana* (Linn.), *Parkia biglobosa* (Jacq.), *Nauclea diderrichii* (Merill), *Mimosops warneckii*, *Terminalia sp.* Food crops like yam, cassava and also grains like rice and maize are grown in large quantities in the area [15, 16, 17].

The experimental materials were twelve samples of honeys collected in the year 2011-2012 from different towns and villages in Ekiti State, South Western Nigeria.

Fig. 1. Map of Ekiti State.

2.2. Collection of Honey Sample

The twelve honey samples were collected randomly from three Senatorial District of Ekiti State. The honey samples were collected from various Apiaries of *Apis melifera* at different locations in the State. The locations were: EK-I (IKERE), EK-2 (IJAN), EK-3 (ILAWE), EK-4 (EFON), EK-5 (IGBEMO), EK-6 (ARAMOKO I), EK-7 (ESURE), EK-8(ADO), EK-9 (IKOLE),EK-10 (IDO), EK-11 (OYE) and EK-12 (ARAMOKO II). The honey samples were collected from domesticated bee-hives and were extracted through pressing, without separating the pollen and honey cells of the combs. The honey samples were later stored in airtight containers at room temperature. The analyses were done at the Palynological Laboratory of the Department of Archeology, University of Ibadan, Oyo-State, Nigeria. Samples were subjected to acetolysis following standard method [18]. Five milliliter of honey per sample was taken from a well reserved stocks of each sample and dissolved in 10ml of distilled water and centrifuged for 10 minutes at 2,500rpm and the supernatant liquid was drawn out. The sediment was treated with 5ml of glacial acetic acid and centrifuged at 2,500rpm for 10minutes. After decanting acetic acid, the sediment was acetolyzed [18]. One part of concentrated Sulphuric acid was added drop by drop to nine parts of Acetic Anhydride acid and warmed in water bath till the liquid turned chest brown color. After cooling, it was again centrifuged at 2,500rpm for 10minutes. The supernatant liquid was decanted off. The sediment was then treated with glacial Acetic acid, later centrifuged at 2,500rpm for 10minutes, followed by two or three rising with distilled water. After each rinsing, it was again centrifuged at 2,500rpm for 10minutes. The 50% aqueous glycerin prepared in distilled water was added and centrifuged for 10minutes at 3,500rpm. The supernatant liquid was decanted off. The pollen sediment was taken on a pellet of glycerin jelly and transferred to the slides, covered with cover slips and sealed with paraffin was [19]. In the microscopic analysis, three slides were prepared for each of the twelve samples and examined under the light microscope. The pollen count (that is the frequency of the pollen of the different representative plants or families in each sample) was taken and recorded [20]. In the evaluation of botanical source of the honeys, specific percentage like: (>45%) for dominant, (16-44%) for secondary, (3-15%) for important minor and <3% for minor pollen were applied in the study. Honey samples containing more than 45% of a single type of pollen were considered as unifloral honey. Plant species with percentage below 1% were not considered in the tables.

3. Results

The types of pollen, the percentage pollen frequency class as well as the quantitative pollen spectra of the honey samples from Ekiti State have been summarized in table 1-3.

Among the different honey samples analyzed, three samples were found to be unifloral while nine samples were multifloral (Table 1). In the group of honeys examined, the contribution of pollen from *Alchornea sp* was found to be the dominant type (>45%) in one sample EK-1 (IKERE) and secondary type (16-44%) in two samples EK-2 (IJAN) and EK-10 (IDO). Other predominant pollen grains identified were *Lannea sp*, Asteraceae, *Elaeis guineesis* and *Pavetta sp* with the *Elaeis guineesis* as the most dominant having dominated in three samples (EK 6, 7 and 8) and secondary in two samples. Secondary pollen types identified from the samples include, Papilionaceae, *Tithonia diversifolia*, *Triplochiton sclerexylon*, *Triumfetta sp*, *Adenodolichas paniculata*, *Oldenlandia corymbosa*, *Mimosops warnekii*, *Spondias mombin*, *Blighia sapinda* and *Euphorbia hirta.* The rest of the pollen types were categorized as important minor and minor pollen types.

Table 1. *Pollen types found in the honey from Ekiti State*

Locality	Type of honey	Pollen types			
		Predominant(>45%)	Secondary(16-44%)	Important minor(3-15%)	Minor Pollen(<3%)
IKERE (EK-1)	Multifloral	*Alchornea sp* (63.3%)	*Elaeis guineensis* (19.2%)	Papilionaceae (12.4%)	*Adenia sp,Zea mays, Solanium sp, Justica flava, Asystasia gangetica,* Poaceae
IJAN (EK-2)	Unifloral	*Lannea sp* (50.2%)	*Alchornea cordifolia* (18.6%)	*Spondias mombis (5.7%), Uapaca sp (3.2%), Entada gigas (8.8%), Blighia sapinda (3.3%), Mimusops warneckii (3.3%)*	Verbenaceae, *Elaeis gueneensis, Syzygium guineensis*
ILAWE (EK-3)	Unifloral	Asteraceae (58.5%)	Papilionaceae (32.4%)	*Mollotus sp* (5.4%)	*Asystasia gangetica, Mesozerum sp, Tridax procumbens*
EFON (EK-4)	Multifloral	-	*Elaeis guineensis* (43.8%), *Tithonia diversifolia* (39.6%)	Combretaceae/ melastomataceae (7.1%), *Asystasia gangetica* (4.3%)	*Nauclea sp,* Asteraceae, *Cardiospermum halicacabum,Zea mays,* Poaceae,*Asystasia gangatica.*
IGBEMO (EK-5)	Multifloral	-	*Triplochiton scleroxylon (39.3%), Triumfetta sp (26.2%), Adenodolichas paniculata (21.0%)*	*Venonia amygdalina (6.6%)*	*Asystasia sp,Ceiba pentandra,Cadiospermum sp,Spondias mombin*
ARAMOKO(I) (EK-6)	Multifloral	*Elaeis guineensis* (68.6%)	-	*Adenia sp (6.3%), Alchornea sp (8.3%), Mansonia sp (4.7%)*	*Choriosa sp,Lannea sp,Morus sp,Spondis mombin*
ESURE (EK-7)	Multifloral	*Elaeis guineensis (46.7%)*	*Oldenlandia corymbosa* (34.0%)	*Adenia sp (13.3%)*	*Alchornea sp,*Asteraceae, *Lantana camara,Spodias mombin,*Rubiaceae.
ADO (EK-8)	Muitifloral	*Elaeis guineesis (61.1%)*	Asteraceae (17.48%)	Rubiaceae (11.46%)	*Justicia flava, Asystasia gangetica, Adenia sp,*Poaceae, *Alchornea sp*
IKOLE (EK-9)	Multifloral	-	*Mimosops warnekii(34.5%),*Rubiaceae (34.5%)	*Piptadenisatrum africanium (23.0%)*	*Lannae welwitschii, Elaeis guineensis* Poaceae,*Chromoleana odorata*
IDO (EK-10)	Multifloral	-	*Elaeis guineesis(37.1%), Alchornea cordifolia (38.9%)*	*Cassia sp (5.7%), Uapaca sp (3.7%), Adenodolichas paniculata (3.3%)*	Asterasceae, *Chromoleana odoranta,Vernonia amygdalina, Triumfetta sp, Erythrinia sp*
OYE (EK-11)	Multifloral	-	*Spondias mombin (16.6%), Alchornea sp (16.6%), Blighia sapida (16.3%)*	*Vernonia amygdalina (10.4%), Pavetta owariensis (3.5%), Brachystegia eurycoma (3.4%), Choris sp (4.2%), Morus sp (3.6%), Celtis sp (3.4%), Phyllanthus discoides (3.7%)*	*Elaeis guineensis,* Asteraceae, *Bombax buonopozense*
ARAMOKO II (EK-12)	Unifloral	*Pavetta sp (62.4%)*	*Euphorbia hirta (25.7%)*	Anacardiaceae *(8.4%),* Combretaceae *(5.3%) Spondias mombin (12.2%).*	*Ceiba pentandra, Daniellia ogea, Trilepisium nadagascariensis*

Table 2. *The percentage pollen frequency class of honey samples from Ekiti State.*

Locality/Sample	Pollen type	Percentage frequency	Frequency class
IKERE	*Asystasia gangetica*	1.2	Sporadic
	Elaeis guineensis	19.2	Frequent
	Papilionaceae	12.4	Rare
	Alchornea sp	63.4	Very frequent
IJAN	*Spondias mombin*	5.7	Rare
	Lannea sp	50.2	Rare
	Uapaca sp	3.2	Rare
	Alchornea cordifolia	18.6	Frequent
	Entada gigas	8.8	Rare
	Blighia sapinda	3.3	Rare
	Mimusops warneckii	3.3	Rare
	Verbenaceae	2.1	Sporadic
ILAWE	*Spondias mombin*	27.0	Frequent
	Lannea sp	3.8	Rare
	Elaeis guineensis	6.0	Rare
	Combretaceae/Melastomataceae	5.4	Rare
	Blighia sapinda	24.2	Frequent
	Rubiaceae	23.0	Frequent
	Pavetta owariensis	2.0	Sporadic
EFON	*Elaeis guineensis*	43.8	Frequent
	Asystasia gangetia	1.3	Sporadic
	Combretaceae/Melastomataceae	7.1	Rare
	Nauclea sp	2.9	Sporadic
	Tithonia diversifolia	39.3	Frequent
IGBEMO	*Triumfetta sp*	26.2	Frequent
	Venonia amygdalina	6.6	Rare
	Triplochiton scleroxylon	39.3	Frequent
	Adenodolichas paniculata	21.0	Frequent
	Ceiba pentandra	2.4	Sporadic
	Cadiospermum sp	1.2	Sporadic
	Vernonia amygdalina	6.6	Rare
ARAMOKO (I)	*Adenia sp*	6.3	Rare
	Alchornea sp	8.3	Rare
	Mansonia sp	4.7	Rare
	Elaeis guineensis	68.6	Very frequent
ESURE	*Oldenlandia corymbosa*	34.0	Frequent
	Adenia sp	13.3	Rare
	Elaeis guineensis	46.7	Very frequent
	Spondias mombin	1.6	Sporadic
ADO	*Elaeis guineensis*	61.0	Very frequent
	Asteraceae	17.5	Frequent
	Rubiaceae	11.5	Rare
IKOLE	*Mimusops warneckii*	34.5	Frequent
	Rubiaceae	34.5	Frequent
	Piptadeniastrum africanum	23.0	Frequent
	Lannea welwetschii	2.3	Sporadic
	Chromoleana odoranta	1.8	Sporadic
IDO	*Cassia sp*	5.7	Rare
	Uapaca sp	3.7	Rare
	Elaeis guineensis	37.1	Frequent
	Alchornea cordifolia	38.9	Frequent
	Adenodolichas paniculata	3.3	Rare
OYE	*Vernonia amygdalina*	10.4	Rare
	Pavetta owariensis	3.5	Rare
	Brachystegia eurycoma	3.4	Rare
	Chorisa sp	4.2	Rare
	Spondias mombin	16.6	Frequent
	Alchornea sp	16.6	Frequent
	Blighia sapinda	16.3	Frequent
	Morus sp	3.6	Rare
	Celtis sp	3.4	Rare
	Phyllanthus discoides	3.7	Rare
ARAMOKO (II)	*Pavetta sp*	62.4	Very frequent
	Euphorbia hirta	25.7	Frequent
	Anacardiaceae	8.4	Rare
	Spondias mombin	12.2	Rare
	Ceiba pentadra	1.4	Sporadic

Table 3. Quantitative summary of pollen spectra identified in the honey samples

Sample	Number of families	Number of pollen types	Quantity of pollen grains/sample
IKERE	23	27	65,589
IJAN	13	20	4,818
ILAWE	12	17	4,818
EFON	20	29	28,086
IGBEMO	18	24	85,087
ARAMOKO (I)	24	28	7,582
ESURE	16	20	8,285
ADO	17	23	14,902
IKOLE	13	15	9,912
IDO	16	27	57,375
OYE	18	24	50,664
ARAMOKO II	21	28	7,582

The range in frequency percentage pollen occurrences of the honey samples varied widely (Table 2). The number of pollen types that occurred as "very frequent" and "frequent" were below 30% while those that occurred as "rare" and "sporadic" were above 70%. *Alchornea sp*, *Elaeis guineensis* and *Pavetta sp* occurred as "very frequent" in six honey samples while *Alchornea cordifolia*, *Spondias mombins*, Rubiaceae, *Tithonia diversifolia*, *Oldenlandia corymbosa*, Asteraceae, *Piptadeniastrum africanum*, *Euphorbia hirta* and *Mimosops warnerkii* occurred frequently in the honey samples. The result of this analysis (Table 3) showed that a range of pollen types varying from a minimum of 15 to a maximum of 29 different plant taxa was identified. Ikole presented the lowest while Efon showed the highest value of pollens. The results show that some of the honey investigated revealed high variety of pollen grains from nectarless plants such as, Amaranthaceae/Chenopodiaceae, *Alchornea sp*, *Alchornea cordifolia*, Combretaceae/Melastomataceae and Cyperaceae which serve as good source of pollen load for the bees. This agreed with the earlier works of [21, 22, 14]. The pollen grains of anemophilous plants of Poaceae were also identified in seven honey samples in the area studied.

4. Discussion

The pollen in Ekiti honeys presented very rich and heterogenous spectra of plant taxa. Pollen count revealed a total of eighty-five plant taxa belonging to thirty-three families among which are Rubiaceae, Fabaceae, Euphorbiaceae, Combretaceae, Asteraceae, Sapindaceae, Arecaceae, Anacadiaceae, Sapotaceae, Poaceae, Passifloraceae and Acanthaceae. In the analyzed honeys, *Elaeis guineesis, Pavetta sp* and *Alchornea sp,* are generally very frequent in the honey samples from the areas, an indication that the pollen and nectar of these botanicals are important bee foods. The abundance and regular occurrence of pollen grains of *Elaeis guineensis* may be attributed to the fact that the indigenous palm trees are not discriminately fell in the area. This finding agreed with the report of [24] who stated that *Elaeis guineensis* pollens are produced abundantly in honey investigated from two vegetation zones of Nigeria. Results point out to a great variety of forest resources and generalized habits of pollen harvesting by the bees, *Apis mellifera*. This study has shown that bees do not respect plant habit or plant stratification. This confirmed the earlier work of [25] in a study carried out in Southwest Nigeria.

5. Conclusion

The areas selected for this study have good potential for sustaining beekeeping activities because of the diversity of nectar and pollen taxa. The honey samples are rich in pollen contents an indication that they are not adulterated. Since *Elaeis guineensis* are major sources of forage for wild honey bees (*Apis mellifera*) efforts should be intensified in increasing their cultivation. More so, plants in the families Rubiaceae, Asteraceae, Euphorbiaceae, Sapotaceae and Acanthaceae which are characteristics of Ekiti vegetation and also reflecting common pollen load and nectar sources should be cultivated the more.

References

[1] Bogdanov S, Matin P, Lullman C.Harmonized methods of the European Honey Commission. *Apidologie*, 1997, (Extral issue) pp.

[2] Breadbear N. Bees and their role in forest livelihoods: A guide to the service provided by bees and the sustainable harvesting, processing and marketing of their products. Non Wood Forest Products 19. Food Agric. Orga. United Nations. Rome, 2009, p. 204.

[3] Louveaux, J, Maurizo A, Vorwohl G. Method of Melissopalynology. International Bee Research Association, Bee World, 1978, 59 (4).139-157.

[4] Ohe W V D, Oddo L, Piana P, MM, Martin P. Harmonized methods of melissopalynology, *Apidologie*, 2004, 35: S18-S25.

[5] Persano Oddo, Land Piro, R. Main European unifloral honeys: descriptive sheets, *Apidologie*, 2004, 35, special issue, S38-S81.

[6] McCanthy J. The antibacterial effects of honey: medical fact of fiction? *American Bee Journal*, 1995, 135: 341-342.

[7] Balasurbramanyam M.V. Chemical characteristics of multifloral wild and apiary honey from western Ghats of Karnataka, *The Bioscan*, 2011 6, 467-467.

[8] Molan P C. Re-introducing honey in the management of wounds and ulcers: theory and practice – Ostomy Wound Manage, 2002, 48: 28-40.

[9] Ramirez- Arriaga, E, Navano-Calvo L A, Diaz-Carbajal E. Botanical characterization of Mexican honeys from a substropical region (Oaxaca) based on pollen analysis- Grana, 2011, 50: 40-54.

[10] Bogdanov S, Kaspar R, Livia, PO. Physico-chemical Methods for the characterization of Unifloral Honeys: A Review, *Apidologie*, 2004, 35: Pp S_4-S_{17}.

[11] Russman, H. Hefen und Glycerin in Blutenenhonigen-Nachweis einer Garung ode einer Abgestoppten Garung Lebensmiltedchemie, 1998, 52, 116-117.

[12] Kerkvhet, J D, Shrestha, M Tuladhar, K, Mandanhar, H, Meijer, H A J. Adulteration of honey relation between microscopic analysis and delta C-13 measurements, *Apidologie*, 2000, 31, 717-726.

[13] Ige O E, Modupe T O. Pollen characterization of Honey samples from North Central Nigeria. *Journal of Biological Sciences*, 2010, 10(1): 43-47.

[14] Kayode J, Oyeyemi S D. Pollen analysis of honey derived from Ekiti State, Nigeria. *Bulletin of Applied Sciences*, 2012,Vol.31B-Botany (No.1):39-47.

[15] EKSG. First Anniversary Celebration of Ekiti State, Nigeria, Nigeria Government Press: 1997, P 22.

[16] Kayode J. Phytosociological investigation of compositae weeds in abandoned farmland in Ekiti State, Nigeria compositae Newslatter, 1999, 34:62-68.

[17] Kayode J. Population dynamics of *Euphorbia heterophylla* (L) after slash and burn agriculture in southwestern Nigeria. *Journal of Biological and Physical Science*, 2000, 1:30-33.

[18] Erdtman G. Handbook of Palynology. An Introduction to the study of Pollen Grain and Spore. Hafner Publishing Co. Inc., New York., 1968.

[19] Agwu C O C, Akanbi T O. A palynological study of honey from four vegetation zones of Nigeria. *Pollen et Spores*, 1985, 25 (3-4), 335-346.

[20] Louveaux J, Maurizo A, Vorwohl G. Methods of Melissopalynology, Bee World, 1970, 1: 125-138.

[21] Sowumi M A. Microscopic analysis of honey. The Nigeria Field, 2001, 66:125-133.

[22] Agwu C O C, Okeke G I. Pollen analytical and thin-layer chromatographic study of honey from three Savanna Zones of Northern Nigeria. *Nig. J. Bot*, 1997, 10:25-36.

[23] Agwu C O C, Uwakwe G O. Melissopalynological study of Abia and Imo States Nigeria honey, *Nigeria Journal of Botany*, 1992, 5:85-91.

[24] Ige OE, Apo K A. Pollen analysis of honey samples from two vegetation zones in Nigeria. *Sci. Focus*, 2007, 13:36-43.

[25] Adeonipekun P A. A palynological study of an apiary in Ibadan, Nigeria. Unpublished B. Sc. Report. Department of Botany and Microbiology, University of Ibadan, Ibadan, Nigeria, 1987, 57pp.

Effects of biochar derived from maize stover and rice straw on the early growth of their seedlings

Alie Kamara[1], Mary Mankutu Mansaray[2], Abibatu Kamara[2], Patrick Andrew Sawyerr[1]

[1]Soil Science Department, School of Agriculture, Njala Campus, Njala University, Sierra Leone
[2]Extension Division, Ministry of Agriculture, Forestry and Food Security, Sierra Leone

Email address:
aliekamara@njala.edu.sl (Alie K.)

Abstract: Crop residues such as maize and rice are important sources of nutrients and their restitution to the soil is an important residue management strategy for maintaining or improving soil and crop productivity. However, most maize and rice farmers in Sierra Leone burn crop residues during land preparation thereby depriving the soil of this vital resource. An alternative approach is the recycling of crop residues through biochar production. This study was therefore carried out to assess the effects of (i) maize stover biochar on the early growth of maize and (ii) rice straw biochar on the early growth of rice. In this study, biochar (0g, 12.5g, 25.0g, 37.5g and 50.0g) was mixed with 3kg of soil (fine sandy loam) and placed in perforated black polythene bags. Maize or rice seeds (3 per pot) were planted to 2cm depth and later thinned to one plant per pot. All pots were placed in the open under direct sunlight and rainfall. No supplemental irrigation was done since there was sufficient rainfall to supply the required moisture. The experiment was laid in completely randomized design with five treatments and three replications. After four weeks, the experiment was terminated. Mean plant height, stem girth, fresh- and dry-shoot weights of both maize and rice plants generally increased significantly ($p<0.05$) relative to the control at biochar applications ≥8.3g/kg soil. The results showed that growing maize and rice on soils treated with biochar derived from their residues (maize stover and rice straw respectively) has the potential to improve crop production. Confirmatory field trials and costs/benefits analysis of converting rice and maize residues to biochar will serve as a good guide for policy makers and end-users (farmers) for increased productivity of maize and rice in Sierra Leone.

Keywords: Biochar, Rice Straw, Maize Stover, Early Growth, Plant Height

1. Introduction

Crop residues are important sources of nutrients and soil organic matter, and their restitution to the soil is an important residue management strategy for maintaining or improving soil and crop productivity. However, most maize and rice farmers in Sierra Leone burn crop residues during land preparation thereby depriving the soil of this vital resource.

Incorporation of maize or rice residues as green manures or compost could offer benefits to maize or rice cropping through the positive effects on soil organic matter, nutrient release, cation exchange capacity, and microbial activity. However, soil organic matter decomposes at a faster rate when incorporated into the soil, particularly on warm humid tropical soils, resulting in rapid loss of soil organic carbon and nutrients through leaching [1, 2]. Rapid decomposition of soil organic matter also leads to the release of biomass carbon in the form of CO_2 [3] making the recycling process carbon neutral. An alternative approach is the recycling of crop residues through biochar production.

Addition of biochar to soil helps in improving soil productivity by improving water and nutrient retention [4]. Biochar increases cation exchange capacity of soil [5, 6], improves porosity [5] and lowers bulk density thus making root penetration easier. Biochar may also supply some amount of nutrients [7, 8] and provide a liming effect to soil [9]. Unlike soil organic matter, biochar does not decompose readily in soils because the biochar carbon is very resistant to microbial attack [10] and as a result it can lasts in the soil for centuries or millennia thereby sequestering carbon in the soil and making the process carbon negative. Thus converting crop residues to biochar is an important strategy to improve soil productivity and mitigate global warming through carbon sequestration.

This study was therefore carried out to assess the effects of (i) maize stover biochar on the early growth of maize and (ii) rice straw biochar on the early growth of rice.

2. Materials and Methods

2.1. Description of the Study Area

The experiment was conducted in Njala Campus, Njala University in Moyamba District in Southern Sierra Leone. Njala is about 255km from Freetown, the capital city of Sierra Leone, and is located at an elevation of 128m above sea level (altitude), on Latitude 8°N and Longitude 12°W. Njala experiences a tropical climate that is characterized by two main seasons; a rainy season and a dry season. The average annual rainfall is about 2500mm and the mean annual temperature ranges from a minimum of 28°C to a maximum of 33°C. The trial was conducted in the rainy season.

2.2. Biochar Production and Processing

Maize stover and rice straw were collected from farms in Mosongo, one of the villages around Njala University Campus. Each plant material was thoroughly dried in the sun and converted to biochar using the ELSA stove designed for the BeBi Project in Njala University. [The Elsa stove is a low-tech Top-Lit-Up-Draft (TLUD) bioenergy stove designed for use in domestic cooking and production of biochar for application to soils under the EU-funded project: "Agricultural and Environmental Benefits of Biochar Use in ACP Countries" – 'BeBi']. The charred material (biochar) was quenched with water (by sprinkling water on the hot char) and dried in the sun for a few days. The dry biochar was crushed and sieved through a 2mm sieve and stored.

2.3. Properties of Soil and Biochar

The soil used in this study had the following characteristics: pH 4.7 (1:1 soil:water ratio), organic carbon 2.1g/kg soil (Walkley-Black), available phosphorus 5.6mg/kg soil (Bray & Kurtz 1), cation exchange capacity 6.2 cmol(+)/kg soil (neutral M NH_4OAc), exchangeable cations 0.28, 0.19, 0.12 and 0.02 cmol+/kg soil Ca, Mg, K and Na respectively, and particle size (70% sand, 16% silt and 14% clay). Maize stover biochar had the following characteristics: pH 8.4, electrical conductivity (1:5 biochar:water) 3.98 dSm^{-1}. Rice straw, on the other hand, had the following characteristics: pH 8.8, electrical conductivity (1:5) 2.82 dSm^{-1}.

2.4. Experimental Setup

The experiment consisted of five treatments in three replications in a completely randomized design (Table 1). For each treatment the appropriate amount of biochar was mixed with 3kg of soil and placed in perforated black polythene bags. The soil in each pot was moistened with water to saturation level and allowed to drain freely for 48 hours. Maize seeds (3 per pot) were planted to 2cm depth and then thinned to one plant per pot after emergence. All pots were placed on an elevated platform under direct sunlight and rainfall. No supplemental irrigation was done since there was sufficient rainfall to supply the required moisture.

Table 1. Details of biochar applications rates used in the trial

Treatment	Biochar Rate (g/3kg soil)	Biochar Rate (g/kg soil)
T1	0.0	0.0
T2	12.5	4.17
T3	25.0	8.33
T4	37.5	12.5
T5	50.0	16.67

Data was collected every week on plant height and stem girth onto the fourth week when the experiment was terminated. Tiller number was also recorded for rice. At the end of the fourth week, each plant was cut from the base and weighed to determine the weight of the fresh biomass for each treatment. The fresh shoots were dried to constant weight in an oven at a temperature of 75°C.

2.5. Data Analysis

The data was analysed for statistical differences using $LSD_{0.05}$ and Analysis of Variance (ANOVA) for Completely Randomized Designs (CRD) in the GenStat 12th Edition computer software.

3. Results

3.1. Effect on Plant Height

Results of weekly plant height measurements for maize and rice (Table 2; also see Figs. 1 and 2)) showed general increases with increasing biochar application levels. However, this increment was not significant ($p<0.05$) for maize in weeks one and two, and for rice in weeks one, two and three. In weeks three and four, maize plant height increased significantly ($p<0.05$) relative to the control above biochar application level of 4.2 g/kg soil. On the other hand, rice plant height increased significantly relative to the control only in week four and beyond biochar application level of 4.2g/kg soil. Generally, further increasing biochar application beyond 4.2g/kg soil did not lead to any significant increase in maize and rice plant heights among biochar treatments.

Figure 1. *Maize plants after four weeks*

Figure 2. *Rice plants after four weeks*

Table 2. *Mean plant height (cm) of maize and rice seedlings grown on a sandy loam soil treated with increasing levels of maize stover and rice straw biochars respectively*

Biochar (g/kg soil)	Maize Plant Height (cm)				Rice Plant Height (cm)			
	Week1	Week2	Week3	Week 4	Week1	Week2	Week3	Week 4
0.0	13.43a	17.87a	15.77a	16.60a	9.57a	14.23a	20.03a	25.00a
4.2	12.50a	17.50a	20.50ab	21.73ab	11.50a	16.37a	18.83a	26.80ab
8.3	15.00a	20.07a	25.33bc	26.80bc	10.83a	18.07a	24.00a	37.20bd
12.5	15.13a	21.13a	25.87bc	27.83bc	15.33a	20.17a	27.23a	42.33cd
16.7	16.37a	19.80a	28.87c	31.73c	11.63a	24.17a	27.43a	42.00cd
LSD	3.21	5.24	6.88	6.99	5.21ns	10.5ns	8.28ns	11.37

Values with the same letters are not significantly different from each other ($p<0.05$)

3.2. Effect on Stem Girth

Maize and rice (main tiller) stem girths were observed to increase with increasing application levels of biochars derived from their respective residues (Table 3). However, except for biochar application level of 16.7g/kg soil in week 2, the increase in maize stem girth was not significant relative to the control in the first and second weeks. In week 3 a significant difference relative to the control was only observed beyond biochar application level of 4.2g/kg soil whereas in week 4, a significant difference relative to the control was observed for all application levels. Rice showed significant increase in stem girth relative to the control only with biochar application levels of 12.5g/kg soil in week1, 12.5g/kg soil and 16.7g/kg soil in week 2, above4.2g/kg soil in week 3 and only at 16.7g/kg soil in week 4. From weeks three and four data, it was generally observed that further biochar applications above 8.3 g/kg soil did not increase maize and rice stem girths significantly.

Table 3. *Mean stem girth (cm) of maize and rice seedlings grown on a sandy loam soil treated with increasing levels of maize stover and rice straw biochars respectively*

Biochar (g/kg soil)	Maize Stem Girth (cm)				Rice Stem Girth (cm)			
	Week1	Week2	Week3	Week 4	Week1	Week2	Week3	Week 4
0.0	1.10a	1.70a	1.43a	1.60a	0.7a	0.93a	1.13a	1.57a
4.2	1.27a	1.80a	1.70a	2.03b	0.7a	1.17ab	1.4a	2.1ab
8.3	1.40a	1.87ab	2.43b	2.47c	0.83ab	1.4ab	1.7b	2.37ab
12.5	1.40a	1.87ab	2.57b	2.87cd	0.9b	1.5b	1.97bc	2.6ab
16.7	1.43a	2.07b	2.67b	3.20d	0.8ab	1.73b	2.29c	3.17b
LSD	0.43ns	0.27ns	0.49	0.43	0.16ns	0.48	0.44	1.23

Values with the same letters are not significantly different from each other ($p<0.05$)

3.3. Effect on Shoot Biomass

The mass of fresh and dry shoot biomass of maize after four weeks of growth are shown in Table 4.1. There was a significant increase in biomass (fresh and dry) relative to the control above biochar application level of 4.2g/kg soil.

Table 4.1. Fresh and dry shoot biomass of four-week old maize seedlings grown on a fine sandy loam soil treated with increasing levels of maize stover biochar

Biochar Rate (g/kg soil)	Fresh Maize Shoot (g)	Dry Maize Shoot (g)
0.0	1.09a	0.14a
4.2	2.69a	0.35a
8.33	4.15b	0.58b
12.5	5.92bc	0.75b
16.67	8.31c	1.08c
LSD (0.05)	2.49	0.32

Values with the same letters are not significantly different from each other ($p<0.05$)

Rice shoot biomass (fresh and dry) as shown in Table 4.2 generally increased with increasing application levels of rice straw biochar. Application of biochar at 4.2g/kg soil did not differ significantly from the control. Rice shoot biomass showed significant difference relative to the control above biochar application level of 4.2g/kg soil for fresh biomass and above 8.33g/kg soil for dry biomass.

Table 4.2. Fresh and dry shoot biomass of four-week old rice seedlings grown on a fine sandy loam soil treated with increasing levels of rice straw biochar

Biochar Rate (g/kg soil)	Fresh Rice Shoot (g)	Dry Rice Shoot (g)
0.0	0.48a	0.14a
4.2	0.63a	0.20a
8.33	2.59b	0.62ab
12.5	3.14bc	0.78b
16.67	4.11c	0.88b
LSD (0.05)	1.88	0.51

3.4. Effect on Rice Tiller Number

Rice started tillering at week 3 for biochar treated pots (Table 5). The control on the other hand only started tillering a week later i.e. at week four. Rice tiller numbers increased with increasing biochar application and differed significantly among all treatments. Thus, application of rice straw biochar resulted in early tillering and increasing application of rice straw biochar generally increased rice tiller numbers.

Table 5. Mean rice tiller numbers produced at three and four weeks after planting on a fine sandy loam soil treated with increasing levels of rice straw biochar.

Biochar Rate	Average Tiller Number	
(g/kg soil)	Week3	Week4
0.0	0.0a	0.7a
4.17	1.0b	1.7b
8.33	2.0c	3.0c
12.5	3.0d	4.0d
16.67	3.3e	5.3e
LSD (0.05)	0.49	0.91

4. Discussion

The overall objective of this study was to demonstrate the usefulness of recycling crop residues through biochar production. The study generally revealed that there is a potential for improving maize and rice production using biochar derived from their residues.

This study has shown that application of biochar derived from maize and rice residues to soil sown to maize and rice respectively could increase crop growth at the early stages (four weeks) relative to the control (where no biochar was applied). Generally, there was an increase in plant growth as measured in terms of plant height, stem girth, and shoot biomass. At the very early stages of growth (one and two weeks) there were no significant differences in plant growth between biochar treatments and the control.

The observed increase in growth for each crop was only significantly different from the control at biochar application levels of ≥8.33g/kg soil and above. However, applying biochar above this level did not lead to any further significant increase in maize and rice growth indicating an optimum application level of about 8.33g/kg soil for biochars derived from maize and rice residues.

The relative increase in crop growth at the early growth stages as observed in this study could be attributed to the soil enhancing properties of the rice straw and maize stover biochars, in particular, the observed increase in pH (from 4.7 to 5.5 by maize stover biochar and to 5.8 by rice straw biochar). Both biochar types had high pH (8.4 for maize biochar and 8.8 for rice straw biochar) and increased soil pH when applied to soil. Soil pH is an important property that influences nutrient retention and availability [11]. Some biochar have been reported to increase soil pH and thus act as liming materials [12, 9]. Several studies have also reported pH increases with biochar application to soil [13, 14]. The increase in soil pH due to application of rice straw and maize biochars could be used to explain the observed differences between the control and the biochar treatments, particularly when no fertilizer was applied. The soil used in this study was very acidic (pH 4.7) and could reduce crop growth in the absence of soil amendment such as liming to counteract the effects of acidity. The increase in soil pH observed in this study may have had positive effects on the growth of maize and rice seedlings.

Application of rice straw biochar was found to increase rice tiller numbers significantly among all treatments. This indicated that biochar had a strong positive effect on rice tiller numbers than other plant growth parameters such as plant height and stem girth as measured in this study.

5. Conclusion

The results of this study have demonstrated that recycling crop residues such as rice straw and maize stover through biochar production and application to soil could be a useful residue management strategy for improving rice and maize production on sandy loams of Sierra Leone.

Generally application of biochar improved maize and rice growth compared to the control where no biochar was applied. However, applications beyond 8.3g/kg soil did not lead to any significant increase in the growth of maize and rice plants. Application of biochar also resulted in early tillering.

Acknowledgements

The authors would like to thank the field technicians Joseph S. Domingo and Luseni Jaward, and the laboratory technician Samuel Jaia all of the Department of Soil Science Njala University, Sierra Leone for their assistance in soil sample collection and preparation. We will also like to thank Mr Tamba in the agricultural engineering Department for his assistance in biochar production. Special thanks also to Mr J.E. Norman for helping in the statistical analysis of rice data.

References

[1] H. Tiessen, E. Cuevas, and P. Chacon. The role of soil organic matter in sustaining soil fertility. Nature, 371: 783-785, 1994.

[2] W. Zech, N. Senesi, G. Guggenberger, K. Kaiser, J. Lehmann, T. M. Miano, A. Miltner, G. Schroth. Factors controlling humification and mineralization of soil organic matter in the tropics. Geoderma 79, 117-161, 1997.

[3] P. M. Fearnside. Global warming and tropical land-use change: greenhouse gas emissions from biomass burning, decomposition and soils in forest conversion, shifting cultivation and secondary vegetation. Climatic Change 46:115–158, 2000.

[4] B. Glaser, J. Lehmann, and W. Zech. Ameliorating physical and chemical properties of highly weathered soils in the tropics with charcoal: A review. Biol Fertil Soils., 35, 219–230, 2002.

[5] M. I. Bird, P. L. Ascough, I. M. Young, C. V. Wood, and A. C. Scott. (2008). X-ray microtomographic imaging of charcoal. J. Archaeol. Sci. 35, 2698–2706.

[6] B. Liang, J. Lehmann, D. Kinyangi, J. Grossman, B. O'Neill, J. O. Skjemstad, J. Thies, F. J. Luizao, J. Peterson, and E.G. Neves. Black carbon increases cation exchange capacity in soils. Soil Sci. Soc. Am. J. 70: 1719–1730, 2006.

[7] K. Chan and Z. Xu. Biochar: Nutrient Properties and Their Enhancement. In 'Biochar for Environmental Management: Science and Technology'. (Eds J. Lehmann and S. Joseph) pp.53-66, 2009. (Earthscan: London, UK).

[8] J. W. Gaskin, R. A. Speir, K. Harris, K. C. Das, R. D. Lee, L. A. Morris and D. S. Fisher. Effect of Peanut Hull and Pine Chip Biochar on Soil Nutrients, Corn Nutrient Status, and Yield. Agronomy Journal. 102:623-633, 2010.

[9] L. Van Zwieten, S. Kimber, S. Morris, K. Y. Chan, A. Downie, J. Rust, S. Joseph, and A. Cowie. Effects of biochar from slow pyrolysis of papermill waste on agronomic performance and soil fertility, Plant and Soil 327(1–2): 235–46, 2010.

[10] D. Granatstein, C. Kruger, H. P. Collins, M. Garcia-Perez, and J. Yoder. Use of biochar from the pyrolysis of waste organic material as a soil amendment. Center for Sustaining Agric. Nat. Res. 2009. Washington State University, Wenatchee, WA. WSDA Interagency Agreement. C0800248. (http://www.ecy.wa.gov/pubs/0907062.pdf).

[11] C. Steiner, W. G. Teixeira, J. Lehmann, T. Nehls, J. L. V. de Macedo, W. E. H. Blum, and W. Zech. Long-term effect of manure, charcoal and mineral fertilization on crop production and fertility on highly weathered central amazonian upland soil. Plant Soil, 291, 275-290, 2007.

[12] J. M. Novak, W. J. Busscher, D. L. Laird, M. Ahmedna, D. W. Watts, and M. A. S. Niandou. Impact of biochar amendment on fertility of a southeastern Coastal Plain soil. Soil Science, 174(2), p.105-112, 2009

[13] K. Y. Chan, B. L. Van Zwieten, I. Meszaros, D. Downie, D. and S. Joseph. Using poultry litter biochars as soil amendments. Australian Journal of Soil Research, 46, 437-444, 2008.

[14] A. Masulili, W. H. Utomo, and Syekhfani. Rice husk biochar for rice based cropping system in acid soil 1. The characteristics of rice husk biochar and its influence on the properties of acid sulfate soils and rice growth in West Kalimantan, Indonesia. Journal of Agriculture Science, 3, 25-33, 2010.

Permissions

All chapters in this book were first published in AJAF, by Science Publishing Group; hereby published with permission under the Creative Commons Attribution License or equivalent. Every chapter published in this book has been scrutinized by our experts. Their significance has been extensively debated. The topics covered herein carry significant findings which will fuel the growth of the discipline. They may even be implemented as practical applications or may be referred to as a beginning point for another development.

The contributors of this book come from diverse backgrounds, making this book a truly international effort. This book will bring forth new frontiers with its revolutionizing research information and detailed analysis of the nascent developments around the world.

We would like to thank all the contributing authors for lending their expertise to make the book truly unique. They have played a crucial role in the development of this book. Without their invaluable contributions this book wouldn't have been possible. They have made vital efforts to compile up to date information on the varied aspects of this subject to make this book a valuable addition to the collection of many professionals and students.

This book was conceptualized with the vision of imparting up-to-date information and advanced data in this field. To ensure the same, a matchless editorial board was set up. Every individual on the board went through rigorous rounds of assessment to prove their worth. After which they invested a large part of their time researching and compiling the most relevant data for our readers.

The editorial board has been involved in producing this book since its inception. They have spent rigorous hours researching and exploring the diverse topics which have resulted in the successful publishing of this book. They have passed on their knowledge of decades through this book. To expedite this challenging task, the publisher supported the team at every step. A small team of assistant editors was also appointed to further simplify the editing procedure and attain best results for the readers.

Apart from the editorial board, the designing team has also invested a significant amount of their time in understanding the subject and creating the most relevant covers. They scrutinized every image to scout for the most suitable representation of the subject and create an appropriate cover for the book.

The publishing team has been an ardent support to the editorial, designing and production team. Their endless efforts to recruit the best for this project, has resulted in the accomplishment of this book. They are a veteran in the field of academics and their pool of knowledge is as vast as their experience in printing. Their expertise and guidance has proved useful at every step. Their uncompromising quality standards have made this book an exceptional effort. Their encouragement from time to time has been an inspiration for everyone.

The publisher and the editorial board hope that this book will prove to be a valuable piece of knowledge for researchers, students, practitioners and scholars across the globe.

List of Contributors

Junjun Guan, Guohao Yang, Haicheng Yin, Feng Jia and Jinshui Wang
College of Biological Engineering, Henan University of Technology, Lianhua Street, Zhengzhou New & High-tech Industrial Development Zone, Zhengzhou 450001, China

Hamed Fathi
Department of soil Sciences, college of Agriculture, Gorgan University of Agriculture Sciences and Natural Resources Gorgan, Iran

Mojan Mirzanejad
Islamic Azad University, Rasht, Iran

Habib Fathi
Academic center for Education, Culture and Research, Environmental Research Instituterasht, Iran

Abdolhossinpari Zangane
Zanjan University Department of chemistry Zanjan, Iran

Hamed Fathi
Department of soil Sciences, college of Agriculture, Gorgan University of agriculture sciences and Natural resources Gorgan, Iran

Hossein moradi
Higher educational complex of saravan, Iran

Tasnia Ahmed, Nusrat Jahan Urmi, Md. Sakil Munna, Kamal Kanta Das, Mrityunjoy Acharjee and Rashed Noor
Department of Microbiology, Stamford University Bangladesh, Dhaka, Bangladesh

M Majibur Rahman
Department of Microbiology, University of Dhaka, Dhaka, Bangladesh

Sunday Brownson Akpan and Ini-Mfon Vincent Patrick
Department of Agricultural Economics and Extension, Akwa Ibom State University, Nigeria

Glory Emmanuel Edet and Daniel Etim John
Department of Agricultural Economics and Extension, University of Uyo, Akwa Ibom State, Nigeria

David Boansi
Department of Economic and Technological Change, Center for Development Research (ZEF), University of Bonn, Germany

Adegun Maria Kikelomo
Department of Animal Production and Health Sciences, Ekiti State University, Ado-Ekiti, Nigeria

Serhat Usanmaz
European University of Lefke, Faculty of Agricultural Sciences and Technologies, Cyprus, Mersin 10 Turkey

İbrahim Kahramanoğlu
European University of Lefke, Faculty of Agricultural Sciences and Technologies, Cyprus, Mersin 10 Turkey

Alnar Pomegranates Ltd., Ataturk Ave. 109/1 Guzelyurt, Cyprus, Mersin 10 Turkey

Nihat Yılmaz
European University of Lefke, Faculty of Agricultural Sciences and Technologies, Cyprus, Mersin 10 Turkey

Erciyes Üniversitesi, Safiye Cıkrıkcıoglu Vocational College, Kayseri, Turkey

Fankem Henri, Nguesseu Njanjouo Ghislain, Ngo Nkot Laurette, and chuisseu Tchakounte Gylaine Vanessa
Department of Plant Biology, Faculty of Science, University of Douala, Douala, Cameroon

Nwaga Dieudonné and Etoa François-Xavier
Department of Microbiology, Faculty of Science, University of Yaoundé I, Yaoundé, Cameroon

Hasna Kadri and Sondes Marouani
Institut National des Sciences et Technologies de la Mer, Centre de Sfax, B.P. 1035, Sfax 3018, Tunisie

Faculte des Sciences de Sfax, B.P. 802, Sfax 3028, Tunisie

Mohamed Nejmeddine Bradai
Institut National des Sciences et Technologies de la Mer, Centre de Sfax, B.P. 1035, Sfax 3018, Tunisie

Abderrahmen Bouaïn
Faculte des Sciences de Sfax, B.P. 802, Sfax 3028, Tunisie

Eric Morize
Lemar/ IRD 230 Broennou, 29870 Landeda, France

Mihir Kanti Sarkar and ATM Jashim Uddin
Bangladesh Agricultural Research Council (BARC), Farmgate, Dhaka- 1215, Bangladesh

Md. Rajib Rahman
Soil Resource Development Institute (SRDI), Farmgate, Dhaka- 1215, Bangladesh

G. M. Faysal Ahmed and Md. Humayan Kabir
Ministry of Agriculture, Dhaka- 1000, Bangladesh

Ringko Kabiraj
Bangladesh Computer Council (BCC), Sher-E-Bangla Nagar, Agargaon, Dhaka-1207, Bangladesh

Ananya Biswas
Faculty of Agriculture, Patuakhali Science and Technology University (PSTU), Dumki, Patuakhali-8602, Bangladesh

Hemant Pathak and Satish Chandra Silawat
Forest Research and Extension Circle, Indore, Madhya Pradesh, India

Saurabh Maru
Laboratory of Pharmacology, Department of Pharmacy, Shri G S Institute of Technology and Science, Indore, Madhya Pradesh, India

Philip Cheriose Nzien Alikwe and Amos Ebiowei Kester
Biochemistry/Biotechnology Research Unit, Animal Science Department, Niger Delta University, Wilberforce Island, Bayelsa State

Elijah Ige Ohimain
Agricultural and Veterinary Microbiology Research Unit, Biological Sciences Department, Niger Delta University, Wilberforce Island, Bayelsa State

Fankem Henri, Ngo Nkot Laurette, Nguesseu Njanjouo Ghislain, Tchuisseu Tchakounte Gylaine Vanessa and Tchiazé Ifoué Alice Virginie
Department of Plant Biology, Faculty of Science, University of Douala, Douala, Cameroon

Nwaga Dieudonné
Department of Plant Biology and Physiology, Faculty of Science, University of Yaoundé I, Yaoundé, Cameroon

Samuel Twumasi Amoah
Department of Environment and Resource Studies, University for Development Studies, WA, Ghana

Irene Akobour Debrah
Faculty of Education, University for Development Studies, WA, Ghana

Razak Abubakari
School of Business and Law, University for Development Studies, WA, Ghana

Simeon Olutoye Olawumi
Department of Animal Production and Health Sciences, Ekiti State University, Ado-Ekiti, Nigeria

Sampson Konlan, Michael Obour Opoku-Agyeman, Kofi Acheampong, Kwabena Opoku-Ameyaw, Gilbert John Anim-Kwapong and Godwin Addo
Cocoa Research Institute of Ghana, P. O. Box 8, New Tafo-Akim

Mohammad Monirul Hasan
Department of Economic and Technological Change, Center for Development Research (ZEF), University of Bonn, Bonn, Germany

Ogisi O'raye Dicta, Begho Toritseju and Ewolor Scholastica Agatha
Department of Agricultural Economics and Extension, Delta State University, Asaba Campus, Asaba Nigeria

Kazi Kamrul Islam
Department of agroforestry, Bangladesh Agricultural University, Mymensingh-2202, Bangladesh

Forest Policy Laboratory, Faculty of Agriculture, Kyushu University, Fukuoka, Japan

Takahiro Fujiwara and Noriko Sato
Forest Policy Laboratory, Faculty of Agriculture, Kyushu University, Fukuoka, Japan

Masakazu Tani
Faculty of Design, Kyushu University, Fukuoka, Japan

Sheikh Helena Bulbul and Md. Redwanur Rahman
Institute of Environmental Science, University of Rajshahi, Rajshahi-6205, Bangladesh

Nsikak-Abasi A. Etim
Department of Agricultural Economics and Extension, University of Uyo, P.M.B. 1017, Uyo, Akwa Ibom State, Nigeria

Abiola Matthew Oladipupo
Department of Agricultural Economics and Extension, Faculty of Agriculture, Federal University of Oye Ekiti State, Nigeria

Omoregbee Friday Egbenayabuwa and Caroline Sede
Department of Agricultural Economics and Extension Services, University of Benin, Edo State, Nigeria

Awal, Mohd. Abdul
Environmental Scientist, Founder & Chief Advisor of Health & Pollution Research Farm, Address: Long Island City, New York, USA

M. A. Rakib
Department of Disaster Management, Begum Rokeya University, Rangpur, Bangladesh

M. S. Akter
Department of Statistics, Jahangirnagar University, Savar, Dhaka, Bangladesh

M. B. Hossain
Department of Zoology, Rangpur Government College, Rangpur, Bangladesh

Binod Ghimire
Agriculture Extension Officer, District Agriculture Development Office, Mustang, Ministry of Agricultural Development, Government of Nepal, Nepal

Shiva Chandra Dhakal
Department of Agricultural Economics, Institute of Agriculture and Animal Science, Tribhuvan University, Rampur, Chitwan, Nepal

Douglas Rodríguez Martínez, Priscilla de Sousa Geraldino Duarte and Antonia dos Reis Figueira
Department of Plant Pathology, Federal University of Lavras, Lavras, Brazil

Justo González Olmedo
Bioplants Center, University of Ciego de Avila, Ciego de Avila, Cuba

Nguyen Thi Hoa, Tang Thi Chinh and Dang Thi Mai Anh
Dept of Environmental Microbiology, Institute of Environmental Technology, Vietnamese Academy of Science and Technology, Hanoi, Vietnam

Ngo Dinh Binh and Le Thi Minh Thanh
Institute of Biotechnology, Vietnam Academy of Science and Technology, Hanoi, Vietnam

Joshua Kayode and Sunday Dele Oyeyemi
Department of Plant Science, Ekiti State University, Ado-Ekiti, Nigeria

Alie Kamara and Patrick Andrew Sawyerr
Soil Science Department, School of Agriculture, Njala Campus, Njala University, Sierra Leone

Mary Mankutu Mansaray and Abibatu Kamara
Extension Division, Ministry of Agriculture, Forestry and Food Security, Sierra Leone

Index